Basic Science Concepts and Applications

PRINCIPLES AND PRACTICES OF WATER SUPPLY OPERATIONS SERIES

Series Editor, Harry Von Huben

Water Sources, Second Edition

Water Treatment, Second Edition

Water Transmission and Distribution, Second Edition

Water Quality, Second Edition

Basic Science Concepts and Applications, Second Edition

Basic Science Concepts and Applications

Second Edition

American Water Works Association

Basic Science Concepts and Applications, Second Edition

Principles and Practices of Water Supply Operations Series

Library of Congress Cataloging-in-Publication Data

Basic science concepts and applications -- 2nd ed.
xv, 670 p. 18x23 cm. -- (Principles and practices of water supply operations)
Includes index.
ISBN 0-89867-796-3
1. Water-supply engineering. 2. Science. I. American Water Works Association. II. Series: Principles and practices of water supply operations (Unnumbered)
TD345.B36 1995
628.1--dc20 95-31429
CIP

Disclaimer

Many of the photographs and illustrative drawings that appear in this book have been furnished through the courtesy of various product distributors and manufacturers. Any mention of trade names, commercial products, or services does not constitute endorsement or recommendation for use by the American Water Works Association or the US Environmental Protection Agency.

Cover and book design by Susan DeSantis

Printed in the United States of America

American Water Works Association
6666 West Quincy Avenue
Denver, CO 80235
(303) 794-7711

Printed on recycled paper.

CONTENTS

Hydraulics Section

Chemistry Section

Electricity Section

Appendixes

FOREWORD

Basic Science Concepts and Applications is a supplementary reference book that accompanies the other four books in the series titled Principles and Practices of Water Supply Operations. The sections of the reference book — Mathematics, Hydraulics, Chemistry, and Electricity — provide extended discussions of the principles and operational calculations related to selected topics in the four books.

The other books in the series are

Water Sources
Water Treatment
Water Transmission and Distribution
Water Quality

A student workbook is available for each of the five books. These may be used by classroom students for completing assignments and reviewing questions and as a convenient method of keeping organized notes from class lectures. The workbooks, which provide a review of important points covered in each chapter, are also useful for self-study.

An instructor guide is also available for each of the books to assist teachers of classes in water supply operations. The guides provide the instructor with additional sources of information for the subject matter of each chapter, as well as an outline of the chapters, suggested visual aids, class demonstrations where applicable, and the answers to review questions provided in the student workbooks.

Acknowledgments

The second edition of *Basic Science Concepts and Applications* has been revised to include metric examples, and material has been updated where necessary. In addition, a student workbook and instructor guide have been created for use with this reference book.

Special thanks are extended to the following individuals who provided technical review of all or portions of the second-edition outline and manuscript:

Bill Hester, Denver Water, Denver, Colo.
Stephen E. Jones, Iowa State University, Ames, Iowa
Kenneth D. Kerri, California State University, Sacramento, Calif.
Gary B. Logsdon, Black & Veatch, Inc., Cincinnati, Ohio
Ken Morrison, M. M. Dillon Limited, London, Ontario
Dennis Mutti, M. M. Dillon Limited, London, Ontario
Robert K. Weir, Denver Water, Denver, Colo.

American Water Works Association staff who reviewed the manuscript included Brian Murphy, Frederick W. Pontius, and Steve Posavec.

Publication of the first edition was made possible through a grant from the US Environmental Protection Agency, Office of Drinking Water, under Grant No. T900632-01.

The principal developers were Joanne Kirkpatrick and Benton C. Price, under contract with VTN Colorado, Inc.

The following individuals are credited with participating in the review of the first edition: Charles R. Beer, Lucille M. Black, James O. Bryant Jr., Susan A. Castle, Jack C. Dice, James T. Harvey, William R. Hill, Jack W. Hoffbuhr, Kenneth D. Kerri, John L. Krantz, Ralph W. Leidholdt, L.H. Lockhart, Andrew J. Piatek Jr., John F. Rieman, Tom F. Staible, Robert L. Wubbena.

The electricity section was reprinted with minor changes from the first edition of AWWA Manual M2, *Automation and Instrumentation*, which was developed by the AWWA Distribution Division Committee on Automation and Instrumentation.

Portions of the text were adapted from *Mathematics for Water and Wastewater Treatment Plant Operators* (1973), published by Ann Arbor Science Publishers, Inc.

INTRODUCTION

Subject matter in this reference book is organized into four main sections — mathematics, hydraulics, chemistry, and electricity — supporting a broad range of topics in the other four books in the series. The chapters in each main section need not be read in the order in which they are presented. Where topics in other sections of this work are related and considered helpful to the reader, a footnote reference is given to the related discussion. A glossary at the end of the book defines important terms.

In general, calculations and answers are carried to two decimal places where the number is less than 999.99 (example: 27.12). Final zeros after the decimal point are not printed (example: 39.7, not 39.70). Numbers greater than 999.99 are given only to the nearest whole number (example: 1,718). Tables and graphs are read with whatever precision possible. These rules are adopted for consistency; they do not reflect the accuracy of the data used in the calculations. In practice, the operator should realize that the accuracy of a final answer can never be better than the accuracy of the data used.

Basic Science Concepts and Applications

Mathematics

MATHEMATICS 1

Powers and Scientific Notation

Two common methods of expressing a number—powers notation and scientific notation—will be discussed in this chapter.

Powers Notation

The most basic form of powers notation is merely a shorthand method of writing multiplication. For example, 4×4 can be written as

$$4^2$$

This is referred to as *4 to the second power*, or *4 squared*. The small 2 is the exponent, or power. It tells you how many 4's are to be multiplied together: two. In expanded form,

$$4^2 = (4)(4)$$

The expression *4 to the third power* (usually called *4 cubed*) is written as

$$4^3$$

In expanded form, this notation means

$$4^3 = (4)(4)(4)$$

The following examples further illustrate this concept of powers notation.

Example 1

How is the term 2^3 written in expanded form?

The power (or exponent) of 3 means that the number is multiplied by itself three times:

$$2^3 = (2)(2)(2)$$

Example 2

How is the term ft^2 written in expanded form?

The power or exponent of 2 means that the term is multiplied by itself two times:

$$ft^2 = (ft)(ft)$$

Example 3

How is the term 10^5 written in expanded form?

The exponent of 5 indicates that 10 is multiplied by itself five times:

$$10^5 = (10)(10)(10)(10)(10)$$

Example 4

How is the term $(2/3)^2$ written in expanded form?

When parentheses are used, the exponent refers to the entire term within the parentheses. Therefore, in this problem, $(2/3)^2$ means

$$\left(\frac{2}{3}\right)^2 = \left(\frac{2}{3}\right)\left(\frac{2}{3}\right)$$

Sometimes a negative exponent is used with a number or term. A number with a negative exponent can be reexpressed using a positive exponent:

$$3^{-2} = \frac{1}{3^2}$$

Another example is

$$12^{-3} = \frac{1}{12^3}$$

Example 5

How is the term 7^{-2} written in expanded form?

$$7^{-2} = \frac{1}{7^2} = \frac{1}{(7)(7)}$$

Example 6

How is the term 10^{-4} written in expanded form?

$$10^{-4} = \frac{1}{10^4} = \frac{1}{(10)\,(10)\,(10)\,(10)}$$

If a term is given in expanded form, you should be able to determine how it would be written in exponential (or power) form. For example,

$$(\text{ft})(\text{ft}) = \text{ft}^2$$

or

$$(3)(3)(3) = 3^3$$

Example 7

How would the following expanded term be rewritten in exponential form?

$$(\text{mm})(\text{mm})(\text{mm})$$

Since the term is multiplied by itself three times, it would be written in exponential form as

$$\text{mm}^3$$

Example 8

Write the following term in exponential form:

$$\frac{(4)\,(4)}{(5)\,(5)\,(5)}$$

The exponent for the numerator of the fraction is 2 and the exponent for the denominator is 3. Therefore, the term would be written as

$$\frac{4^2}{5^3}$$

Since the exponents are not the same, parentheses cannot be placed around the fraction and a single exponent cannot be used.

Example 9

Write the following term in exponential form:

$$\frac{(\text{ft})\,(\text{ft})}{(\text{in.})\,(\text{in.})}$$

The exponent of both the numerator and denominator is 2:

$$\frac{\text{ft}^2}{\text{in.}^2}$$

Since the exponents are the same, parentheses can be used to express this term, if desired:

$$\left(\frac{\text{ft}}{\text{in.}}\right)^2$$

Perhaps the two most common situations in which you may see powers used with a number or term are in denoting area or volume units (in.^2, ft^2, in.^3, ft^3) and in scientific notation.

Scientific Notation

Scientific notation is a method by which any number can be expressed as a term multiplied by a power of 10. The term itself is greater than or equal to 1 but less than 10. Examples of numbers written in scientific notation are

$$5.4 \times 10^1$$

$$1.2 \times 10^3$$

$$9.789 \times 10^4$$

$$3.62 \times 10^{-2}$$

The numbers can be taken out of scientific notation by performing the indicated multiplication. For example,

$$\begin{aligned} 5.4 \times 10^1 &= (5.4)(10) \\ &= 54 \end{aligned}$$

$$\begin{aligned} 1.2 \times 10^3 &= (1.2)(10)(10)(10) \\ &= 1{,}200 \end{aligned}$$

$$9.789 \times 10^4 = (9.789)(10)(10)(10)(10)$$
$$= 97{,}890$$

$$3.62 \times 10^{-2} = (3.62)\frac{1}{10^2}$$

$$= (3.62)\frac{(1)}{(10)\,(10)}$$

$$= \frac{3.62}{(10)\,(10)}$$

$$= 0.0362$$

An easier way to take a number out of scientific notation is by moving the decimal point the number of places indicated by the exponent.

Rule 1
When a number is taken *out* of scientific notation, a *positive* exponent indicates a decimal point move to the *right*, and a *negative* exponent indicates a decimal point move to the *left*.

Let's look again at the examples above, using the decimal point move rather than the multiplication method. The first example is

$$5.4 \times 10^1$$

The *positive* exponent of 1 indicates that the decimal point in 5.4 should be moved one place to the *right*:

$$5.4 = 54$$

The next example is

$$1.2 \times 10^3$$

The *positive* exponent of 3 indicates that the decimal point in 1.2 should be moved three places to the *right*:

$$1.200 = 1{,}200$$

The next example is

$$9.789 \times 10^4$$

The *positive* exponent of 4 indicates that the decimal point should be moved four places to the *right*:

$$9.7890_{\wedge} \quad = \quad 97{,}890$$

The final example is

$$3.62 \times 10^{-2}$$

The *negative* exponent of 2 indicates that the decimal point should be moved two places to the *left*:

$$_{\wedge}03.62 \quad = \quad 0.0362$$

Example 10

Take the following number out of scientific notation:

$$7.992 \times 10^5$$

The positive exponent of 5 indicates that the decimal point should be moved five places to the right:

$$7.99200_{\wedge} \quad = \quad 799{,}200$$

Example 11

Take the following number out of scientific notation:

$$2.119 \times 10^{-3}$$

The negative exponent of 3 indicates that the decimal point should be moved three places to the left:

$$_{\wedge}002.119 \quad = \quad 0.002119$$

Although there will be very few instances in which you will need to put a number *into* scientific notation, the method is discussed below.

To put a number into scientific notation, the decimal point is moved the number of places necessary to result in a number between 1 and 9. This number is multiplied by a power of 10, with the exponent equal to the number of places that the decimal point was moved. (Remember that if no decimal point is shown in the number to be converted, it is assumed to be at the end of the number.)

Rule 2
When a number is put *into* scientific notation, a decimal point move to the *left* indicates a *positive* exponent; a decimal point move to the *right* indicates a *negative* exponent.

Now let's try converting a few numbers into scientific notation, using the same numbers as in the previous examples.

First, let's convert

$$54$$

To obtain a number between 1 and 9, the decimal point should be moved one place to the left. This move of one place gives the exponent, and the move to the *left* means that the exponent is *positive*:

$$54 = 5.4 \times 10^1$$

The next number to be put into scientific notation is

$$1,200$$

To obtain a number between 1 and 9, the decimal point should be moved three places to the left. The number of place moves (3) becomes the exponent of the power of 10, and the move to the *left* indicates a *positive* exponent:

$$1,200 = 1.2 \times 10^3$$

The next example is

$$97,890$$

To obtain a number between 1 and 9, the decimal point should be moved four places to the *left*, resulting in a *positive* exponent of 4:

$$97,890 = 9.789 \times 10^4$$

The final example is

$$0.0362$$

To obtain a number between 1 and 9, the decimal point must be moved two places to the right. This indicates an exponent of 2, and the move to the *right* requires a *negative* exponent:

$$0.0362 = 3.62 \times 10^{-2}$$

Example 12

Put the following number into scientific notation:

$$4{,}573{,}000$$

To obtain a number between 1 and 9, the decimal point should be moved six places to the left, resulting in a positive exponent of 6:

$$4{,}573{,}000 = 4.573 \times 10^{6}$$

Example 13

Convert the following decimal to scientific notation:

$$0.000375$$

To obtain a number between 1 and 9, the decimal point should be moved four places to the right, resulting in a negative exponent of 4:

$$0.000375 = 3.75 \times 10^{-4}$$

MATHEMATICS 2

Dimensional Analysis

Dimensional analysis is a tool that you can use to determine whether you have set up a problem correctly. In checking a math setup using dimensional analysis, you work only with the dimensions or units of measure and not with the numbers themselves.

To use the dimensional analysis method, you must know three things:

- how to express a horizontal fraction (such as gal/ft^3) as a vertical fraction (such as $\frac{\text{gal}}{\text{ft}^3}$)
- how to divide by a fraction
- how to divide out or cancel terms in the numerator and denominator of a fraction

These techniques are reviewed briefly below.

When you are using dimensional analysis to check a problem, it is often desirable to write any horizontal fractions as vertical fractions, thus

$$\text{ft}^3/\text{min} = \frac{\text{ft}^3}{\text{min}}$$

$$\text{s}/\text{min} = \frac{\text{s}}{\text{min}}$$

$$\frac{\text{gal}/\text{min}}{\text{gal}/\text{ft}^3} = \frac{\dfrac{\text{gal}}{\text{min}}}{\dfrac{\text{gal}}{\text{ft}^3}}$$

When a problem involves division by a fraction, the rule is to invert (or turn over) the terms in the denominator and then multiply. For example,

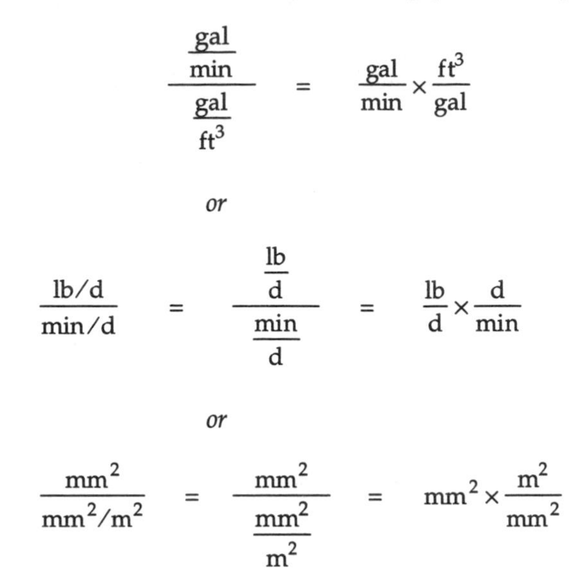

Once the fractions in a problem, if any, have been rewritten in the vertical form, and division by a fraction has been reexpressed as multiplication as shown above, then the terms can be divided out or canceled. For every term canceled in the numerator of a fraction, a similar term must be canceled in the denominator, and vice versa, as shown below:

$$\frac{\cancel{\text{gal}}}{\text{min}} \times \frac{\text{ft}^3}{\cancel{\text{gal}}} = \frac{\text{ft}^3}{\text{min}}$$

$$\frac{\text{kg}}{\cancel{\text{d}}} \times \frac{\cancel{\text{d}}}{\text{min}} = \frac{\text{kg}}{\text{min}}$$

$$\cancel{\text{mm}^2} \times \frac{\text{m}^2}{\cancel{\text{mm}^2}} = \text{m}^2$$

$$\frac{\cancel{\text{ft}^3}}{\cancel{\text{s}}} \times \frac{\text{gal}}{\cancel{\text{ft}^3}} \times \frac{\cancel{\text{s}}}{\cancel{\text{min}}} \times \frac{\cancel{\text{min}}}{\text{d}} = \frac{\text{gal}}{\text{d}}$$

You may wish to review the concept of powers[1] before continuing with the following examples in dimensional analysis.

Suppose you wish to convert 1,200 ft^3 volume to gallons, and suppose that you know you will use 7.48 gal/ft^3 in the conversion but that you don't know whether to multiply or divide by 7.48. Let's look at both possible ways and see how dimensional analysis can be used to choose the correct way. *Only the dimensions* will be used to determine if the math setup is correct.

First, try multiplying the dimensions:

$$(ft^3)\,(gal/ft^3) \quad = \quad (ft^3)\left(\frac{gal}{ft^3}\right)$$

Then multiply the numerators and denominators to get

$$= \quad \frac{(ft^3)\,(gal)}{ft^3}$$

and cancel common terms to get

$$= \quad \frac{\cancel{(ft^3)}\,(gal)}{\cancel{ft^3}}$$

$$= \quad gal$$

So, by dimensional analysis you know that if you *multiply* the two dimensions (ft^3 and gal/ft^3), the answer you get will be in gallons, which is what you want. Therefore, since the math setup is correct, you would then multiply the numbers to obtain the number of gallons.

$$(1{,}200\ ft^3)(7.48\ gal/ft^3) \quad = \quad 8{,}976\ gal$$

What would have happened if you had divided the dimensions instead of multiplying?

$$\frac{ft^3}{gal/ft^3} \quad = \quad \frac{ft^3}{\frac{gal}{ft^3}}$$

---

[1]Mathematics 1, Powers and Scientific Notation.

$$= \quad ft^3\left(\frac{ft^3}{gal}\right)$$

Then multiply the numerators and denominators of the fraction to get

$$= \quad \frac{ft^6}{gal}$$

So had you *divided* the two dimensions (ft^3 and gal/ft^3), the units of the answer would have been ft^6/gal, *not* gal. Clearly you do not want to divide in making this conversion.

Example 1

You wish to obtain an answer in square feet. If you are given the two terms — 80 ft^3/s and 3.5 ft/s — is the following math setup correct?

$$(80\ ft^3/s)(3.5\ ft/s)$$

First, only the dimensions are used to determine if the math setup is correct. By multiplying the two dimensions, you get

$$(ft^3/s)\,(ft/s) \quad = \quad \left(\frac{ft^3}{s}\right)\left(\frac{ft}{s}\right)$$

Then multiply the terms in the numerators and denominators of the fraction:

$$= \quad \frac{(ft^3)\,(ft)}{(s)\,(s)}$$

$$= \quad \frac{ft^4}{s^2}$$

The math setup is wrong since the dimensions of the answer are not square feet. Therefore, if you multiply the numbers just as you did the dimensions, the answer will be wrong.

Let's try division of the two dimensions instead.

$$\frac{ft^3/s}{ft/s} \quad = \quad \frac{\dfrac{ft^3}{s}}{\dfrac{ft}{s}}$$

Then invert the denominator and multiply to get

$$= \left(\frac{ft^3}{s}\right)\left(\frac{s}{ft}\right)$$

$$= \frac{(ft)\,(ft)\,(ft)\,(s)}{(s)\,(ft)}$$

$$= \frac{(ft)\,(ft)\,\cancel{(ft)}\,\cancel{(s)}}{\cancel{(s)}\,\cancel{(ft)}}$$

$$= ft^2$$

This math setup is correct since the dimensions of the answer are square feet. Therefore, if you *divide* the numbers as you did the units, the answer will also be correct:

$$\frac{80\ ft^3/s}{3.5\ ft/s} = 22.86\ ft^2$$

Example 2

Suppose you have two terms — 3 m/s and 6 m^2 — and you wish to obtain an answer in cubic meters per second (m^3/s). Is multiplying the two terms the correct math setup?

$$(m/s)\,(m^2) = \frac{m}{s} \times m^2$$

Then multiply the numerators and denominators of the fraction:

$$= \frac{(m)\,(m^2)}{s}$$

$$= \frac{m^3}{s}$$

The math setup is correct since the dimensions of the answer are cubic meters per second (m^3/s). So if you multiply the numbers, just as you did the dimensions, you will get the correct answer:

$$(3\ m/s)\,(6\ m^2) = 18\ m^3/s$$

Example 3

Suppose you have been given the following problem: "The flow rate in a water line is 2.3 ft^3/s. What is the flow rate expressed as gallons per minute?" You then set up the math problem as shown below. Use dimensional analysis to determine if this math setup is correct.

$$(2.3 \text{ ft}^3/\text{s})\,(7.48 \text{ gal/ft}^3)\,(60 \text{ s/min})$$

Dimensional analysis is used to check the math setup:

$$(\text{ft}^3/\text{s})\,(\text{gal/ft}^3)\,(\text{s/min}) = \left(\frac{\text{ft}^3}{\text{s}}\right)\left(\frac{\text{gal}}{\text{ft}^3}\right)\left(\frac{\text{s}}{\text{min}}\right)$$

$$= \left(\frac{\cancel{\text{ft}^3}}{\cancel{\text{s}}}\right)\left(\frac{\text{gal}}{\cancel{\text{ft}^3}}\right)\left(\frac{\cancel{\text{s}}}{\text{min}}\right)$$

$$= \frac{\text{gal}}{\text{min}}$$

This analysis indicates that the math setup is correct as shown above.

Example 4

You have been given the following problem: "A channel is 3 ft wide with water flowing to a depth of 2 ft. The velocity in the channel is found to be 1.8 ft/s. What is the flow rate in the channel in cubic feet per second?" You then set up the math problem as shown below. Use dimensional analysis to determine if this math setup is correct.

$$\frac{(3 \text{ ft})\,(2 \text{ ft})}{1.8 \text{ ft/s}}$$

Dimensional analysis is used to check the math setup:

$$\frac{(\text{ft})\,(\text{ft})}{\text{ft/s}} = \frac{(\text{ft})\,(\text{ft})}{\frac{\text{ft}}{\text{s}}}$$

$$= (\text{ft})\,(\text{ft})\,\frac{\text{s}}{\text{ft}}$$

$$= \frac{(\text{ft})\,(\text{ft})\,(\text{s})}{\text{ft}}$$

$$= \frac{\cancel{(\text{ft})}\,(\text{ft})\,(\text{s})}{\cancel{\text{ft}}}$$

$$= \text{ft s}$$

Since the dimensions of the answer are incorrect, the math setup shown in example 4 is also incorrect. Had the math setup instead been (3 ft)(2 ft)(1.8 ft/s), the dimensions of the answer would have been correct.

MATHEMATICS 3

Rounding and Estimating

The practice of estimating the size of the answer to a calculation helps you to avoid reporting an answer as 5,000, for example, when it should be 50,000, or 40,000 when it should be 4,000,000. Suppose when you are solving a problem that you forget to multiply or divide by a certain number, or suppose you punch a wrong button on the calculator. If you have an idea of what the approximate answer should be, then you can recognize an incorrect answer and recheck your arithmetic.

In general, the process of estimating the size of an answer involves two steps:

- rounding the numbers
- completing the calculation using the rounded numbers

Rounding means replacing the final digits of a number with zeros, thus expressing the number as tens, hundreds, thousands, or tenths, hundredths, thousandths, etc. (for example, expressing 498 as 500; 0.19 as 0.2; or 5,754,193 as 6,000,000). Equations using rounded numbers are often written with an approximate sign (8 or ≈) instead of an equals sign (=) to show that the answer is an estimate.

Rounding

The technique of rounding numbers is based on a particular place value in the decimal system. The various place values are reviewed below to emphasize the importance of understanding this concept before proceeding further.

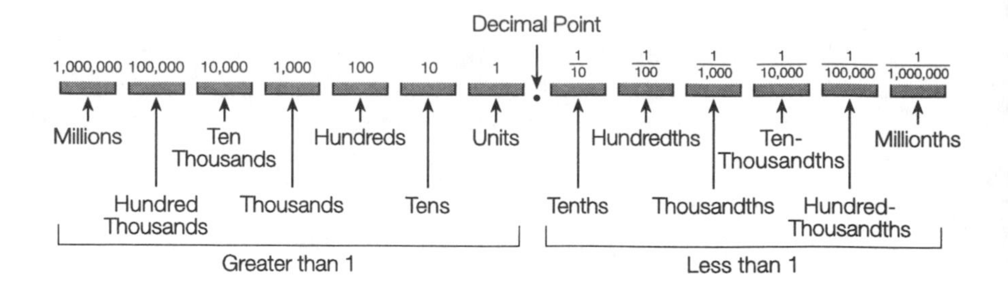

Suppose you want to round the number 2,427 to the nearest hundred. Rounding this number depends only on the *size of the digit just to the right* of the hundreds place:

hundreds place
↓
2,427
↑
digit that determines rounding

Similarly, if you want to round the number 323,772 to the nearest thousand, this rounding depends only on the *size of the digit just to the right* of the thousands place:

thousands place
↓
323,772
↑
digit that determines rounding

Now consider the rules for *rounding whole numbers*. The procedure depends on whether the digit just to the right of the rounding place is less than 5, is 5, or is greater than 5.

Rule 1
If the digit is less than 5: When rounding to any desired place, if the digit to the right of that place is less than 5, replace all digits to the right of the rounding place with zeros.

Rule 2
If the digit is 5 or greater than 5: When rounding to any desired place, if the digit to the right of that place is 5 or greater than 5, increase the digit in the rounding place by 1 and replace all digits to the right of the increase with zeros.

The best way to understand rounding rules is to look at a few examples of rounding.

Example 1

Round 37,926 to the nearest hundred.

The procedure used in this rounding depends on the digit just to the right of the hundreds place:

37,926
↑
hundreds place

Since the digit to the right of the hundreds place is less than 5, the 9 is not changed and all the digits to the right of the 9 are replaced with zeros:

37,926 ≈ 37,900 (rounded to the nearest hundred)

Example 2

Round 248,722 to the nearest thousand.

The procedure used depends on the digit just to the right of the thousands place:

248,722
↑
thousands place

Since the digit to the right of the thousands place is greater than 5, the 8 in the thousands place is increased by 1 and all the digits to the right of the 8 are replaced with zeros:

248,722 ≈ 249,000 (rounded to the nearest thousand)

Example 3

Round 25,675 to the nearest ten thousand.

The digit just to the right of the ten-thousands place determines how the number is rounded:

25,675
↑
ten-thousands place

Since the digit to the right of the ten-thousands place is 5, the 2 is increased by 1 and all digits to the right of the 2 are replaced with zeros:

25,675 ≈ 30,000 (rounded to the nearest ten thousand)

Example 4

Round 14,974 to the nearest hundred.

The digit to the right of the hundreds place determines how the number is rounded:

14,974
↑
hundreds place

Since the digit to the right of the hundreds place is 7, the 9 should be increased by 1 and all digits to the right replaced with zeros. But notice that increasing the 9 by 1 changes it to 10. The zero in the 10 replaces the 9, and the 1 is carried and added to the 4 in the thousands place:

14,974 ≈ 15,000 (rounded to the nearest hundred)

The rules for rounding decimal numbers are basically the same as those for rounding whole numbers, with the following modification:

Rule 3
When rounding decimal numbers to a place to the right of the decimal point, instead of replacing the rounded digits with zeros, drop the rounded digits.

The following examples illustrate this procedure.

Example 5

Round 37.18236 to the nearest thousandth.

As in rounding whole numbers, the procedure used depends on the size of the digit just to the right of the rounding place:

37.18236
↑
thousandths place

According to the basic rounding rules, since the digit to the right of the thousandths place is less than 5, the 2 remains unchanged and all digits to the right of the 2 are dropped:

37.18236 ≈ 37.182 (rounded to the nearest thousandth)

Example 6

Round 5.654 to the nearest tenth.

5.654
↑
tenths place

The digit to the right of the tenths place is 5. Therefore the 6 is increased by 1 and all digits to the right are dropped:

5.654 ≈ 5.7 (rounded to the nearest tenth)

Example 7

Round 483.16 to units.

483.16
↑
units place

Even though the decimal point falls between the rounding place and the digit just to the right, the procedure still follows the basic rounding rules. The digit to the right of the units place is less than 5; therefore it (and all the digits after it) should be replaced with zeros. But, rounded digits on the right side of the decimal point are dropped. Therefore, the decimal number is rounded to a whole number.

483.16 ≈ 483 (rounded to the nearest unit)

Estimating

When estimating the answer to a problem, you will find it helpful to round each number in the calculation so that only one digit remains, with the rest of the digits in the number either changed to zeros or dropped in accordance with the rounding rules. In this way the estimation can often be done in your head, or at least with a minimum of computing with pencil or calculator.

However, you should be aware that the more rounding you do in an estimation, the greater the difference may be between the digits of your estimated answer and those of the actual answer. Keep this principle in mind:

> Rounding and estimating indicate the approximate size (place value) of a calculated answer but do not necessarily indicate the numerical value of the answer.

For instance, if the estimate to a problem is 40,000, you can expect the calculated answer to be in the tens of thousands (not hundreds or millions); but the value of the calculated answer could fall between 10,000 and 90,000.

Also consider what the place value of your answer could be if the digit in the estimated answer is 1 or 9. With an estimate of 900, the actual answer will probably be in the hundreds (somewhere between 100 and 900) — but it could also be a little more than 900 (such as 1,020). On the other hand, if the estimate is 1,000, then the calculated answer could be in the thousands, or a little less than 1,000 (in the high hundreds).

Now let's look at the mechanics of estimating. Suppose you have rounded two numbers so that your estimate of the answer is 20×30. In making this estimate, first multiply the two digits (2 and 3):

$$\begin{array}{c} (20)\ (30) \\ \downarrow \quad \downarrow \\ 2 \times 3 \end{array} = 6$$

Then count all the zeros in the calculation and put that number of zeros after the 6:

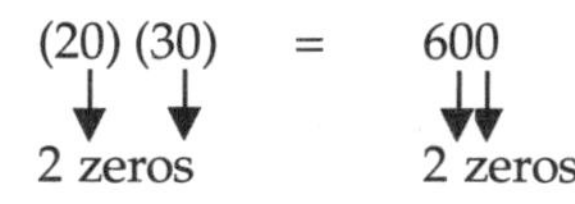

The estimated answer is 600, so the actual answer should be in the hundreds.

Let's look at a few more examples of this procedure.

Example 8

A calculation has been rounded so that it can be estimated. What is the estimated answer for the calculation?

(600) (3,000)

First the 6 and the 3 are multiplied, then the total number of zeros are added:

$$(600)\ (3,000) = \underbrace{18}_{6 \times 3}\ \underbrace{00000}_{5 \text{ zeros}}$$

Therefore, the estimated answer for the calculation is 1,800,000; the actual answer should be in the millions or high hundred thousands.

Example 9

A calculation has been rounded to the numbers shown below. What is the estimated answer for the calculation?

(400) (70,000)

First the 4 and 7 are multiplied, then the total number of zeros are added:

$$(400)\ (70,000) = \underbrace{28}_{4 \times 7}\ \underbrace{000000}_{6 \text{ zeros}}$$

The estimated answer is therefore 28,000,000; the actual answer should be in the ten millions.

Example 10

What is the estimated answer for the following rounded calculation?

(20) (300) (40)

As in the two preceding examples, the 2, 3, and 4 are multiplied and then the total number of zeros are added:

$$(20)\ (300)\ (40) \quad = \quad \underbrace{24}_{2 \times 3 \times 4} \quad \underbrace{0000}_{4\ \text{zeros}}$$

The estimated answer is therefore 240,000; the actual answer should be in the hundred thousands.

When three or more digits must be multiplied, rounding is sometimes required during the multiplication process. In the last example, the three digits could be multiplied easily, so no rounding was needed. Suppose, however, that you wanted to estimate the answer to

$$(800)\ (7{,}000)\ (800)$$

In this case the multiplication of the digits is not quite so obvious. Therefore, rounding during multiplication is permissible:

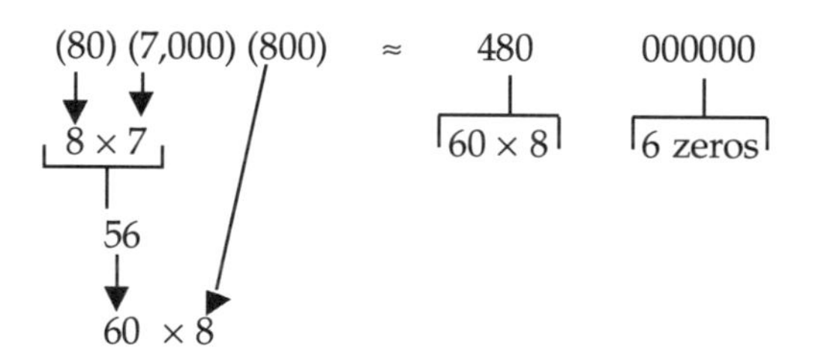

The following two examples illustrate the technique of rounding during multiplication of digits.

Example 11

The estimate of a calculation is shown below. What is the estimated answer?

$$(50)\ (50)\ (3{,}000)$$

As in the other examples, the 5, 5, and 3 are multiplied; then the zeros are added:

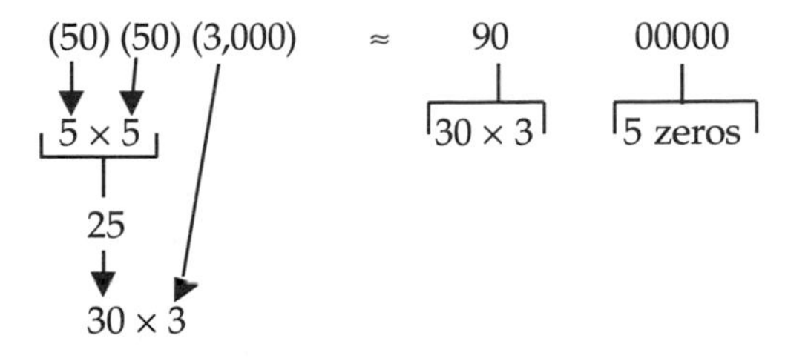

The estimated answer is therefore 9,000,000; the actual answer should be in the millions or ten millions.

Example 12

What is the estimated answer to the following calculation?

(9) (700) (60) (70)

First the 9, 7, 6, and 7 are multiplied; then the zeros are added. This problem requires more rounding during the multiplication step than was required in the previous examples:

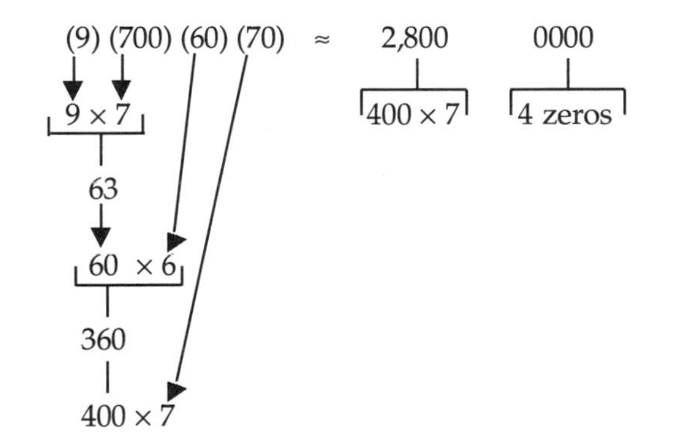

The estimated answer is 28,000,000; so the actual answer should be in the ten millions.

Sometimes the calculation being estimated involves division. When this occurs, the estimate can often be simplified by dividing out (or crossing off) final zeros; that is, zeros at the end of a whole number. The next few examples illustrate this technique.

Example 13

What is the estimated answer to the following calculation?

$$\frac{40{,}000}{200}$$

In this problem the estimate may first be simplified. When a final zero is crossed off in the denominator (bottom) of the fraction, a final zero must be crossed off in the numerator (top) of the fraction. The simplified estimate is therefore

$$\frac{40{,}0\not{0}\not{0}}{2\not{0}\not{0}} = \frac{400}{2}$$

Then the estimate can be completed:

$$\frac{400}{2} = 200$$

Next, let's consider problems where the *estimated answer is a decimal less than 1.* For estimating purposes, it is necessary to obtain only the first nonzero digit to the right of the decimal point. If an estimated answer is 0.01, for example, the actual answer could be 0.014, or 0.02, or even 0.009. The estimate of 0.01 merely shows that the answer will probably be in the *low* hundredths (although it could be in the high thousandths). The point is that the estimate should give you a "ball-park" answer, but *you must take particular care in determining where to place the decimal point* in the calculated answer.

Example 14

Complete the following estimate:

$$\frac{700}{6{,}000}$$

First the estimate is simplified by crossing out final zeros, and then the problem is completed by division. As previously stated, for estimated decimal answers less than 1, it is necessary to obtain only the first nonzero digit to the right of the decimal point.

$$\frac{7\not{0}\not{0}}{6{,}0\not{0}\not{0}} = \frac{7}{60}$$

$$\begin{array}{r} .1 \\ 60\ \overline{)\,7.0} \\ \underline{6.0} \end{array}$$

Therefore, the estimated answer is 0.1; the actual answer should be in the tenths (0.1 to 0.9) or high hundredths (such as 0.09 or 0.08).

Example 15

The estimate of a calculation is shown below. What is the estimated answer?

$$\frac{2{,}000}{7{,}000{,}000}$$

This estimate is first simplified by crossing out final zeros:

$$\frac{2{,}\not{0}\not{0}\not{0}}{7{,}000{,}\not{0}\not{0}\not{0}} = \frac{2}{7{,}000}$$

Then the problem is completed by division. As in the last example, it is necessary to obtain only the first nonzero digit (after the required number of zeros) to the right of the decimal point.

$$\begin{array}{r} .0002 \\ 7{,}000\ \overline{)\,2.0000} \\ \underline{1.4000} \end{array}$$

Therefore, the answer to this estimate is 0.0002; the actual answer should be in the ten thousandths.

In many practical applications involving rounding and estimating, the problem requires both multiplication and division. When this is the case, first simplify as much as possible by crossing out zeros. Next, multiply the numbers in the numerator, then those in the denominator. Finally, complete the problem by division. The next two examples illustrate this technique.

Example 16

Calculate the estimated answer to

$$\frac{(20)\,(400)}{(50)\,(80)}$$

First, the problem is simplified by crossing out final zeros where possible:

$$\frac{(20)\,(4\cancel{0}\cancel{0})}{(5\cancel{0})\,(8\cancel{0})} = \frac{(20)\,(4)}{(5)\,(8)}$$

The numerator and denominator are each multiplied:

$$\frac{(20)\,(4)}{(5)\,(8)} = \frac{80}{40}$$

Additional zeros may be crossed out and the problem completed by division:

$$\frac{(8\cancel{0})}{(4\cancel{0})} = \frac{8}{4}$$

$$= 2$$

The estimated answer is 2, so the actual answer should be in the units.

Example 17

Calculate the answer to the following estimated problem:

$$\frac{(4{,}000)\,(40)}{(60)\,(3)}$$

First, the problem is simplified by crossing out zeros:

$$\frac{(4{,}00\cancel{0})\,(40)}{(6\cancel{0})\,(3)} = \frac{(400)\,(40)}{(6)\,(3)}$$

The numerator and denominator are each multiplied:

$$\frac{(400)\,(40)}{(6)\,(3)} = \frac{16{,}000}{18}$$

Before the problem is completed by division, the numbers in the fraction can be rounded and additional zeros crossed out:

$$\frac{16{,}000}{18} \approx \frac{20{,}00\cancel{0}}{2\cancel{0}}$$

$$\approx \frac{2{,}000}{2}$$

$$\approx 1{,}000$$

The estimated answer is therefore 1,000, and the actual answer should be in the thousands or high hundreds.

In examples 16 and 17, the numbers were already in rounded form. In practical applications, however, you must round the numbers in the problem before continuing with the estimate. And as noted at the beginning of this section, the more rounding you do, the greater the difference may be between the digits of your estimated answer and the actual answer. The estimated answer will show you what the approximate place value of the calculated answer should be, but the numerical values of the two answers may vary considerably.

Let's look at a few examples that require you to round the numbers before estimating. In each case the estimate is calculated according to the rules given in this section. Then the actual answer, based on the numbers before rounding, is given to show the comparison between the estimated and actual answers.

Example 18

Estimate the answer to the following problem:

$$\frac{(29)\,(244)}{(12)\,(32)}$$

First, each term in the calculation is rounded:

$$\frac{(29)\,(244)}{(12)\,(32)} \approx \frac{(30)\,(200)}{(10)\,(30)}$$

The estimate can now be completed as in the previous examples. The problem is first simplified by crossing out zeros; then the numerator and denominator are each multiplied, and finally the estimated answer is obtained by division:

$$\frac{(3\cancel{0})\,(20\cancel{0})}{(1\cancel{0})\,(3\cancel{0})} = \frac{60}{3}$$

$$= 20$$

The estimated answer is 20; the actual answer is 18.43.

Example 19

Estimate the answer to the following problem:

$$\frac{(430)\,(61{,}702)}{(65)\,(65)}$$

Each term in the calculation is first rounded; then the estimation is completed as in previous examples.

$$\frac{(430)\,(61{,}702)}{(65)\,(65)} \approx \frac{(4\cancel{0}\cancel{0})\,(60{,}000)}{(7\cancel{0})\,(7\cancel{0})}$$

$$\approx \frac{240{,}000}{49}$$

Then round the terms before division, cancel zeros, and complete the estimate:

$$\approx \frac{200{,}00\cancel{0}}{5\cancel{0}}$$

$$\approx \quad 4{,}000$$

The estimated answer is 4,000; the actual answer is 6,280.

Example 20

Estimate the answer to the following problem:

$$\frac{(291)\,(34)\,(419)}{(21)\,(169)}$$

First, each term is rounded and the problem is simplified by crossing out final zeros. Then the numerator and denominator are each multiplied; finally, the answer is obtained by division.

$$\frac{(291)\,(34)\,(419)}{(21)\,(169)} \approx \frac{(3\cancel{0}\cancel{0})\,(3\cancel{0})\,(400)}{(2\cancel{0})\,(2\cancel{0}\cancel{0})}$$

$$\approx \frac{(3)\,(3)\,(400)}{(2)\,(2)}$$

$$\approx \quad \frac{3{,}600}{4}$$

$$\approx \quad 900$$

The estimated answer is 900; the actual answer is 1,168. This problem shows how the place value of the estimated and actual answers can vary.

Up to this point, we have been considering estimates using primarily rounded *whole* numbers. In some practical problems, however, you will be working with decimals less than 1 or with combinations of whole numbers and decimals less than 1. The rounding rules for decimals have already been illustrated. But in performing the necessary multiplication and division steps in an estimate, you may have to do some of the arithmetic in longhand in order to determine the position of the decimal point (and therefore the place value) of the answer.

Suppose, for example, you must round and estimate the following calculation:

$$\frac{(2.375)\,(4.75)\,(247)}{(61)\,(92)\,(0.785)}$$

First, round each term to one digit (plus the necessary number of zeros); then cancel the final zeros where possible:

$$\frac{(2.375)\,(4.75)\,(247)}{(61)\,(92)\,(0.785)} \approx \frac{(2)\,(5)\,(2\not{0}\not{0})}{(6\not{0})\,(9\not{0})\,(0.8)}$$

Next, multiply the terms in the numerator, then those in the denominator. Round and cancel final zeros again.

$$\frac{(2)\,(5)\,(2)}{(6)\,(9)\,(0.8)} \approx \frac{20}{(54)\,(0.8)}$$

$$\approx \frac{2\not{0}}{(5\not{0})\,(0.8)}$$

$$\approx \frac{2}{4}$$

Finally, divide to obtain the estimated answer:

$$4\overline{)2.0}\quad = .5$$

The estimated answer is 0.5; the actual answer is 0.63.

Now consider another example:

$$\frac{(271)(32)(0.06725)}{(725)(0.0024)}$$

Round each term to one digit plus the required number of zeros:

$$\frac{(300)(30)(0.07)}{(700)(0.002)}$$

Now cancel the final zeros of whole numbers, as possible. Note that *the only zeros that can be canceled are the final zeros of whole numbers.*

$$\frac{(3\cancel{00})(30)(0.07)}{(7\cancel{00})(0.002)} = \frac{(3)(30)(0.07)}{(7)(0.002)}$$

Next, multiply the numerator, then the denominator. You will probably want to do the arithmetic in longhand to place the decimal points properly.

$$= \frac{(90)(0.07)}{(7)(0.002)}$$

$$= \frac{6.3}{0.014}$$

Round each term again, then divide to complete the estimate:

$$\frac{6.3}{0.014} \approx \frac{6}{0.01}$$

$$0.01\overline{)6.00}\quad = 600.$$

The estimated answer is 600; the actual answer is 335.17.

As you can see from these two examples, estimates involving decimal numbers less than 1 can be time consuming, and if you misplace the decimal point, your answer could be wrong by one or more place values. You need to be particularly careful in estimating this type of problem because the possibility of error greatly increases.

An easier way to estimate problems that have decimal numbers less than 1 is to put all the rounded numbers of the estimate into scientific notation. The rules for converting numbers to and from scientific notation are given in another section[1] and are not discussed here. However, there are some basic rules for multiplying and dividing numbers in scientific notation that you must know.

> **Rule 1**
> When you *multiply* the numbers in scientific notation, multiply the numbers but *add the exponents.*

> **Rule 2**
> When you *divide* in scientific notation, divide the numbers but *subtract the exponent* in the denominator from the exponent in the numerator.

First let's look at some examples that demonstrate the multiplication rule.

Example 21

Calculate the following:

$$(20)\,(400)$$

You can see immediately that the answer is 8,000. But, for practice, put the calculation into scientific notation:

$$(20)\,(400) \quad = \quad (2 \times 10^1)\,(4 \times 10^2)$$

---

[1]Mathematics 1, Powers and Scientific Notation.

Now multiply the numbers, then add the exponents:

$$(2 \times 10^1)\,(4 \times 10^2) = (2)\,(4) \times 10^{1+2}$$

$$= 8 \times 10^3$$

Finally, take the answer out of scientific notation by moving the decimal point. The *positive* exponent of 3 indicates that the decimal point should be moved three places to the *right*.

$$8.000 = 8{,}000$$

Example 22

Multiply the following whole number and decimal, using scientific notation:

$$(0.0003)\,(300)$$

In scientific notation this becomes

$$(0.0003)\,(300) = (3 \times 10^{-4})\,(3 \times 10^2)$$

Now multiply the numbers and add the exponents:

$$(3 \times 10^{-4})\,(3 \times 10^2) = (3)\,(3) \times 10^{-4+2}$$

$$= 9 \times 10^{-2}$$

Finally, take the answer out of scientific notation by moving the decimal point. The *negative* exponent of 2 indicates a decimal point move two places to the *left*.

$$09. = 0.09$$

The next example shows how to multiply three decimals and not lose track of the decimal point.

Example 23

Calculate the following, using scientific notation:

$$(0.003)\,(0.2)\,(0.0006)$$

First, convert to scientific notation; then multiply the numbers and add the exponents:

$$(0.003)\ (0.2)\ (0.0006) = (3 \times 10^{-3})\ (2 \times 10^{-1})\ (6 \times 10^{-4})$$

$$= (3)\ (2)\ (6) \times 10^{-3-1-4}$$

$$= 36 \times 10^{-8}$$

Finally, take the answer out of scientific notation by moving the decimal point. The *negative* exponent of 8 indicates a decimal point move of eight places to the *left*.

$$00000036. = 0.00000036$$

For estimating purposes, a *whole* number between 1 and 9 need not be converted to scientific notation. It would merely be the same as multiplying the number by 10^0, which is simply 1. Consider how this can happen:

2×10^2 means (2) (10) (10) or 200

2×10^1 means (2) (10) or 20

2×10^0 means (2)[*not multiplied by 10*] or 2

However, in using scientific notation to perform a calculation, you may sometimes get an answer where the positive and negative exponents cancel each other and the answer becomes a number multiplied by 10^0. How do you take such a number out of scientific notation? Since 10^0 means that the number is *not* raised to a power of 10, the answer is, in effect, already out of scientific notation, and the number stands "as is." Another way to state this is that any number times 10^0 equals the number itself.

Let's consider an example:

$$(0.002)\ (3{,}000)$$

First, convert to scientific notation; then multiply the numbers and add the exponents:

$$(0.002)\ (3{,}000) = (2 \times 10^{-3})\ (3 \times 10^{3})$$

$$= (2)\ (3) \times 10^{-3+3}$$

$$= 6 \times 10^0$$

$$= 6 \times 1$$

$$= 6$$

Now let's look at some examples using the division rule for scientific notation; that is, divide the numbers and subtract the exponent in the denominator from the exponent in the numerator.

Example 24

Calculate the following:

$$\frac{800}{20}$$

You know immediately that the answer is 40. But let's put it into scientific notation:

$$\frac{800}{20} = \frac{8 \times 10^2}{2 \times 10^1}$$

Now divide the numbers, but subtract the denominator exponent from the numerator exponent:

$$\frac{8 \times 10^2}{2 \times 10^1} = \left(\frac{8}{2}\right) \times 10^{2-1}$$

$$= 4 \times 10^1$$

Finally, take the number out of scientific notation by moving the decimal point. The *positive* exponent of 1 indicates a decimal point move of one place to the *right*.

$$4.0 = 40$$

Example 25

Calculate the following, using scientific notation:

$$\frac{800}{0.2} = \frac{8 \times 10^2}{2 \times 10^{-1}}$$

Divide the numbers, and subtract the denominator exponent from the numerator exponent:

$$\frac{8 \times 10^2}{2 \times 10^{-1}} = \left(\frac{8}{2}\right) \times 10^{2-(-1)}$$

Here, the two minus signs become a plus sign (two negatives make a positive).

$$= \left(\frac{8}{2}\right) \times 10^{2+1}$$

So, the answer becomes

$$= 4 \times 10^3$$

And taking the answer out of scientific notation,

$$4 \times 10^3 = 4{,}000$$

The next example shows how you can divide decimals and keep track of the decimal point for the answer.

Example 26

Make the following calculation, using scientific notation:

$$\frac{0.006}{0.3} = \frac{6 \times 10^{-3}}{3 \times 10^{-1}}$$

Divide the numbers, and subtract the exponent in the denominator from the exponent in the numerator:

$$\frac{6 \times 10^{-3}}{3 \times 10^{-1}} = \left(\frac{6}{3}\right) \times 10^{(-3)-(-1)}$$

Again, the two minus signs become a plus:

$$= \left(\frac{6}{3}\right) \times 10^{-3+1}$$

And taking the answer out of scientific notation

$$2 \times 10^{-2} = 0.02$$

Now consider the estimating problem we used previously, in which there are whole numbers and decimals less than 1 in both the numerator and denominator:

$$\frac{(271)\,(32)\,(0.06725)}{(725)\,(0.0024)}$$

First, round each term to one digit plus the required number of zeros; then cancel the final zeros of whole numbers where possible:

$$\frac{(271)\,(32)\,(0.06725)}{(725)\,(0.0024)} \approx \frac{(3\cancel{00})\,(30)\,(0.07)}{(7\cancel{00})\,(0.002)}$$

$$\approx \frac{(3)\,(30)\,(0.07)}{(7)\,(0.002)}$$

Next, convert the numbers of both the numerator and denominator to scientific notation. (Remember that a whole number between 1 and 9 need not be converted.)

$$\frac{(3)\,(30)\,(0.07)}{(7)\,(0.002)} = \frac{(3)\,(3 \times 10^{1})\,(7 \times 10^{-2})}{(7)\,(2 \times 10^{-3})}$$

Multiply the numerator, then the denominator, following the rule for multiplication of numbers in scientific notation. (Multiply the numbers but add the exponents.)

$$= \frac{(3)\,(3)\,(7) \times 10^{1-2}}{(7)\,(2) \times 10^{-3}}$$

$$= \frac{63 \times 10^{-1}}{14 \times 10^{-3}}$$

Round the numbers in the numerator and denominator, and cancel zeros where possible. Then, perform the division according to the division rule for scientific notation. (Divide the numbers, but subtract the denominator exponent from the numerator exponent.)

$$\frac{63 \times 10^{-1}}{14 \times 10^{-3}} \approx \frac{6\cancel{0} \times 10^{-1}}{1\cancel{0} \times 10^{-3}}$$

$$\approx \left(\frac{6}{1}\right) \times 10^{(-1)-(-3)}$$

Again, two minus signs become a plus:

$$\approx 6 \times 10^{-1+3}$$

$$\approx 6 \times 10^{2}$$

Finally, take the number out of scientific notation. A *positive* exponent of 2 indicates a decimal point move two places to the *right*.

$$6.00 = 600$$

The estimated answer is 600, so the calculated answer should be in the hundreds. (The actual answer is 355.17.)

MATHEMATICS 4

Solving for the Unknown Value

In treatment plant operations, you may use equations for such calculations as the detention time of a tank, flow rate in a channel, filter loading rate, or chlorine dosage. To make these calculations, you must first know the values for all but one of the terms of the equation to be used.

For example, in making a flow rate calculation,[1] you would use the equation

$$Q = AV$$

Where:

Q = flow rate
A = area
V = velocity

The terms of the equation are Q, A, and V. In solving problems using this equation, you would need to be given values to substitute for any two of the three terms. The term for which you do not have information is called the *unknown value*, or merely the *unknown*. The unknown value is often denoted by a letter such as x, but may be any other letter such as Q, A, V, etc.

Suppose you have this problem:

$$58 = (x)(25)$$

How can you determine the value of x? The rules that allow you to solve for x are discussed in this section. The discussion includes only *what* the rules are

[1]Hydraulics 7, Flow Rate Problems.

and *how* they work. (If you are interested in *why* these rules work and how they came about, you should consult a book on elementary algebra.)

Almost all problems in water treatment plant calculations have equations that involve only multiplication (such as the one just shown) and/or division. There are occasional problems, however, that use equations involving addition and subtraction. The way to solve for the unknown in each of these types of equations is discussed in this section. Equations involving all four operations of multiplication, division, addition, and subtraction are encountered only in more advanced calculations and are thus beyond the scope of this text.

Equations Using Multiplication and Division

To solve for the unknown value in an equation using multiplication and division, you must rearrange the terms so that x is by itself. This often involves two steps:

1. Move x to the numerator (top) of the fraction, if it is not already there.
2. Move any other terms away from x, to the other side of the equal sign, so that x stands alone.

To accomplish these two steps, follow this rule:

Rule 1
In equations using multiplication and division, to move a term from one side of the equation to the other, move from the top of one side to the bottom of the other side, or from the bottom of one side to the top of the other side.

Although you rearrange only *some* of the terms in an equation when you solve for the unknown value, any of the terms in the equation may be moved if they are moved according to the rule just stated. Examples 1 and 2 show how to rearrange terms.

Example 1

Given the equation below, move the 8 to another position.

$$\frac{(x)(2)}{8} = \frac{3}{7}$$

According to the rearrangement rule, there is only one possible way to move the 8. It must be moved to the *top* of the right side of the equation.

$$\frac{(x)(2)}{⑧} = \frac{\bigcirc(3)}{7}$$

After making this move, the equation is

$$(x)(2) = \frac{(8)(3)}{7}$$

Example 2

Given the following equation, move the 4 to another position.

$$\frac{(5)(3)}{6} = \frac{(13)(4)}{(7)(x)}$$

There is only one possible move for the 4. According to the rearrangement rule, it must be moved to the *bottom* of the left side of the equation.

$$\frac{(5)(3)}{(6)\bigcirc} = \frac{(13)④}{(7)(x)}$$

After making this move, the equation is

$$\frac{(5)(3)}{(6)(4)} = \frac{13}{(7)(x)}$$

Rearranging the terms *always* involves this diagonal pattern. Any other movement is not permissible. The following three examples illustrate how terms can be rearranged to solve for an unknown value.

Example 3

Solve for the unknown x in this equation:

$$\frac{x}{7} = \frac{(2)(3)}{13}$$

You must first ask yourself if x is in the numerator. The answer is yes. You should then ask yourself if x is by itself. The answer is no. Since x is not by itself, some rearranging is necessary.

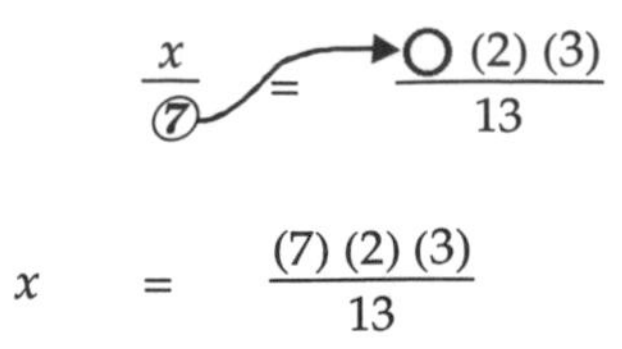

$$x = \frac{(7)(2)(3)}{13}$$

Now that x is by itself, arithmetic completes the problem.

$$x = \frac{42}{13}$$

$$x = 3.23$$

Example 4

Solve for the unknown b in this equation:

$$\frac{(3)(b)}{4} = \frac{(6)(5)}{7}$$

Again you must consider two questions: First, is b in the numerator? The answer is yes; so then ask, is b by itself? The answer is no. Since b is not by itself, some rearranging is necessary. The 3 and 4 must be moved away from b.

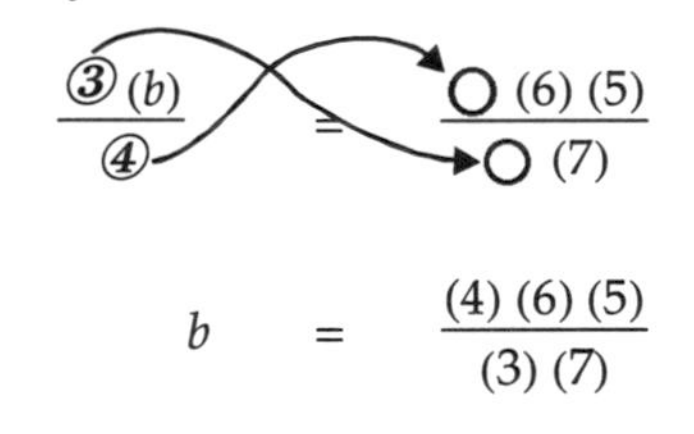

$$b = \frac{(4)(6)(5)}{(3)(7)}$$

The unknown b is now by itself, and arithmetic completes the problem.

$$b = \frac{120}{21}$$

$$b = 5.71$$

Example 5

Solve for the unknown y in this equation:

$$\frac{2}{(3)(y)} = (7)(3)$$

Consider again the first of the two questions: Is the unknown y in the numerator? The answer is no, and so it must be moved.

Before rearranging terms, make the right side of the equation into a fraction by putting a 1 as the denominator. This does not change the value of that side of the equation, and it makes the crisscross method of rearranging more apparent.

$$\frac{2}{(3)\,(y)} = \frac{(7)\,(3)}{1}$$

Now move the y according to the rearranging rule:

$$\frac{2}{(3)\,(y)} = \frac{\bigcirc\,(7)\,(3)}{1}$$

The resulting equation is

$$\frac{2}{3} = \frac{(y)\,(7)\,(3)}{1}$$

The unknown is in the numerator, so ask the second question: Is y by itself? The answer is no. Since y is not by itself, some rearranging is necessary. The 7, 3, and 1 must be moved away from the y.

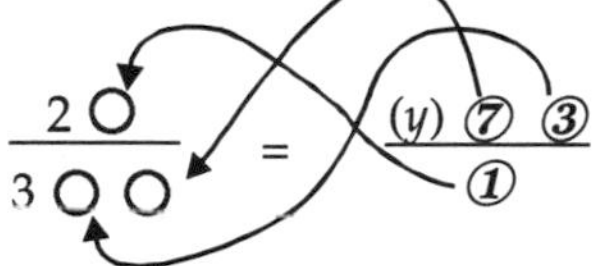

The resulting equation is

$$\frac{(2)\,(1)}{(3)\,(3)\,(7)} = y$$

The unknown is by itself, and the problem can be completed by arithmetic:

$$\frac{2}{63} = y$$

$$0.03 = y$$

Equations Using Addition and Subtraction

As previously mentioned, most water treatment calculations involve equations using only multiplication and division. There are some, however, such as chlorine dosage problems, that involve addition and subtraction. To solve for the unknown value in an equation using addition and subtraction, you must rearrange the terms so that x is positive and is by itself. However, in this type of equation there is neither a numerator nor a denominator, so the crisscross method of rearranging *cannot be used*. Instead, the terms are moved directly from one side of the equation to the other side, according to the following rule:

> **Rule 2**
> In equations using addition and subtraction, when a term is moved from one side of the equation to the other side, the sign of the term must be changed.

This rule means that if a term is positive on one side of the equation, it becomes negative when moved to the other side. Conversely, if a term is negative on one side of the equation, it becomes positive when moved to the other side.

If a term does not show either a plus or minus sign in front, it is assumed to be positive. For example, in the equation

$$7 - 2 - 5 + x = 19 - 3$$

both the 7 and 19 are considered positive although no plus sign is shown in front of them.

To solve an equation involving addition and subtraction, you must ask yourself two questions:

- Is x positive?
- Is x by itself?

The terms of the equation must be rearranged so that the answer to both questions is yes. It is important to understand, however, that although x as a term in the equation should be *positive*, the *value* of x in the answer could be positive or negative (see example 8).

Example 6

Solve for x:

$$7 - 2 - 5 + x = 19 - 3$$

In answer to the first question, x is positive. However, x is not by itself. In this problem, then, you must move the 7, 2, and 5 to the other side of the equation. To do this, all you have to remember is to *change the sign* of the terms moved:

$$x = 19 - 3 - 7 + 2 + 5$$

The unknown has been solved for, and now arithmetic is used to complete the problem:

$$x = 16$$

Example 7

Solve for x:

$$-18 - 2 + 5 = 7 - x$$

First of all, x is not positive. To make it positive, you merely have to move it to the other side of the equation (because a negative changes to a positive in crossing to the other side of the equation):

$$x - 18 - 2 + 5 = 7$$

Since x is not by itself, the 18, 2, and 5 must be put on the other side of the equation:

$$x = 7 + 18 + 2 - 5$$

The unknown x has been solved for, and arithmetic is used to complete the problem:

$$x = 22$$

Example 8

Solve for x:

$$x + 4 - 9 = 6 - 23$$

In answer to the first question, the x term is positive. Since it is not by itself, however, the 4 and 9 must be moved to the other side of the equation:

$$x = 6 - 23 - 4 + 9$$

The unknown has been solved for, and arithmetic completes the problem:

$$x = -12$$

Here, although the x term in the equation is positive, the *value* of x is negative, because the sum of the negative numbers is larger than the sum of the positive numbers.

Although the method discussed above may be used in solving for the unknown in addition and subtraction problems, most water treatment calculations you will encounter involving addition and subtraction are far simpler than the examples just given. In fact, the answer may be obvious without any rearranging of terms. To illustrate, let's look at perhaps the most common type of problem using addition and subtraction — chlorine dosage.[2] This problem is expressed by the equation

$$\text{chlorine dosage} = \text{chlorine demand} + \text{chlorine residual}$$

In such problems, two of the terms are given and the third term is the unknown. As you can see in the example below, the value of x can be determined without rearranging the terms:

$$7 \text{ mg/L} = x \text{ mg/L} + 0.2 \text{ mg/L}$$

The answer is 6.8 mg/L, because 6.8 must be added to 0.2 to make 7.

[2]Chemistry 7, Chemical Dosage Calculations.

Mathematics 5

Ratios and Proportions

A *ratio* is the relationship between two numbers and may be written using a colon (1:2, 4:7, 3:5) or as a fraction ($\frac{1}{2}$, $\frac{4}{7}$, $\frac{3}{5}$). When the relationship between two numbers in a ratio is the same as that between two other numbers in another ratio, the ratios are said to be in *proportion*, or *proportionate*. Stated another way, proportionate ratios are mathematically equal.

A method used to determine if two ratios are in proportion is *cross multiplication*. If the answers are the same when numbers diagonally across from each other are multiplied, then the ratios are proportionate.

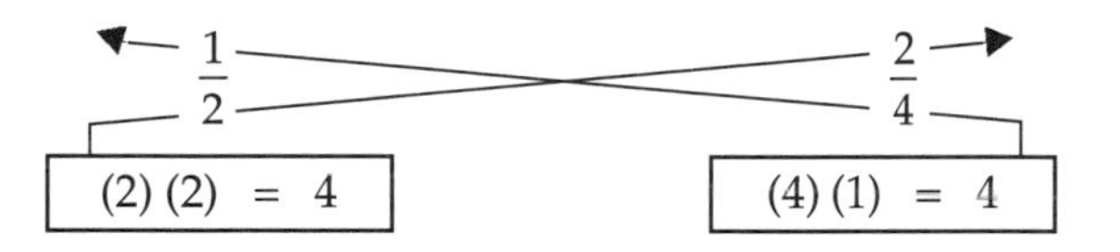

Thus, $\frac{1}{2}$ is equal to $\frac{2}{4}$.

The next two examples use the cross-multiplication method to determine whether ratios are proportionate.

Example 1

Are $\frac{3}{7}$ and $\frac{42}{98}$ proportionate?

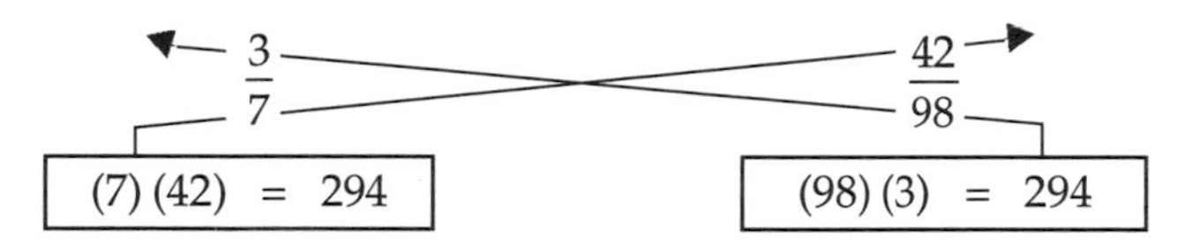

Since the answers (products) of the cross multiplication are the same (294), the ratios are proportionate.

Example 2

Are 5:13 and 22:95 proportionate?

First the ratios are written as fractions, then the cross-multiplication method is used to determine if the two ratios are proportionate:

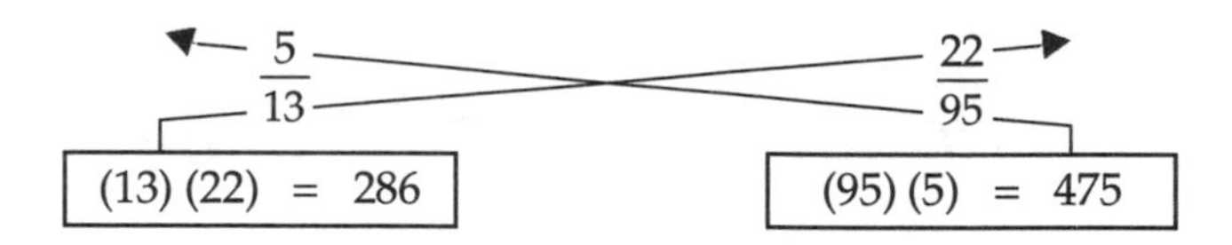

The products of cross multiplication (286 and 475) are different; therefore the ratios are not proportionate.

In examples 1 and 2, cross multiplication has been used to determine whether two ratios are proportionate. Sometimes only three of the four numbers in a proportion are known. In such cases, by solving for the unknown value,[1] you can determine the fourth number.

Suppose, for example, you are given the proportion below. What is the value of the unknown number?

$$\frac{6}{13} = \frac{x}{40}$$

To determine the value of x, solve for the unknown value:

$$\frac{(40)(6)}{13} = x$$

$$18.46 = x$$

Example 3

In the following proportion, solve for the unknown value:

$$\frac{6.5}{x} = \frac{13}{20}$$

$$6.5 = \frac{(13)(x)}{20}$$

$$(20)(6.5) = (13)(x)$$

[1]Mathematics 4, Solving for the Unknown Value.

$$\frac{(20)(6.5)}{13} = x$$

$$10 = x$$

Example 4

Solve for the unknown value in the proportion:

$$\frac{x}{72} = \frac{14}{15}$$

$$x = \frac{(14)(72)}{15}$$

$$x = 67.2$$

Practical Applications

In practical application problems, the ratios will not be arranged as in the examples just given. You must know how to set up the proportion before you can solve for the unknown value. As previously explained, a proportion establishes a simple relation between two ratios. If any number in one of the ratios changes, then another must also change to maintain the equality. In performing various treatment plant operations, you will find that a change in one operating condition produces a proportionate change in some other condition. For example, changing the speed of a metering pump changes the quantity metered, or changing the amount of chemical added to a certain volume of water changes the strength of the solution. The point is this: If you can set up a relation (ratio) for a particular operating condition, then when one of the two quantities of that ratio changes, you can calculate what the new value of the other quantity should be. For example, suppose you know the speed of a metering pump and the corresponding quantity of water metered. If the speed is changed to a different (but known) value, you may wish to determine the new quantity of water metered.

To set up a proportion, first analyze the problem to decide what is unknown. Then decide whether you would expect the unknown value to be larger or smaller than the known value of the same unit. For example, if the unknown in a particular problem is pounds, would you expect the amount of pounds to be larger or smaller than the number of pounds given in the problem? Now write all values (including the unknown x) in fraction form, grouping values having the same units (pounds go with pounds, time with time, liters with liters, and so on) and putting the smaller value in the numerator (top) of each fraction and the larger value in the denominator (bottom).

Use this arrangement:

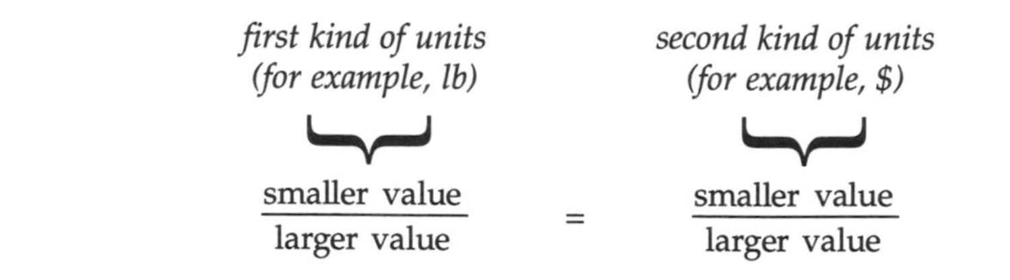

And follow these rules:

Rule 1
If the unknown is expected to be smaller than the known value, put an *x* in the numerator of the first fraction, and put the known value of the same unit in the denominator.

Rule 2
If the unknown is expected to be larger than the known value, put an *x* in the denominator of the first fraction, and put the known value of the same unit in the numerator.

Rule 3
Make the two remaining values of the problem into the second fraction (the smaller value in the numerator, the larger in the denominator).

The next examples demonstrate this method of setting up proportions to solve typical problems.

Example 5

If 3 people can complete a job in 11 days, how many days will it take 5 people to complete it?

First, decide what is unknown. In this problem, a number of days is the unknown. If it takes 3 people 11 days to complete a job, then 5 people should complete the job in less time, or fewer than 11 days. Therefore, the unknown x should be smaller than the known (11 days). Now set up the proportion. Applying rule 1, set up a fraction with x as the numerator and 11 days as the denominator. Then set up the other side of the proportion according to rule 3, with 3 people as the numerator and 5 people as the denominator:

$$\frac{x \text{ days}}{11 \text{ days}} = \frac{3 \text{ people}}{5 \text{ people}}$$

Now solve for the unknown value:

$$x = \frac{(3)(11)}{5}$$

$$x = 6.6 \text{ days}$$

Example 6

If 5 lb of chemical are mixed with 2,000 gal of water to obtain a desired solution, how many pounds of chemical would be mixed with 10,000 gal of water to obtain a solution of the same concentration?

Here, the unknown is some number of pounds of chemical. Analyzing the problem, you would expect that the more water you used, the more chemical would be needed to keep the same concentration. Therefore, the unknown number of pounds (x) should be greater than the known number (5 lb). Now set up the proportion using the rules given above (in this case, rules 2 and 3):

$$\frac{5 \text{ lb}}{x \text{ lb}} = \frac{2{,}000 \text{ gal}}{10{,}000 \text{ gal}}$$

And solve for the unknown value:

$$5 = \frac{(2{,}000)(x)}{10{,}000}$$

$$(5)(10{,}000) = (2{,}000)(x)$$

$$\frac{(5)(10{,}000)}{2{,}000} = x$$

$$25 \text{ lb} = x$$

Example 7

If a pump will fill a tank in 13 hours at 6 gpm (gallons per minute), how long will it take a 15-gpm pump to fill the same tank?

Again, analyze the problem. Here the unknown is some number of hours. But should the answer be larger or smaller than 13 hours? If a 6-gpm pump can fill the tank in 13 hours, a larger pump (15-gpm) should be able to complete the filling in less than 13 hours. Therefore, the answer should be less than 13 hours. Now set up the proportion, following rules 1 and 3:

$$\frac{x \text{ hours}}{13 \text{ hours}} = \frac{6 \text{ gpm}}{15 \text{ gpm}}$$

$$x = \frac{(6)\,(13)}{15}$$

$$x = 5.2 \text{ hours}$$

After you have worked with proportion problems for a while, you will gain an understanding that will allow you to skip some of the intermediate steps. The following examples show one "short-cut" approach that an experienced operator might use to analyze and solve proportion problems.

Example 8

To make up a certain solution, 57.2 mg of a chemical must be added to 100 L of water. How much of the chemical should be added to 22 L to make up the same strength solution?

To solve this problem, first decide what is unknown, and whether you expect the unknown value to be larger or smaller than the known value of the same unit. The amount of chemical to be added to 22 L is the unknown, and you would expect this to be smaller than the 57.2 mg needed for 100 L.

Now take the two known quantities of the same unit (22 L and 100 L) and make a fraction to multiply the third known quantity (57.2 mg) by. Notice that there are two possible fractions you can make with 22 and 100:

$$\frac{22}{100} \quad \textit{or} \quad \frac{100}{22}$$

Choose the fraction that will make the unknown number of milligrams less than the known (57.2 mg). Multiplying 57.2 by the fraction $^{22}/_{100}$ would result in a number *smaller* than 57.2. Multiplying 57.2 by $^{100}/_{22}$, however, would result in a number larger than 57.2.

You wish to obtain a number smaller than 57.2, so multiply by $^{22}/_{100}$; then complete the arithmetic to solve the problem:

$$\left(\frac{22}{100}\right)(57.2) = x$$

$$\frac{(22)\,(57.2)}{100} = x$$

$$12.58 \text{ mg} = x$$

The key to this method is arranging the two known values of like units into a fraction that, when multiplied by the third known value, will give a result that is larger or smaller as needed.

Example 9

Operating at an 86 percent efficiency, a pump can fill a tank in 49 min. How long will it take the pump to fill the tank if the pump efficiency is 59 percent?

The unknown value is the time it will take the pump to fill the tank while operating at reduced efficiency. At the reduced pumping efficiency, the unknown number of minutes should be larger than the known (49 min).

The known values of the other unit (percent) are 59 percent and 86 percent. Arrange these numbers into a fraction by which to multiply 49 min. The answer should be larger than 49 min, so the fraction should be larger than 1 ($^{86}/_{59}$ rather than $^{59}/_{86}$).

$$\left(\frac{86\%}{59\%}\right)(49 \text{ min}) = x$$

Arithmetic completes the problem:

$$\frac{(86)\,(49)}{59} = x$$

$$71 \text{ min} = x$$

Using this shortcut method then, you need to remember only two things:

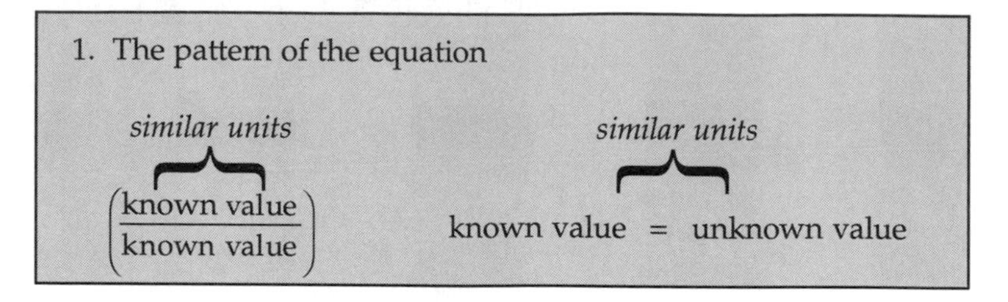

2. Arrangement of the fraction so that the multiplication will result in a larger or smaller number as expected.

MATHEMATICS 6

Averages

To assess the performance of a water treatment plant, much data must be collected and evaluated. Because there may be considerable variation in the information, it is often difficult to determine *trends* in performance.

The calculation of an average is a method to *group the information* so that trends in the information may be determined. When evaluating information based on averages, you must keep in mind that the "average" reflects the *general* nature of the group and does not necessarily reflect any one element of that group.

The *arithmetic mean* is the most commonly used measurement of average value. It is calculated as follows:

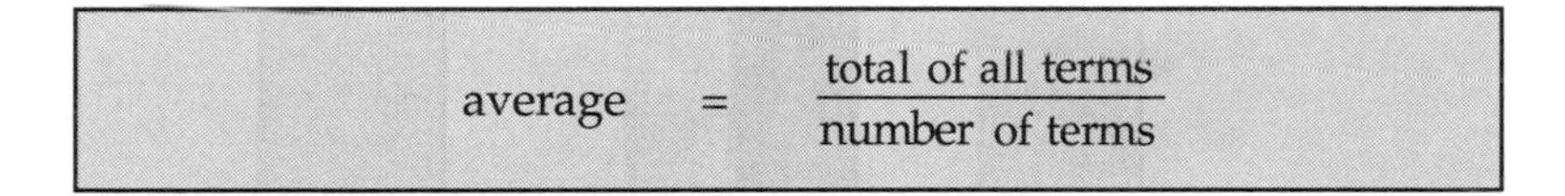

$$\text{average} = \frac{\text{total of all terms}}{\text{number of terms}}$$

Example 1

The following raw-water turbidities (measured in turbidity units, or TU) were recorded for a week: Monday, 8.2 TU; Tuesday, 7.9 TU; Wednesday, 6.3 TU; Thursday, 6.5 TU; Friday, 7.4 TU; Saturday, 6.2 TU; Sunday, 5.9 TU. What was the average daily turbidity?

Monday	8.2
Tuesday	7.9
Wednesday	6.3
Thursday	6.5

Friday	7.4
Saturday	6.2
Sunday	5.9
Total	48.4 TU

$$\text{average} = \frac{\text{total of all terms}}{\text{number of terms}}$$

$$= \frac{48.4 \text{ TU}}{7}$$

$$= 6.91 \text{ TU}$$

Example 2

The sick days taken during the year by each of five operators were recorded as follows: operator A, 6 days; operator B, 3 days; operator C, 7 days; operator D, 0 days; operator E, 1 day. For these five operators, what was the average number of days of sick leave taken?

Operator A	6 days
Operator B	3 days
Operator C	7 days
Operator D	0 days
Operator E	1 day
Total	17 days

$$\text{average} = \frac{\text{total of all terms}}{\text{number of terms}}$$

$$= \frac{17 \text{ days}}{5}$$

$$= 3.4 \text{ days of sick leave}$$

MATHEMATICS 7

Percent

The word *percent* (symbolized %) comes from the Latin words *per centum*, meaning "per one hundred." For example, if 62 percent of the voters are in favor of passing a bond issue, then 62 voters out of every 100 voters are in favor of the issue. Or if a student scores 87 percent on a test with 100 questions, that student has answered 87 of the 100 questions correctly.

There is a direct relationship between percents, fractions, and decimal numbers. They are merely different ways of expressing the same mathematical proportions, as shown in the following examples.

$$20\% = \frac{20}{100} = 0.20$$

$$95\% = \frac{95}{100} = 0.95$$

$$5\% = \frac{5}{100} = 0.05$$

$$14.5\% = \frac{14.5}{100} = 0.145$$

$$0.5\% = \frac{0.5}{100} = 0.005$$

Many problems involving percent require converting from a fraction to a decimal number to a percent. For example,

$$\frac{20}{100} = 0.20 = 20\%$$

$$\frac{95}{100} = 0.95 = 95\%$$

$$\frac{18}{60} = 0.30 = 30\%$$

$$\frac{4}{32} = 0.125 = 12.5\%$$

Note that to change from a decimal number to a percent, the decimal point is moved two places to the right. (The result is the same as multiplying by 100.) To convert a percent to a decimal, the decimal point is moved two places to the left (the same as dividing by 100).

A fraction is the key to calculating percent. The fraction is divided out, resulting in a decimal number that is then multiplied by 100 to be expressed as a percent. The formula for all percent problems is

$$\text{percent} = \frac{\text{part}}{\text{whole}} \times 100$$

Remember that the *whole* is always the entire quantity; 100 percent of anything is the whole thing. For percentages less than 100, the *part* is less than the whole. For example, 50¢ is less than $1.00; in fact, it is ½ (or 50 percent) of the dollar. And $25.00 is less than $100.00; it is ¼, or 25 percent, of the whole amount.

Examples 1 through 3 illustrate how to use the percent formula in solving some water treatment plant problems.

Example 1

A certain piece of equipment is having mechanical difficulties. If the equipment fails 6 times out of 25 tests, what percent failure does this represent?

The equation to be used in solving percent problems is

$$\text{percent} = \frac{\text{part}}{\text{whole}} \times 100$$

In this problem, *percent* is unknown, but information is given regarding the *part* and the *whole*. Fill this information into the equation:

$$\text{percent failure} = \frac{\text{6 failed tests}}{\text{25 tests}} \times 100$$

$$= 0.24 \times 100$$

$$= 24\% \text{ failure}$$

Example 2

A treatment plant operator has been ill during the past month and has had to miss a number of days' work. If she missed an average of 3 days out of 7, what was her percent absence?

The problem gives information regarding the *part* and the *whole* — 3 days absent (part) out of a total of 7 days (whole). This information is used in the percent absence calculation as follows:

$$\text{percent} = \frac{\text{part}}{\text{whole}} \times 100$$

$$\text{percent absence} = \frac{\text{3 days absent}}{\text{7 days total}} \times 100$$

$$= 0.43 \times 100$$

$$= 43\% \text{ absence}$$

Example 3

The planning commission of a community voted on a bond issue concerning the construction of a new treatment plant. If 8 were in favor of the bond issue and 5 were opposed, what percent of the commission was in favor of the bond?

$$\text{percent} = \frac{\text{part}}{\text{whole}} \times 100$$

$$\text{percent in favor} = \frac{\text{8 favor}}{\text{13 total}} \times 100$$

$$= 0.62 \times 100$$

$$= 62\% \text{ in favor}$$

In the three examples just given, the unknown value was always *percent*. However, *any one* of the three terms (percent, part, or whole) can be the unknown value. The same general method is used in solving any percent problem. Fill the information that is given into the percent equation, and then solve for the unknown value.[1]

Example 4

The flow rate at a treatment plant is 7.2 mgd, which is 60 percent of the plant's capacity. What is the capacity of the plant?

In solving this problem, first write down the equation for percent, then determine what information is given.

$$\text{percent} = \frac{\text{part}}{\text{whole}} \times 100$$

The 7.2 mgd is not the total capacity of the plant, so it must be the *part* in this problem. The 60 is the *percent*. Fill in the information in the problem:

$$60\% = \frac{7.2 \text{ mgd}}{x \text{ mgd}} \times 100$$

Then solve for the unknown x:

$$60 = \frac{(7.2)\,(100)}{x}$$

$$(60)\,(x) = (7.2)\,(100)$$

$$x = \frac{(7.2)\,(100)}{60}$$

$$x = 12 \text{ mgd}$$

Example 5

A tank is filled with water to 55 percent of its capacity. If the capacity of the tank is 2,000 gal, how many gallons of water are in the tank?

---

[1]Mathematics 4, Solving for the Unknown Value.

First write down the equation for percent, then determine what information is given:

$$\text{percent} = \frac{\text{part}}{\text{whole}} \times 100$$

The *percent* is 55. The question now is whether the 2,000 gal is the *part* or the *whole*. Since the tank will hold only 2,000 gal, that number must be the *whole*. Fill this information into the percent equation:

$$55\% = \frac{x \text{ gal}}{2{,}000 \text{ gal}} \times 100$$

And solve for x:

$$55 = \frac{(x)(100)}{2{,}000}$$

$$(2{,}000)(55) = (x)(100)$$

$$\frac{(2{,}000)(55)}{100} = x$$

$$1{,}100 \text{ gal} = x$$

Example 6

The raw water entering a treatment plant has a turbidity of 10 ntu. If the turbidity of the finished water is 0.5 ntu, what is the turbidity removal efficiency of the treatment plant?

In math problems, the term *efficiency* refers to a percent. Therefore, another way of stating this problem is "What percent of the turbidity has been removed?"

$$\text{percent} = \frac{\text{part}}{\text{whole}} \times 100$$

Here, *percent* is the unknown and 10 ntu is the *whole*. However, 0.5 ntu is not the *part* removed. It is the amount of turbidity *still in* the water. The amount of turbidity removed must therefore be 10 ntu – 0.5 ntu, or 9.5 ntu.

$$\text{percent turbidity remove} = \frac{9.5 \text{ ntu}}{10 \text{ ntu}} \times 100$$

$$= 0.95 \times 100$$

$$= 95\% \text{ turbidity removal efficiency}$$

Although most percentages are less than 100 percent, in certain circumstances it is possible to have a percentage greater than 100 percent. Suppose, for example, you had $1.00 and found another dollar bill on the street. You would then have $2.00 — double what you had to start, or 200 percent. If you had a dollar and found a dime, you would have $1.10, or 110 percent of what you had to start. Example 7 shows one way in which this concept of percents greater than 100 percent applies to water treatment operations.

Example 7

A treatment plant was designed to treat 60 ML/d. On one particular day the plant treated 66 ML. What percent of the design capacity does this represent?

As in the other examples, first write down the equation for percent, then determine what information is given.

$$\text{percent} = \frac{\text{part}}{\text{whole}} \times 100$$

In this problem the *percent* of design capacity is the unknown. The *whole* is 60 ML/d, since the whole refers to the design capacity of the plant. So what is the part? The *part* is the amount of water actually treated (66 ML). And since the plant treated more water than it was designed to treat, the percent of the design capacity should be *greater than 100 percent*.

$$\text{percent of design capacity} = \frac{\text{part actually treated}}{\text{total design capacity}} \times 100$$

$$= \frac{66 \text{ ML/d}}{60 \text{ ML/d}} \times 100$$

$$= 1.1 \times 100$$

$$= 110\%$$

The rule to remember is

Rule 1
In calculations of percents greater than 100 percent, the numerator of the percent equation must always be larger than the denominator.

MATHEMATICS 8

Linear Measurements

Linear measurements define the distance (or length) along a line. They can be expressed in customary or metric units. The customary units of linear measurement are inches, feet, yards, and miles; the metric units are centimeters, meters, and kilometers.

The focus of this section is on one particular linear measurement: the distance around the outer edge of a shape, called the *perimeter*. To determine the distance around the outer edge of an *angular shape*, such as a square, rectangle, or any other shape *with sides that are straight lines*, merely add the length of each side, as shown in the examples below.

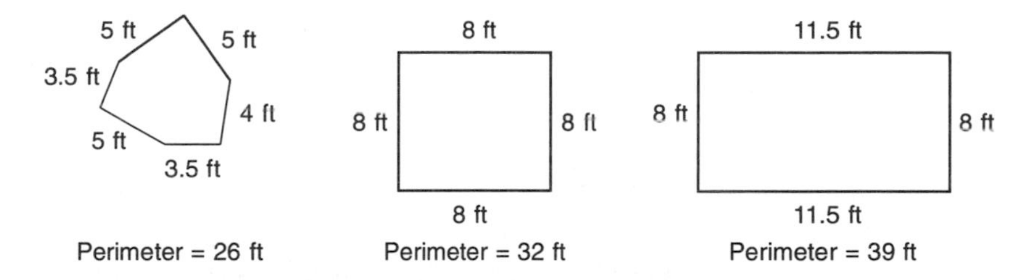

The mathematical equation for perimeter is

$$\text{perimeter} = \text{side}_1 \text{ length} + \text{side}_2 \text{ length} + \text{side}_3 \text{ length} \ldots \text{etc.}$$

Use this type of calculation if, for example, you must determine the length of wire needed for a fence.

Because the outside of a circle is not made up of easily measured straight lines, perimeter calculations for a circle are approached differently. The following diagram illustrates two of the linear terms associated with the circle.

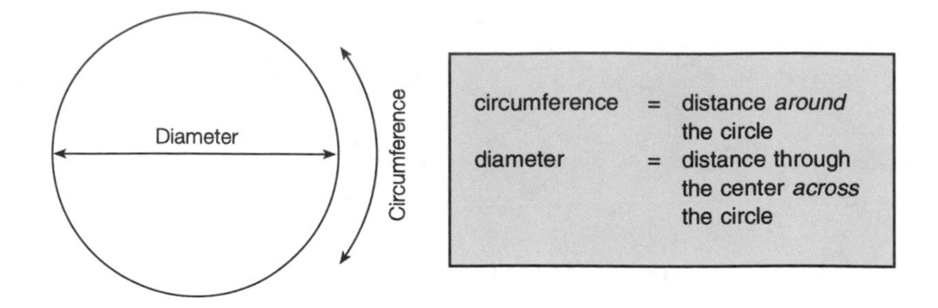

As shown, the distance measured around the outside edge of a circle is called the circumference; this is just a special name for the perimeter of a circle. The diameter is a straight line drawn from one side of the circle *through the center* to the other side.

If you compare the diameter of any circle with the circumference of that same circle, you will find that the circumference is just a little more than three times the length of the diameter. In mathematics, this length comparison (or ratio) of circumference to diameter is indicated by the Greek letter pi (π), and has a value of approximately 3.14.

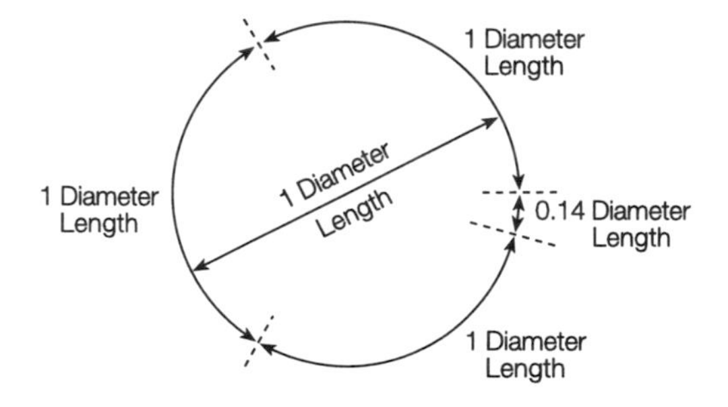

Because this relationship between the diameter length and circumference length is true for all circles, you can calculate the circumference of any circle if you know the circle's diameter. For example, if you know the diameter of a circle is 20 ft, the distance around the circle (circumference) is roughly 3×20 ft, or about 60 ft. The exact circumference of this circle is

$$(3.14)\ (20 \text{ ft}) \quad = \quad 62.8 \text{ ft}$$

The mathematical equation for circumference is

$$\text{circumference} \quad = \quad (3.14)\ (\text{diameter})$$

Use this type of calculation if, for example, you must determine the circumference of a circular tank.

Example 1

You wish to put a fence around one of the storage areas at your treatment plant. It has the dimensions shown below. Determine the perimeter of this area so you can order the correct length of fencing material.

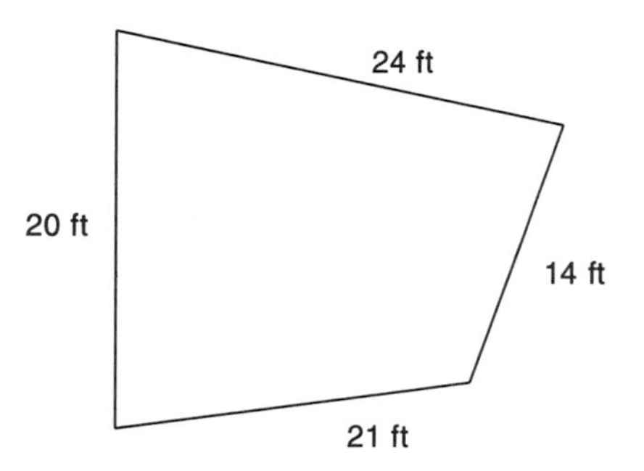

To calculate the perimeter, or distance around the area, add the lengths of the four sides:

$$\text{perimeter} = \text{side}_1 \text{ length} + \text{side}_2 \text{ length} + \text{side}_3 \text{ length} + \text{side}_4 \text{ length}$$

$$= 20 \text{ ft} + 24 \text{ ft} + 14 \text{ ft} + 21 \text{ ft}$$

$$= 79 \text{ ft}$$

Example 2

A circular clarifier has a diameter of 15 m. What is the circumference of this tank?

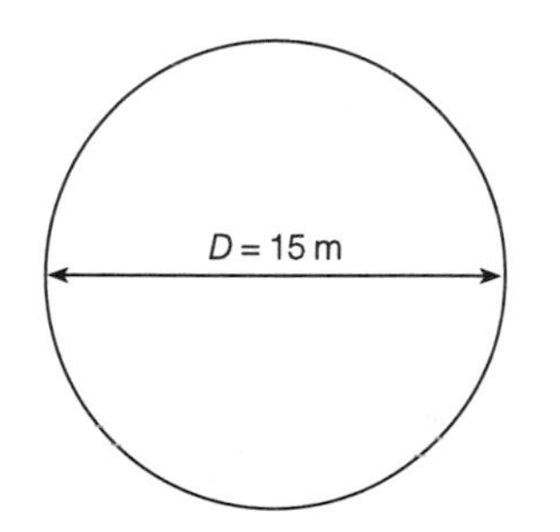

From the relationship between the diameter and circumference of a circle, you know that the circumference is approximately three times the length of the diameter. More accurately, it is

$$\text{circumference} = (3.14)\,(\text{diameter})$$

$$= (3.14)\,(15\text{ m})$$

$$= 47.1\text{ m}$$

Example 3

The circumference of a circular tank is 235.5 ft. What is the diameter of the tank?

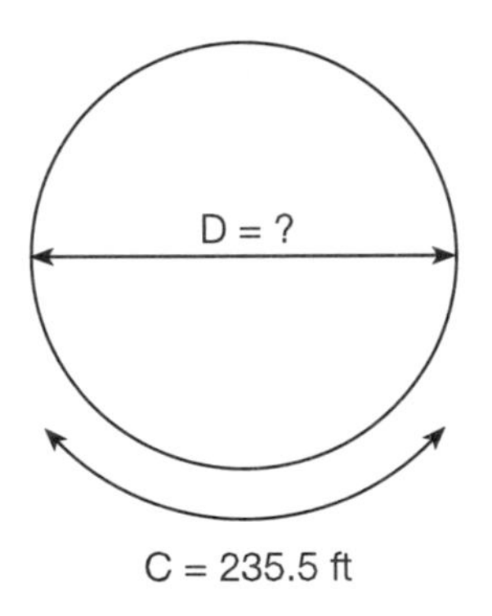

In this problem the circumference is known, and the diameter is unknown. Use the same formula for circumference and fill in the given information.

$$\text{circumference} = (3.14)\,(\text{diameter})$$

$$235.5\text{ ft} = (3.14)\,(D)$$

Solve for the unknown value:[1]

$$\frac{235.5\text{ ft}}{3.14} = D$$

$$75\text{ ft} = D$$

[1] Mathematics 4, Solving for the Unknown Value.

MATHEMATICS 9

Area Measurements

Area measurements define the size of the *surface* of an object. The customary units most frequently used to express this surface space are square inches, square feet, and square yards; the metric units are square millimeters, square centimeters, and square meters.

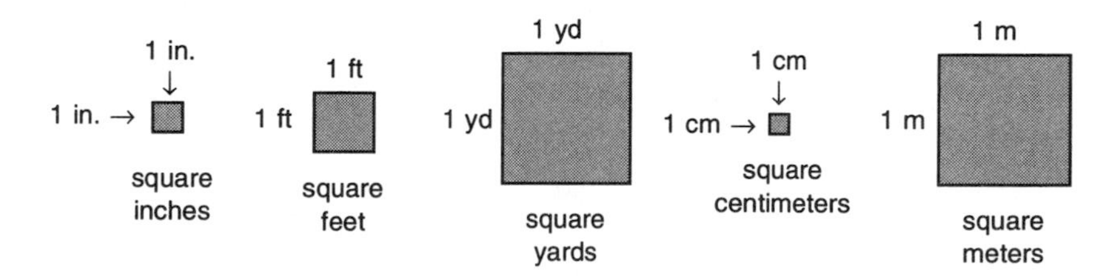

The fact that area measurements are expressed in square units does *not* mean that the surface must be a square in order to measure it. The surface of any shape can be measured. Although the shapes differ in the illustration below, the total surface area of each is the same.

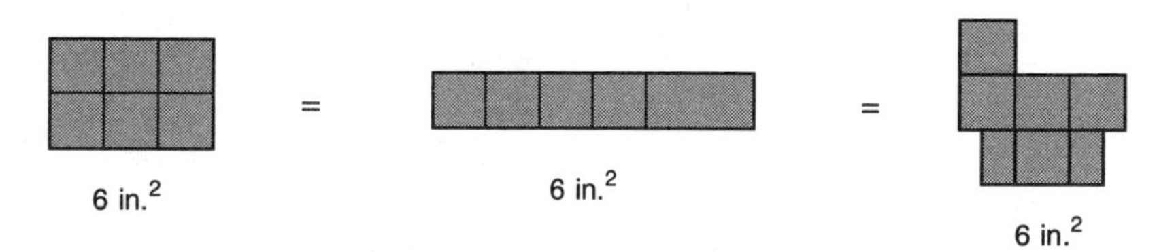

For area measurements in water treatment plant calculations, three shapes are particularly important, namely rectangles, triangles, and circles. Most problems involve one or two combinations of these shapes. Equations (formulas) for each of the three basic shapes are presented on page 74.

Because these formulas are used so often in treatment plant calculations, you should memorize them.

Surface Area Formulas

Rectangle: $$\text{Area} = (\text{length})(\text{width}) = lw$$

Triangle: $$\text{Area} = \frac{(\text{base})(\text{height})}{2} = \frac{bh}{2}$$

Circle: $$\text{Area} = (0.785)(\text{diameter}^2) = (0.785)(D^2)$$

or

$$= (3.14)(\text{radius}^2) = (3.14)(r^2)$$

Use of these formulas is demonstrated in the examples that follow.

Area of a Rectangle

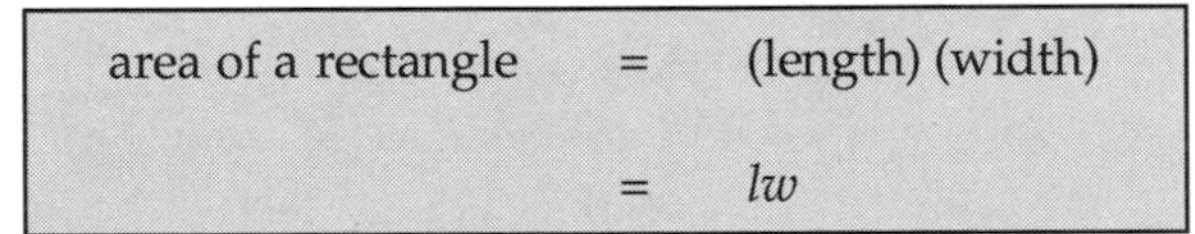

Example 1

What is the area of the rectangle shown below?

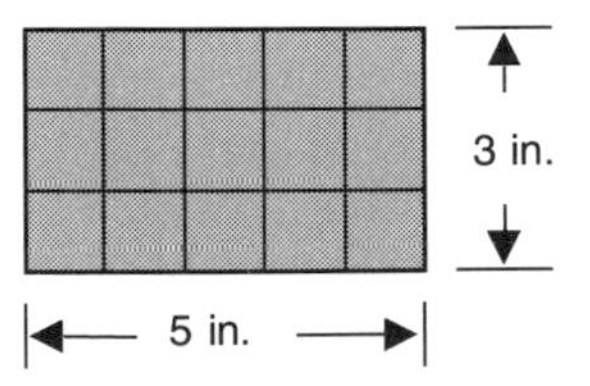

area of rectangle = (length) (width)

= (5 in.) (3 in.)

= 15 in.^{2} surface area

Example 2

What is the area of the rectangle in the diagram below?

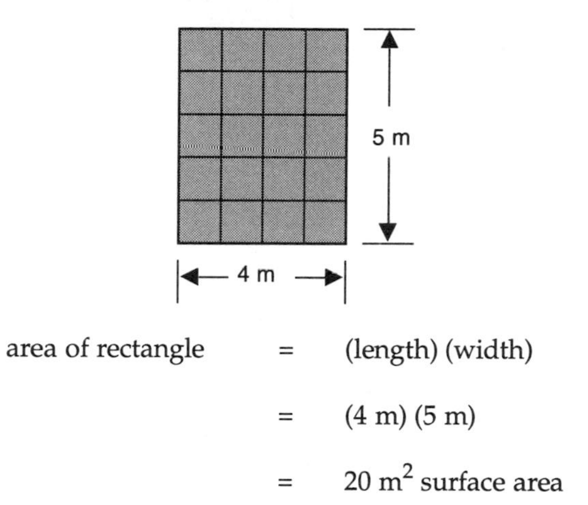

area of rectangle = (length) (width)

= (4 m) (5 m)

= 20 m^2 surface area

Example 3

A tank is 10 ft long, 10 ft wide. What is the area of the water surface of thc tank?

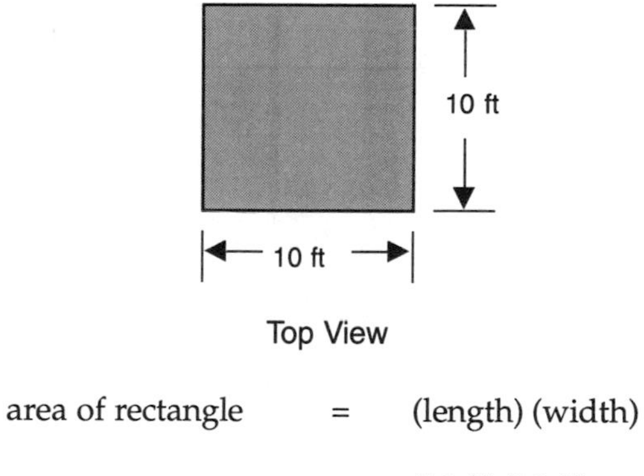

Top View

area of rectangle = (length) (width)

= (10 ft) (10 ft)

= 100 ft^2 surface area

Example 4

A sedimentation tank is 20 m long and 7 m wide. What is the area of the water surface in the tank?

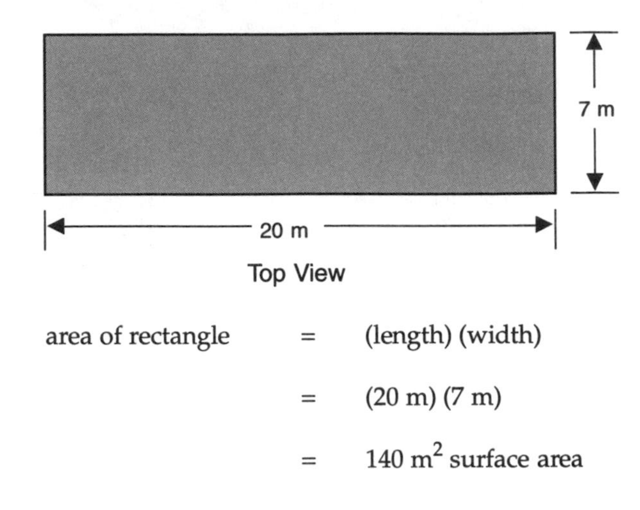

area of rectangle = (length) (width)

= (20 m) (7 m)

= 140 m^2 surface area

Example 5

A piece of land 800 ft by 1,200 ft is to be purchased for expansion of water treatment facilities. How many acres is this?

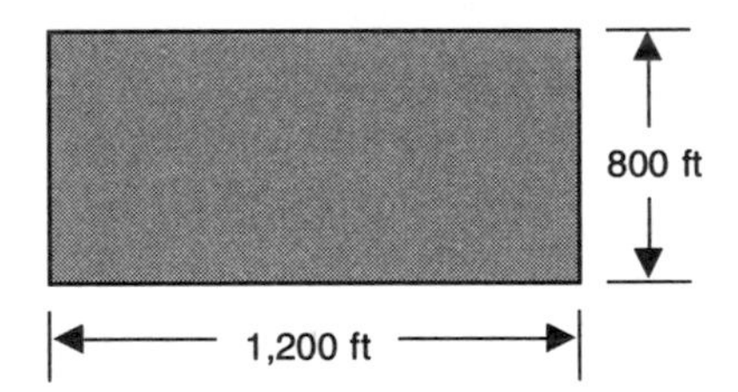

First calculate the number of square feet in the area, then convert the number of square feet to acres.[1]

area of rectangle = (length) (width)

= (800 ft) (1,200 ft)

= 960,000 ft^2

[1]Mathematics 11, Conversions.

Since 1 acre $=$ 43,560 ft^2, the number of acres in this area is

$$\frac{960{,}000 \ ft^2}{43{,}560 \ ft^2/acre} = 22 \text{ acres}$$

Area of a Triangle

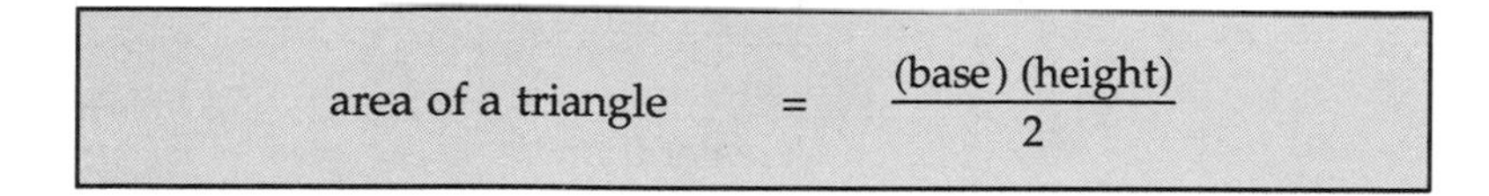

$$\text{area of a triangle} = \frac{(\text{base})(\text{height})}{2}$$

As shown in the following three examples, the *height* of the triangle must be measured *vertically* from the horizontal base.

Example 6

What is the area of the triangle in the diagram below?

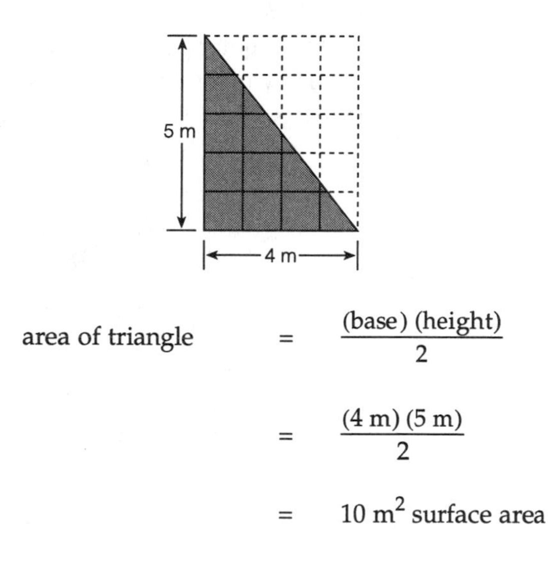

$$\text{area of triangle} = \frac{(\text{base})(\text{height})}{2}$$

$$= \frac{(4 \text{ m})(5 \text{ m})}{2}$$

$$= 10 \text{ m}^2 \text{ surface area}$$

Example 7

What is the area of the triangle shown below?

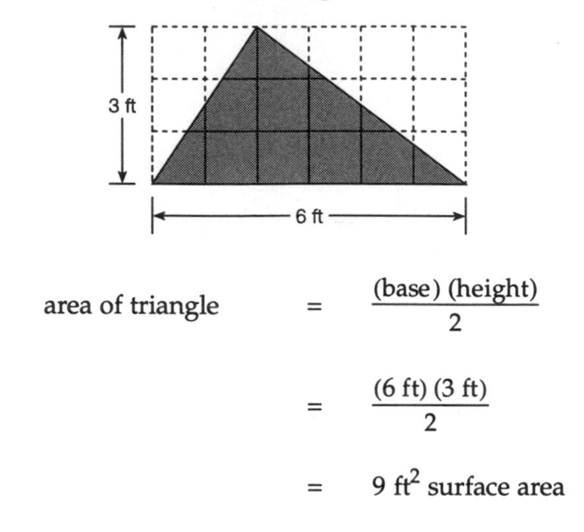

$$\text{area of triangle} = \frac{(\text{base})(\text{height})}{2}$$

$$= \frac{(6\text{ ft})(3\text{ ft})}{2}$$

$$= 9\text{ ft}^2 \text{ surface area}$$

Example 8

A triangular portion of the treatment plant grounds is not being used. How many square feet does this represent if the height of the triangle is 170 ft and the base is 200 ft?

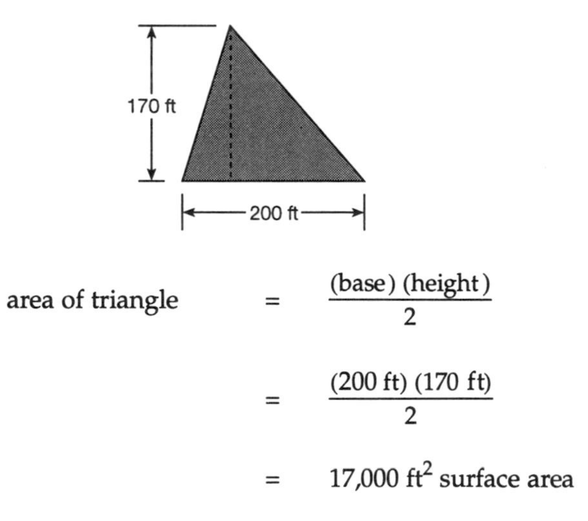

$$\text{area of triangle} = \frac{(\text{base})(\text{height})}{2}$$

$$= \frac{(200\text{ ft})(170\text{ ft})}{2}$$

$$= 17{,}000\text{ ft}^2 \text{ surface area}$$

Area of a Circle

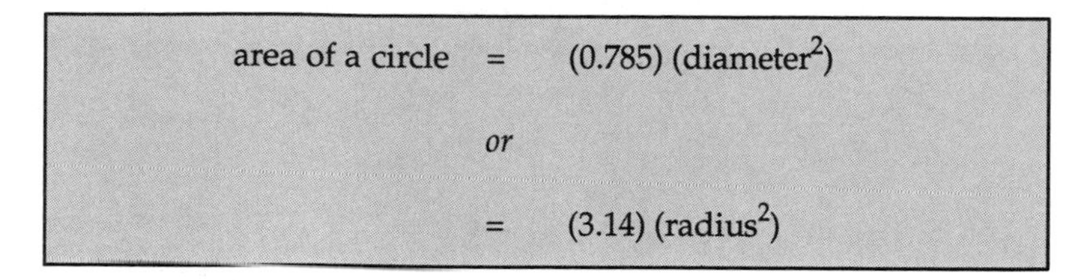

$$\text{area of a circle} = (0.785)(\text{diameter}^2)$$

or

$$= (3.14)(\text{radius}^2)$$

The more familiar formula for the area of a circle is πr^2. The r stands for the *radius* of the circle; that is, the distance from the circle's center to its edge. The radius of any circle is exactly *half* of the diameter.

However, since the diameter rather than the radius is generally given for circular tanks and basins, the formula using diameter is preferred for water treatment plant calculations. The relationship between the two formulas is shown in the following equations:

$$\text{area} = \pi r^2$$

Since the radius is half the diameter, r^2 in the formula may be expressed as $(D/2)^2$, where D is the diameter:

$$\text{area} = (\pi)\left(\frac{D}{2}\right)^2$$

$$= (\pi)\left(\frac{D}{2}\right)\left(\frac{D}{2}\right)$$

or

$$= \frac{\pi D^2}{4}$$

And, because $\pi = 3.14$,

$$\text{area} = \left(\frac{3.14}{4}\right)(D^2)$$

The formula may be reexpressed as

$$\text{area} = (0.785)(D^2)$$

Another advantage of the formula using diameter is that it can be better understood by diagram than the formula using radius. The diameter formula describes a surface area that can be thought of as *a square with the corners cut off.*

Let's examine the relationship between a square and a circle to better understand the formula for circle area:

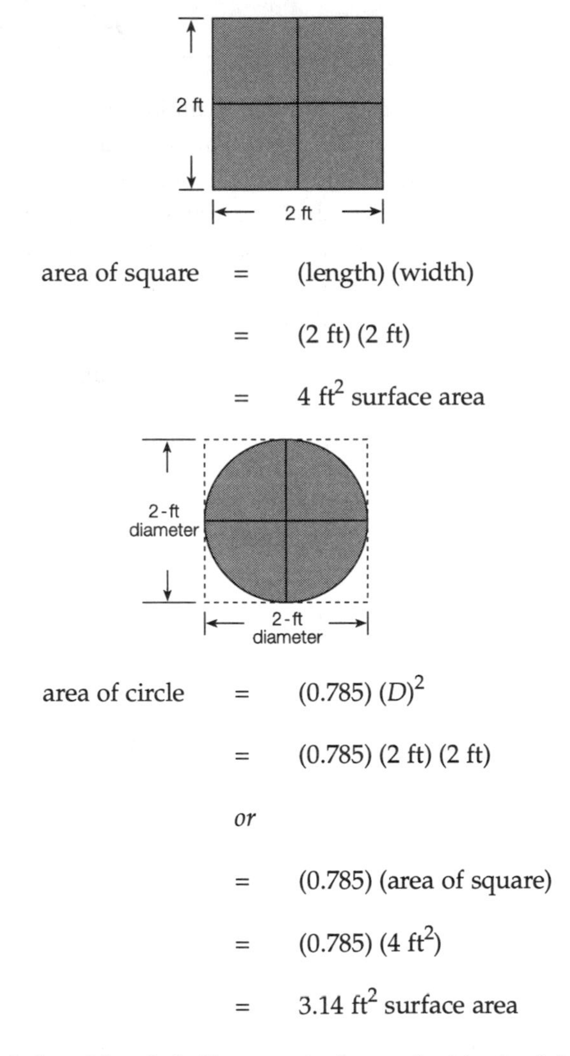

area of square = (length) (width)

= (2 ft) (2 ft)

= $4\ ft^2$ surface area

area of circle = $(0.785)\ (D)^2$

= (0.785) (2 ft) (2 ft)

or

= (0.785) (area of square)

= $(0.785)\ (4\ ft^2)$

= $3.14\ ft^2$ surface area

When a circle with a 2-ft diameter is drawn inside the 2-ft square, you can see that the surface area of the circle is less than that of the square. However, it is not necessary to construct a square around each circle that is being measured. Mathematically, the D^2 (diameter squared) of the formula represents the square. Therefore, in finding the area of the circle, 0.785 is essential to the formula, because it is this factor that "cuts off the corners" of the square.

Example 9

Calculate the area of the circle shown below.

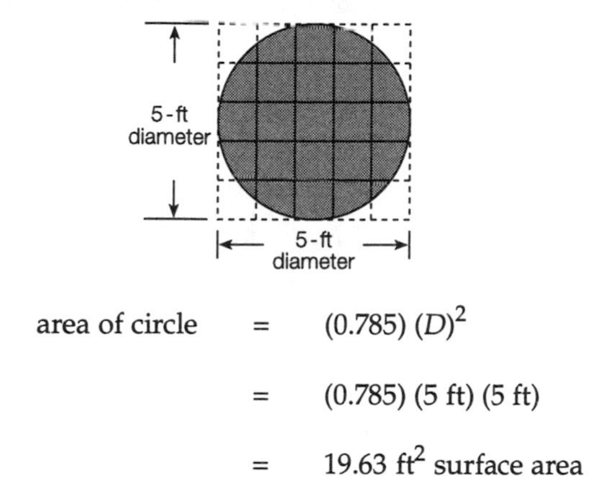

$$\text{area of circle} = (0.785)\,(D)^2$$

$$= (0.785)\,(5\text{ ft})\,(5\text{ ft})$$

$$= 19.63\text{ ft}^2\text{ surface area}$$

Example 10

What is the area of the circle shown in the diagram below?

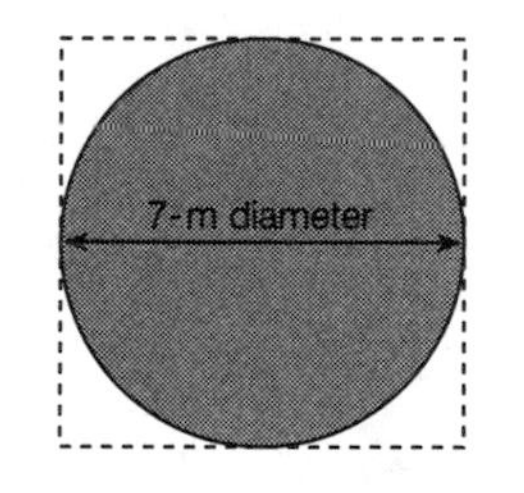

$$\text{area of circle} = (0.785)\,(D)^2$$

$$= (0.785)\,(7\text{ m})\,(7\text{ m})$$

$$= 38.47\text{ m}^2\text{ surface area}$$

Example 11

A circular clarifier has a diameter of 40 ft. What is the surface area of the clarifier?

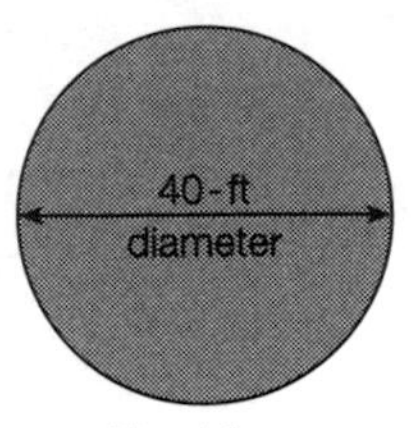

Top View

$$\text{area of circle} = (0.785)\,(D)^2$$

$$= (0.785)\,(40 \text{ ft})\,(40 \text{ ft})$$

$$= 1{,}256 \text{ ft}^2 \text{ surface area}$$

MATHEMATICS 10

Volume Measurements

A volume measurement defines the amount of space that an object occupies. The basis of this measurement is the *cube,* a square-sided box with all edges of equal length, as shown in the diagram below. The customary units commonly used in volume measurements are cubic inches, cubic feet, cubic yards, gallons, and acre-feet. The metric units commonly used to express volume are cubic centimeters, cubic meters, and liters.

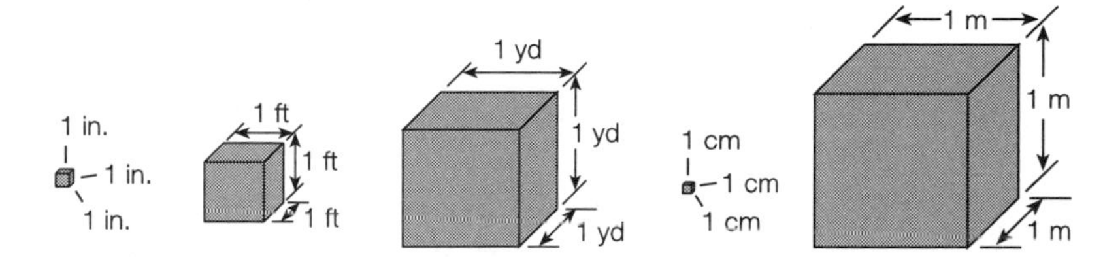

The calculations of surface area and volume are closely related. For example, to calculate the surface area of one of the cubes above, you would multiply two of the dimensions (length and width) together. To calculate the volume of that cube, however, a *third dimension* (depth) is used in the multiplication. The concept of volume can be simplified as

$$\text{volume} = (\text{area of surface})(\text{third dimension})$$

The *area of surface* to be used in the volume calculation is the *representative surface area,* the side that gives the object its basic shape. For example, suppose you begin with a rectangular area as shown on the next page. Notice the shape that would be created by stacking a number of those same rectangles one on top of the other.

Because the rectangle gives the object its basic shape in this example, it is considered the representative area.

The same volume could have been created by stacking a number of smaller rectangles one behind the other.

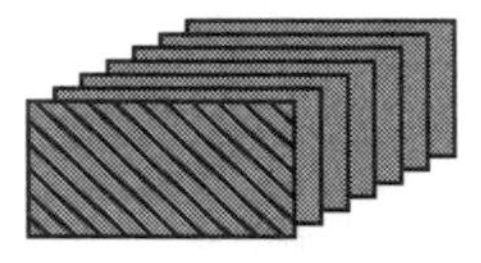

Although an object may have more than one representative surface area as illustrated in the two preceding diagrams, sometimes only one surface is the representative area. Consider, for instance, a shape like this:

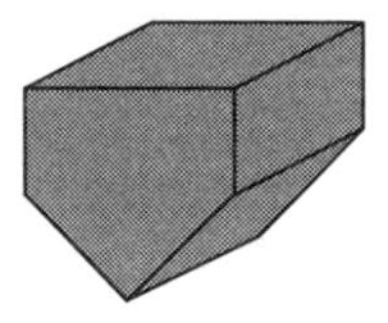

Let's compare two different sides of this shape (the top and the front) to determine if they are representative areas. In the first case, a number of the top rectangles stacked together does not result in the same shape volume. Therefore, this rectangular area is not a representative area. In the second case, however, a number of front shapes stacked one behind the other results in the same shape volume as the original object. Therefore, this front area may be considered a representative surface area.

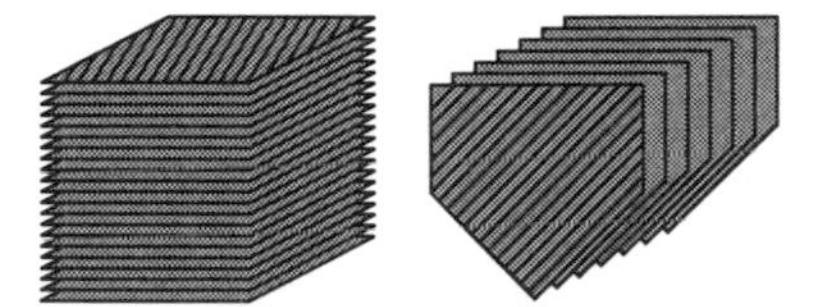

Rectangles, Triangles, and Circles

For treatment plant calculations, representative surface areas are most often rectangles, triangles, circles, or a combination of these. The following diagrams illustrate the three basic shapes for which volume calculations are made.

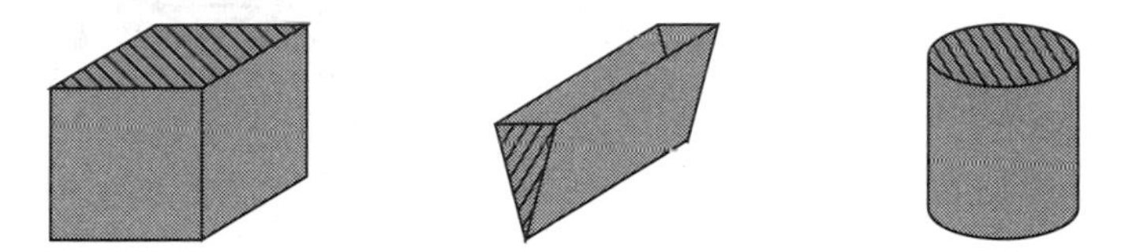

In the first diagram the rectangle defines the shape of the object; in the second diagram the triangle, rather than the rectangle, gives the trough its basic shape; in the third, the surface that defines the shape of the cylinder is the circle.

The formulas for calculating the volume of each of these three shapes are given below. Note that they are closely associated with the area formulas given previously.

Volume Formulas

Rectangular Tank Volume $= \left(\begin{array}{c}\text{area of}\\ \text{rectangle}\end{array}\right)\left(\begin{array}{c}\text{third}\\ \text{dimension}\end{array}\right)$

$= (lw)\left(\begin{array}{c}\text{third}\\ \text{dimension}\end{array}\right)$

Trough Volume $= \left(\begin{array}{c}\text{area of}\\ \text{triangle}\end{array}\right)\left(\begin{array}{c}\text{third}\\ \text{dimension}\end{array}\right)$

$= \left(\frac{bh}{2}\right)\left(\begin{array}{c}\text{third}\\ \text{dimension}\end{array}\right)$

Cylinder Volume $= \left(\begin{array}{c}\text{area of}\\ \text{circle}\end{array}\right)\left(\begin{array}{c}\text{third}\\ \text{dimension}\end{array}\right)$

$= (0.785\ D^2)\left(\begin{array}{c}\text{third}\\ \text{dimension}\end{array}\right)$

Cone Volume $= \frac{1}{3}$ (volume of a cylinder)

Sphere Volume $= \left(\frac{\pi}{6}\right)(\text{diameter})^3$

Use of the volume formulas is demonstrated in the examples that follow.

Example 1

Calculate the volume of water contained in the tank illustrated below if the depth of water (called *side water depth*, SWD) in the tank is 10 ft.

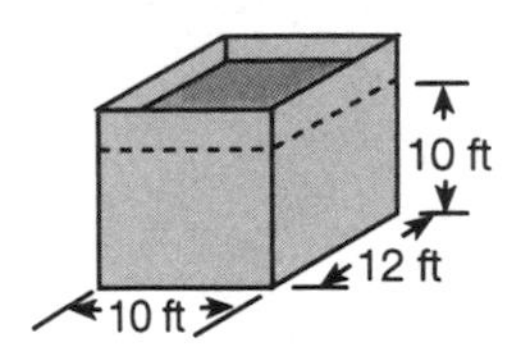

volume = (area of surface) (third dimension)

In this example, the *rectangle* is the representative surface, and the dimension not used in the area calculation is the *depth*.

$$\text{volume} = (\text{length})\ (\text{width})\ (\text{depth})$$

$$= (12\ \text{ft})\ (10\ \text{ft})\ (10\ \text{ft})$$

$$= 1{,}200\ \text{ft}^3$$

Example 2

What is the volume (in cubic inches) of water in the trough shown below if the depth of water is 8 in.?

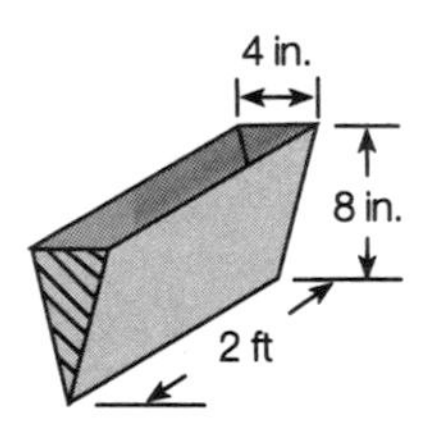

First, all dimensions must be expressed in the same terms.[1] Since the answer is desired in cubic inches, the 2-ft dimension should be converted to inches:

$$(2\ \text{ft})\ (12\ \text{in./ft}) = 24\ \text{in.}$$

[1]Mathematics 11, Conversions.

Then

$$\text{volume} = (\text{area of surface})\,(\text{third dimension})$$

The triangle is the representative surface; the third dimension may be considered length or width — a difference in terminology.

$$\text{volume} = \left(\frac{bh}{2}\right)(\text{length})$$

$$= \frac{(4\text{ in.})\,(8\text{ in.})}{2}\,(24\text{ in.})$$

$$= \frac{(4\text{ in.})\,(8\text{ in.})\,(24\text{ in.})}{2}$$

$$= 384\text{ in.}^3$$

Example 3

What is the volume of water contained in the tank shown below if the depth of water (SDW) is 12 ft?

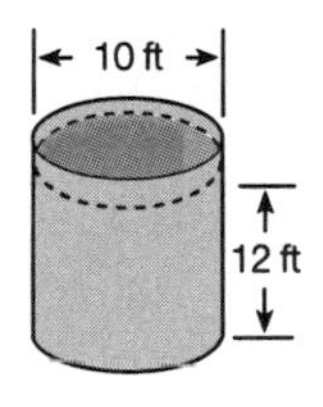

$$\text{volume} = (\text{area of surface})\,(\text{third dimension})$$

The circle is the representative surface, and the third dimension is depth.

$$\text{volume} = (0.785)\,(D^2)\,(\text{depth})$$

$$= (0.785)\,(10\text{ ft})^2\,(12\text{ ft})$$

$$= (0.785)\,(10\text{ ft})\,(10\text{ ft})\,(12\text{ ft})$$

$$= 942\text{ ft}^3$$

Example 4

A tank with a cylindrical bottom has dimensions as shown below. What is the capacity of the tank? (Assume that the cross section of the bottom of the tank is a half circle.)

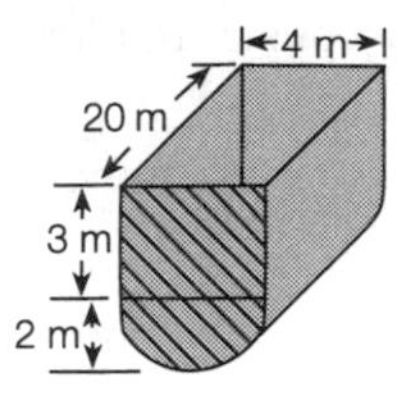

In problems involving a representative surface area that is a combination of shapes, it is often easier to calculate the representative surface area first, then calculate the volume:

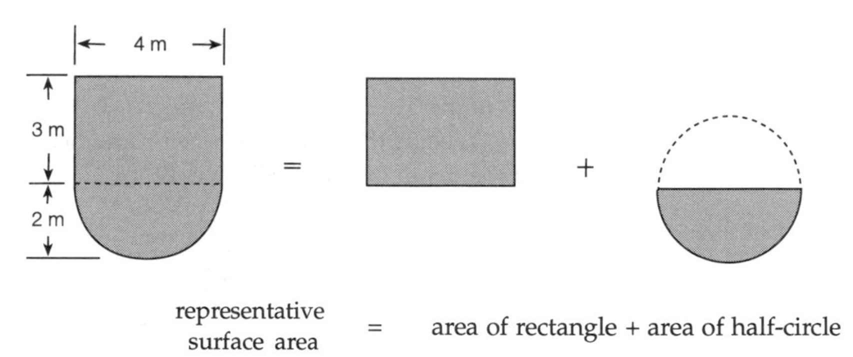

$$\text{representative surface area} = \text{area of rectangle} + \text{area of half-circle}$$

$$= (4 \text{ m})(3 \text{ m}) + \frac{(0.785)\,(4 \text{ m})\,(4 \text{ m})}{2}$$

$$= 12 \text{ m}^2 + 6.28 \text{ m}^2$$

$$= 18.28 \text{ m}^2$$

And now calculate the volume of the tank:

$$\text{volume} = (\text{area of surface})\,(\text{third dimension})$$

$$= (18.28 \text{ m}^2)\,(20 \text{ m})$$

$$= 365.6 \text{ m}^3$$

Cones and Spheres

There are many shapes (though very few in water treatment calculations) for which the concept of a "representative surface" does not apply. The cone and sphere are notable examples of this. That is, we cannot "stack" areas of the same size on top of one another to obtain a cone or sphere.

Calculating the volume of a cylinder was discussed earlier. The volume of a cone represents ⅓ of that volume.

$$\text{volume of a cone} = \tfrac{1}{3}\,(\text{volume of a cylinder})$$

or

$$= \frac{(0.785)\,(D^2)\,(\text{Depth})}{3}$$

The volume of a sphere is more difficult to relate to the other calculations or even to a diagram. In this case the formula should be memorized. To express the volume of a sphere mathematically, use the following formula:

$$\text{volume of a sphere} = \left(\frac{\pi}{6}\right)(\text{diameter})^3$$

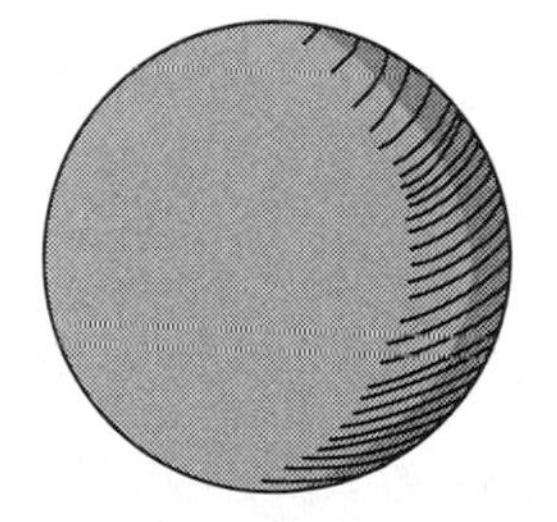

(π is the relationship between the circumference and diameter of a circle. The number 3.14 is used for pi.)[2] The equation may be reexpressed as

$$\text{volume of a sphere} = \left(\frac{3.14}{6}\right)(\text{diameter})^3$$

Example 5

Calculate the volume of a cone that is 3 m tall and has a base diameter of 2 m.

$$\text{volume of a cone} = \tfrac{1}{3}\,(\text{volume of a cylinder})$$

or

$$= \frac{(0.785)\,(D^2)\,(\text{third dimension})}{3}$$

$$= \frac{(0.785)\,(2\text{ m})\,(2\text{ m})\,(3\text{ m})}{3}$$

$$= 3.14\text{ m}^3$$

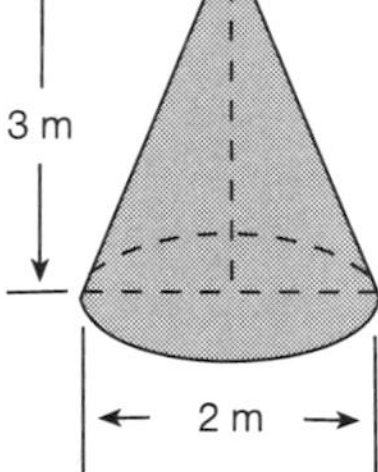

Example 6

If a spherical tank is 30 ft in diameter, what is its capacity?

$$\text{volume of a sphere} = \left(\frac{\pi}{6}\right)(\text{diameter})^3$$

---

[2]Mathematics 8, Linear Measurements.

$$= \left(\frac{3.14}{6}\right)(30\text{ ft})\,(30\text{ ft})\,(30\text{ ft})$$

$$= 14{,}130\text{ ft}^3$$

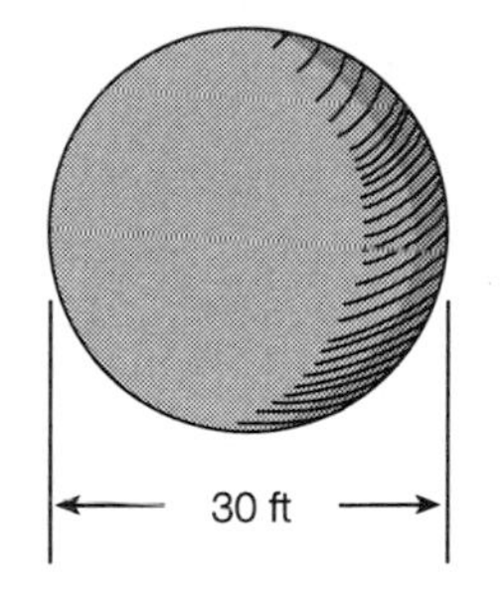

Occasionally it is necessary to calculate the volume of a tank that consists of two distinct shapes. In other words, there is no representative surface area for the entire shape. In this case, the volumes should be calculated separately, then the two volumes added. The diagrams below illustrate this method.

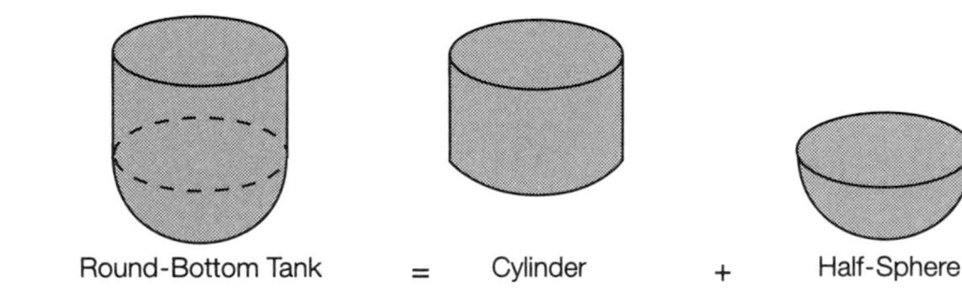

MATHEMATICS 11

Conversions

In making the conversion from one unit to another, you must know

- the number that relates the two units
- whether to multiply or divide by that number

For example, in converting from feet to inches, you must know that in 1 ft there are 12 in., and you must know whether to multiply or divide the number of feet by 12.

Although the number that relates the two units of a conversion is usually known or can be looked up, there is often confusion about whether to multiply or divide. One method to help decide whether to multiply or divide for a particular conversion is called *dimensional analysis,* discussed in a previous section.[1]

Conversion Tables

Usually the fastest method of converting units is to use a conversion table, such as the one included in appendix A, and to follow the instructions indicated by the table headings. For example, if you want to convert from feet to inches, look in the *Conversion* column of the table for *From* "feet" *To* "inches." Read across this line and perform the operation indicated by the headings of the other columns; that is, multiply the number of feet by 12 to get the number of inches.

Suppose, however, that you want to convert inches to feet. Look in the *Conversion* column for *From* "inches" *To* "feet," and read across this line. The

[1]Mathematics 2, Dimensional Analysis.

headings tell you to multiply the number of inches by 0.08333 (which is the decimal equivalent of 1⁄12) to get the number of feet. *Multiplying by either 1⁄12 or 0.08333 is the same as dividing by 12.*

The instruction to *multiply* by certain numbers (called conversion factors) is used throughout the conversion table. There is no column headed *Divide by* because the fractions representing division (such as 1⁄12) were converted to decimal numbers (such as 0.08333) when the table was prepared.

To use the conversion table, remember the following three steps:

1. In the *Conversion* column, find the units you want to change *From* and *To*. (Go *From* what you have *To* what you want.)
2. Multiply the *From* number you have by the conversion factor given.
3. Read the answer in *To* units.

Example 1

Convert 288 in. to feet.

In the *Conversion* column of the table, find *From* "inches" *To* "feet." Reading across the line, perform the multiplication indicated; that is, multiply the number of inches (288) by 0.08333 to get the number of feet.

$$(288 \text{ in.})\,(0.08333) \quad = \quad 24 \text{ ft}$$

Example 2

A tank holds 50 gal of water. How many cubic feet of water is this, and what does it weigh?

First, convert gallons to cubic feet. Using the table, you find that to convert *From* "gallons" *To* "cubic feet," you must multiply by 0.1337 to get the number of cubic feet:

$$(50 \text{ gal})\,(0.1337) \quad = \quad 6.69 \text{ ft}^3$$

Note that this number of cubic feet is actually a rounded value (6.685 is the actual calculated number).[2] Rounding helps simplify calculations.

Next, convert gallons to pounds of water. Using the table, you find that to convert *From* "gallons" *To* "pounds of water," you must multiply by 8.34 to get the number of pounds of water:

$$(50 \text{ gal})\,(8.34) \quad = \quad 417 \text{ lb of water}$$

---

[2]Mathematics 3, Rounding and Estimating.

Notice that you could have arrived at approximately the same weight by converting 6.69 ft^3 to pounds of water. Using the table, we get

$$(6.69\ ft^3)\ (62.4) \quad = \quad 417.46\ \text{lb of water}$$

This slight difference in the two answers is due to rounding[3] numbers both when the conversion table was prepared and when the numbers are used in solving the problem. You may notice the same sort of slight difference in answers if you have to convert from one kind of units to two or three other units, depending on whether you round intermediate steps in the conversions.

Box Method

Another method that may be used to determine whether multiplication or division is required for a particular conversion is called the *box method* and is based on the relative sizes of different squares ("boxes"). The box method can be used when a conversion table is not available (such as during a certification exam). This method of conversion is often slower than using a conversion table, but many people find it simpler.

Because multiplication is usually associated with an *increase* in size, moving from a smaller box to a larger box corresponds to using multiplication in the conversion:

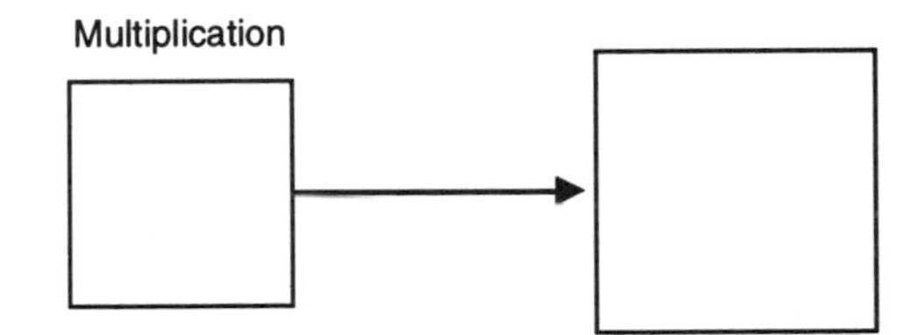

Division, on the other hand, is usually associated with a *decrease* in size. Therefore, moving from a larger box to a smaller box corresponds to using division in the conversion:

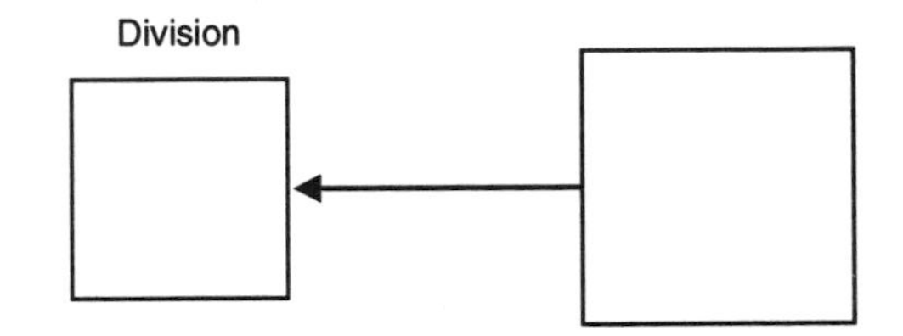

---

[3]Mathematics 3, Rounding and Estimating.

To use the box method to determine whether to multiply or divide in making a conversion, set up and label the boxes according to the following procedure:

1. Write the equation that relates the two types of units involved in the conversion. (One of the two numbers in the equation must be a 1; for example, 1 ft = 12 in., or 1 ft = 0.305 m.)
2. Draw a small box on the left, a large one on the right, and connect them with a line (as in the drawings at the beginning of this section).
3. In the *smaller* box, write the name of the units associated with the 1 (for example, 1 *ft* = 12 in.—*ft* should be written in the smaller box). Note that the name of the units next to the 1 must be written in the *smaller* box, *or the box method will give incorrect results.*
4. In the larger box, write the name of the remaining units. Those units will also have a number next to them, a number that is not 1. Write that number over the line between the boxes.

Suppose, for example, that you want to make a box diagram for feet-to-inches conversions. First, write the equation that relates feet to inches:

1 ft = 12 in.

Next, draw the conversion boxes (smaller box on the left) and the connecting line:

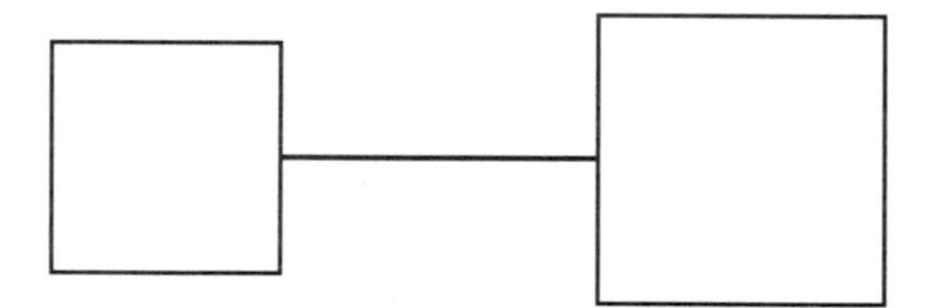

Now label the diagram. Because the number 1 is next to the units of *feet* (1 ft), write *ft* in the smaller box. Write the name of the other units, inches (*in.*), in the larger box. And write the number that is next to inches, 12, over the line between the boxes.

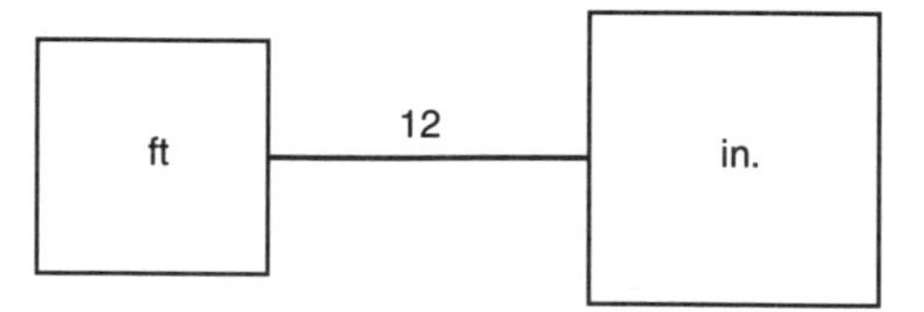

To convert from feet to inches, then *multiply by 12* because you are moving from a smaller box to a larger box. And to convert from inches to feet, *divide by 12* because you are moving from a larger box to a smaller box.

Let's look at another example of making and using the box diagram. Suppose you want to convert cubic feet to gallons. First write down the equation that relates these two units:

$$1 \text{ ft}^3 = 7.48 \text{ gal}$$

Then draw the smaller and larger boxes and the connecting line; label the boxes, and write in the conversion number:

The smaller box corresponds to cubic feet and the larger box to gallons. To convert from cubic feet to gallons according to this box diagram, *multiply* by 7.48 because you are moving from a smaller to a larger box. And to convert from gallons to cubic feet, *divide* by 7.48 because you are moving from a larger to a smaller box.

Conversions of US Customary Units

This section discusses important conversions between terms expressed in US customary units (based on the box method).

Conversions From Cubic Feet to Gallons to Pounds

In making the conversion from cubic feet to gallons to pounds of water, you must know the following relationships:

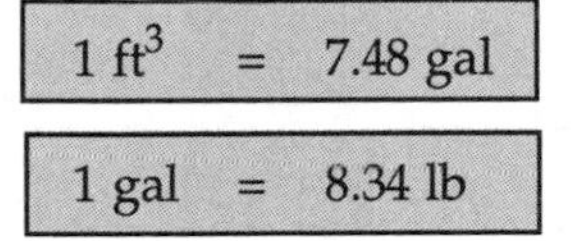

You must also know whether to multiply or divide, and which of the above numbers are used in the conversion. The following box diagram should assist in making these decisions.

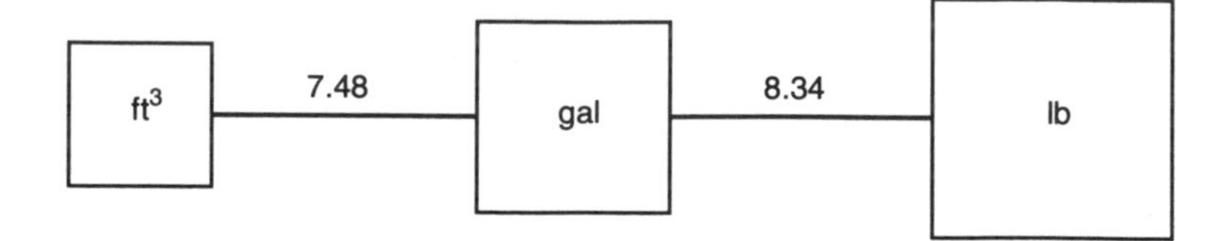

Example 3

Convert 1 ft^3 to pounds.

First write down the diagram to aid in the conversion:

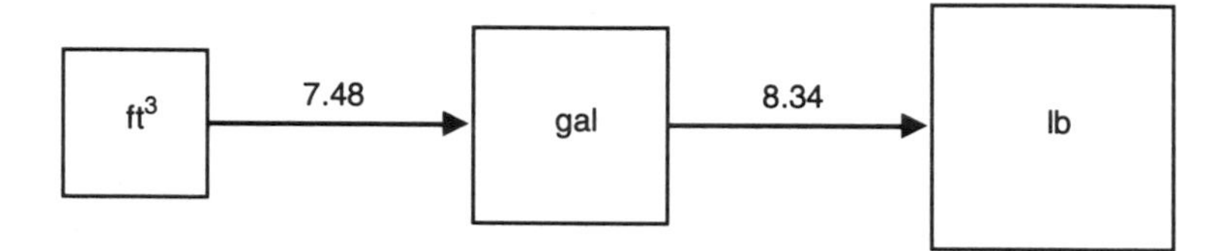

When you are converting from cubic feet to pounds, you are moving from smaller to larger boxes. Therefore *multiplication* is indicated in both conversions:

$$(1\ ft^3)\ (7.48\ gal/ft^3)\ (8.34\ lb/gal) = 62.38\ lb$$

or

$$= 62.4\ lb$$ (rounded to the number commonly used for water treatment calculations)

Example 4

A tank has a capacity of 60,000 ft^3. What is the capacity of the tank in gallons?

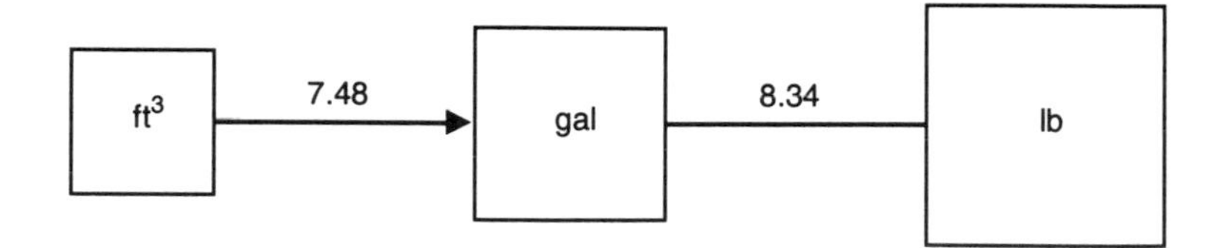

When converting from cubic feet to gallons, you are moving from a smaller to a larger box. Therefore, *multiplication* by 7.48 is indicated:

$$(60{,}000 \text{ ft}^3)\ (7.48 \text{ gal/ft}^3) \quad = \quad 448{,}800 \text{ gal}$$

Example 5

If a tank will hold 1,500,000 lb of water, how many cubic feet of water will it hold?

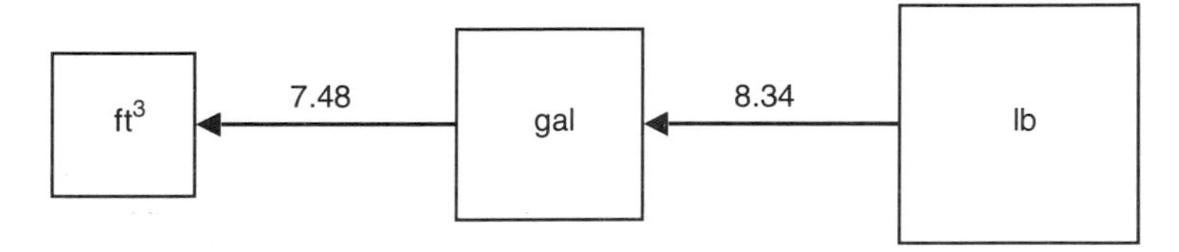

A move from larger to smaller boxes indicates *division* in both conversions:

$$\frac{1{,}500{,}000 \text{ lb}}{(8.34 \text{ lb/gal})\ (7.48 \text{ gal/ft}^3)} \quad = \quad 24{,}045 \text{ ft}^3$$

Flow Conversions

The relationships among the various US customary flow units are shown by the following diagram:

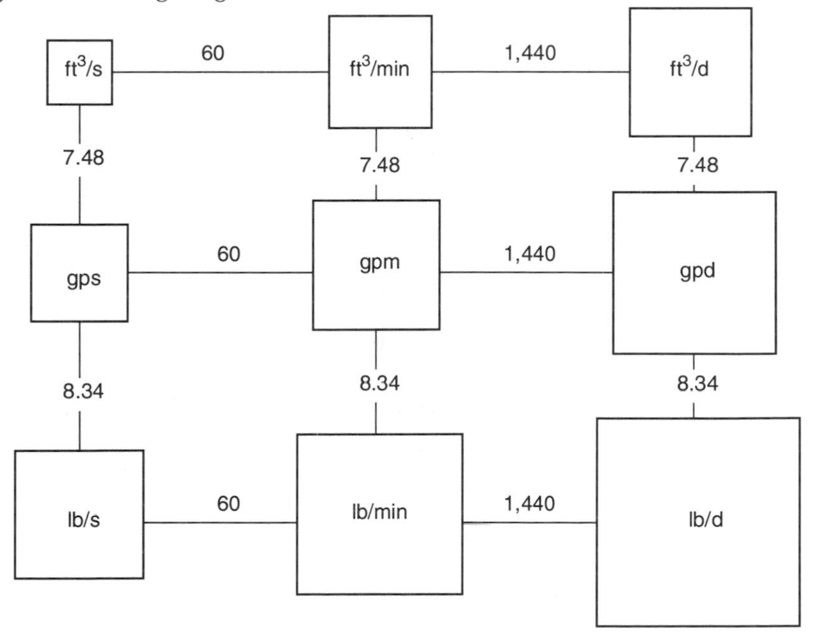

Note that

gps = gallons per second

gpm = gallons per minute

gpd = gallons per day

Since flows of cubic feet per hour, gallons per hour, and pounds per hour are less frequently used, they have not been included in the diagram. However, some chemical feed rate calculations require converting from or to these units.[4]

The lines that connect the boxes and the numbers associated with them can be thought of as *bridges* that relate two units directly and all other units in the diagram indirectly.

The relative sizes of the boxes are an aid in deciding whether multiplication or division is appropriate for the desired conversion.

The relationship among the boxes should be understood, not merely memorized. The principle is basically the same as that described in the preceding section. For example, looking at just part of the above diagram, notice how every box in a single vertical column has the same *time* units; a conversion in this direction corresponds to a change in volume units. Every box in a single horizontal row has the same *volume* units; a conversion in this direction corresponds to a change in time units.

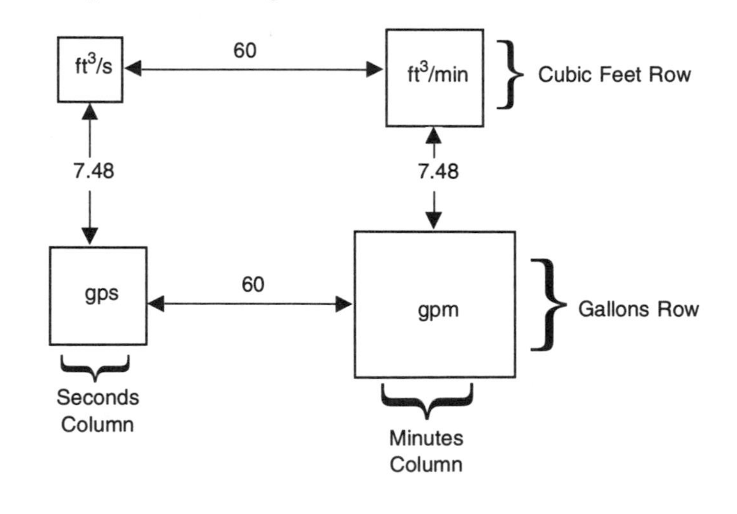

---

[4]Chemistry 7, Chemical Dosage Problems.

Although you need not draw the nine boxes each time you make a flow conversion, it is useful to have a mental image of these boxes to make the calculations. For the examples that follow, however, the boxes are used in analyzing the conversions.

Example 6

If the flow rate in a water line is 2.3 ft^3/s, what is this rate expressed as gallons per minute? (Assume the flow is steady and continuous.)

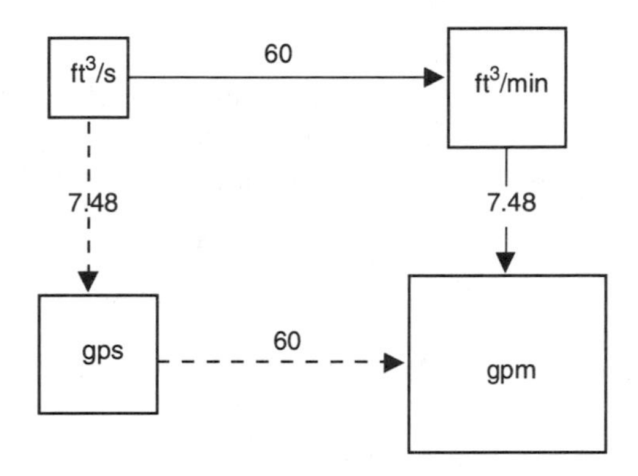

There are two possible paths from cubic feet per second to gallons per minute. Either will give the correct answer. Notice that each path has factors of 60 and 7.48, with only a difference in order. In each case, you are moving from a smaller to a larger box, and thus *multiplication* by both 60 and 7.48 is indicated:

$$(2.3 \text{ ft}^3/\text{s})\ (60 \text{ s/min})\ (7.48 \text{ gal/ft}^3) \quad = \quad 1{,}032 \text{ gpm}$$

Notice that you can write both multiplication factors into the same equation — you do not need to write one equation for converting cubic feet per second to cubic feet per minute and another for converting cubic feet per minute to gallons per minute.

Example 7

The flow rate to a sedimentation basin is 2,450,000 gpd. At this rate, what is the average flow in cubic feet per second?

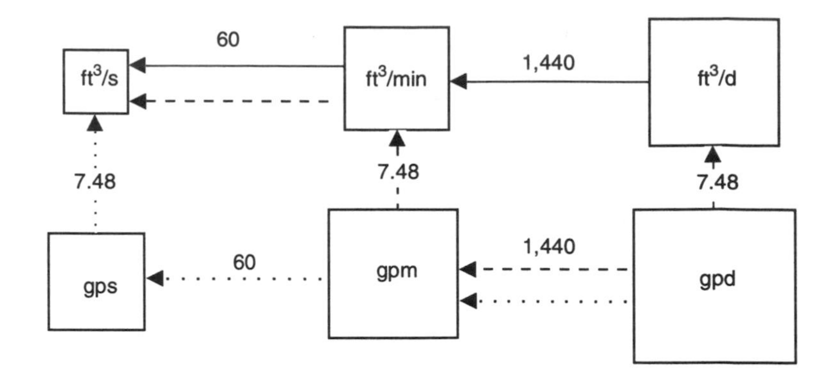

There are three possible paths from gallons per day to cubic feet per second. In each case, you would be moving from a larger to a smaller box, thus indicating *division* by 7.48, 1,440, and 60 (in any order).

$$\frac{2{,}450{,}000 \text{ gpd}}{(7.48 \text{ gal/ft}^3)\ (1{,}440 \text{ min/d})\ (60 \text{ s/min})} = 3.79 \text{ ft}^3\text{/s}$$

Again, the divisions are all written into one equation.

Example 8

If a flow rate is 200,000 gpd, what is this flow expressed as pounds per minute?

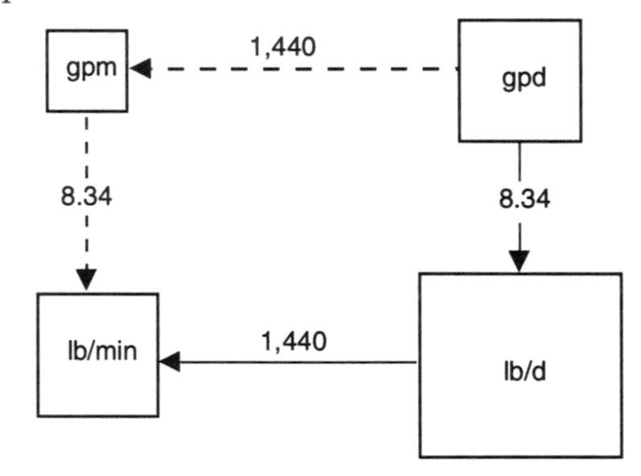

There are two possible paths from gallons per day to pounds per minute. The only difference in these paths is the order in which the numbers appear. The answer is the same in either case. In the following explanation, the solid-line path of the diagram is used.

Converting from gallons per day to pounds per day, you are moving from a smaller box to a larger box. Therefore, *multiplication* by 8.34 is indicated. Then from pounds per day to pounds per minute, you are moving from a larger to a smaller box, which indicates *division* by 1,440. These multiplication and division steps are combined into one equation:

$$\frac{(200{,}000 \text{ gpd})\,(8.34 \text{ lb/gal})}{1{,}440 \text{ min/d}} = 1{,}158 \text{ lb/min}$$

Linear Measurement Conversions

Linear measurement defines the distance along a line; it is the measurement between two points. The US customary units of linear measurement include the inch, foot, yard, and mile. In most treatment plant calculations, however, the mile is not used. Therefore, this section discusses conversions of inches, feet, and yards only. The box diagram associated with these conversions is

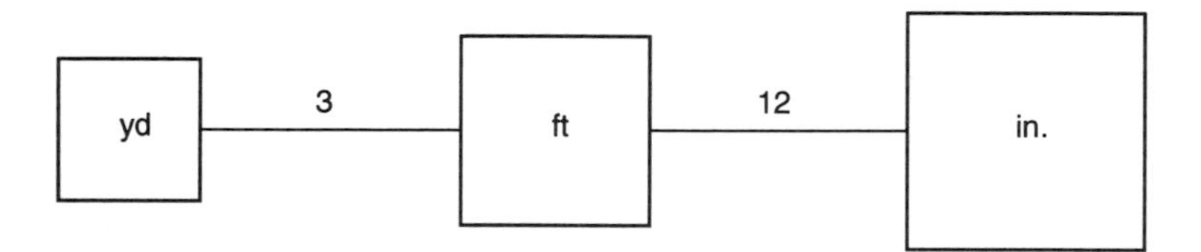

Example 9

The maximum depth of sludge drying beds is 14 in. How many feet is this?

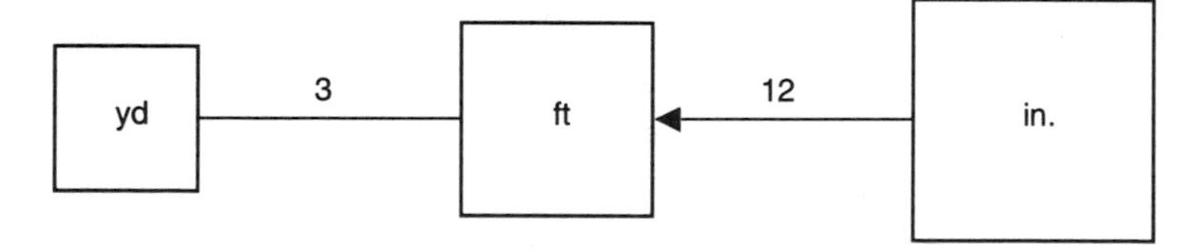

In converting from inches to feet, you are moving from a larger to a smaller box. Therefore, *division* by 12 is indicated.

$$\frac{14 \text{ in.}}{12 \text{ in./ft}} = 1.17 \text{ ft}$$

Example 10

During backwashing, the water level drops 0.6 yd during a given time interval. How many feet has it dropped?

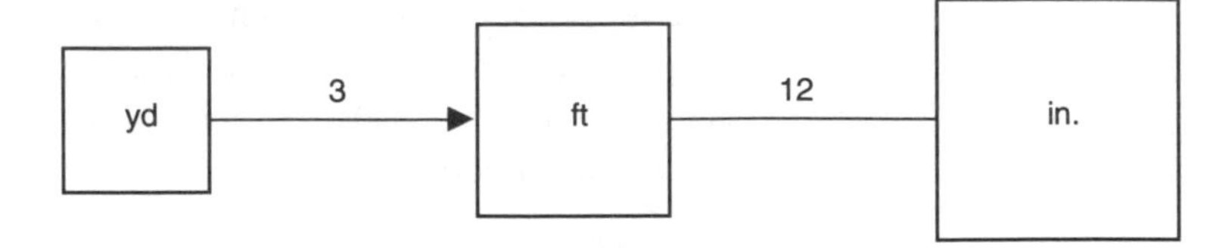

Moving from a smaller to a larger box indicates *multiplication* by 3:

$$(0.6 \text{ yd})\,(3 \text{ ft/yd}) = 1.8 \text{ ft}$$

Area Measurement Conversions

To make area conversions in US customary units, you work with units such as square yards, square feet, or square inches. These are derived from the multiplications

(yards) (yards) = square yards

(feet) (feet) = square feet

(inches) (inches) = square inches

By examining the relationship of yards, feet, and inches in linear terms, you can recognize the relationship between yards, feet, and inches in square terms. For example,

$$1 \text{ yd} = 3 \text{ ft}$$

$$(1 \text{ yd})\,(1 \text{ yd}) = (3 \text{ ft})\,(3 \text{ ft})$$

$$1 \text{ yd}^2 = 9 \text{ ft}^2$$

This method of comparison may be used whenever you wish to compare linear terms with square terms.

Compare the diagram used for linear conversions with that used for square measurement conversions:

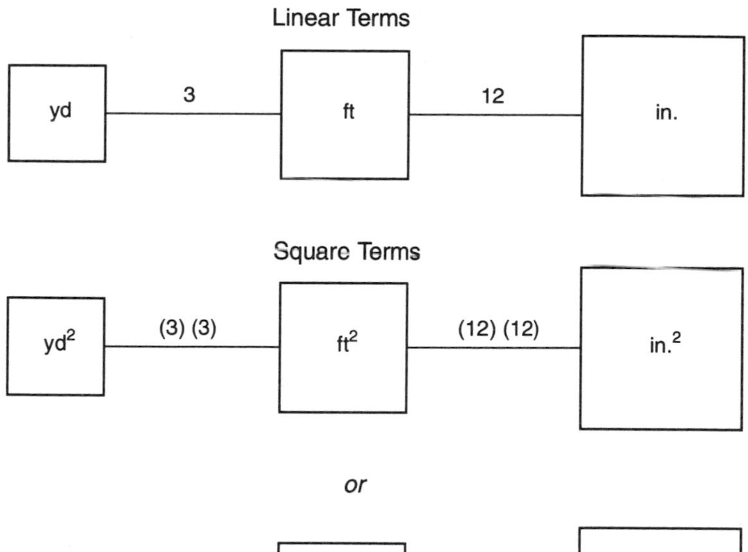

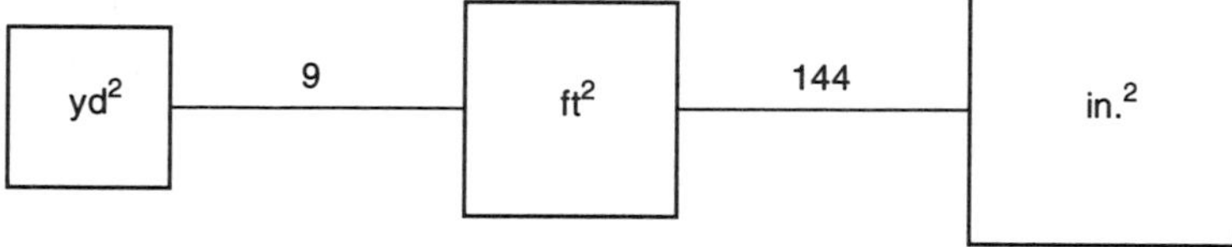

Example 11

The surface area of a sedimentation basin is 170 yd^2. How many square feet is this?

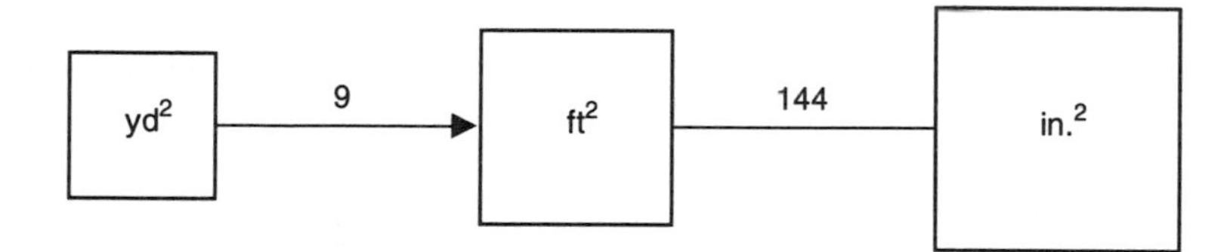

When converting from square yards to square feet, you are moving from a smaller to a larger box. Therefore, *multiplication* by 9 is indicated.

$$(170\ yd^2)\,(9\ ft^2/yd^2) \quad = \quad 1{,}530\ ft^2$$

Example 12

The cross-sectional area of a pipe is 64 in^2. How many square feet is this?

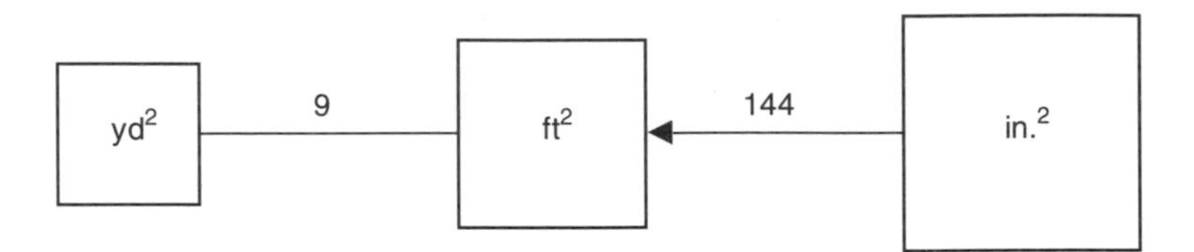

Converting from square inches to square feet, you are moving from a larger to a smaller box. *Division* by 144 is indicated.

$$\frac{64 \text{ in.}^2}{144 \text{ in.}^2/\text{ft}^2} = 0.44 \text{ ft}^2$$

One other area conversion important in treatment plant calculations is that between square feet and acres. This relationship is expressed mathematically as

$$1 \text{ acre} = 43{,}560 \text{ ft}^2$$

A box diagram can be devised for this relationship. However, the diagram should be separate from the diagram relating square yards, square feet, and square inches because you usually wish to relate directly with square feet. As in the other diagrams, the relative sizes of the boxes are important.

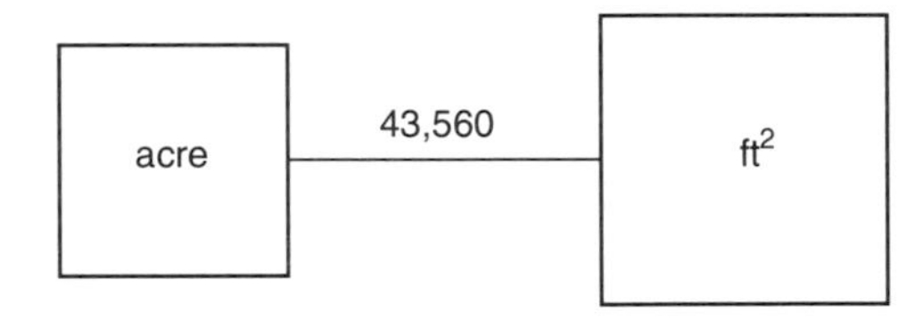

Example 13

A treatment plant requires 0.2 acre for drying beds. How many square feet are required?

Converting acres to square feet, you are moving from a smaller to a larger box. *Multiplication* by 43,560 is therefore indicated:

$$(0.2 \text{ acre})\,(43{,}560 \text{ ft}^2/\text{acre}) = 8{,}712 \text{ ft}^2$$

Volume Measurement Conversions

To make volume conversions in US customary unit terms, you work with such units as cubic yards, cubic feet, and cubic inches. These units are derived from the multiplications

(yards) (yards) (yards)	=	cubic yards
(feet) (feet) (feet)	=	cubic feet
(inches) (inches) (inches)	=	cubic inches

By examining the relationship of yards, feet, and inches in linear terms, you can recognize the relationship between yards, feet, and inches in cubic terms. For example,

1 yd	=	3 ft
(1 yd) (1 yd) (1 yd)	=	(3 ft) (3 ft) (3 ft)
1 yd^3	=	27 ft^3

The box diagram associated with these cubic conversions is

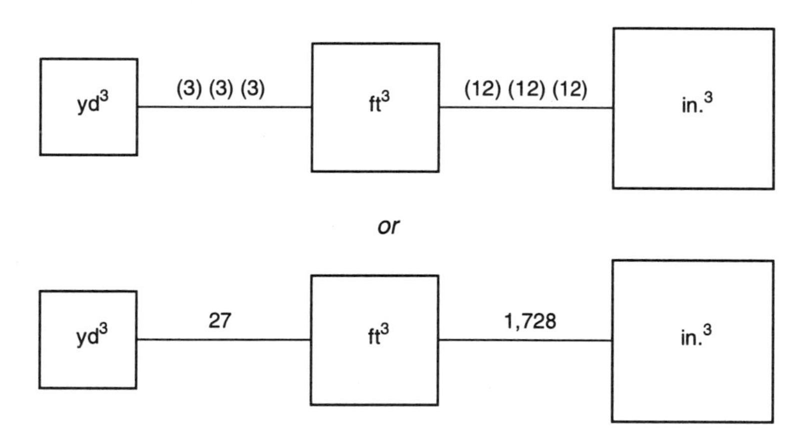

Example 14

Convert 15 yd^3 to cubic inches.

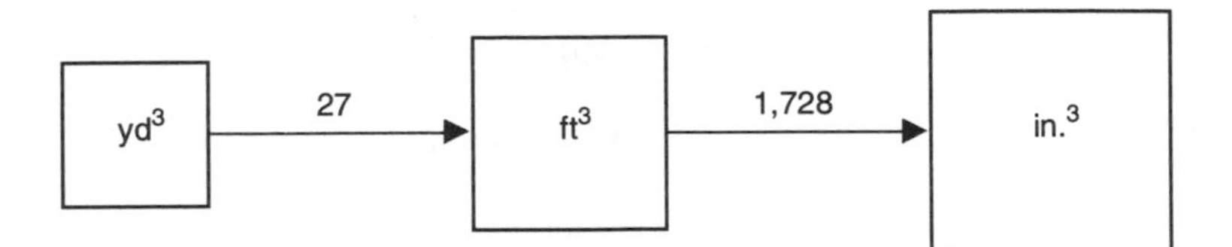

In converting from cubic yards to cubic feet and from cubic feet to cubic inches, you are moving from smaller to larger boxes. Thus, *multiplication* by 27 and 1,728 is indicated:

$$(15 \text{ yd}^3)(27 \text{ ft}^3/\text{yd}^3)(1{,}728 \text{ in.}^3/\text{ft}^3) = 699{,}840 \text{ in.}^3$$

Example 15

The required volume for a chemical is 325 ft^3. What is this volume expressed as cubic yards?

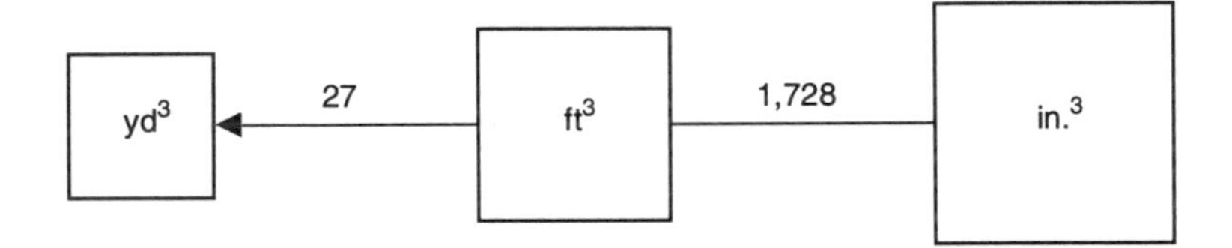

When you move from a larger to a smaller box, *division* is indicated:

$$\frac{325 \text{ ft}^3}{27 \text{ ft}^3/\text{yd}^3} = 12.04 \text{ yd}^3$$

Another volume measurement important in treatment plant calculations is that of acre-feet. A reservoir with a surface area of 1 acre and a depth of 1 ft holds exactly 1 acre-ft.

$$1 \text{ acre-ft} = 43{,}560 \text{ ft}^3$$

The relative sizes of the boxes are again important:

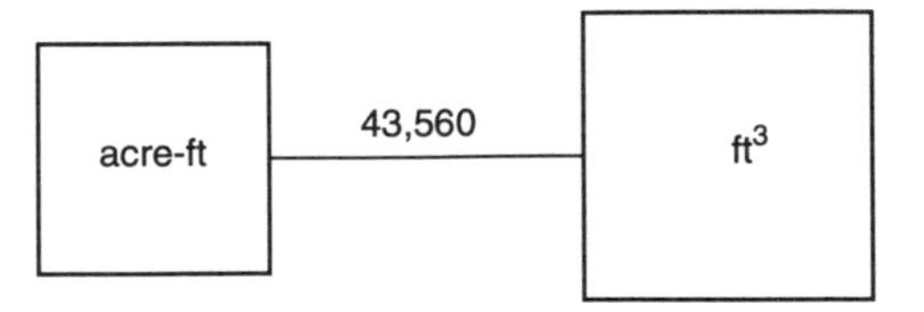

Example 16

The available capacity of a reservoir is 220,000 ft^3. What is this volume expressed in terms of acre-feet?

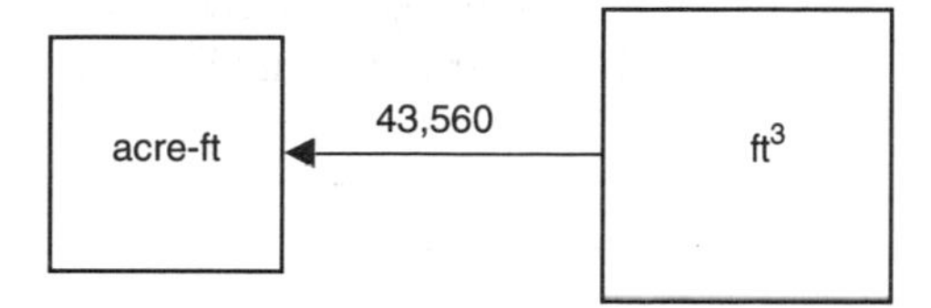

When you move from the larger to the smaller box, *division* by 43,560 is indicated:

$$\frac{220{,}000\ \text{ft}^3}{43{,}560\ \text{ft}^3/\text{acre-ft}} = 5.05\ \text{acre-ft}$$

Concentration Conversions

A milligrams-per-liter (mg/L) concentration can also be expressed in terms of grains per gallon (gpg) or parts per million (ppm). However, of the three, the preferred unit of concentration is milligrams per liter. To convert milligrams per liter to percent or to weight per day (lb/d), refer to the Chemistry portion of this handbook.[5]

Milligrams per Liter to Grains per Gallon

Conversions between milligrams per liter and grains per gallon are based on the relationship

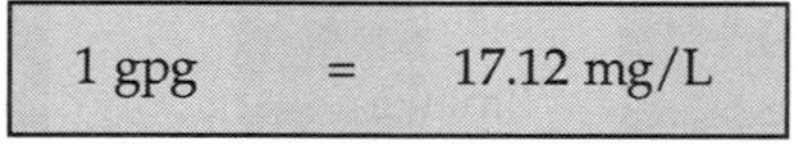

As with any other conversion, often the greatest difficulty in converting from one term to another is deciding whether to multiply or divide by the number given. The box diagram below should help in making this decision.

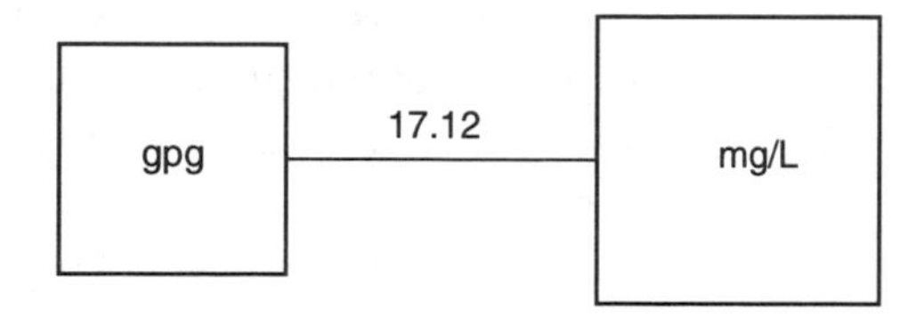

[5]Chemistry 7, Chemical Dosage Problems.

Example 17

Convert 25 mg/L to grains per gallon.

In this example, you are converting from milligrams per liter to grains per gallon. Therefore, as shown below, you are moving from the larger to the smaller box:

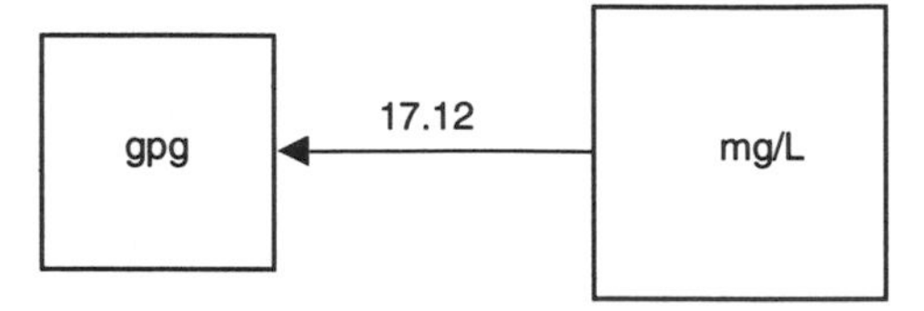

Larger to smaller indicates *division* by 17.12:

$$\frac{25 \text{ mg/L}}{17.12 \text{ mg/L/gpg}} = 1.46 \text{ gpg}$$

Example 18

Express a 20-gpg concentration in terms of milligrams per liter.

In this example, you are converting from grains per gallon to milligrams per liter. Therefore, you are moving from the smaller to the larger box:

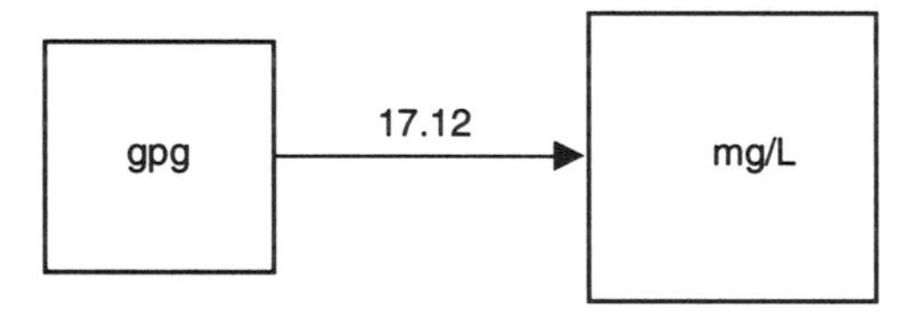

Smaller to larger indicates *multiplication* by 17.12:

$$(20 \text{ gpg})\,(17.12 \text{ mg/L/gpg}) = 342.4 \text{ mg/L}$$

Example 19

If the dosage rate of alum is 1.5 gpg, what is the dosage rate expressed in milligrams per liter?

The desired conversion is from grains per gallon to milligrams per liter:

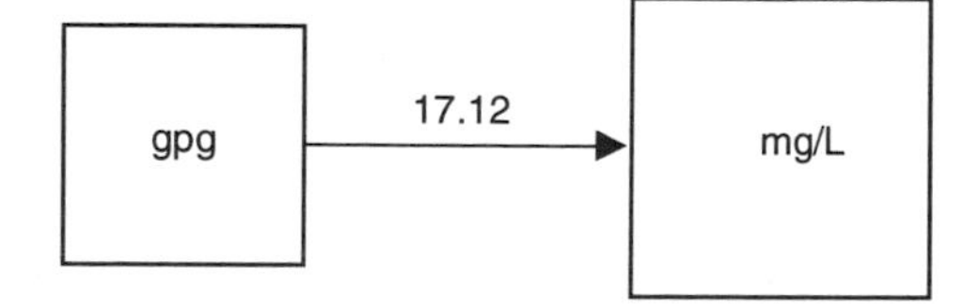

Smaller to larger indicates *multiplication* by 17.12:

$$(1.5 \text{ gpg}) (17.12 \text{ mg/L/gpg}) = 25.68 \text{ mg/L}$$

Milligrams per Liter to Parts per Million

The concentration of impurities in water is usually so small that it is measured in milligrams per liter. This means that the impurities in a standard volume (a liter) of water are measured by weight (milligrams). Concentrations in the range of 0–2,000 mg/L are roughly equivalent to concentrations expressed as the same number of parts per million (ppm). For example, "12 mg/L of calcium in water" expresses roughly the same concentration as "12 ppm calcium in water." However, milligrams per liter are the preferred units of concentration.

Metric System Conversions

In order to convert from the system of US customary units to the metric system, or vice versa, you must understand how to convert within the metric system. This requires a knowledge of the common metric prefixes. *These prefixes should be learned before any conversions are attempted.*

As shown in Table M11-1, on page 112, the metric system is based on *powers*, or multiples of 10,[6] just like the decimal system is. These prefixes may be associated with positions in the place value system.

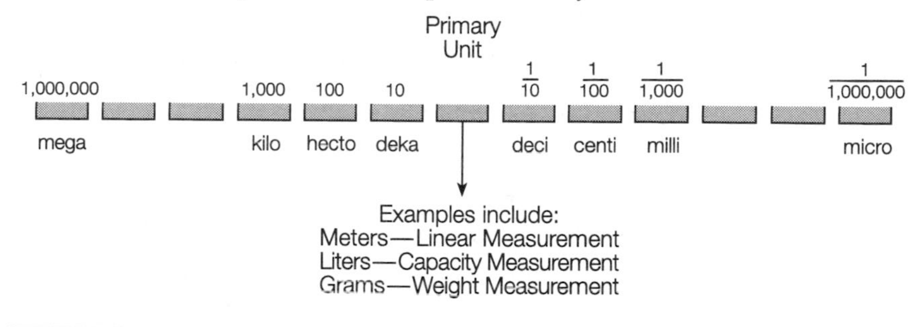

[6]Mathematics 1, Powers and Scientific Notation.

TABLE M11-1 Metric system notations

Prefix	Abbreviation	Mathematical Value	Powers Notation
giga	G	1,000,000,000	10^9
mega	M	1,000,000	10^6
kilo	k	1,000	10^3
hecto*	h	100	10^2
deka*	da	10	10^1
(none†)	(none)	1	10^0
deci*	d	$\frac{1}{10}$ *or* 0.1	10^{-1}
centi*	c	$\frac{1}{100}$ *or* 0.01	10^{-2}
milli	m	$\frac{1}{1,000}$ *or* 0.001	10^{-3}
micro	μ	$\frac{1}{1,000,000}$ *or* 0.000001	10^{-6}
nano	n	$\frac{1}{1,000,000,000}$ *or* 0.000000001	10^{-9}

*Use of these units should be avoided when possible.
†Primary units, such as meters, liters, grams.

Understanding the position of these prefixes in the place value system is important because the method discussed below for metric-to-metric conversions is based on this system.

It is also important to understand the abbreviations used for metric terms. The basic measurement terms and their abbreviations are meters (m), liters (L), and grams (g). The prefixes added to the basic measurement terms may also be abbreviated (as shown in Table M11-1). For example,

1 megaliter = 1 ML

1 millimeter = 1 mm

1 kilogram = 1 kg

Use of these abbreviations greatly simplifies expressions of measurement.

Metric-to-Metric Conversions

When conversions are being made for linear measurement (meters), volume measurement (liters), and weight measurement (grams), each change in prefix place value represents one decimal point move. This system of conversion is demonstrated by the following examples.

Example 20

Convert 1 m to decimeters (dm).

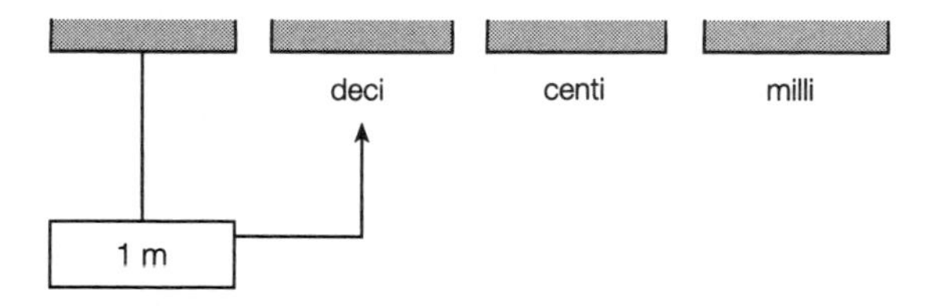

Converting from meters to decimeters requires moving one place to the right. Therefore, move the decimal point from its present position *one place to the right.*

1.0 = 10 decimeters

Example 21

Convert 1 g to (a) decigrams; (b) centigrams; and (c) milligrams.

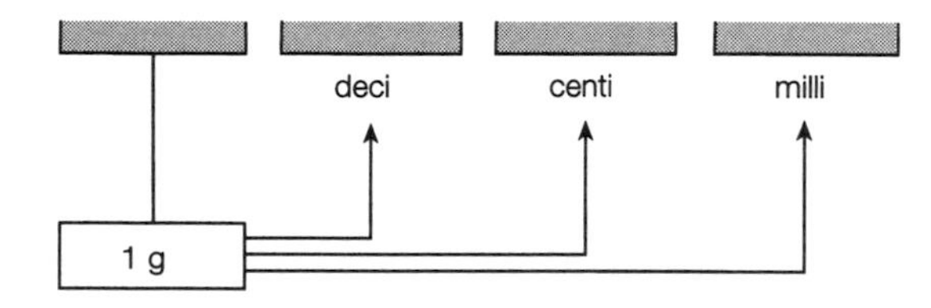

(a) Move the decimal point one place to the right.

1.0 = 10 decigrams

(b) Move the decimal point two places to the right.

1.00 = 100 centigrams

(c) Move the decimal point three places to the right.

1.000 = 1,000 milligrams

This system of conversion applies whether you are converting from the primary unit, as in examples 22 and 23, or from any other unit.

Example 22

Convert 1 dL to milliliters.

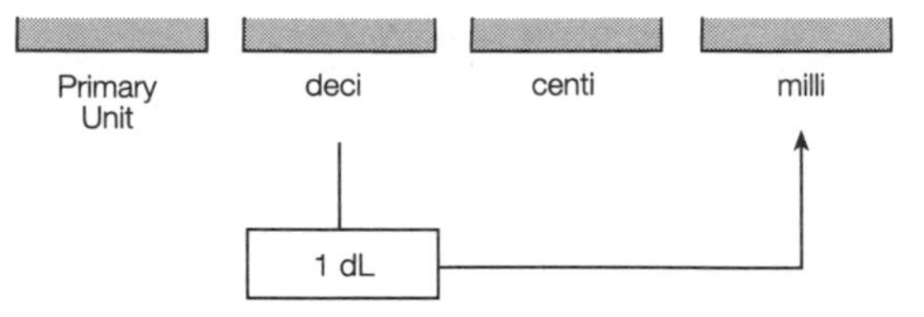

Move the decimal point two places to the right.

1.00 = 100 milliliters

This system of conversion also applies whether the number you are converting is 1.0, as in examples 20 through 22, or any other number.

Example 23

Convert 3.5 kg to grams.

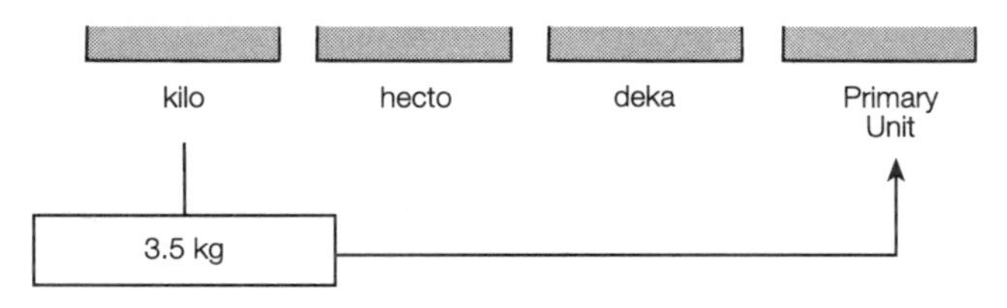

Move the decimal point three places to the right.

3.500 = 3,500 grams

Example 24

Convert 0.28 cm to meters.

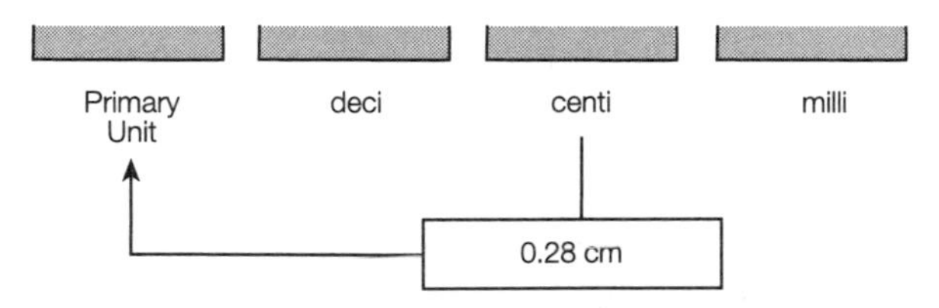

Move the decimal point two places to the left.

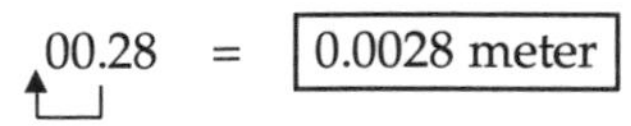

00.28 = 0.0028 meter

Most metric conversion errors are made in moving the decimal point to the left. You must be very careful in moving the decimal point from its present position, counting every number (including zeros) to the left as a decimal point move.

Example 25

Convert 1,750 L to kiloliters.

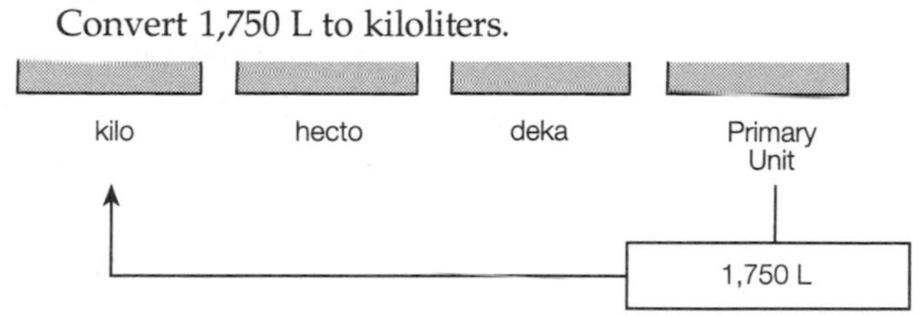

Move the decimal point three places to the left.

$$1750. = \boxed{1.75 \text{ kiloliters}}$$

In the examples just given, there were no conversions of square or cubic terms. However, area and volume measurements can be expressed as square and cubic meters, centimeters, kilometers, and so on. The following discussion shows the special techniques needed for converting between these units.

Square meters indicates the mathematical operation:

$$\text{(meter) (meter)} = \text{square meters}$$

Square meters may also be written as m^2. The exponent of 2 indicates that *meter* appears twice in the multiplication, as shown above.[7]

In conversions, the term *square* (or exponent of 2) indicates that *each prefix place value move requires 2 decimal point moves*. All other aspects of the conversions are similar to the preceding examples.

Example 26

Convert 1 m^2 to square decimeters.

Primary Unit^2 deci^2 centi^2 milli^2

1 m^2

---

[7]Mathematics 1, Powers and Scientific Notation.

Converting from square meters to square decimeters requires moving one place value to the right. In square terms, *each prefix place move requires 2 decimal point moves.* Making the move in groups of two may be easier:

$$1.00 \quad = \quad \boxed{\text{100 square decimeters}}$$

Now check this conversion. From example 20 you know that 1 m = 10 dm. Squaring both sides of the equation, we get

$$(1\ \text{m})\ (1\ \text{m}) \quad = \quad (10\ \text{dm})(10\ \text{dm})$$

$$1\ \text{m}^2 \quad = \quad 100\ \text{dm}^2$$

Example 27

Convert 32,000 m^2 to square kilometers.

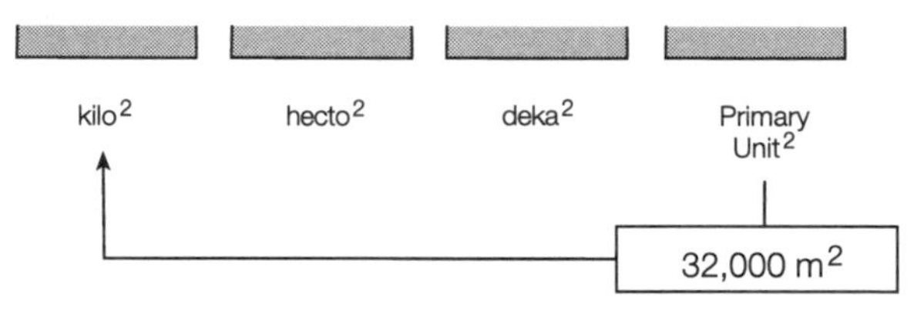

Three place value moves to the left correspond to six total decimal point moves to the left:

$$032{,}000. \quad = \quad \boxed{\text{0.032 square kilometer}}$$

Cubic meters indicates the following mathematical operation:

$$(\text{meters})\ (\text{meters})\ (\text{meters}) \quad = \quad \text{cubic meters}$$

Cubic meters may also be written as m^3. The exponent of 3 indicates that *meter* appears three times in the multiplication. When you are converting cubic terms, *each prefix place value move requires 3 decimal point moves.* Again, it may be easier to make the decimal point moves in groups — groups of three for cubic-term conversions as opposed to groups of two for square-term conversions.

Example 28

Convert 1 m^3 to cubic decimeters.

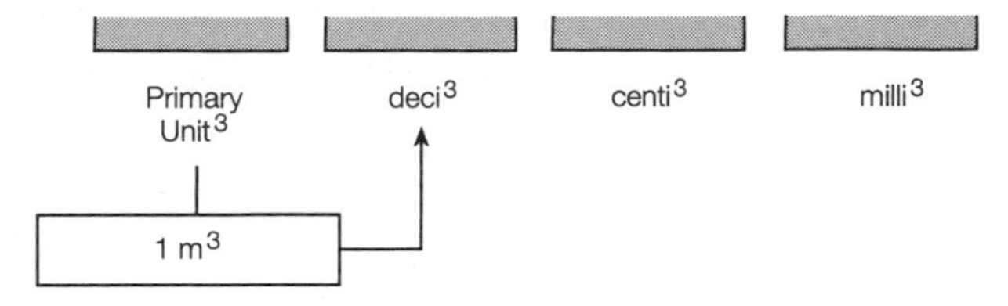

Converting from cubic meters to cubic decimeters requires moving one place value to the right. In cubic terms, *each prefix place value move requires 3 decimal point moves*:

1.000 = 1,000 cubic decimeters

Now check this conversion. From example 20 you know that 1 m = 10 dm. Cubing both sides of the equation, we get

$$(1 \text{ m})(1 \text{ m})(1 \text{ m}) = (10 \text{ dm})(10 \text{ dm})(10 \text{ dm})$$

$$1 \text{ m}^3 = 1{,}000 \text{ dm}^3$$

Example 29

Convert 155,000 mm^3 to cubic meters.

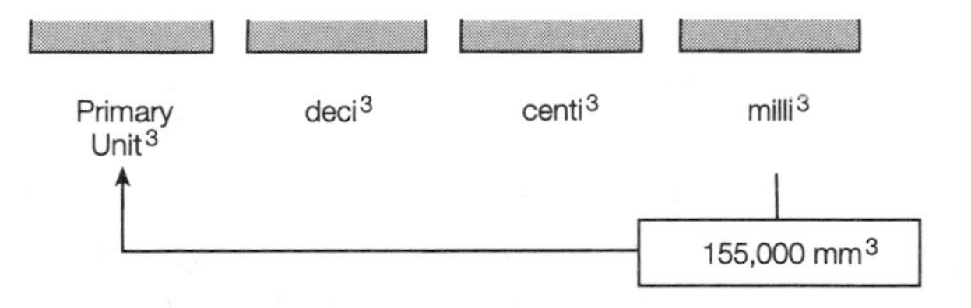

Three place value moves to the left indicate three *groups of 3* decimal point moves to the left:

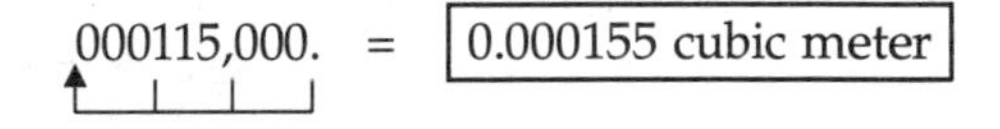

Cross-System Conversions

For conversions from the US customary unit system to the metric system, or vice versa, the conversion table appearing in appendix A and discussed at the beginning of this chapter is useful.

Example 30

Convert 20 gal to liters.

The factor given in the table to convert *From* "gallons" *To* "liters" is 3.785. This means that there are 3.785 L in 1 gal. Therefore, the conversion is

$$(20 \text{ gal})(3.785) = 75.7 \text{ L}$$

Example 31

Convert 3.7 acres to square meters.

The factor given in the table for converting *From* "acres" *To* "square meters" is 4,047. The conversion is therefore

$$(3.7 \text{ acre})(4{,}047) = 14{,}974 \text{ m}^2$$

Example 32

Convert 0.8 m/s to ft/min.

The factor given in the table is 196.8. Therefore, the conversion is

$$(0.8 \text{ m/s})(196.8) = 157.44 \text{ ft/min}$$

Occasionally, when making cross-system conversions, you may not find the factor in the table for the two terms of interest. For example, you may wish to convert from inches to decimeters and the table gives factors only for converting from inches to centimeters or inches to millimeters. Or you may wish to convert cubic centimeters to gallons and the table gives factors only for converting cubic meters to gallons.

In situations such as these, it is usually easiest to make sure that the US customary system unit is in the desired form and then make any necessary changes to the metric unit (e.g., changing inches to centimeters, then centimeters to decimeters). As shown in the example problems for metric-to-metric conversions, changing units in the metric system requires only a decimal point move.

The following two examples illustrate how the cross-system conversion may be made when the table does not give the precise units you need.

Example 33

The water depth in a channel is 1.2 ft. How many decimeters is this?

First check the conversion table in appendix A to see if a factor is given for converting feet to decimeters. The conversion factor is given only for feet to kilometers, meters, centimeters, or millimeters.

To make the conversion from feet to decimeters then, first convert from feet to the closest metric unit to decimeters given in the table (centimeters); then convert that answer to decimeters. The conversion from feet to centimeters is

$$(1.2 \text{ ft}) (30.48) = 36.58 \text{ cm}$$

Then converting from centimeters to decimeters, we get

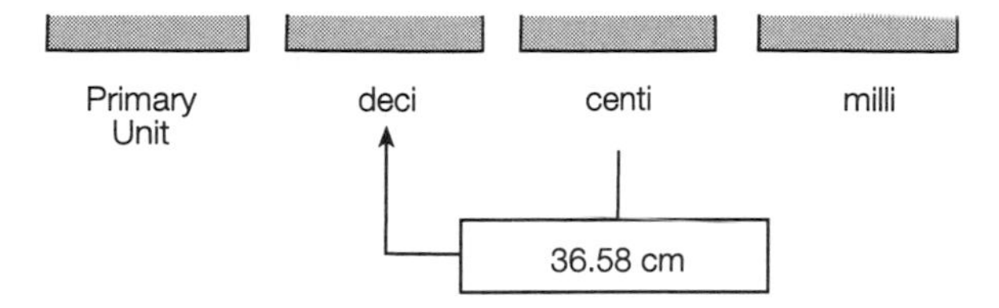

Move the decimal point one place to the left:

$$36.58 \text{ cm} = 3.658 \text{ dm}$$

Example 34

If you use 0.12 kg of a chemical to make up a particular solution, how many ounces of that chemical are used?

First, check the conversion table in appendix A to determine if a factor is given for converting kilograms to ounces. No such conversion factor is given.

Try to find a conversion in the table from some other metric unit (such as milligrams or grams) to ounces. The conversion from grams to ounces is given in the table. Therefore, first convert the 0.12 kg to grams, then convert grams to ounces:

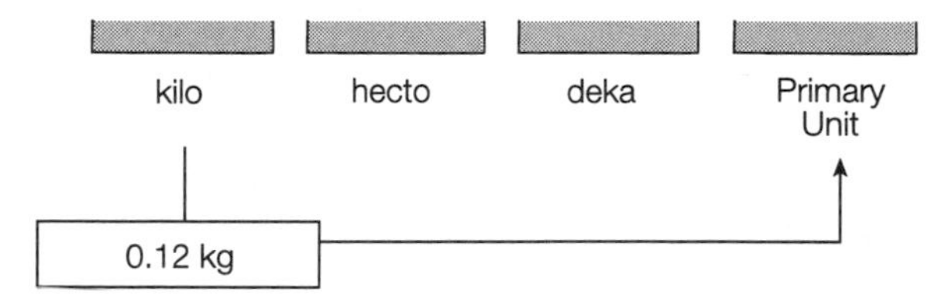

Move the decimal point three places to the right:

$$0.12 \text{ kg} = 120 \text{ g}$$

Then

$$(120 \text{ g})(0.03527) = 4.23 \text{ oz}$$

Temperature Conversions

The formulas used for Fahrenheit and Celsius temperature conversions are

$$°C = \frac{5}{9}(°F - 32°)$$

$$°F = \frac{9}{5}(°C) + 32°$$

These formulas are difficult to remember unless used frequently. There is, however, another method of conversion that is perhaps easier to remember because the following three steps are used for both Fahrenheit and Celsius conversions:

1. Add 40°.
2. Multiply by the appropriate fraction ($\frac{5}{9}$ or $\frac{9}{5}$).
3. Subtract 40°.

The only variable in this method is the choice of $\frac{5}{9}$ or $\frac{9}{5}$ in the multiplication step. To make this choice, you must know something about the two scales. As shown in Figure M11-1, on the Fahrenheit scale the freezing point of water is 32°, whereas it is 0° on the Celsius scale. The boiling point of water is 212° on the Fahrenheit scale and 100° on the Celsius scale.

Thus for the same temperature, higher numbers are associated with the Fahrenheit scale and lower numbers with the Celsius scale. This information helps you decide whether to multiply by $\frac{5}{9}$ or $\frac{9}{5}$. Let's look at a few conversion problems to see how the three-step process works.

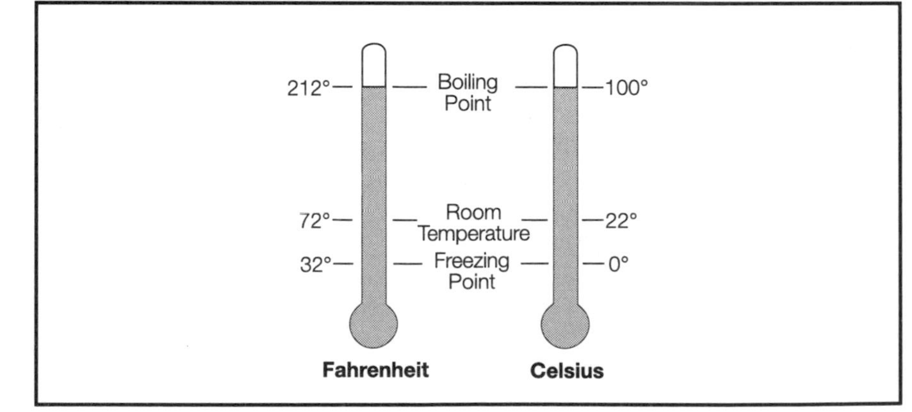

FIGURE M11-1 Fahrenheit and Celsius temperature scales

Example 35

Suppose that you wish to convert 212°F to Celsius. From the sketch of the two scales, you know that the answer should be 100°C. But let's verify it using the three-step process.

The first step is to add 40°:

$$\begin{array}{r} 212^\circ \\ +\ 40^\circ \\ \hline 252^\circ \end{array}$$

Next, 252° must be multiplied by either 5⁄9 or 9⁄5. Since the conversion is to the *Celsius* scale, you will be moving to a number *smaller* than 252. Multiplying by 9⁄5 is roughly the same as multiplying by 2, which would double 252 rather than make it smaller. On the other hand, multiplying by 5⁄9 is about the same as multiplying by ½, which would cut 252 in half.

In this problem, since you wish to move to a smaller number, you should multiply by 5⁄9:

$$\left(\frac{5}{9}\right)(252^\circ) = \frac{1{,}260^\circ}{9}$$

$$= 140^\circ$$

The problem can now be completed using step 3 (subtract 40°):

$$\begin{array}{r} 140^\circ \\ -\ 40^\circ \\ \hline 100^\circ \end{array}$$

Therefore, 212°F = 100°C.

Example 36

Convert 0°C to Fahrenheit.

The sketch of the two scales indicates that 0°C = 32°F, but this can be verified using the three-step method of conversion:
First, add 40°.

$$\begin{array}{r} 0^\circ \\ +\ 40^\circ \\ \hline 40^\circ \end{array}$$

In this problem, you are going from Celsius to Fahrenheit. Therefore you will be moving from a smaller number to a larger number, and 9⁄5 should be used in the multiplication:

$$\left(\frac{9}{5}\right)(40°) = \frac{360°}{5}$$

$$= 72°$$

Subtract 40°.

$$\begin{array}{r} 72° \\ -\ 40° \\ \hline 32° \end{array}$$

Thus, 0°C = 32°F.

Example 37

A thermometer indicates that the water temperature is 15°C. What is this temperature expressed in degrees Fahrenheit?

First, add 40°.

$$\begin{array}{r} 15° \\ +\ 40° \\ \hline 55° \end{array}$$

Moving from a smaller number (Celsius) to a larger number (Fahrenheit) indicates multiplication by 9⁄5:

$$\left(\frac{9}{5}\right)(55°) = \frac{495°}{5}$$

$$= 99°$$

Subtract 40°.

$$\begin{array}{r} 99° \\ -\ 40° \\ \hline 59° \end{array}$$

Therefore, 15°C = 59°F.

Although it is useful to know how to make these temperature conversion calculations, in practical applications you may wish to use a temperature conversion table such as the one in appendix A. Let's look at a couple of example conversions using the table.

Example 38

Normal room temperature is considered to be 68°F. What is this expressed in degrees Celsius?

Use the Fahrenheit-to-Celsius temperature conversion table. Coming down the Fahrenheit column to 68°, you can see that 68°F = 20°C.

Example 39

Convert 90°C to degrees Fahrenheit.

Use the Celsius-to-Fahrenheit temperature conversion table. Coming down the Celsius column to 90°, you can see that 90°C = 194°F.

MATHEMATICS 12

Graphs and Tables

Often the operator of a water treatment plant is responsible for collecting and analyzing data about many aspects of treatment. These data and their analyses are important in making operation modifications and in summarizing operating trends.

Graphs and tables are a means of sorting and organizing data so that trends in information can be identified and interpreted. Graphs and tables may also be regarded as tools for decision making. This section explains how to read the graphs and tables that you may encounter in treatment plant operation, and also how to construct some of the simpler graphs.

Graphs

There are four types of graphs — bar, circle, line, and nomographs — frequently encountered in water treatment data presentation and analysis. The first three (bar, circle, and line graphs) are commonly used to summarize water treatment data, whereas nomographs are usually used to solve mathematical problems pertaining to chemical feed rates.

Before considering the distinctive features of each type of graph, you should have a general understanding of the scales used for graphs, and of how to read or interpret information using the scales.

Scales are series of intervals, usually marked along the side and bottom of a graph, that represent the range of values of the data. On a typical graph, the bottom (horizontal) scale indicates one kind of information, and the side (vertical) scale indicates another kind. For example, the horizontal scale may show the months or years during which a certain water treatment was performed, and the vertical scale of the same graph may show the number of gallons of water treated.

A scale in which the intervals (marks or lines) are equally spaced is called an *arithmetic scale*. An example of a grid in which both the horizontal and vertical scales are arithmetic (equally spaced) is shown in Figure M12-1.

A *logarithmic scale*, or *log scale*, is one in which the intervals are varied logarithmically and therefore not equally spaced. Discussions of logarithms and of how to construct a log scale are beyond the scope of this text. However, Figures M12-2 and M12-3 illustrate log scales. Figure M12-2 is an example of a grid having an arithmetic horizontal scale and a logarithmic

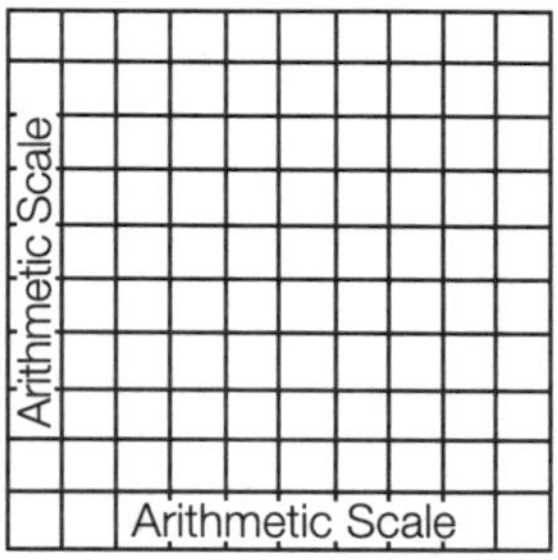

FIGURE M12-1
Grid with two arithmetic scales

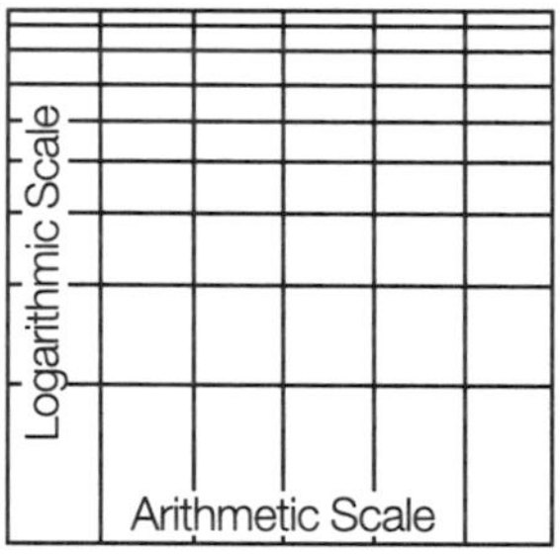

FIGURE M12-2
Grid with logarithmic and arithmetic scales

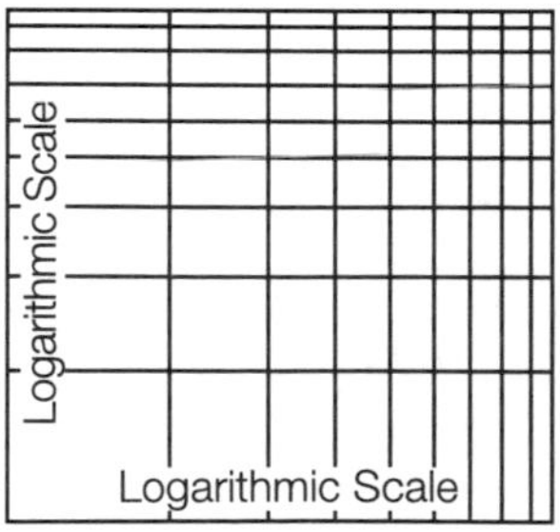

FIGURE M12-3
Grid with two logarithmic scales

vertical scale; Figure M12-3 is an example of a grid in which both the horizontal and vertical scales are logarithmic.

The technique used to determine values that fall between the marked intervals on a scale is called *interpolation*. A good example of everyday interpolation is the way that you use an ordinary ruler. Although only the inches are numbered on a ruler, the spaces between the marks are divided into equal intervals (½ in., ¼ in., ⅛ in., etc.; or, on some rulers, 0.1 in., 0.2 in., 0.3 in., etc.) so that you have no difficulty measuring dimensions between the inches.

Interpolation on arithmetic scales of a graph is similar to using a ruler. For example, where would you read a value of 3 on an arithmetic scale like this?

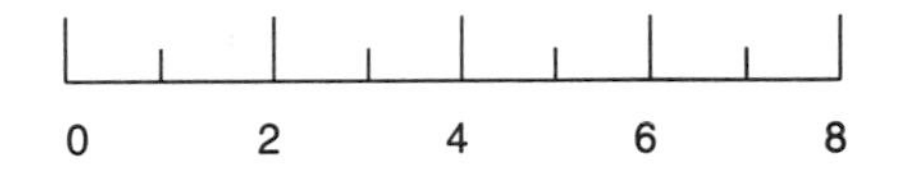

Every other interval is numbered, and the unnumbered intervals fall halfway between the numbers, so 3 is read where the arrow indicates:

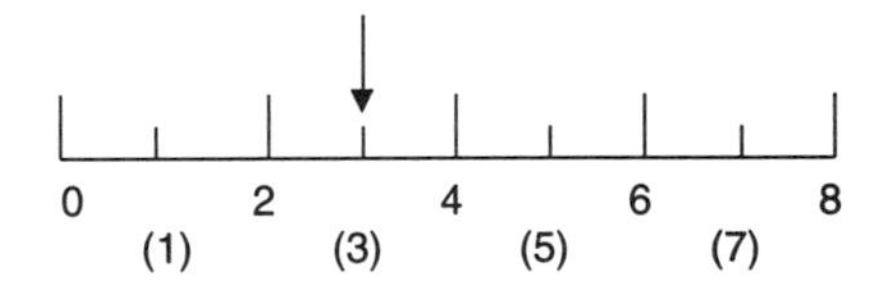

If every fourth interval is numbered, then the space between numbers is divided into four equal parts:

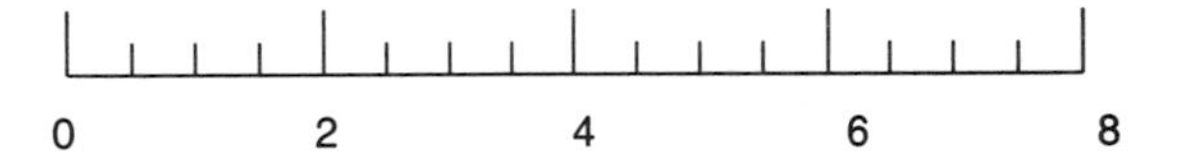

And the values assigned to the in-between divisions on the scale are

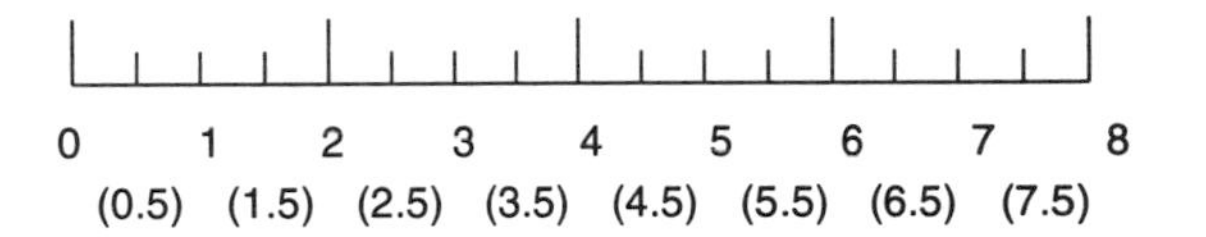

Suppose every fourth interval is marked like this:

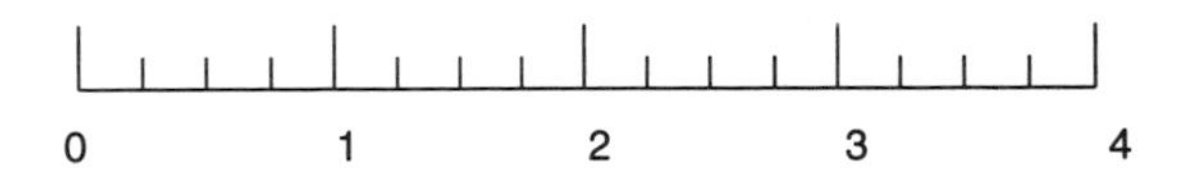

Then the in-between divisions are

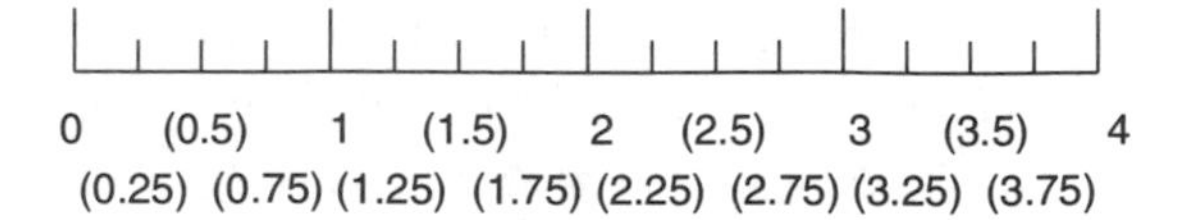

As the examples above illustrate, when you interpolate on an arithmetic scale, you must divide the space between two known points into equal intervals. On a logarithmic scale, however, the space is not divided evenly — the divisions are larger in the first part of the space, becoming smaller in the last part of the space. Let's compare the two types of scales, using 2 and 3 as the known points on both.

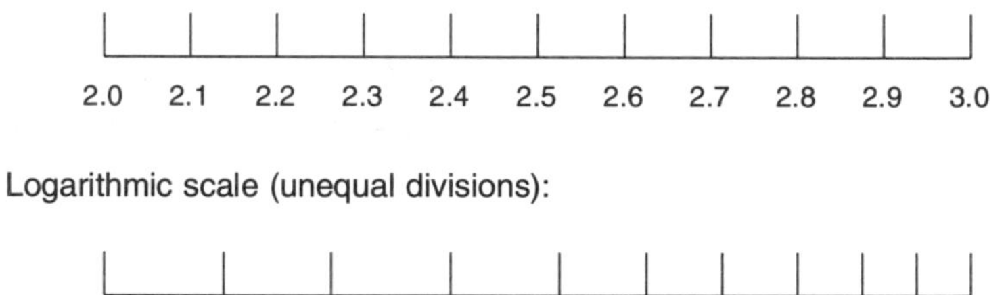

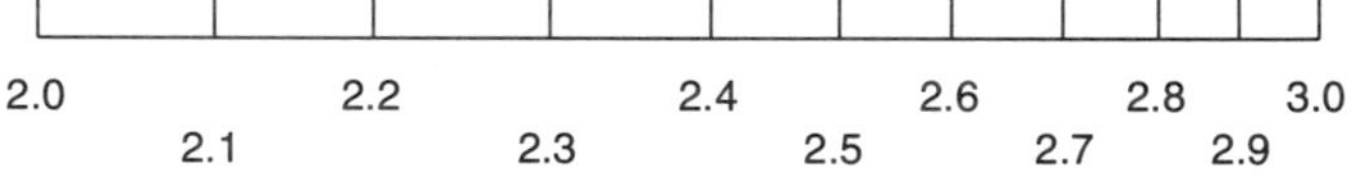

Explaining the construction of a log scale is beyond the scope of this section, so when you need to interpolate on such a scale, you will have to estimate the logarithmic divisions between two known numbers.

Let's look now at how to number the 10 subdivisions between two known numbers on a log scale. Suppose the two known points are 7 and 8, as shown below. How would each of the subdivisions be numbered?

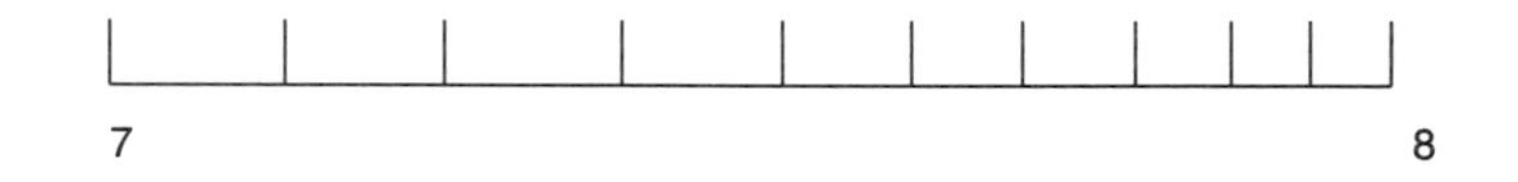

Since the two known numbers represent *units* in the decimal system, the subdivisions are numbered in *tenths*. The starting point is 7.0. The next lines are therefore numbered 7.1, 7.2, 7.3, and so on, up to 7.9 and 8.0:

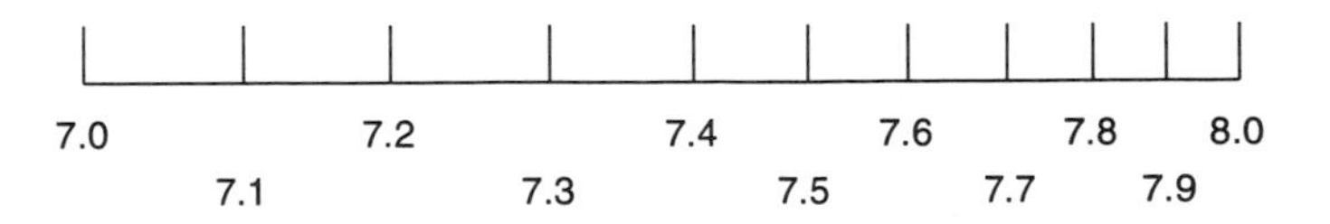

What would the numbering be if the two known numbers were 70 and 80?

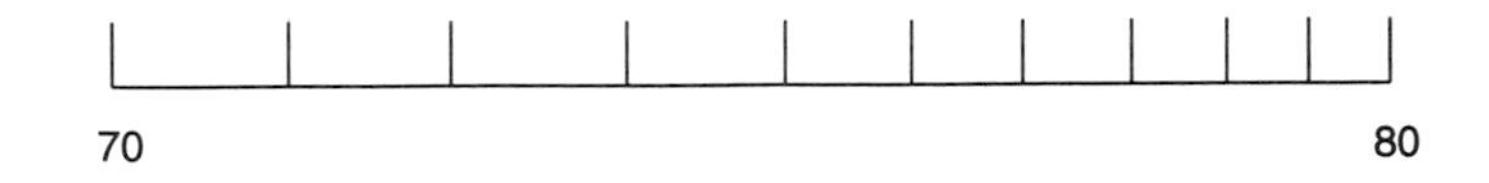

Since the two known numbers represent *tens* in the decimal system, the subdivisions are numbered in *units*. The starting point is 70 and the numbering continues, with 71, 72, 73, up to 79 and 80:

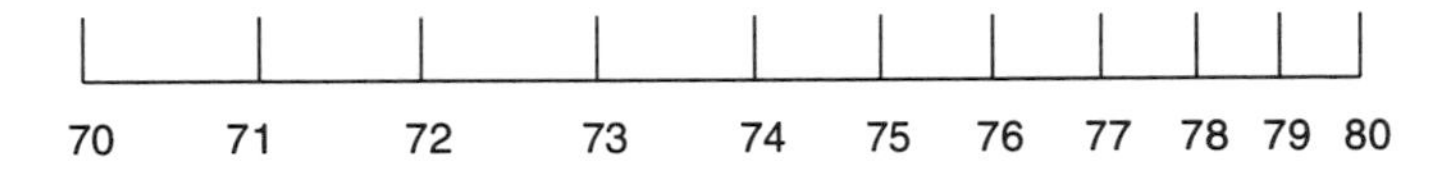

Let's look at a last example using whole numbers. Suppose the two known points this time are 700 and 800. How would the subdivisions be numbered then?

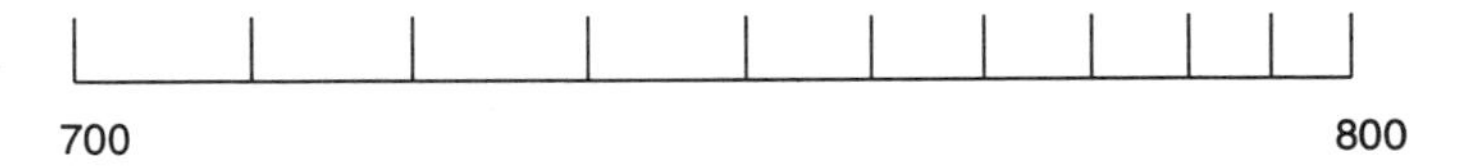

In this example the two known numbers represent *hundreds* in the decimal system. Therefore, the subdivisions are numbered in *tens*. The starting point is 700. The next lines are numbered 710, 720, 730, and so on, up to 790 and 800:

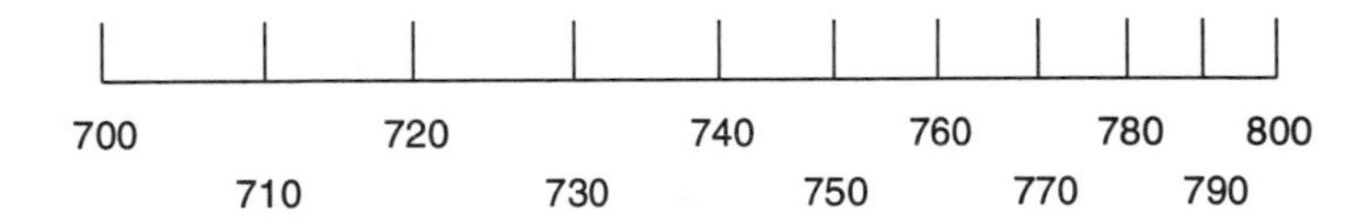

What happens in numbering the subdivisions if the two known numbers are decimal numbers less than 1? For example, suppose the two known numbers are 0.07 and 0.08.

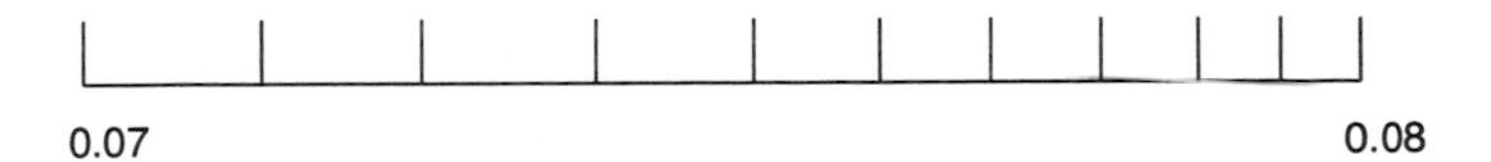

Since the two known numbers represent *hundredths* in the decimal system, the subdivisions are numbered in *thousandths*. Therefore, the starting point is 0.070 and the succeeding lines are numbered 0.071, 0.072, 0.073, and so on, up to 0.079 and 0.080:

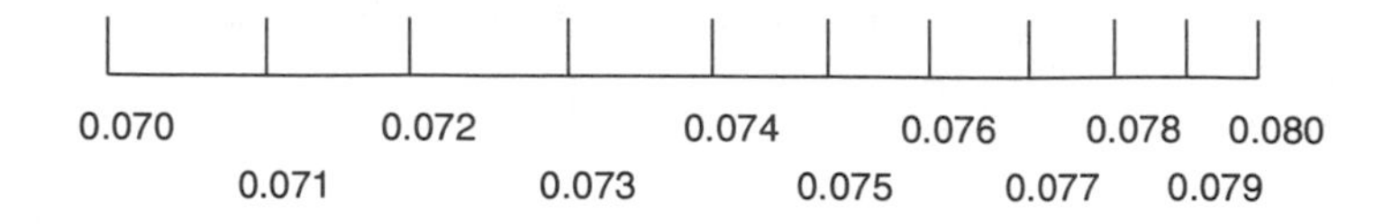

Now let's consider the four types of graphs you are most likely to use in analyzing water treatment data. (Several examples in the following discussion of graphs require interpolation to determine the answer.)

Bar Graphs

In a bar graph the data being compared are represented by bars, which may be drawn vertically or horizontally, as shown in Figures M12-4 and M12-5. The information that the graph represents is usually given along the bottom and left side of the graph. In addition, information is sometimes given at the end of each bar to make reading the graph a little easier. The title of the graph indicates the general type of information being presented.

Two special types of bar graphs are the *historical* and the *divided*. On the historical bar graph, which represents data at various *times*, the bars are drawn vertically. The time factor is given on the horizontal scale, and other information is represented on the vertical scale. In Figure M12-6, for example, time is represented on the horizontal scale and billions of gallons on the vertical scale.

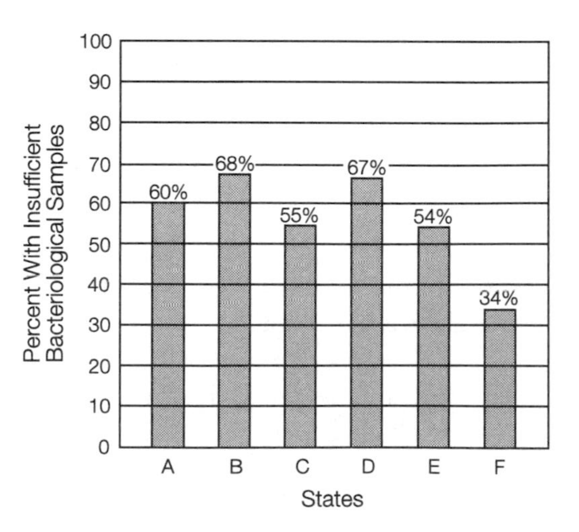

FIGURE M12-4 Vertical bar graph depicting community water supplies with insufficient bacteriological monitoring

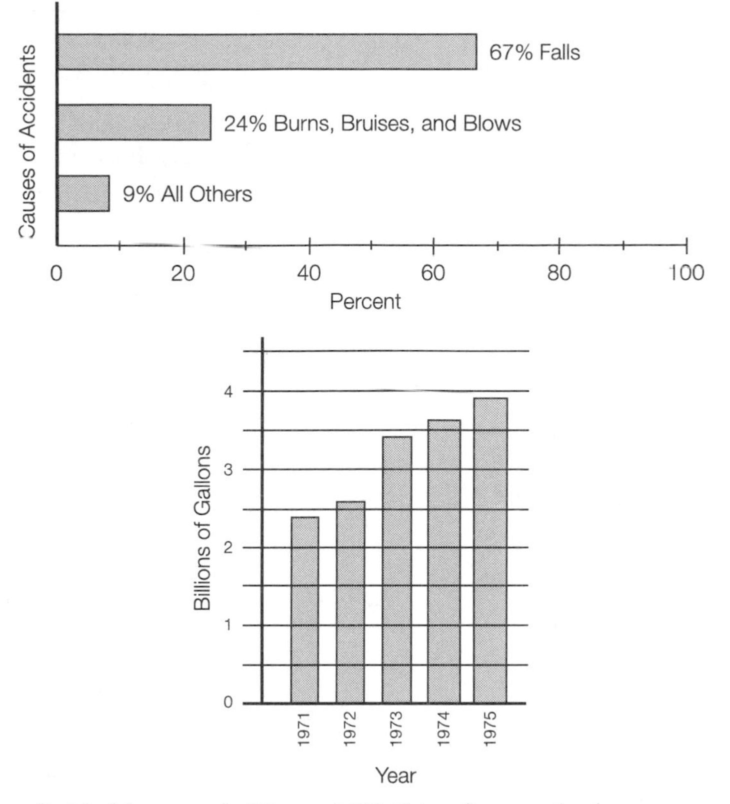

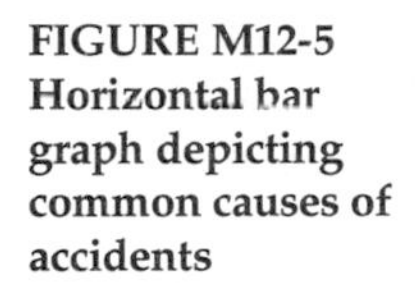

FIGURE M12-5 Horizontal bar graph depicting common causes of accidents

FIGURE M12-6 Historical bar graph depicting total annual water production

The divided bar graph (Figure M12-7) is often used when summarizing data pertaining to percent. The sum of the percents must equal 100.

Occasionally you may see a bar graph that has a broken scale, such as that shown in Figure M12-8. The break, usually on the vertical scale and denoted by a zigzag line, is made to avoid unnecessarily long bars.

Example 1

Given the bar graph in Figure M12-9 (p. 133), answer the following questions.

(a) What was the chemical cost per million gallons of treated water for the month of March?

(b) If the total water treated during July was 137 mil gal, what was the total chemical cost for that month?

(c) How much more were the chemical costs (per million gallons of treated water) for December than for February?

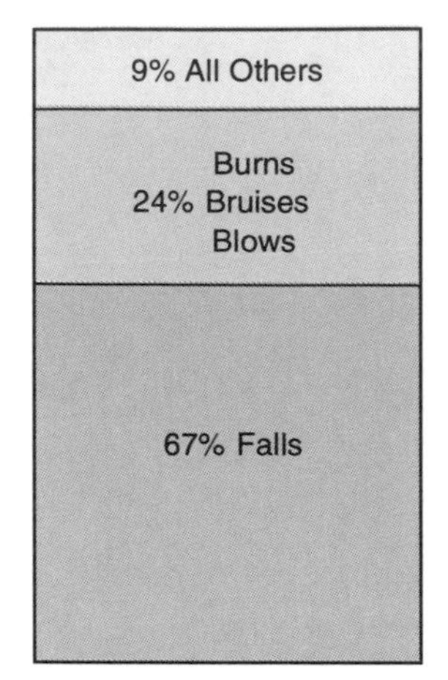

FIGURE M12-7
Divided bar graph depicting common causes of accidents

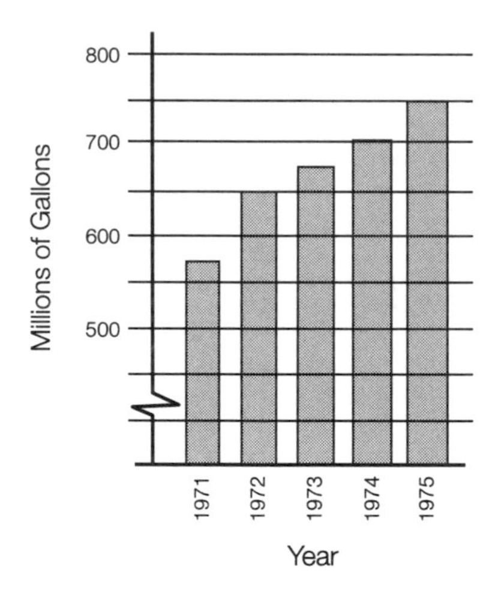

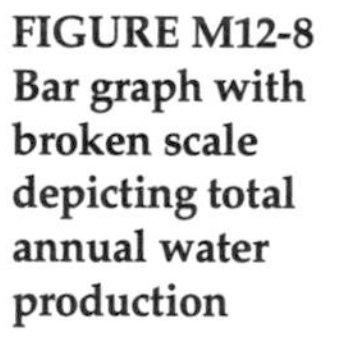

FIGURE M12-8
Bar graph with broken scale depicting total annual water production

(a) The bar height corresponding to March indicates that the chemical costs for the month were approximately \$4.50 per million gallons treated water.

(b) To calculate the total chemical costs for the month, first determine the cost per million gallons treated water. The bar height corresponding to July indicates the cost was about \$7.90 for each million gallons of water treated. Since a total of 137 mil gal of water was treated during the month, the total chemical cost was

$$(\$7.90/\text{mil gal})\ (137\ \text{mil gal}) = \$1{,}082.30$$

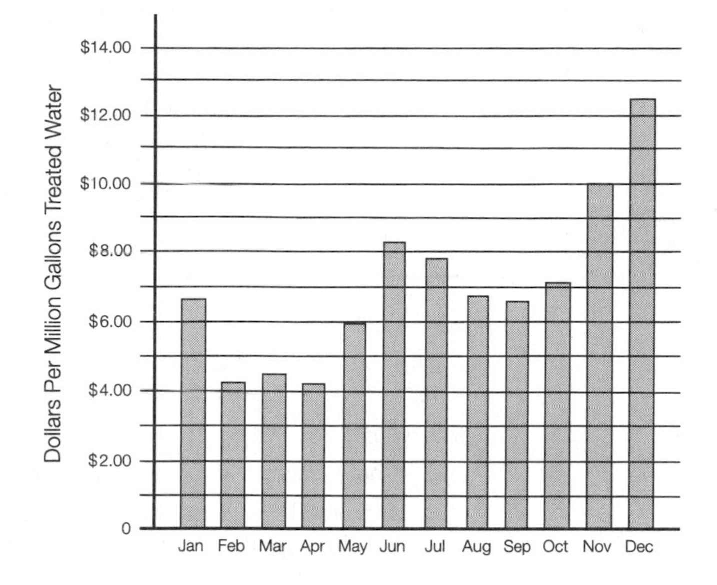

FIGURE M12-9
Chemical costs for 1977

(c) The chemical costs for the month of December were approximately $12.50 per million gallons treated water, whereas the costs for February were only about $4.30 per million gallons treated water. The difference between these costs is

$12.50	December cost
− 4.30	February cost
$ 8.20	difference in cost

Therefore, the chemical costs for December were about $8.20 per million gallons *more* than the chemical costs for February.

Circle Graphs

The circle graph, sometimes referred to as a circle chart or pie chart, is similar to the divided bar graph mentioned above in that the sum of the different parts must equal 100. The circle graph is often used when summarizing data pertaining to percentages because it is an effective way of showing the relationship of the whole to its parts, as in Figure M12-10.

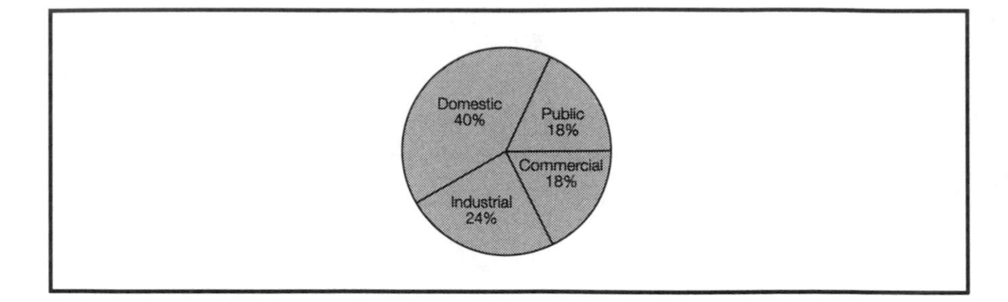

FIGURE M12-10 Water use by category for 1977

Example 2

Using the circle graph in Figure M12-10 as representing the water use by category for a particular water system, answer the following questions:

(a) What percent of the total water use was for domestic purposes?

(b) If the total water use for 1977 was 980 mgd, how much water (in million gallons) was used for domestic purposes?

(a) According to the circle graph in Figure M12-10, 40 percent of the total water use was for domestic purposes.

(b) Since both the *percent domestic use* and the *total water used* are known, the actual amount of water used for domestic purposes can be calculated:[1]

$$\text{percent domestic water use} = \frac{\text{domestic water use}}{\text{total water use}} \times 100$$

Filling in the information,

$$40 = \frac{x \text{ mil gal domestic water use}}{980 \text{ mil gal total water use}} \times 100$$

This may also be expressed as

$$40 = \frac{(x)(100)}{980}$$

Then solve for the unknown value:[2]

$$(980)(40) = (x)(100)$$

[1]Mathematics 7, Percent.

[2]Mathematics 4, Solving for the Unknown Value.

$$\frac{(980)\,(40)}{100} = x$$

$$392 \text{ mil gal domestic use} = x$$

Line Graphs

Line graphs are perhaps the most common type you will encounter in water treatment data presentation. As shown in Figures M12-11 and M12-12, line graphs fall into two general categories: broken-line and smooth-line. Of the two, smooth-line graphs are more often found in water treatment literature.

Broken-line graphs are merely dots (data points) connected by straight line segments. Connecting the data points helps highlight the trends in the data.

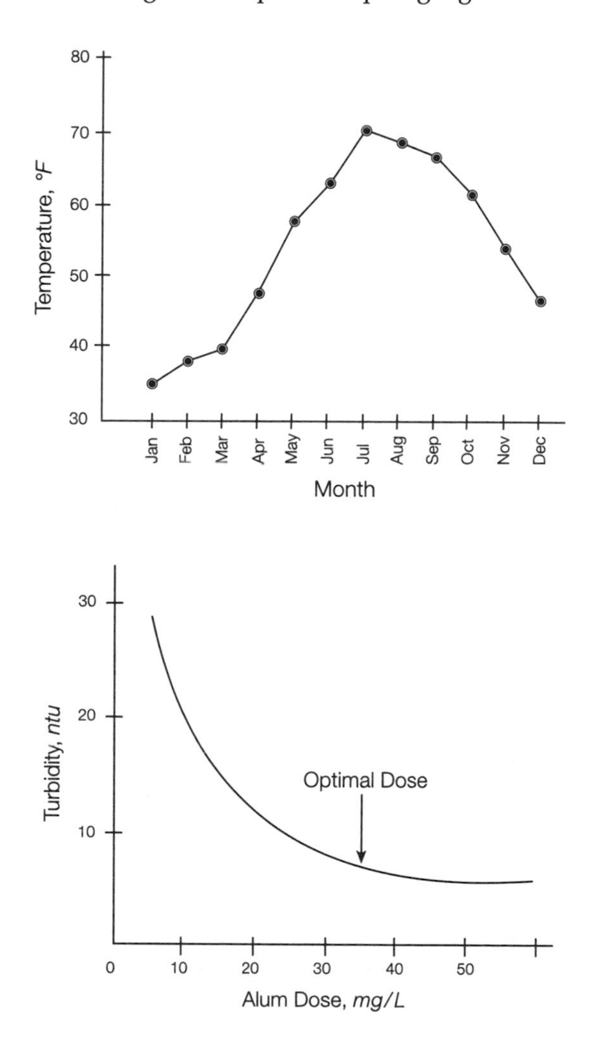

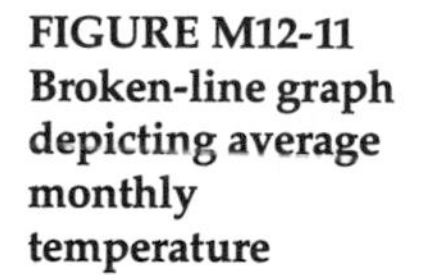

FIGURE M12-11 Broken-line graph depicting average monthly temperature

FIGURE M12-12 Smooth-line graph depicting jar test results

When the in-between points are analyzed and found to be predictable, a smooth-line graph can be constructed, using advanced mathematical calculations to determine the proper shape of the curve.

The scales on a line graph are usually numbered along the bottom and left side of the graph, with additional numbering occasionally found along the right side and top. The scales on a line graph are most often arithmetic (equally spaced), although for some types of data logarithmic (graduated interval) scales are used.

In Figure M12-13, a line graph is shown on a circular scale — somewhat like arithmetic graph paper stretched around into a circle. Reading this type of scale is not much different from reading the more conventional arithmetic scales.

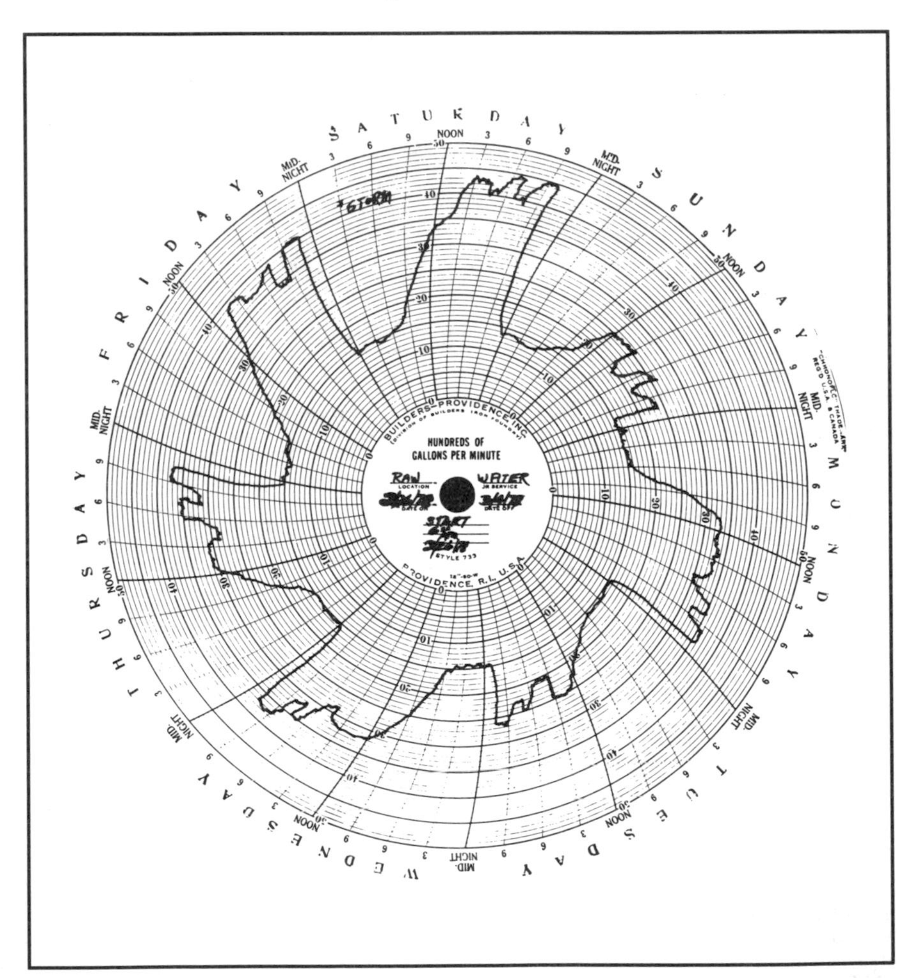

FIGURE M12-13
Recorder chart with circular scale

Courtesy of Leeds & Northrup (BIF)

The following sample problems include some line graphs with arithmetic scales and some with logarithmic scales.

Example 3

Given the graph in Figure M12-14, answer the following questions.

(a) What is represented by the vertical scale?

(b) What is represented by the horizontal scale?

(c) About how many waterborne outbreaks were reported in the 1951–1955 period? In the 1971–1974 period?

(a) As shown on the graph, the vertical scale represents the average annual number of waterborne disease outbreaks.

(b) The horizontal scale indicates the years in which those outbreaks occurred.

(c) As shown at point A, there were 10 outbreaks reported during the 1951–1955 period. At point B it is necessary to interpolate to determine that there were about 23 outbreaks reported during the 1971–1974 period.

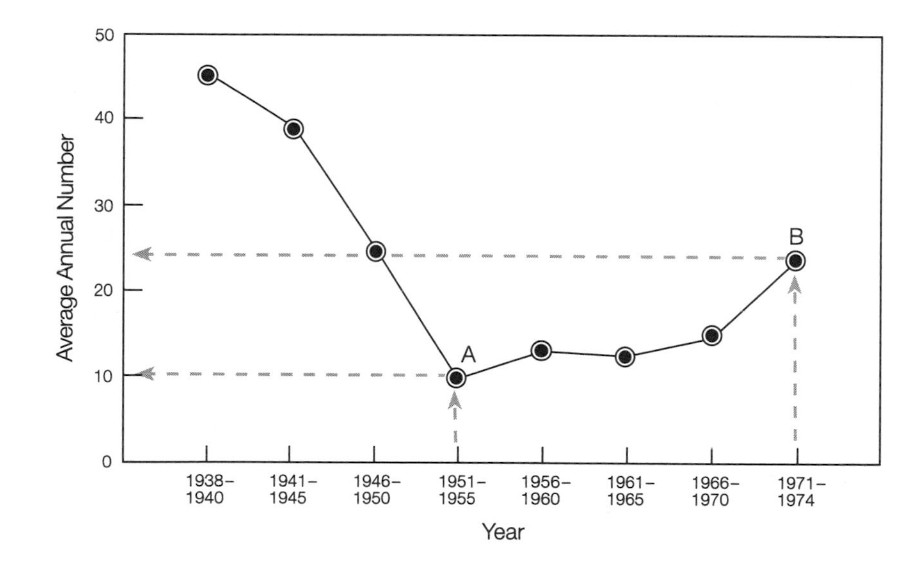

FIGURE M12-14 Average annual number of waterborne disease outbreaks, 1938–1974

Example 4

Given the graph in Figure M12-15, answer the following questions.

(a) If the staff gauge on a 12-in. Parshall flume indicates a head of 18 in., what is the flow in million gallons per day?

(b) If the staff gauge on a 12-in. Parshall flume indicates a head of 7½ in., what is the flow in million gallons per day?

(c) If there is a head of 15 in. on a 12-in. Parshall flume, what is the flow in million gallons per day?

Note that, since all of the answers fall between numbers marked on the scale, interpolation is necessary to answer all of the questions.

(a) As shown by point A on the graph, the flow corresponding to a head of 18 in. would be halfway between 4.5 mgd and 5.0 mgd. By interpolation, the indicated flow is 4.75 mgd.

(b) Interpolation is again necessary to locate point B (7½-in. head). At this head, the flow would be 1.25 mgd.

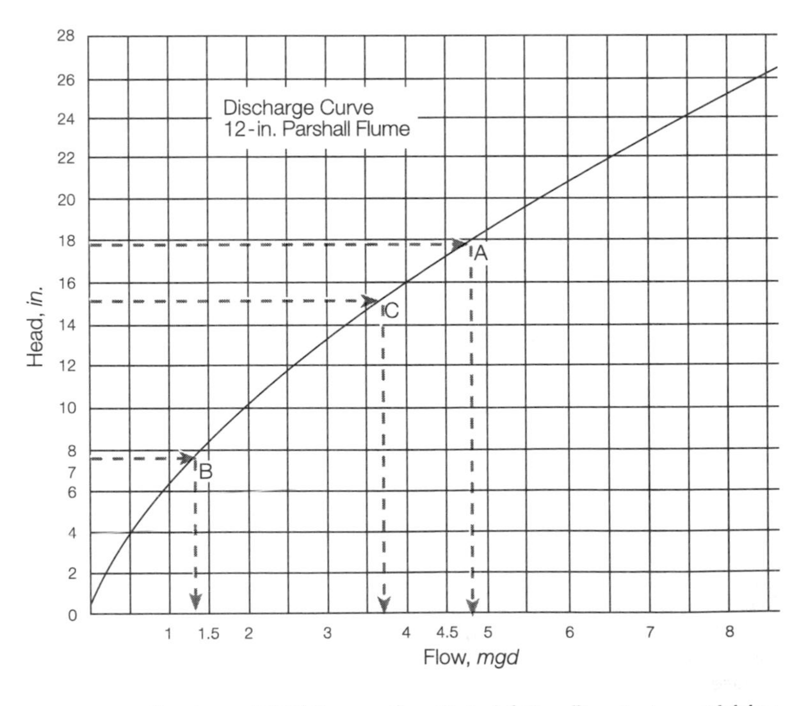

FIGURE M12-15 Discharge curve for 12-in. Parshall flume

Courtesy of FMC Corporation, Material Handling Systems Division.

(c) At a head of 15 in. (point C), the flow is somewhere between 3.5 mgd and 3.75 mgd. Let's take a closer look at the scale between 3 and 4 mgd to determine the indicated flow.

Halfway between 3 and 4 mgd is 3.5 mgd:

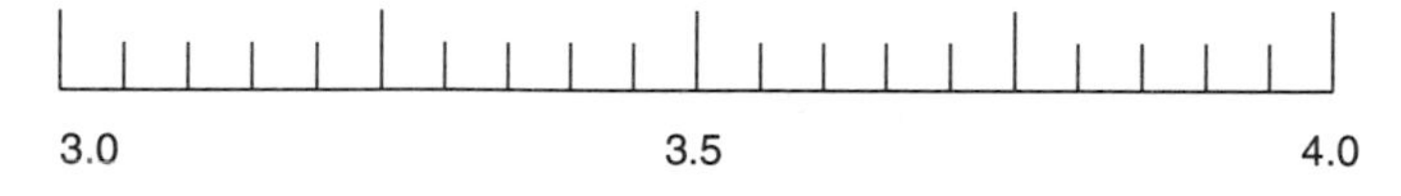

And 3.25 mgd is halfway between 3.0 mgd and 3.5 mgd. Similarly, 3.75 mgd is halfway between 3.5 mgd and 4.0 mgd:

The question in this problem is *what flow rate is indicated by the arrow?*

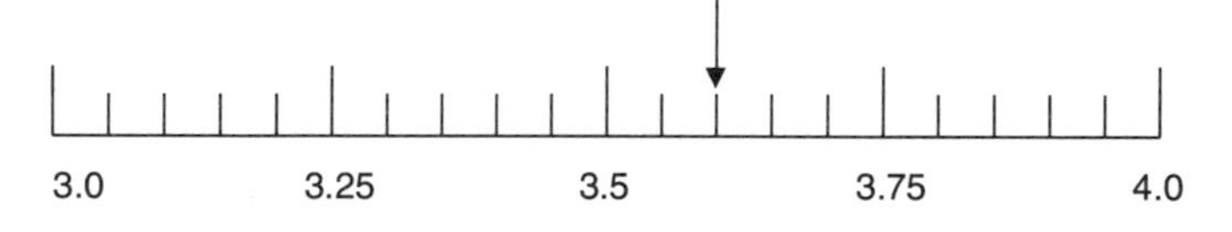

Notice that from 3.5 to 3.75, the space has been divided into five parts, or fifths, and that the arrow lies ⅖ of the distance from 3.5 to 3.75. The distance from 3.5 to 3.75 is

$$\begin{array}{r} 3.75 \\ -3.50 \\ \hline 0.25 \end{array}$$

And now ⅖ of the distance can be calculated:

$$\left(\frac{2}{5}\right)(0.25) = \frac{0.50}{5}$$

$$= 0.10$$

Note that ⅖ could also have been converted to a decimal and then multiplied to get the same answer:

$$(0.4)\,(0.25) = 0.10$$

The arrow lies 0.10 further on the scale than 3.5:

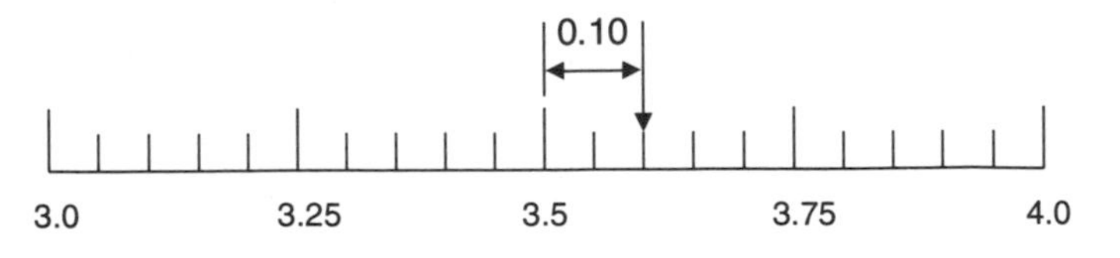

Therefore, the arrow must be at 3.5 + 0.10, or 3.60 mgd. A head of 15 in. indicates a flow of 3.60 mgd.

Example 5

Given the graph in Figure M12-16, answer the following questions. (Notice that both the horizontal and vertical scales in this graph are logarithmic.)

(a) If the staff gauge shows a head of 5½ in. on a 90° V-notch weir, what is the flow in cubic feet per second?

(b) With a head of 0.6 ft on a 90° V-notch weir, what is the flow in cubic feet per second?

(c) Using the answer to part b, what is this flow expressed in gallons per day?

(d) If the staff gauge shows a head of 1½ in. on a 90° V-notch weir, what is the flow in cubic feet per second?

(a) As shown by point A, a head of 5½ in. on a 90° V-notch weir indicates a flow of 0.4 ft^3/s.

(b) Point B shows that a head of 0.6 ft on a 90° V-notch weir indicates a flow of 0.7 ft^3/s.

(c) Convert 0.7 ft^3/s to gallons per day:[3]

$$(0.7\ ft^3/s)\ (60\ s/min)\ (1{,}440\ min/d)\ (7.48\ gal/ft^3) = 452{,}390\ gpd$$

(d) A head of 1½ in. (point C) indicates the flow is somewhere between 0.01 and 0.02 ft^3/s. As in other example problems above, interpolation is needed to determine the indicated flow. However, interpolation in this problem is a little different.

---

[3]Mathematics 11, Conversions.

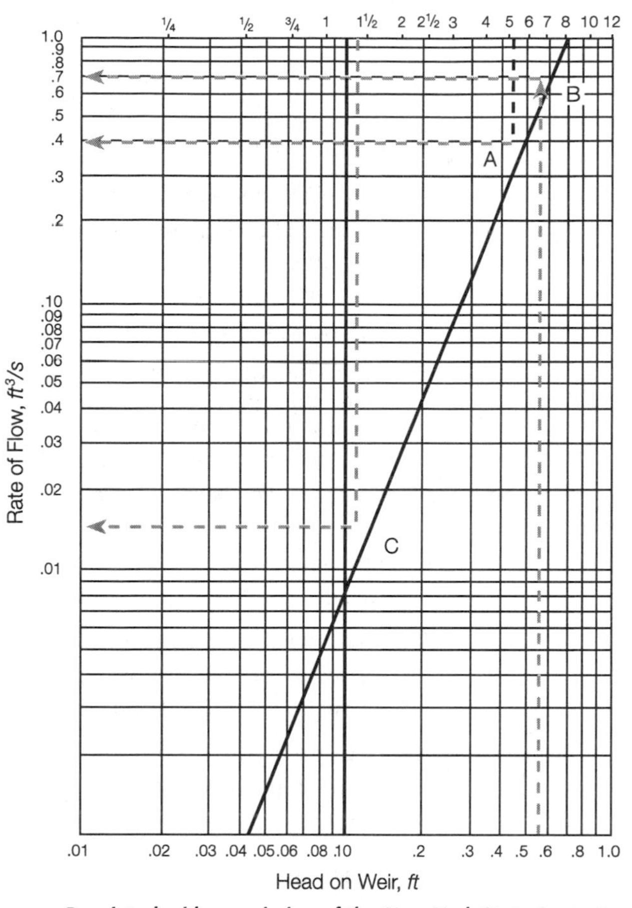

FIGURE M12-16 Flow rate through 90° V-notch weir

Reprinted with permission of the New York State Department of Health

In Example 4(c) the interpolation was on an *arithmetic* scale (the space between two known points was equally divided). This problem uses a logarithmic scale, so the space between two known points must be divided logarithmically. (There is a general discussion of *logarithmic* scales at the beginning of this section.)

Consider the part of the scale needed for this problem; that is, the section between 0.01 and 0.02 ft^3/s:

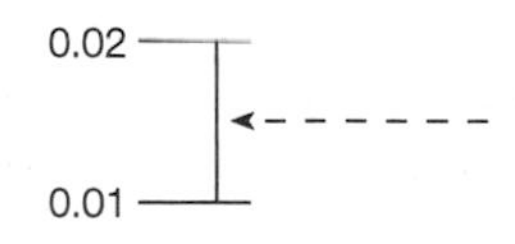

Estimate and mark off the logarithmic scale between the two known numbers:

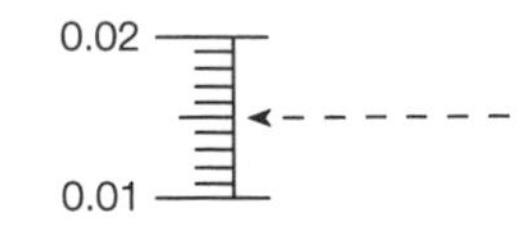

On the estimated scale the arrow falls on the halfway mark. The estimated scale begins at the point 0.01 (equal to 0.010) and is divided into 10 parts, so the first subdivision line is 0.011; the second is 0.012; and the halfway line is 0.015. Therefore, a head of 1½ in. on a 90° V-notch weir indicates a flow rate of approximately 0.015 ft^3/s.

Sometimes a graph will have more than one line plotted on it. The lines may be interrelated, or they may be unrelated lines that have been grouped on one graph for convenience. The next two problems involve such multilined graphs.

Example 6

Use the graph in Figure M12-17 to answer the following questions:

(a) You wish to pump a particular solution that has a specific gravity[4] of 0.90. If you are pumping against a head of 70 psi, how many feet of head are you pumping against?

(b) Suppose you are pumping a liquid that has a specific gravity of 1.62. If you are pumping against a total head of 140 ft, how many pounds per square inch are you pumping against?

There are five data lines on this graph. Each is *unrelated* to the next except that each reflects data for a particular specific gravity.

(a) In this part of the problem you are interested only in the line depicting data for a specific gravity of 0.90. Point A on the graph indicates where 70 psi intersects this line. Reading the scale directly to the left of this point indicates that 70 psi is equivalent to a head of 180 ft.

(b) For this part of the problem, only the line relating data for a specific gravity of 1.62 is needed. Following the line across from 140 ft to point B and from there downward indicates that

---

[4]Hydraulics 1, Density and Specific Gravity.

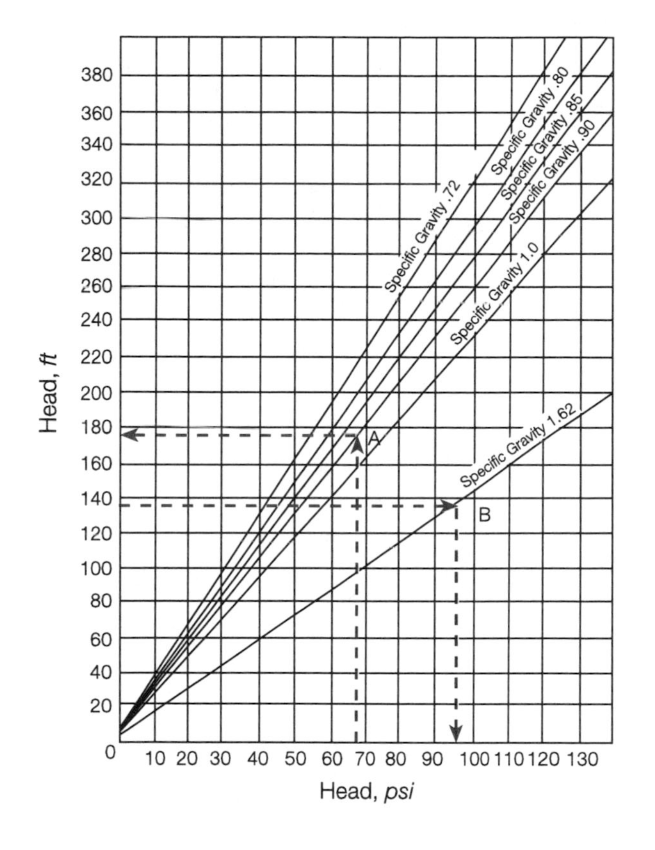

FIGURE M12-17 Conversion chart: head in feet versus head in pounds per square inch

a head of 140 ft is equivalent to a pressure between 90 and 100 psi. Since the arrow does not fall directly on a numbered line, interpolation must be used to determine the answer. The arrow falls somewhere between 95 and 100 psi, at about 98 psi. Therefore, for a liquid with a specific gravity of 1.62, a head of 140 ft is approximately equal to 98 psi.

In pump characteristics curves, such as the one shown in Figure M12-18A, four types of information are given by the scales:

- pump capacity Q
- pump head H
- pump efficiency E (sometimes labeled η)
- brake horsepower P

The H–Q curve (the *head curve*) shows the relationship between head and capacity (flow rate); the P–Q curve (the *power curve*) shows the relationship

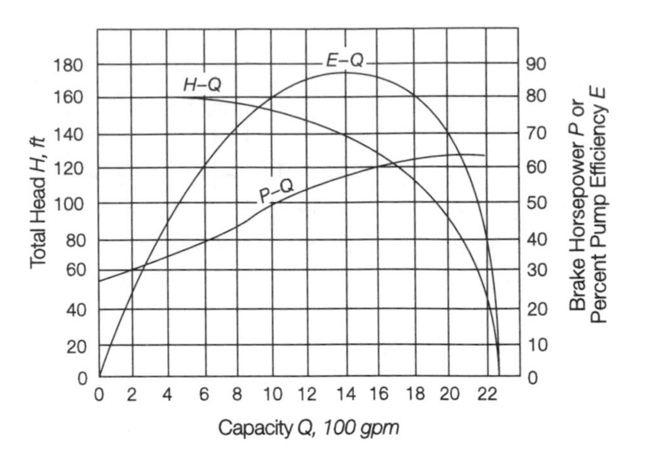

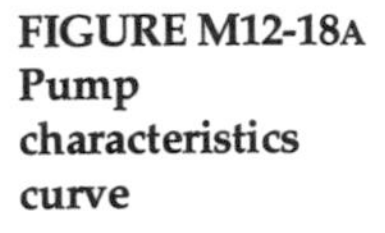
FIGURE M12-18A Pump characteristics curve

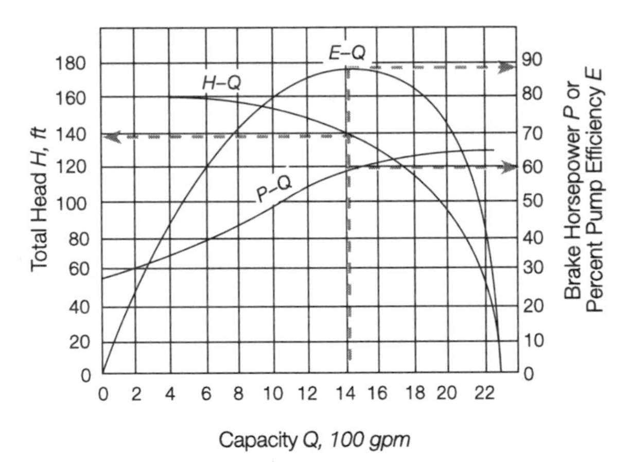

FIGURE M12-18B Pump characteristics curve showing values of *E*, *H*, and *P* when capacity is known

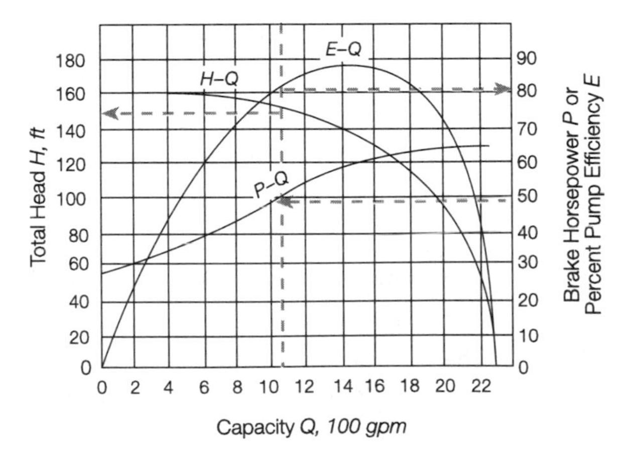

FIGURE M12-18C Pump characteristics curve showing values of *Q*, *H*, and *E* when brake horsepower is known

between brake horsepower and capacity; and the *E–Q* curve (the *efficiency curve*) shows the relationship between pump efficiency and capacity.

By using pump characteristics curves,[5] you can determine *any three* of the characteristics when given information about the fourth. (Either the head or the capacity is normally the known characteristic.) The key to solving this kind of pump problem is to use the known characteristic to determine the location of a specific vertical line and to determine the points where it intersects all three pump curves.

Example 7

Use the graph in Figure M12-18A to answer the following questions:

(a) Assume the capacity of the pump to be 1,400 gpm. Determine the brake horsepower (bhp), total head in feet, and efficiency of the pump at this capacity.

(b) Determine the capacity, total head, and efficiency of the pump at 50 bhp.

(a) In the first part of this example, the known characteristic is *capacity* (1,400 gpm). How can head, efficiency, and power be determined? Of the four types of pump characteristics problems, this is probably the easiest to solve. First, draw a vertical line upward from 14 on the horizontal (capacity) scale (Figure M12-18B). This is equivalent to a capacity of 1,400 gpm.

Note that the vertical line intersects each of the three curves. As you follow the vertical line upward, it first intersects the *P–Q* (power) curve. To determine what the power is at this point, move horizontally to the right toward the brake horsepower scale. You should hit this scale at about 57 bhp.

Now look at the point where the vertical line (1,400 gpm) intersects the *H–Q* (head) curve. To determine the head at this point, move horizontally to the left toward the total head scale. You should hit this scale at about 137 ft.

Finally, look at the point where the vertical line intersects the *E–Q* (efficiency) curve. To determine the efficiency at this point, move horizontally to the right to the efficiency scale. The reading is about 86 percent.

---

[5]Hydraulics 6, Pumping Problems.

Knowing just the pump capacity of 1,400 gpm, you were able to determine the following characteristics:

brake horsepower P = 57 bhp
pump head H = 137 ft
pump efficiency E = 86 percent

(b) In the second part of this example, 50 bhp is the known characteristic. This problem is approached in much the same manner as part a. Locate the value 50 on the brake horsepower scale, then draw a horizontal line until you reach the P–Q (power) curve. This intersection point indicates where the vertical line should be drawn, as shown in Figure M12-18C.

The vertical line enables you to determine the three unknown characteristics. At its base on the capacity scale, the line falls somewhere between 10 and 12, not quite halfway. Therefore, you can estimate a capacity of about 1,090 gpm.

As you move upward, the vertical line first intersects the P–Q curve. This point corresponds with 50 bhp, the information given in the problem.

As you move further upward, the vertical line intersects the H–Q curve. To determine the head value at this point, draw a line horizontally to the left toward the total head scale. The indicated head is about 150 ft.

The last point on the vertical line is on the E–Q curve. Draw a line horizontally to the right, to the efficiency scale. The pump efficiency is about 82 percent.

Therefore, given a brake horsepower of 50, you were able to determine the following characteristics:

pump capacity Q = 1,090 gpm
pump head H = 150 ft
pump efficiency E = 82 percent

Nomographs

A *nomograph* is a graph in which three or more scales are used to solve mathematical problems. The most commonly used nomographs in the water treatment field are those that pertain to chemical feed rates. In such problems, usually the plant flow and desired chemical concentrations are known and the chemical feed rate must be determined. Let's look at a few example problems using nomographs.

Example 8

Assuming that a treatment plant flow is 600 gpm and the desired fluoride concentration is 0.9 ppm, use the nomograph in Figure M12-19 to answer the following questions.

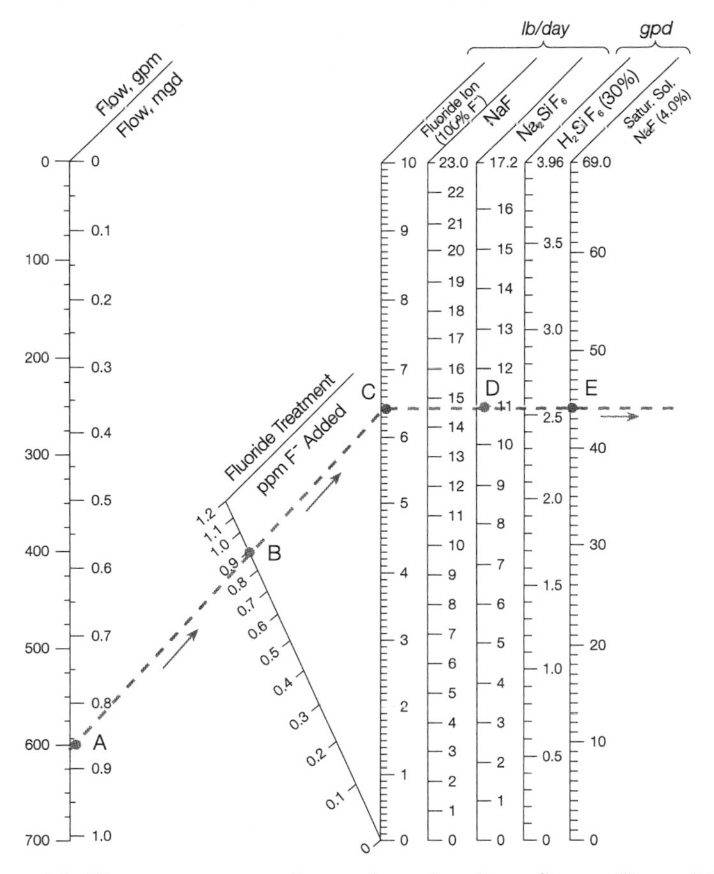

FIGURE M12-19 Fluoridation nomograph

(a) How many pounds per day of sodium fluorosilicate (Na_2SiF_6) should be added to obtain the desired fluoride concentration?

(b) How many gallons per day of a 4 percent saturated solution of sodium fluoride (NaF) should be added to obtain the desired fluoride concentration?

First, locate one end of a straightedge at 600 gpm (point A) on the nomograph. From there, draw a line through 0.9 ppm (point B), extending the line until it intersects with the fluoride ion line (point C). Then, to obtain the pounds-per-day or gallons-per-day feed rate for each type of fluoride listed, draw a horizontal line from point C to the right.

(a) The indicated feed rate for Na_2SiF_6 (point D) is 11 lb/d.

(b) The rate at which a 4 percent saturated solution of NaF must be added to obtain the desired fluoride concentration is 44.3 gpd (point E).

Example 9

A dosage of 2 gpg (34 ppm) of alum liquor is required for treatment of 20-mgd water flow. Using the nomograph in Figure M12-20, determine how many gallons per day of alum liquor must be fed.

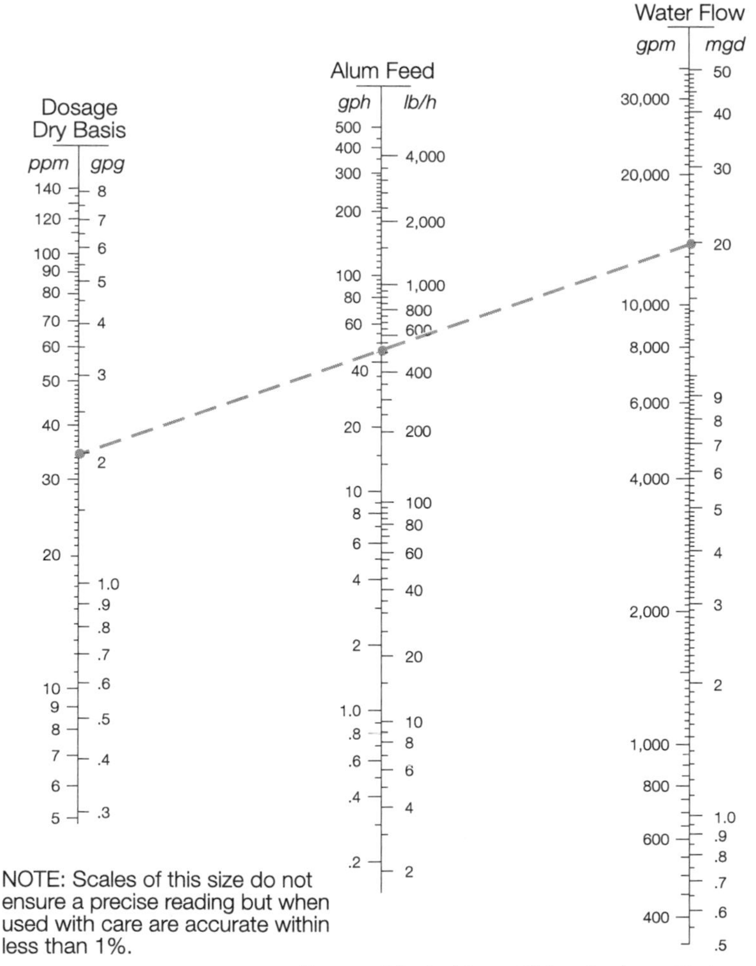

FIGURE M12-20 Alum liquor feed nomograph

Source: Adapted from Rhône-Poulenc, Shelton, Conn.

First, place one end of a straightedge at 2 gpg on the dosage dry basis scale and place the other edge or 20 mgd on the water flow scale. Then read the dosage rate from the alum feed scale. In this case the feed rate is 45 gallons per *hour*. But the question asked for gallons per *day*.

Convert gallons per hour to gallons per day:

$$(44.4 \text{ gph})\,(24 \text{ h/d}) \quad = \quad 1{,}060 \text{ gpd}$$

Constructing Simple Graphs

In summaries of water treatment data, the three types of graphs most commonly used are bar, circle, and broken-line.

Bar Graphs

When constructing a *bar graph,* you should include the following components:

- a title
- clearly labeled horizontal and vertical scales (except when constructing a divided bar graph)
- descriptions of the bars (when needed for clarity)

The bars on the graph should be the same width and should be spaced an equal distance apart. Let's look at an example of constructing a bar graph.

Example 10

Given the following monthly turbidity data, construct a bar graph.

Month	Average Turbidity Value, *ntu*
January	0.70
February	0.61
March	0.64
April	0.78
May	0.91
June	0.71
July	0.59
August	0.68
September	0.72
October	0.83
November	0.78
December	0.71

First the horizontal and vertical scales must be established. Since *time* is part of this graph, a *vertical bar graph* should be constructed with the time factor along the horizontal scale. There is no set rule for the distance between the bars. However, they should be of equal width and spaced a visually pleasing, equal distance apart.

In this example, it is especially important to give some thought to the spacing on the vertical scale. This scale will determine the lengths of the bars. As shown in Figure M12-21, if the scale is too small all the bars will be short, with little distinguishable difference between the bar heights. If the scale is too large, as in Figure M12-22, bars will be longer than necessary.

Once an appropriate vertical scale has been selected, the bar graph can be completed. As you can see from Figure M12-23, although some of the details of the data cannot be determined (for instance, it would be difficult to distinguish between a bar height of 0.77 and 0.78), it is much easier to see at a glance the trend in turbidity values over the 12-month period.

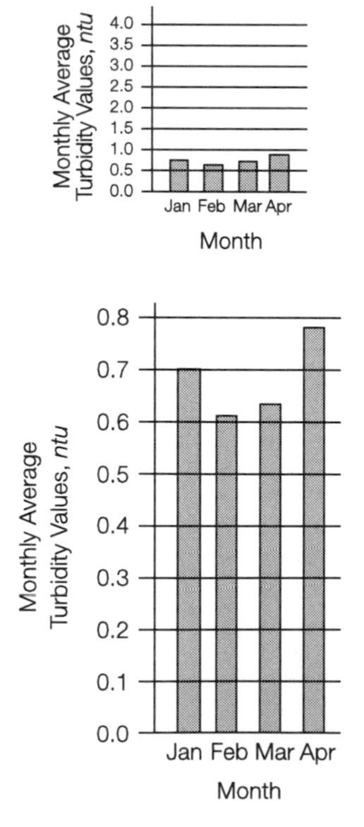

FIGURE M12-21 Bar graph with vertical scale too small

FIGURE M12-22 Bar graph with vertical scale too large

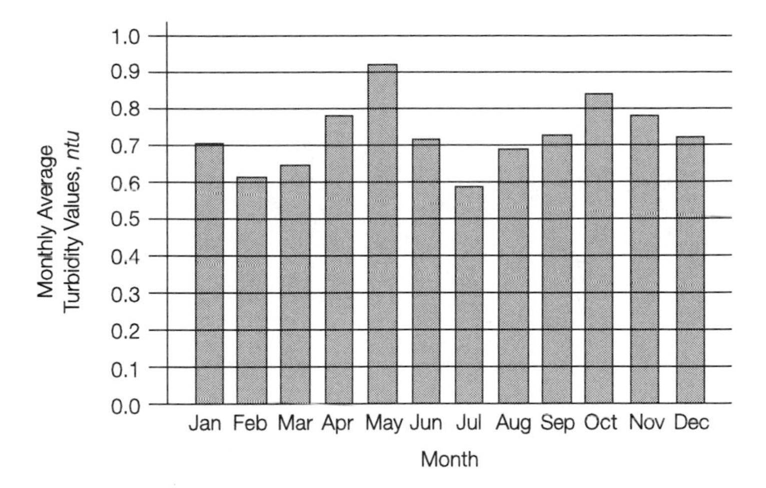

FIGURE M12-23 Completed bar graph for example 10

Circle Graphs

When constructing a *circle graph*, you should include the following components:

- a title
- a description of each part of the circle, including the percentage it represents

The most important thing you will need to know in constructing a circle graph is how to determine the correct size for each part. The following examples illustrate how the size of each part is determined.

Example 11

Suppose you are given information that, of all accidents, 65 percent are due to falls; 25 percent are due to burns, bruises, and blows; and 10 percent are due to all other causes. Construct a circle graph depicting this information.

First you must determine what size to make each part. A full circle is 360°. Each part, therefore, will represent a percentage[6] of the 360°:

(0.65) (360°)	=	234°	(represents falls)
(0.25) (360°)	=	90°	(represents burns, bruises, and blows)
(0.10) (360°)	=	36°	(represents all other causes)
		360°	(total)

[6]Mathematics 7, Percent.

Draw these angles using a protractor.

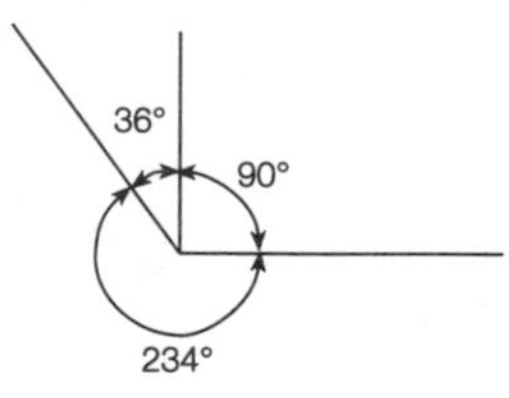

And then draw the circle and label the graph (Figure M12-24).

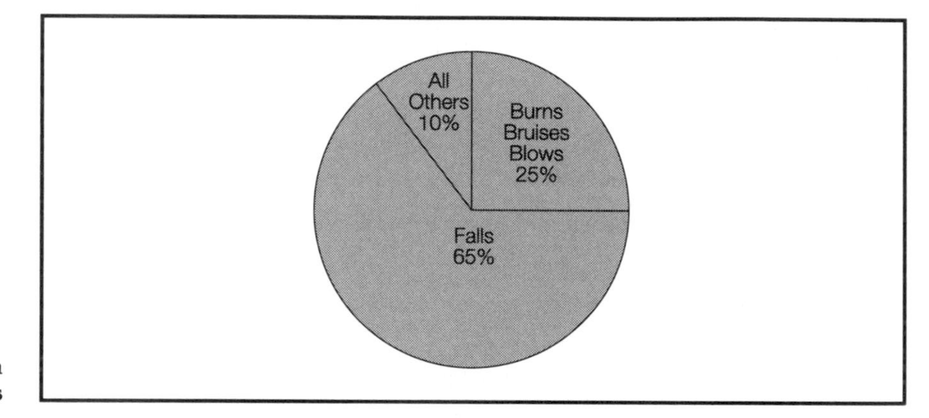

FIGURE M12-24 Circle graph depicting common causes of accidents

Example 12

In 1975 the national average water use was 414 bgd. Suppose that 29 bgd were for public water supplies, 140 bgd for irrigation, 5 bgd for rural uses, and 240 bgd for self-supplied industry. Report this information in terms of a circle graph.

First determine the percent of the whole that each part represents:

$$\frac{\text{29 bgd public water supplies}}{\text{414 bgd total}} \times 100 = 7\% \quad \text{(public water supplies)}$$

$$\frac{\text{140 bgd irrigation}}{\text{414 bgd total}} \times 100 = 34\% \quad \text{(irrigation)}$$

$$\frac{\text{5 bgd rural uses}}{\text{414 bgd total}} \times 100 = 1\% \quad \text{(rural uses)}$$

$$\frac{\text{240 bgd self-supplied industry}}{\text{414 bgd total}} \times 100 = \frac{58\%}{100\%} \quad \text{(self-supplied industry)}$$

Once the percentages have been established, calculate the angle that each part represents:

(0.07) (360°)	=	25°	(public water supplies)
(0.34) (360°)	=	122°	(irrigation)
(0.01) (360°)	=	4°	(rural uses)
(0.58) (360°)	=	209°	(self-supplied industry)
		360°	(total)

From this information draw the angles using a protractor.

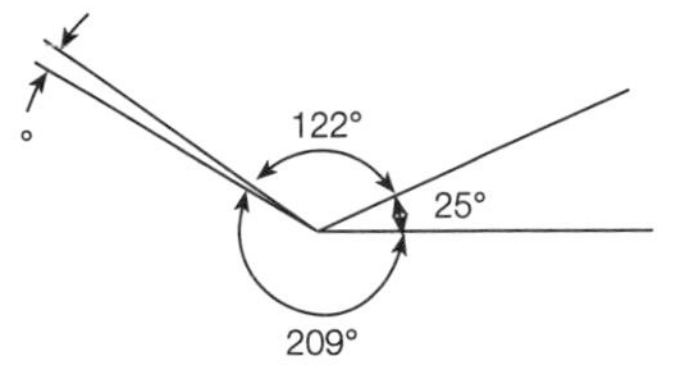

And then draw the circle and label the graph (Figure M12-25).

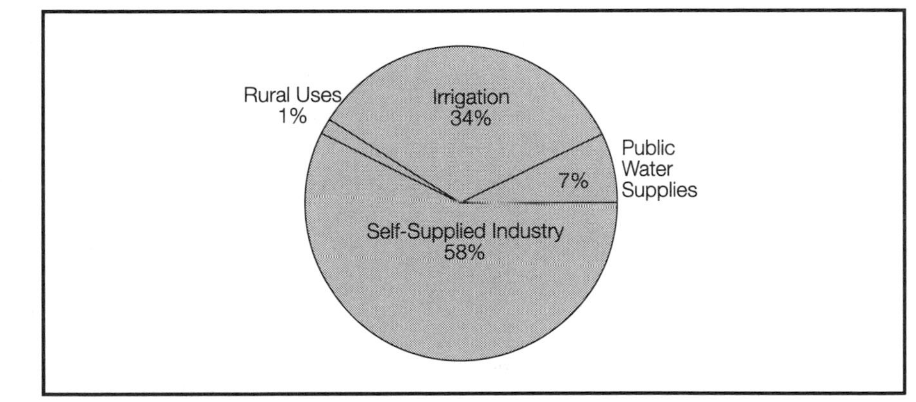

FIGURE M12-25 Circle graph depicting 1975 national daily average water use

Broken-Line Graphs

When constructing a *broken-line graph*, you should include the following components:

- a title
- at least one vertical and one horizontal scale, clearly labeled

Plot each data point on the graph, and draw a connecting line between the points. The following example illustrates the construction of a broken-line graph.

Example 13

Given the following average monthly flows, construct a broken-line graph.

Month	Average Monthly Flow, *mgd*
January	1.6
February	1.4
March	2.0
April	2.6
May	4.5
June	5.4
July	6.3
August	5.7
September	3.8
October	2.8
November	2.4
December	1.8

As in constructing a bar graph, you must first select a suitable scale. The scale should be one that is neither blown out of proportion nor unnecessarily compressed. The same basic principles apply as those discussed with respect to bar graph scales.

Since *time* is part of this graph, the time information should be placed along the horizontal scale. Flow, in million gallons per day, will therefore go along the vertical scale. Then plot the points on the graph and draw the connecting lines (Figure M12-26). Notice that

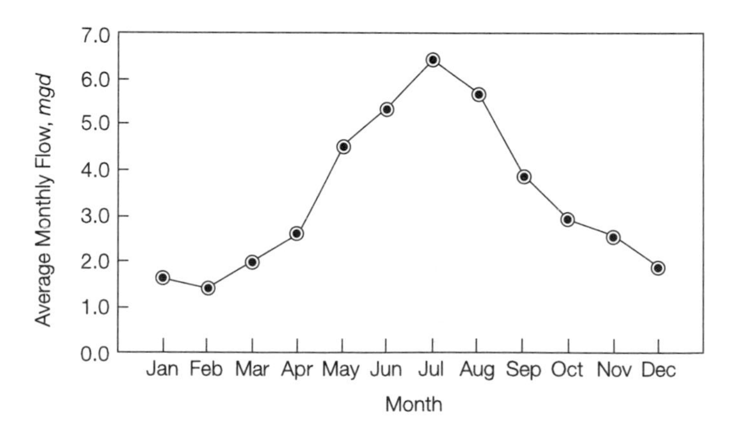

FIGURE M12-26 Average monthly flow

the small dots on the graph are circled so that data points are not obscured by the lines connecting them.

Tables

For the most part, tables that you will use in water treatment plant operations are plainly marked and need no explanation regarding use. However, the use of a table that has numerical scales for column and row headings can be clarified by example. This type of table is more complex than most others you may encounter.

Example 14

Suppose the staff gauge measuring the depth of flow over a 90° V-notch weir indicates a head of 1.35 ft. Using Table M12-1, determine the flow rate in cubic feet per second and million gallons per day.

To read Table M12-1 you must know that part of the head is listed on the vertical, or side, scale and part on the horizontal, or top, scale. In this problem, 1.3 ft of the head is read on the side scale. The remaining portion, 0.05 ft, is read on the top scale. These two readings pinpoint the cubic-feet-per-second and million-gallons-per-day discharge from the weir. Moving to the right of the 1.3 reading until you reach the column headed "0.05," you can see that the indicated discharge is 5.29 ft^3/s or 3.42 mgd.

Let's look at two more head readings on Table M12-1 to illustrate how the scales should be used in determining discharge. Suppose the staff gauge indicated a head of 0.88 ft. Where on the top and side scales would this be read?

$$\underset{}{0.88} = \underset{\substack{\text{side}\\ \text{scale}}}{0.8} + \underset{\substack{\text{top}\\ \text{scale}}}{0.08}$$

The flow for 0.88 ft of head is shown on the table as 1.82 ft^3/s, or 1.17 mgd. How would the reading be located if the state gauge indicated a head of 1.9 ft?

$$1.9 = 1.90 = \underset{\substack{\text{side}\\ \text{scale}}}{1.9} + \underset{\substack{\text{top}\\ \text{scale}}}{0.00}$$

The flow for 1.9 ft of head is shown as 12.4 ft^3/s, or 8.04 mgd.

TABLE M12-1 Discharge rates for 90° V-notch weirs

Head,	.00		.01		.02		.03		.04		.05		.06		.07		.08		.09	
ft	*ft*^{3}/*s*	*mgd*	*ft*^{3}/*s*	*mgd*	*ft*^{3}/*s*	*mgd*	*ft*^{3}/*s*	*mgd*	*ft*^{3}/*s*	*mgd*	*ft*^{3}/*s*	*mgd*	*ft*^{3}/*s*	*mgd*	*ft*^{3}/*s*	*mgd*	*ft*^{3}/*s*	*mgd*	*ft*^{3}/*s*	*mgd*
0.1	.008	.005	.010	.006	.012	.008	.015	.010	.018	.012	.022	.014	.026	.017	.030	.019	.034	.022	.039	.025
0.2	.045	.029	.051	.033	.057	.037	.063	.041	.071	.046	.078	.050	.086	.056	.095	.061	.104	.067	.113	.073
0.3	.123	.080	.134	.086	.145	.094	.156	.101	.169	.109	.181	.117	.194	.126	.208	.135	.223	.144	.237	.153
0.4	.253	.163	.269	.174	.286	.185	.303	.196	.321	.207	.340	.219	.359	.232	.379	.245	.399	.258	.420	.272
0.5	.442	.286	.464	.300	.487	.315	.511	.330	.536	.346	.561	.362	.587	.379	.613	.396	.640	.414	.668	.432
0.6	.697	.451	.727	.470	.757	.489	.788	.509	.819	.529	.852	.550	.885	.579	.919	.594	.953	.616	.989	.639
0.7	1.02	.662	1.06	.686	1.10	.711	1.14	.736	1.18	.761	1.22	.787	1.26	.814	1.30	.841	1.34	.868	1.39	.896
0.8	1.43	.925	1.48	.954	1.52	.984	1.57	1.01	1.62	1.04	1.66	1.08	1.71	1.11	1.76	1.14	1.82	1.17	1.87	1.21
0.9	1.92	1.24	1.97	1.28	2.03	1.31	2.08	1.35	2.14	1.38	2.20	1.42	2.26	1.46	2.32	1.50	2.38	1.54	2.44	1.58
1.0	2.50	1.62	2.56	1.66	2.63	1.70	2.69	1.74	2.76	1.78	2.82	1.82	2.89	1.87	2.96	1.91	3.03	1.96	3.10	2.00
1.1	3.17	2.05	3.24	2.10	3.32	2.14	3.39	2.19	3.47	2.24	3.55	2.29	3.62	2.34	3.70	2.39	3.78	2.44	3.86	2.50
1.2	3.94	2.55	4.03	2.60	4.11	2.66	4.19	2.71	4.28	2.77	4.37	2.82	4.45	2.88	4.54	2.94	4.63	2.99	4.72	3.05
1.3	4.82	3.11	4.91	3.17	5.00	3.23	5.10	3.30	5.20	3.36	5.29	3.42	5.39	3.48	5.49	3.55	5.59	3.61	5.69	3.68
1.4	5.80	3.75	5.90	3.81	6.01	3.88	6.11	3.95	6.22	4.02	6.33	4.09	6.44	4.16	6.55	4.23	6.66	4.30	6.77	4.38
1.5	6.89	4.45	7.00	4.53	7.12	4.60	7.24	4.68	7.36	4.75	7.48	4.83	7.60	4.91	7.72	4.99	7.84	5.07	7.97	5.15
1.6	8.09	5.23	8.22	5.31	8.35	5.40	8.48	5.48	8.61	5.56	8.74	5.65	8.88	5.74	9.01	5.82	9.15	5.91	9.28	6.00
1.7	9.42	6.09	9.56	6.18	9.70	6.27	9.84	6.36	9.98	6.45	10.1	6.55	10.3	6.64	10.4	6.73	10.6	6.83	10.7	6.93
1.8	10.9	7.02	11.0	7.12	11.2	7.22	11.3	7.32	11.5	7.42	11.6	7.52	11.8	7.62	11.9	7.73	12.1	7.83	12.3	7.93
1.9	12.4	8.04	12.6	8.15	12.7	8.25	12.9	8.36	13.1	8.47	13.3	8.58	13.4	8.69	13.6	8.80	13.8	8.91	14.0	9.03
2.0	14.1	9.14	14.3	9.25	14.5	9.37	14.7	9.49	14.9	9.60	15.0	9.72	15.2	9.84	15.4	9.96	15.6	10.1	15.8	10.2

Source: Adapted from Leupold & Stevens, Inc., P.O. Box 688, Beaverton, OR 97005, from Stevens Water Resources Book. *Formula:* $ft^3/s = 2.50\, H^{2.5}$ *and* $mgd = ft^3/s \times 0.646$.

TABLE M12-2 Flow rate through Parshall flumes

Head, ft	Discharge Through Throat Width *W* of —													
	1 in.		2 in.		3 in.		6 in.		9 in.		12 in.		18 in.	
	ft^3/s	*mgd*	ft^3/s	*mgd*	ft^3/s	*mgd*	ft^3/s	*mgd*	ft^3/s	*mgd*	ft^3/s	*mgd*	ft^3/s	*mgd*
.46	.101	.065	.203	.131	.299	.193	.61	.39	.94	.61	1.23	.79	1.82	1.18
.47	.105	.068	.210	.136	.309	.200	.63	.41	.97	.63	1.27	.82	1.88	1.22
.48	.108	.070	.217	.140	.319	.206	.65	.42	1.00	.65	1.31	.85	1.94	1.25
.49	.112	.072	.224	.145	.329	.213	.67	.43	1.03	.67	1.35	.87	2.00	1.29
.50	.115	.074	.230	.149	.339	.219	.69	.45	1.06	.69	1.39	.90	2.06	1.33
.51	.119	.077	.238	.154	.350	.226	.71	.46	1.10	.71	1.44	.93	2.13	1.38
.52	.123	.079	.245	.158	.361	.233	.73	.47	1.13	.73	1.48	.96	2.19	1.42
.53	.126	.081	.253	.164	.371	.240	.76	.49	1.16	.75	1.52	.98	2.25	1.45
.54	.130	.084	.260	.168	.382	.247	.78	.50	1.20	.78	1.57	1.01	2.32	1.50
.55	.134	.087	.268	.173	.393	.254	.80	.52	1.23	.79	1.62	1.05	2.39	1.54
.56	.138	.089	.275	.178	.404	.261	.82	.53	1.26	.81	1.66	1.07	2.45	1.58
.57	.141	.091	.283	.183	.415	.268	.85	.55	1.30	.84	1.70	1.10	2.52	1.63
.58	.145	.094	.290	.187	.427	.276	.87	.56	1.33	.86	1.75	1.13	2.59	1.67
.59	.149	.096	.298	.193	.438	.283	.89	.58	1.37	.89	1.80	1.16	2.66	1.72
.60	.153	.099	.306	.198	.450	.291	.92	.59	1.40	.90	1.84	1.19	2.73	1.76
.61	.157	.101	.314	.203	.462	.299	.94	.61	1.44	.93	1.88	1.22	2.80	1.81
.62	.161	.104	.322	.208	.474	.306	.97	.63	1.48	.96	1.93	1.25	2.87	1.85
.63	.165	.107	.330	.213	.485	.313	.99	.64	1.51	.98	1.98	1.28	2.95	1.91
.64	.169	.109	.338	.218	.497	.321	1.02	.66	1.55	1.00	2.03	1.31	3.02	1.95
.65	.173	.112	.347	.224	.509	.329	1.04	.67	1.59	1.03	2.08	1.34	3.09	2.00

Source: Adapted from Leupold & Stevens, Inc., P.O. Box 688, Beaverton, OR 97005, from Stevens Water Resources Book. *Formula:* $ft^3/s = 2.50\ H^{5/2}$ *and* $mgd = ft^3/s \times 0.646$.

Example 15

Assume that you have a Parshall flume with a 3-in. throat width, and a staff gauge water depth reading of 0.62 ft. Using Table M12-2, determine the cubic-foot-per-second and million-gallons-per-day flow rates.

As in the last problem, this table has both vertical and horizontal scales. However, the scales here are somewhat different. The vertical scale shows head, and the horizontal scale shows the Parshall flume throat width. Once these two variables are known, the corresponding flow can be determined.

In this problem the head is given as 0.62 ft and the throat width as 3 in. Move to the right of 0.62 ft until you come under the "3 in." column heading. The indicated discharge is 0.474 ft^3/s, or 0.306 mgd.

MATHEMATICS 13

Per Capita Water Use

The amount of water a community uses is, of course, of primary importance in water treatment. The water use establishes the amount of water that must be treated. One method of expressing the water use of a community is gallons per capita per day (gpcd). The term *per capita* means "per person." Calculating gallons per capita per day merely determines the average water use *per person*. The mathematical formula used in calculating gallons per capita per day is

$$\text{gpcd} = \frac{\text{water used (gpd)}}{\text{total number of people}}$$

Note that the "total number of people" refers to the total number of people served by the water system; the water system service population is not always the same as the population of the city or town in which the system is located. The following examples illustrate the calculation of gallons per capita per day.

Example 1

During 1977, the average daily water use at a particular water system was 8.4 mgd. If the water system served a population of 26,900, what was this average daily water use expressed in gallons per capita per day?

$$\text{water use} = \frac{\text{water used (gallons per day)}}{\text{total number of people}}$$

$$= \frac{8{,}400{,}000\text{-gpd flow}}{26{,}900 \text{ people}}$$

$$= 312.27 \text{ gpcd}$$

Example 2

The average daily water use during 1995 for a certain water system serving a population of 10,300 was 2.7 mgd. The percent of water used in each of four categories is given below. Use this information to determine the 1995 gallons-per-capita-per-day water use for each of the four categories.

Domestic uses	46%
Commercial uses	16%
Industrial uses	25%
Public uses	13%

First, calculate the total gallons-per-capita-per-day water use for 1995:

$$\text{water use} = \frac{\text{water used (gallons per day)}}{\text{total number of people}}$$

$$= \frac{2{,}700{,}000\text{-gpd flow}}{10{,}300 \text{ people}}$$

$$= 262.14 \text{ gpcd}$$

Next, use the percents given to calculate the water use by category:[1]

$$(262.14 \text{ gpcd})(0.46) = 120.58 \text{ gpcd for domestic uses}$$

$$(262.14 \text{ gpcd})(0.16) = 41.94 \text{ gpcd for commercial uses}$$

$$(262.14 \text{ gpcd})(0.25) = 65.54 \text{ gpcd for industrial uses}$$

$$(262.14 \text{ gpcd})(0.13) = 34.08 \text{ gpcd for public uses}$$

---

[1]Mathematics 7, Percent.

Example 3

The total water use on a particular day was 4,503,000 L. If the water system served a population of 9,800, what was this water use expressed in liters per day per capita?

$$\text{water use} = \frac{\text{water used (liters per day)}}{\text{total number of people}}$$

$$= \frac{4{,}503{,}000\text{-L/d flow}}{9{,}800 \text{ people}}$$

$$= 459.5 \text{ L/d per capita}$$

MATHEMATICS 14

Domestic Water Use Based on Household Fixture Rates

Of the four principal uses of community water supplies — domestic, commercial, industrial, and public — domestic consumption often accounts for the largest portion. The approximate amounts of water required for typical household activities are tabulated in Table M14-1. Wasteful practices will, of course, greatly increase the quantities used.

This information, although not directly applicable to daily operation of a water treatment plant, helps to identify the factors that contribute to domestic water consumption. The following examples illustrate how to use

TABLE M14-1 Household fixture use rate

Use	Approximate Rate
Laundry	20 to 45 gal per load
Shower	20 to 30 gal per shower
Tub bath	30 to 40 gal per bath
Dishwashing	15 to 30 gal per load
Toilet	3.5 to 7 gal per flush
Drinking water	1–2 qt/day/person
Garbage disposal	5 gpm*
Car washing	5 gpm*
Lawn watering	7 to 43 gal per 100 ft^2 per wk†

*Assuming that a faucet draws 5 gpm fully open.

†The lower value applies in humid areas, the higher value in arid areas.

these household fixture rates to estimate the water demand of an individual residence.

Example 1

The major water uses of a family of four are listed below. Based on the approximate values given in Table M14-1, determine the average amount of water used each day by this family. Express the answer in gallons per capita per day, and identify the one activity that represents the highest water consumption per day. (When a range of values is given in the table, use a median value.)

Laundry	5 loads per week
Showers	2 per day
Tub baths	2 per day
Dishwashing	1 load per day
Toilet	12 flushes per day
Drinking water	4 people
Garbage disposal	2 min per day
Car washing	10 min per week
Lawn watering	5,000 ft^2

To calculate the average amount of water used daily by this family, first determine the amount for each activity, then add these amounts. (Notice that some uses are per week and others are per day.)

Laundry: In this problem, you were instructed to use a median value when a range is shown for the approximate use rate. For laundry the range is 20 to 45 gal. To determine the middle value, add the two numbers and divide by 2:

Step 1:

$$\begin{array}{r} 20 \\ +\ 45 \\ \hline 65 \end{array}$$

Step 2:

$$\frac{65}{2} = 32.5 \text{ (middle value)}$$

Now continue with the water use calculation:

$$(32.5 \text{ gal/ load}) (5 \text{ loads/ week}) = 162.5 \text{ gal/ week}$$

Expressing this in terms of gallons per day,

$$\frac{162.5 \text{ gal/week}}{7 \text{ d/week}} = 23.21 \text{ gpd}$$

Showers: First determine the middle value of the range given for shower use rate.

Step 1:

$$\begin{array}{r} 20 \\ +\,30 \\ \hline 50 \end{array}$$

Step 2:

$$\frac{50}{2} = 25 \text{ (middle value)}$$

Next, calculate the daily water use:

$$(25 \text{ gal/shower})\ (2 \text{ showers/d}) = 50 \text{ gpd}$$

Tub baths: A range of 30 to 40 gal/bath is given in the rate table. The middle value is 35 gal. The daily water use for tub baths is

$$(35 \text{ gal/bath})\ (2 \text{ baths/d}) = 70 \text{ gpd}$$

Dishwashing: The table indicates that the approximate water use rate for dishwashing is 15 to 30 gal/load. Using the midpoint calculation,

Step 1:

$$\begin{array}{r} 15 \\ +\,30 \\ \hline 45 \end{array}$$

Step 2:

$$\frac{45}{2} = 22.5 \text{ (middle value)}$$

The water used daily for dishwashing can now be determined:

$$(22.5 \text{ gal/load})\ (1 \text{ load/d}) = 22.5 \text{ gpd}$$

Toilet: The range given for the approximate water use rate is 3.5 to 7 gal. Using the midpoint calculation,

Step 1:
$$\begin{array}{r} 3.5 \\ +\,7.0 \\ \hline 10.5 \end{array}$$

Step 2:
$$\frac{10.5}{2} = 5.25 \text{ (middle value)}$$

Then calculate the water used daily:

$$(5.25 \text{ gal/flush})\ (12 \text{ flushes/d}) = 63 \text{ gpd}$$

Drinking water: Since there are four in the family, the water used daily for drinking purposes (using the middle value of the range) is

$$(1.5 \text{ qt/person})\ (4 \text{ people}) = 6 \text{ qt}$$

To be consistent with the other calculations, this water use must be expressed in gallons:

$$\frac{6 \text{ qt/d}}{4 \text{ qt/gal}} = 1.5 \text{ gpd}$$

Garbage disposal: Since no range in use rates is given, the water used daily can be calculated directly for 2 min of use:

$$(5 \text{ gpm})\ (2 \text{ min/d}) = 10 \text{ gpd}$$

Car washing: No range in use is given. For 10 min/week of use, the water used weekly is

$$(5 \text{ gpm})\ (10 \text{ min/week}) = 50 \text{ gal/week}$$

Express this water in gallons per day:

$$\frac{50 \text{ gal/week}}{7 \text{ d/week}} = 7.14 \text{ gpd}$$

Lawn watering: Calculate the midpoint of the range:

Step 1:

$$\begin{array}{r} 7 \\ +\ 43 \\ \hline 50 \end{array}$$

Step 2:

$$\frac{50}{2} = 25 \text{ (middle value)}$$

Therefore, the water used weekly is

$$(25\ \text{gal}/100\ \text{ft}^2)\ (50\ \text{segments of}\ 100\ \text{ft}^2) = 1{,}250\ \text{gal/week}$$

Express this in terms of gallons per day:

$$\frac{1{,}250\ \text{gal/week}}{7\ \text{d/week}} = 178.57\ \text{gpd}$$

Now add all the daily amounts calculated above:

Laundry		23.21	gpd
Showers		50	gpd
Tub baths		70	gpd
Dishwashing		22.5	gpd
Toilet		63	gpd
Drinking water		1.5	gpd
Garbage disposal		10	gpd
Car washing		7.14	gpd
Lawn watering		178.57	gpd
	Total	425.92	gpd

In this problem, you were asked to express the water use in terms of gallons per capita per day.[1]

$$\text{water use} = \frac{\text{total water used (gallons per day)}}{\text{total number of people}}$$

$$= \frac{425.92\ /\text{gpd}}{4\ \text{people}}$$

[1]Mathematics 13, Per Capita Water Use.

= 106.48 gpcd

= 106 gpcd (rounded to the nearest whole number)

In the second part of the problem, you were asked to determine which activity represents the highest water use per day. According to the calculations above, lawn watering constitutes the highest daily water use for this family.

Example 2

The actual uses of water by a particular family of three are described by the list below. With this information and the approximate values given in Table M14-1, calculate the average amount of water used daily by this household. (When a range is given for the approximate rate, use the lower value; e.g., 75 L per shower.)

Laundry	4 loads per week
Showers	2 per day
Tub baths	1 per day
Dishwashing	1 load per day
Toilet	10 flushes per day
Drinking water	3 people
Garbage disposal	1 min per day
Car washing	5 min per week
Lawn watering	390 m^2 (once per week)

To calculate the average amount of water used daily by this family, first determine the amount for each activity, then add these individual amounts.

Laundry: The approximate use rate for laundry is 75 L/load. For 4 loads/week, the total water use for the week is

$$(75 \text{ L/load})(4 \text{ loads/week}) = 300 \text{ L/week}$$

The daily water use for laundry is

$$\frac{300 \text{ L/week}}{7 \text{ d/week}} = 42.9 \text{ L/d}$$

Shower: The approximate use rate for showers is 75 L/shower. For 2 showers/d the daily water use is

$$(75 \text{ L/shower})\ (2 \text{ showers/d}) = 150 \text{ L/d}$$

Tub bath: The approximate use rate is 115 L/bath. In this problem there is 1 bath/d, so the water use is

$$(115 \text{ L/bath})\ (1 \text{ bath/d}) = 115 \text{ L/d}$$

Dishwashing: The approximate use rate is 55 L/load. At 1 load/d then,

$$(55 \text{ L/load})\ (1 \text{ load/d}) = 55 \text{ L/d}$$

Toilet: The use rate for toilets is 13 L/flush. At 10 flushes/d the water used is therefore

$$(13 \text{ L/flush})\ (10 \text{ flushes/d}) = 130 \text{ L/d}$$

Drinking: At a use rate of 0.9 L per day per person, the daily water use for a family of three is

$$(0.9 \text{ L/person/d})\ (3 \text{ people}) = 2.7 \text{ L/d}$$

Garbage disposal: The approximate use rate is 19 L/min. In this problem, assume the garbage disposal is used 1 min/d. Therefore the daily water use is

$$(19 \text{ L/min})\ (1 \text{ min/d}) = 19 \text{ L/d}$$

Car washing: In this problem, assume water is used for car washing 5 min/week. At an approximate rate of 19 L/min, the water used is

$$(19 \text{ L/min})\ (5 \text{ min/week}) = 95 \text{ L/week}$$

Converting this to liters used per day,

$$\frac{95 \text{ L/week}}{7 \text{ d/week}} = 13.6 \text{ L/d}$$

Lawn watering: The use rate is 0.9 L/m^2/week. Therefore the water use during one week is

$$(0.9 \text{ L/m}^2\text{/week})\ (390 \text{ m}^2) = 351 \text{ L/week}$$

and the daily water use is

$$\frac{351 \text{ L/week}}{7 \text{ d/week}} = 50.1 \text{ L/d}$$

Now add all the daily amounts calculated above to determine the average amount of water used by this household per day:

Laundry		42.9	L/d
Shower		150	L/d
Tub bath		115	L/d
Dishwashing		55	L/d
Toilet		130	L/d
Drinking water		2.7	L/d
Garbage disposal		19	L/d
Car washing		13.6	L/d
Lawn watering		50.1	L/d
	Total	578.3	L/d

MATHEMATICS 15

Water Use per Unit of Industrial Product Produced

Water used by industry is commonly measured in terms of the products produced. For example, the water required by a green bean cannery is typically 20,000 gal for each ton of green beans processed; for the paper industry, 47,000 gal of water is typically required for each ton of paper produced. Table M15-1 summarizes water requirements for some typical industries. The following examples illustrate how this information can be used in calculating water requirements for a particular industrial plant.

Example 1

A new paper mill is being planned for construction in your area. The planners have requested that water from your treatment plant be used in meeting their paper-processing needs. If the paper mill is planning to produce 25 tons of paper daily at full production, how much water will the mill require daily? (Use Table M15-1 for water requirements per product produced.)

According to Table M15-1, the water requirement for each ton of paper produced is 47,000 gal. Therefore, the daily water requirement for 25 tons of paper is

$$(47{,}000 \text{ gal water/ton})(25 \text{ tons}) = 1{,}175{,}000 \text{ gal water required per day}$$

Example 2

In 1992, the total production of the sulfur manufacturing industry was 1,095 tons of sulfur. On the average, how much water

TABLE M15-1 Water requirements in selected industries

Process	Water Required (in gallons per unit of product noted)
Canneries	
Green beans, per ton	20,000
Peaches and pears, per ton	5,300
Other fruits and vegetables, per ton	2,000–10,000
Chemical industries	
Ammonia, per ton	37,500
Carbon dioxide, per ton	24,500
Gasoline, per 1,000 gal	7,000–34,000
Lactose, per ton	235,000
Sulfur, per ton	3,000
Food and beverage industries	
Beer, per 1,000 gal	15,000
Bread, per ton	600–1,200
Meat packing, per ton live weight	5,000
Milk products, per ton	4,000–5,000
Whiskey, per 1,000 gal	80,000
Pulp and paper	
Pulp, per ton	82,000–230,000
Paper, per ton	47,000
Textiles	
Bleaching, per ton cotton	72,000–96,000
Dyeing, per ton cotton	9,500–19,000
Mineral products	
Aluminum (electrolytic smelting), per ton	56,000 (maximum)
Petroleum, per barrel of crude oil	800–3,000
Steel, per ton	1,500–50,000

Adapted from Metcalf and Eddy. 1972. Wastewater Engineering. *New York: McGraw-Hill and Fair, G.M., J.G. Geyer, and D.A. Okun. 1971.* Elements of Water Supply and Wastewater Disposal. *2d ed. New York: Wiley.*

NOTE: *A ton is approximately 900 kg.*

was required each day, assuming a 7-day work week? (Use Table M15-1 for water requirements per product produced.)

First calculate the amount of water required for the full year, then calculate the amount required per day. According to Table M15-1, the water requirement for each ton of sulfur produced is 3,000 gal. Therefore, the water requirement for the year's production was

$$(3{,}000 \text{ gal/ton})\ (1{,}095 \text{ tons}) \quad = \quad 3{,}285{,}000 \text{ gal}$$

Since sulfur was produced each day of the year, there were 365 days of sulfur production. With this information, the average amount of water required for each day can be determined:

$$\frac{3{,}285{,}000 \text{ gal/year}}{365 \text{ day/year}} \quad = \quad 9{,}000 \text{ gpd}$$

Mathematics 16

Average Daily Flow

The amount of water a community uses every day can be expressed in terms of an *average daily flow* (ADF); that is, the average[1] of the actual daily flows that occur within a period of time, such as a week, a month, or a year. This is expressed mathematically as

$$\text{ADF} = \frac{\text{sum of all daily flows}}{\text{total number of daily flows used}}$$

The average daily flow can reflect a week's data used in the averaging (weekly ADF), a month's data used in the averaging (monthly ADF), or a year's data used in the averaging (annual ADF, or AADF — the most commonly calculated average).

Average daily flow is important because it is used in several treatment plant calculations. The following examples illustrate the calculation of average daily flow.

Example 1

A water treatment plant reported that the total volume of water treated for the calendar year 1995 was 152,655,000 gal. What was the annual average daily flow for 1995?

In this problem, the sum of all daily flows has already been determined — a total of 152,655,000 gal was treated during the year. Knowing that there are 365 days in a year, calculate the average daily flow:

---

[1]Mathematics 6, Averages.

$$ADF = \frac{\text{sum of all daily flows}}{\text{total number of daily flows used}}$$

$$= \frac{152{,}655{,}000 \text{ gal}}{365 \text{ days}}$$

$$= 418{,}233 \text{ gpd}$$

Example 2

In January 1993, a total of 68,920,000 L of water was treated at a plant. What was the average daily flow at the treatment plant for this period?

As in the previous example, the sum of all daily flows has already been determined — a *total* of 68,920,000 L was treated for the month. January has 31 days; fill the information into the ADF formula:

$$ADF = \frac{\text{sum of all daily flows}}{\text{total number of daily flows used}}$$

$$= \frac{68{,}920{,}000 \text{ L}}{31 \text{ days}}$$

$$= 2{,}223{,}226 \text{ L/d}$$

Example 3

The following daily flows (in million gallons per day) were treated during June 1994 at a water treatment plant. What was the average daily flow for this period?

June	1	6.21	June	11	7.59	June	21	6.43
	2	6.68		12	7.01		22	6.26
	3	7.31		13	6.85		23	6.87
	4	7.80		14	6.43		24	7.27
	5	6.77		15	6.52		25	7.95
	6	6.32		16	6.79		26	7.33
	7	5.96		17	6.91		27	6.72
	8	5.83		18	7.37		28	6.51
	9	6.09		19	7.02		29	5.92
	10	7.22		20	6.88		30	5.90

To calculate the average daily flow for this time period, first add all daily flows. The total of these flows is 202.72 mil gal. With this information, calculate the average daily flow for the period:

$$\text{ADF} = \frac{\text{sum of all daily flows}}{\text{total number of daily flows used}}$$

$$= \frac{202.72 \text{ mil gal}}{30 \text{ days}}$$

$$= 6.76 \text{ mgd}$$

In some problems, you will not know *actual* daily flows for each day during a particular period, but you will know the ADF for that time period. ADF information can be used in calculating other ADFs. The following example contrasts the two methods of calculating ADFs.

Example 4

The volume of water (in megaliters) treated for each day during a 2-week period is listed below. What is the average daily flow for this 2-week period?

Week 1		Week 2	
Sunday	2.41	Sunday	2.52
Monday	3.37	Monday	3.39
Tuesday	3.44	Tuesday	3.48
Wednesday	3.61	Wednesday	3.88
Thursday	3.23	Thursday	3.19
Friday	2.86	Friday	2.82
Saturday	2.75	Saturday	2.70

Since you are given the actual flows for each day in the 2-week period, you could calculate the average daily flow in a similar way to the previous example problems in this chapter. That is,

$$\text{ADF} = \frac{\text{sum of all daily flows}}{\text{total number of daily flows used}}$$

$$= \frac{43.65 \text{ ML}}{14 \text{ days}}$$

$$= 3.12 \text{ ML/d}$$

Suppose, however, that in this problem you did not know the actual flow for each day in the week, but you knew the ADF for week 1 (3.10 ML/d) and the ADF for week 2 (3.14 ML/d). The ADF for the 2-week period can still be calculated, but *not* using the ADF equation given above (because in this case you do not know the *sum* of all daily flows). However, a similar formula is used:

$$\text{ADF} = \frac{\text{sum of all weekly ADFs}}{\text{total number of weekly ADFs used}}$$

In this case,

$$\text{ADF} = \frac{3.10 \text{ ML/d} + 3.14 \text{ ML/d}}{2}$$

$$= \frac{6.24 \text{ ML/d}}{2}$$

$$= 3.12 \text{ ML/d}$$

Notice that, although only ADF information was used, the answer was the same as when all 14 actual daily flows were used in the calculation.

In a similar manner, ADFs for each of 12 months can be used to determine the average daily flow (AADF) for a year. The equation that is used is

$$\text{AADF} = \frac{\text{sum of all monthly ADFs}}{\text{total number of monthly ADFs used}}$$

Example 5

The average daily flow (in million gallons per day) at a treatment plant for each month in the year is given below. Using this information, calculate the annual average daily flow.

January	10.71	July	11.96
February	9.89	August	12.24
March	10.32	September	11.88
April	10.87	October	11.53
May	11.24	November	11.36
June	11.58	December	10.98

If you knew all 365 flows for the year, the annual average daily flow (AADF) would be calculated using the formula

$$\text{AADF} = \frac{\text{sum of all daily flows}}{\text{total number of daily flows used}}$$

In this problem, however, average daily flows for each *month* of the year are given. Therefore, the average daily flow is calculated using the formula

$$\text{AADF} = \frac{\text{sum of all monthly ADFs}}{\text{total number of monthly ADFs used}}$$

Filling in the information given in this problem,

$$\text{AADF} = \frac{134.56 \text{ mgd}}{12}$$

$$= 11.21 \text{ mgd}$$

MATHEMATICS 17

Surface Overflow Rate

The faster the water leaves a sedimentation tank, or clarifier, the more turbulence is created and the more suspended solids are carried out in the effluent. Overflow rate — the speed with which water leaves the sedimentation tank — is controlled by an increase or decrease in the flow rate into the tank.

The *surface overflow rate* measures the amount of water leaving a sedimentation tank per square foot of tank surface area. The treatment plant

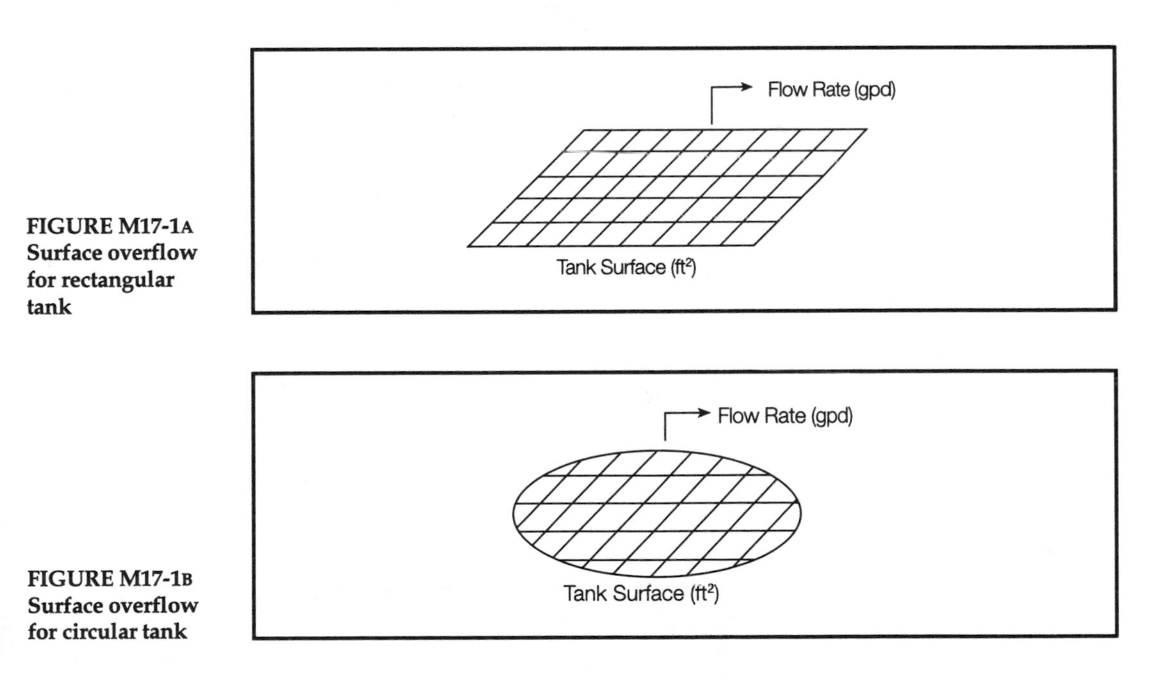

FIGURE M17-1A Surface overflow for rectangular tank

FIGURE M17-1B Surface overflow for circular tank

operator must be able to determine the surface overflow rate that produces the best-quality effluent leaving the tank.

Since surface overflow is the gallons per day flow *up and over* each square foot of tank surface, the corresponding mathematical equation is

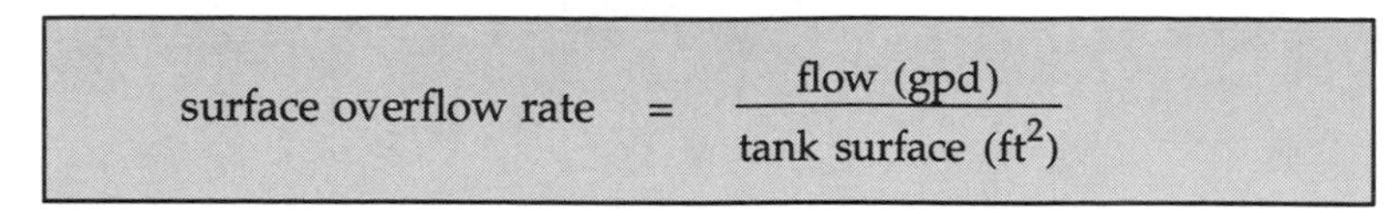

$$\text{surface overflow rate} = \frac{\text{flow (gpd)}}{\text{tank surface (ft}^2\text{)}}$$

Notice that the depth of the sedimentation tank is not a consideration in the calculation of surface overflow rate. Figure M17-1 depicts surface overflow for rectangular and circular tanks.

Example 1

The flow to a treatment plant is 1.2 mgd. If the sedimentation tank is 70 ft long, 15 ft wide, and 7 ft deep, what is the surface overflow rate?

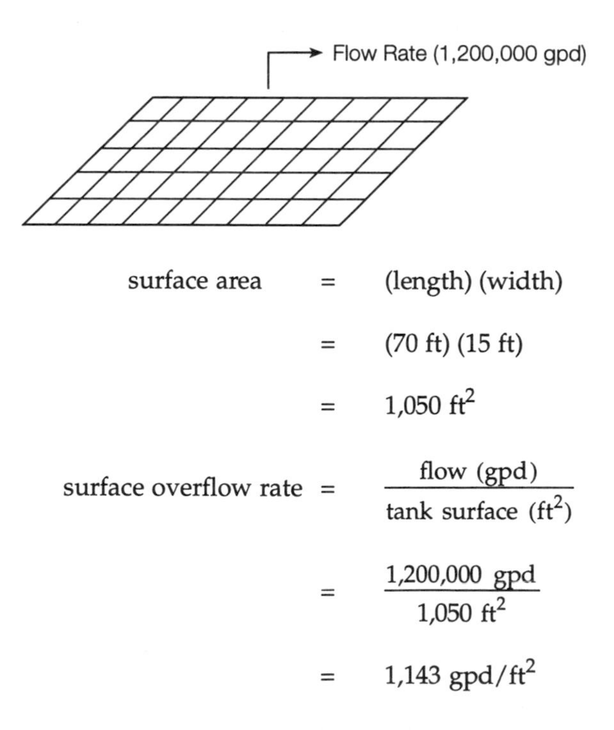

$$\text{surface area} = \text{(length) (width)}$$

$$= \text{(70 ft) (15 ft)}$$

$$= 1{,}050 \text{ ft}^2$$

$$\text{surface overflow rate} = \frac{\text{flow (gpd)}}{\text{tank surface (ft}^2\text{)}}$$

$$= \frac{1{,}200{,}000 \text{ gpd}}{1{,}050 \text{ ft}^2}$$

$$= 1{,}143 \text{ gpd/ft}^2$$

Example 2

A circular sedimentation basin with a 55-ft diameter receives a flow of 2,075,000 gpd. What is the surface overflow rate?

$$\text{surface area} = 0.785\ D^2$$

$$= (0.785)\ (55\ \text{ft})\ (55\ \text{ft})$$

$$= 2{,}375\ \text{ft}^2$$

$$\text{surface overflow rate} = \frac{\text{flow (gpd)}}{\text{tank surface (ft}^2)}$$

$$= \frac{2{,}075{,}000\ \text{gpd}}{2{,}375\ \text{ft}^2}$$

$$= 873.68\ \text{gpd/ft}^2$$

In the previous two examples, the surface overflow rate was the unknown factor. However, *any one* of three factors (surface overflow rate, water flow rate, or surface area) can be unknown. If the other two factors are known, the same mathematical setup can be used to solve for the unknown value.

Example 3

A 20-m diameter tank has a surface overflow area of 0.35 $L/m^2/s$. What is the daily flow to the tank?

$$\text{surface area} = 0.785\ D^2$$

$$= (0.785)\ (20\ \text{m})\ (20\ \text{m})$$

$$= 314\ \text{m}^2$$

$$\text{surface overflow rate} = \frac{\text{flow (L/s)}}{\text{tank surface (m}^2)}$$

$$0.35\ \text{L/m}^2\text{/s} = \frac{x\ \text{L/s}}{314\ \text{m}^2}$$

$$(0.35)\ (314) = x$$

$$109.9\ \text{L/s} = x$$

Example 4

A sedimentation tank receives a flow of 5.4 mgd. If the surface overflow rate is 689 gpd/ft^2, what is the surface area of the sedimentation tank?

$$\text{surface overflow rate} = \frac{\text{flow (gpd)}}{\text{tank surface (ft}^2\text{)}}$$

$$689 \text{ gpd/ft}^2 = \frac{5{,}400{,}000 \text{ gpd}}{x \text{ ft}^2}$$

$$(689)(x) = 5{,}400{,}000$$

$$x = \frac{5{,}400{,}000}{689}$$

$$x = 7{,}837 \text{ ft}^2$$

MATHEMATICS 18

Weir Overflow Rate

The calculation of weir overflow rate (Figure 18-1) is important in detecting high velocities near the weir, which adversely affect the efficiency of the sedimentation process. With excessively high velocities, the settling solids are pulled over the weirs and into the effluent troughs.

Since weir overflow rate is gallons-per-day flow over each foot of weir length, the corresponding mathematical equation for the weir overflow rate is

$$\text{weir overflow rate} = \frac{\text{flow (gpd)}}{\text{weir length (ft)}}$$

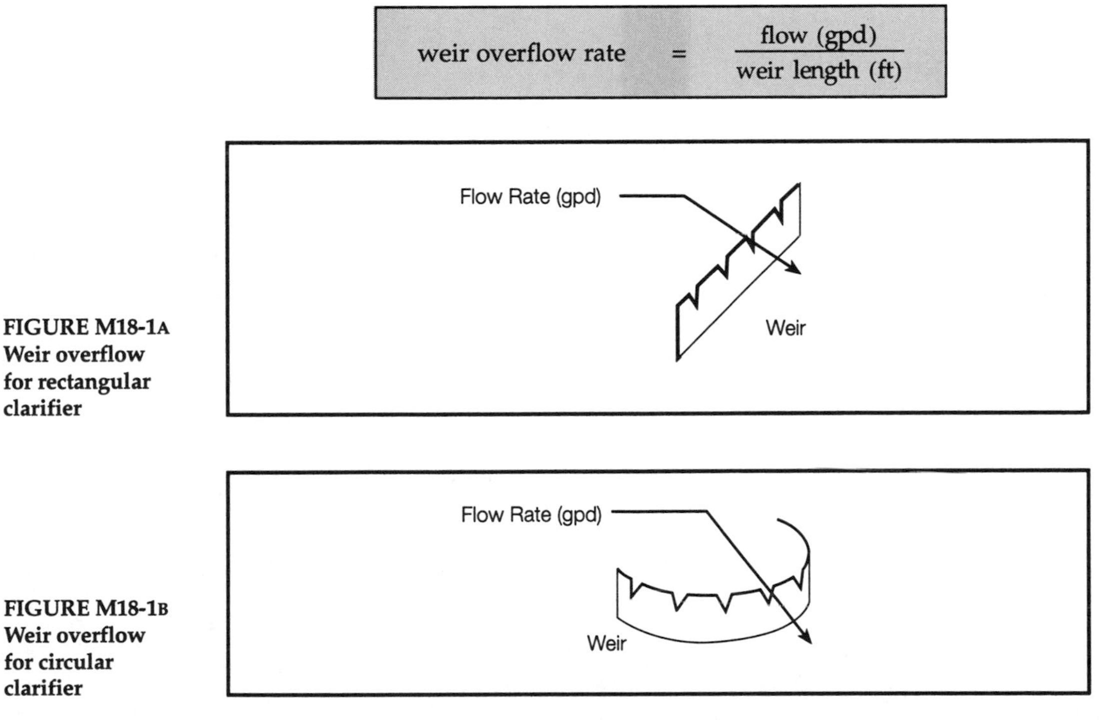

FIGURE M18-1A Weir overflow for rectangular clarifier

FIGURE M18-1B Weir overflow for circular clarifier

Example 1

A circular clarifier, which has a continuous weir around its circumference, receives a flow of 3.6 mgd. If the diameter is 80 ft, what is the weir overflow rate?

Before you can calculate the weir overflow rate, you must know the total length of the weir. The relationship of the diameter and circumference of a circle is the key to determining this.[1]

$$\text{circumference} = (3.14)\,(\text{diameter})$$

In this problem, the diameter is 80 ft. Therefore, the length of weir (circumference) is

$$\text{circumference} = (3.14)\,(80\text{ ft})$$

$$= 251.2\text{ ft}$$

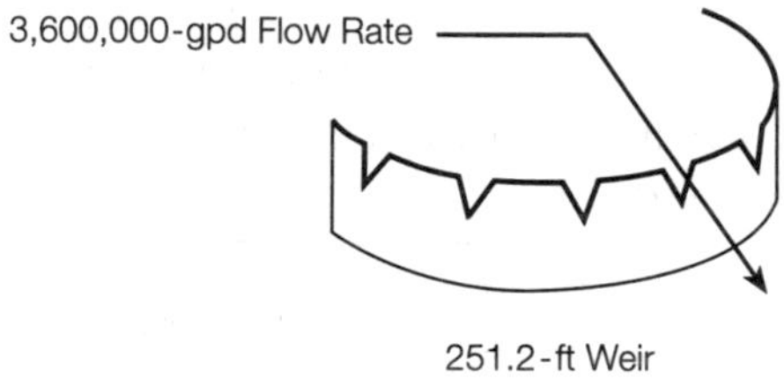

3,600,000-gpd Flow Rate

251.2-ft Weir

Now solve for the weir overflow rate:

$$\text{weir overflow rate} = \frac{\text{flow (gpd)}}{\text{weir length (ft)}}$$

$$= \frac{3{,}600{,}000\text{ gpd}}{251.2\text{ ft}}$$

$$= 14{,}331\text{ gpd/ft}$$

Example 2

If a sedimentation tank has a total of 25 m of weir over which the water flows, what is the weir overflow rate when the flow is 40.5 L/s?

---

[1]Mathematics 8, Linear Measurements.

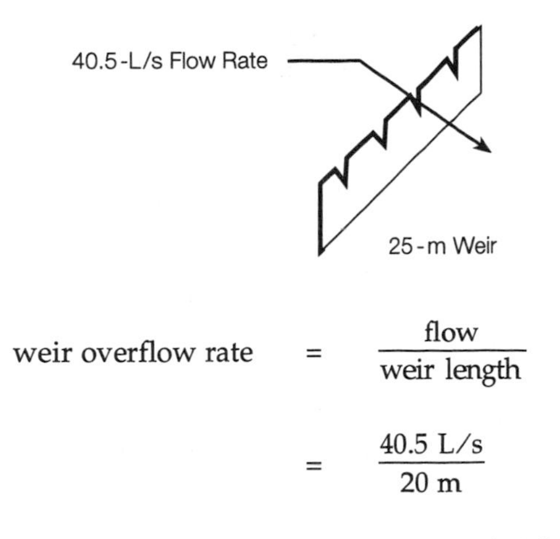

$$\text{weir overflow rate} = \frac{\text{flow}}{\text{weir length}}$$

$$= \frac{40.5 \text{ L/s}}{20 \text{ m}}$$

$$= 2.025 \text{ (L/m)/s}$$

Example 3

A clarifier has a continuous weir around the perimeter and a diameter of 60 ft. It receives a flow of 1.87 mgd. What is the weir overflow rate?

If the diameter is 60 ft, then the weir length (circumference) is

$$\text{circumference} = (3.14)\ (60 \text{ ft})$$

$$= 188.4 \text{ ft}$$

Now solve for the weir overflow rate:

$$\text{weir overflow rate} = \frac{\text{flow (gpd)}}{\text{weir length (ft)}}$$

$$= \frac{1{,}870{,}000 \text{ gpd}}{188.4 \text{ ft}}$$

$$= 9{,}926 \text{ gpd/ft}$$

MATHEMATICS 19

Filter Loading Rate

The filter loading rate is measured in gallons of water applied to each square foot of surface area. Or, stated more precisely, filter loading rate measures the amount of water flowing down through each square foot of filter area. Whereas the surface overflow rate is measured in gallons per *day* per square foot, the filter loading rate is measured in gallons per *minute* per square foot.

Figure M19-1 depicts filter loading.

The mathematical equation associated with filter loading rate is

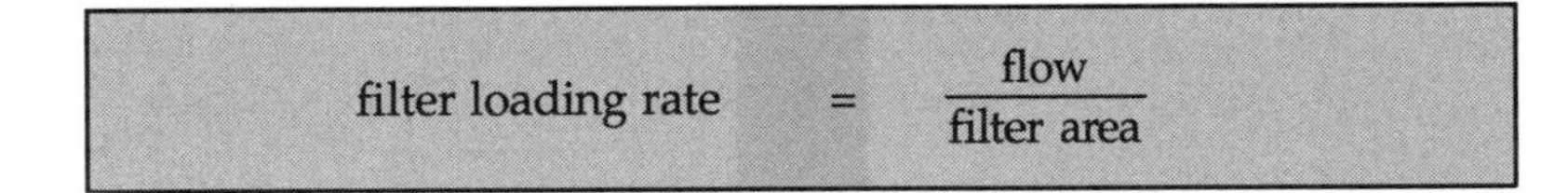

$$\text{filter loading rate} = \frac{\text{flow}}{\text{filter area}}$$

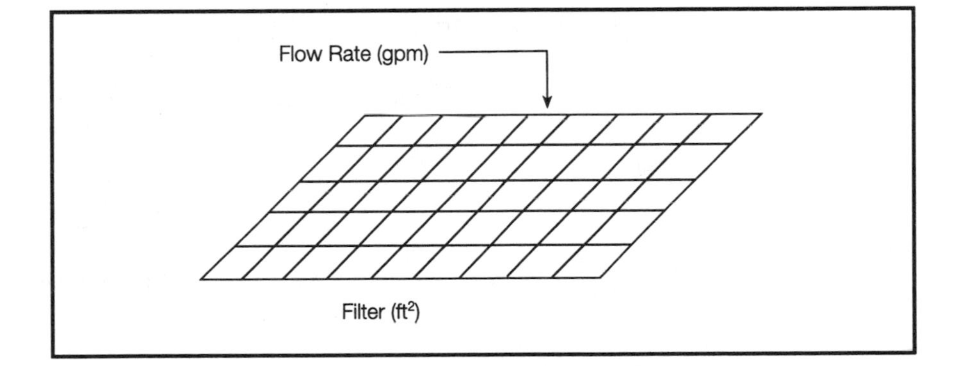

FIGURE M19-1
Filter loading

Sometimes the filter loading rate is measured as the fall of water (in inches) through a filter per minute. This is expressed mathematically as

$$\text{filter loading rate} = \frac{\text{inches of water fall}}{\text{minutes}}$$

Some typical loading rates for various types of filters are tabulated below.

Type of Filter	Common Loading Rate
Slow sand	0.016 to 0.16 gpm/ft^2
Rapid sand	2 gpm/ft^2
Dual media (coal, sand)	2 to 4 gpm/ft^2
Multimedia (coal, sand, and garnet; or coal, sand, and ilmenite)	5 to 10 gpm/ft^2

Example 1

A slow sand filter is 25 ft wide and 40 ft long. If the filter receives a flow of 173,000 gpd, what is the filter loading rate in gallons per minute per square foot?

Convert the flow from gallons per day to gallons per minute:[1]

$$\frac{173{,}000 \text{ gpd}}{1{,}440 \text{ min/d}} = 120.14 \text{ gpm}$$

Fill this information into the equation:

$$\text{filter loading rate} = \frac{\text{flow (gpm)}}{\text{filter area (ft}^2\text{)}}$$

$$= \frac{120.14 \text{ gpm}}{(25 \text{ ft})(40 \text{ ft})}$$

$$= 0.12 \text{ gpm/ft}^2$$

[1]Mathematics 11, Conversions (Flow Conversions).

Filter loading rates, as previously noted, are sometimes expressed as the vertical movement of water in the filter in one minute, that is, as inches-per-minute (in./min) fall in the water level.

The gallons-per-minute-per-square-foot and inches-per-minute measurements are directly related as noted by the following equation:

$$1 \text{ gpm/ft}^2 = 1.6 \text{ in./min}$$

You can use the box method of conversion when converting from one term to another.[2]

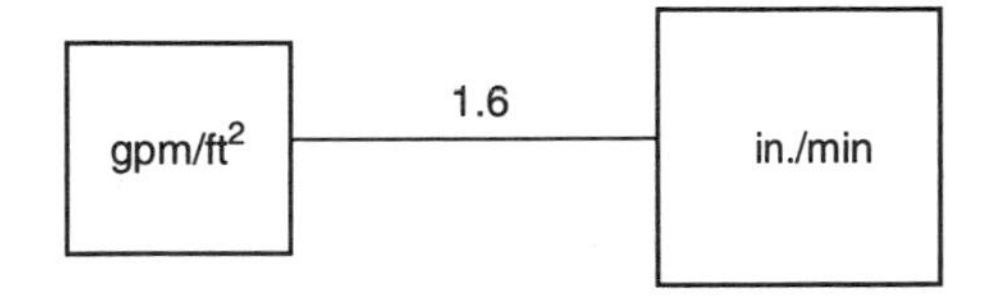

The following two examples illustrate conversion between these two terms.

Example 2

A drop of how many inches per minute corresponds to a filter loading rate of 2.6 gpm/ft^2?

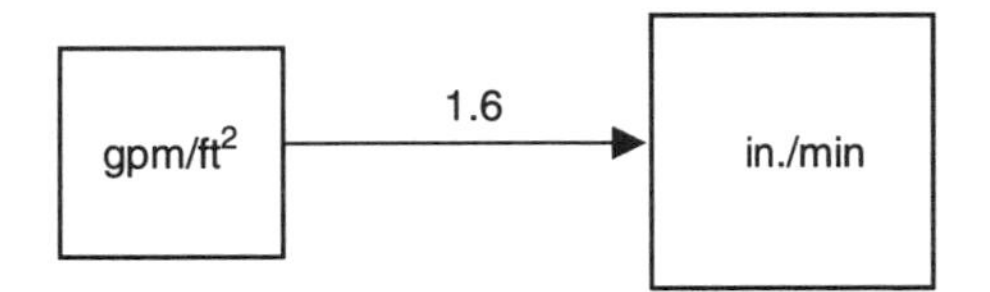

In converting from gallons per minute per square foot to inches per minute filter loading rate, you are moving from a smaller box to a larger box. Therefore, multiplication by 1.6 is indicated:

$$(2.6 \text{ gpm/ft}^2)\,(1.6) = 4.16 \text{ in./min}$$

Example 3

What is the filter loading rate in gallons per minute per square foot of a sand filter in which the water level dropped 15 in. in 3 min after the influent valve was closed?

[2]Mathematics 11, Conversions.

First, determine the inches per minute drop in water level:

$$\frac{15 \text{ in.}}{3 \text{ min}} = 5 \text{ in./min}$$

Next, convert to gallons per minute per square foot.

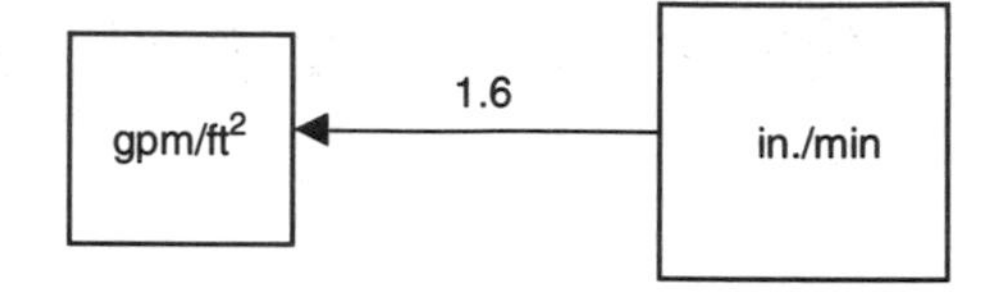

Converting from inches per minute to gallons per day per square foot, you move from a larger box to a smaller box and divide by 1.6:

$$\frac{5 \text{ in./min}}{1.6} = 3.13 \text{ gpm/ft}^2$$

Example 4

A rapid sand filter is 15 ft wide and 40 ft long. If the flow through the filter is 1.6 mgd, what is the filter loading rate in gallons per minute per square foot?

First, convert the gallons per day to gallons per minute:

$$\frac{1{,}600{,}000 \text{ gpd}}{1{,}440 \text{ min/d}} = 1{,}111 \text{ gpm}$$

Then express the filter loading rate mathematically as

$$\text{filter loading rate} = \frac{\text{flow (gpm)}}{\text{filter area (ft}^2\text{)}}$$

$$= \frac{1{,}111 \text{ gpm}}{(15 \text{ ft})(40 \text{ ft})}$$

$$= 1.85 \text{ gpm/ft}^2$$

MATHEMATICS 20

Filter Backwash Rate

The filter backwash rate is measured in gallons of water flowing upward (backward) each minute through a square foot of filter surface area. The units of measure for backwash rate are the same as those for filter loading rate; that is, gallons per *minute* per square foot of filter area.

Figure M20-1 depicts filter backwash.

The mathematical equation associated with the filter backwash rate is

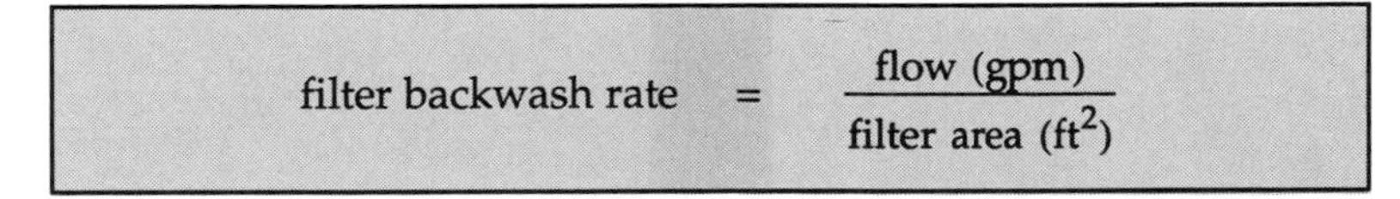

$$\text{filter backwash rate} = \frac{\text{flow (gpm)}}{\text{filter area (ft}^2\text{)}}$$

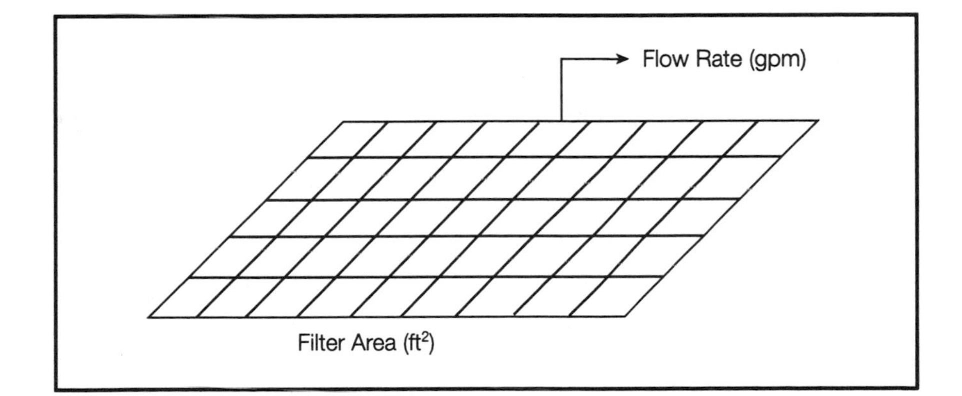

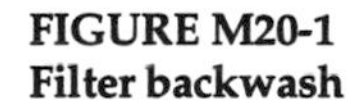
FIGURE M20-1
Filter backwash

Sometimes the backwash rate is measured as the rise of water in *inches* that occurs each minute. This can be expressed mathematically as

$$\text{filter backwash rate} = \frac{\text{inches of water rise}}{\text{minutes}}$$

Typically, backwash rates will vary from 15 gpm/ft^2 to 22.5 gpm/ft^2. This is equivalent to a rise in the water level from 24 to 36 in./min.

Example 1

A mixed-media filter is 25 ft wide and 32 ft long. If the filter receives a backwash flow of 17,300,000 gpd, what is the filter backwash flow rate in gallons per minute per square foot?

First convert the backwash flow from gallons per day to gallons per minute:

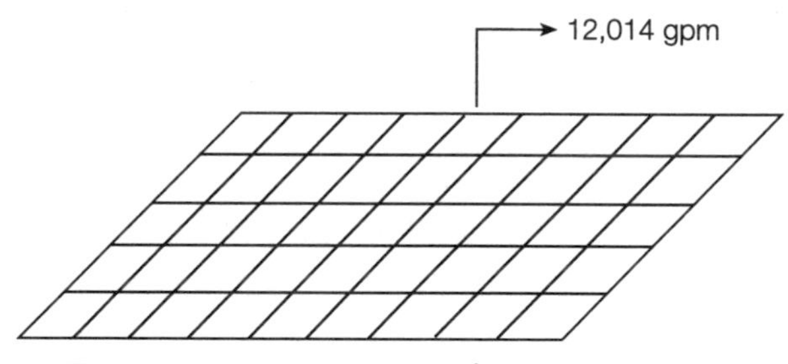

Filter Area = (25 ft)(32 ft) = 800 ft^2

$$\frac{17{,}300{,}000 \text{ gpd}}{1{,}440 \text{ min/d}} = 12{,}014 \text{ gpm}$$

With this information, calculate filter backwash rate:

$$\text{filter backwash rate} = \frac{\text{flow (gpm)}}{\text{filter area (ft}^2\text{)}}$$

$$= \frac{12{,}014 \text{ gpm}}{(25 \text{ ft})\ (32 \text{ ft})}$$

$$= \frac{12{,}014 \text{ gpm}}{800 \text{ ft}^2}$$

$$= 15.02 \text{ gpm/ft}^2$$

Filter backwash rates, as noted earlier, are sometimes expressed in terms of vertical rise of water in a time interval measured in minutes, for example, inches per minute (in./min). The gallons per minute per square foot and inches per minute units of measure are directly related to each other as shown in Example 2.

Example 2

How many inches rise per minute corresponds to a filter backwash rate of 18 gpm/ft^2?

Use the box method to convert from gallons per minute per square feet to inches per minute. Moving from a smaller to a larger box indicates multiplication:[1]

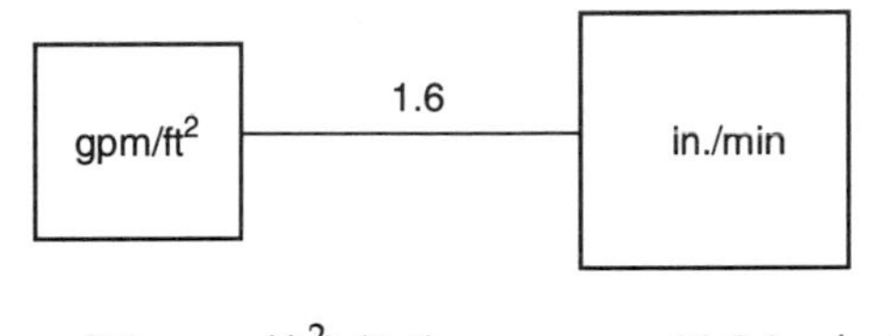

$$(18 \text{ gpm/ft}^2)(1.6) \quad = \quad 28.8 \text{ in./min}$$

Example 3

A rapid sand filter is 3 m wide and 5 m long. If backwash water is flowing upward at a rate of 18 ML/d, what is the backwash rate in liters per square meter per second?

First convert flow from liters per day to liters per second:[2]

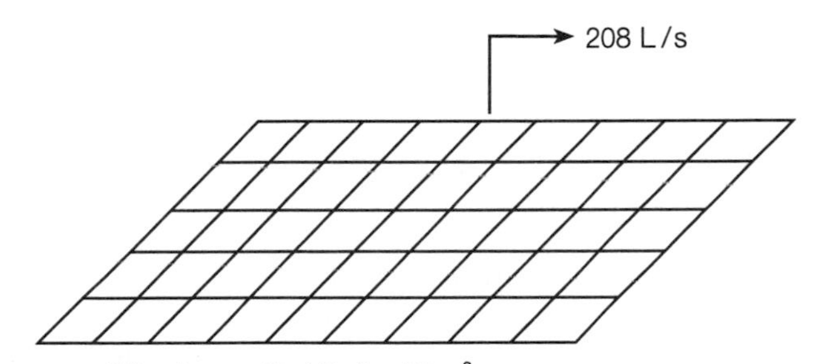

[1]Mathematics 11, Conversions.

[2]Mathematics 11, Conversions.

$$\frac{18{,}000{,}000 \text{ L/d}}{86{,}400 \text{ s/d}} = 208 \text{ L/s}$$

The filter backwash rate can be written mathematically as

$$\text{filter backwash rate} = \frac{\text{flow (L/s)}}{\text{filter area (m}^2\text{)}}$$

$$= \frac{208 \text{ L/s}}{15 \text{ m}^2}$$

$$= 13.9 \text{ (L/m}^2\text{)/s}$$

MATHEMATICS 21

Mudball Calculation

Mudballs, a conglomeration of floc particles and sand, are a common problem in sand filters. The presence of mudballs is the basic cause of many filter malfunctions. Therefore, measuring and controlling the amount of accumulated mudballs are of prime importance in sand filter operation.

To measure the amount of mudballs in a filter, a tube about 3 in. in diameter and 6 in. long is pushed down into the sand of the filter and then lifted, full of sand (Figure M21-1). A sieve is used to separate the sand from any mudballs present.

The volume of mudballs can then be determined by placing the balls in a graduated cylinder containing a known amount of water. As mudballs are put in, the water level rises. The volume of mudballs, therefore, is the *increase*

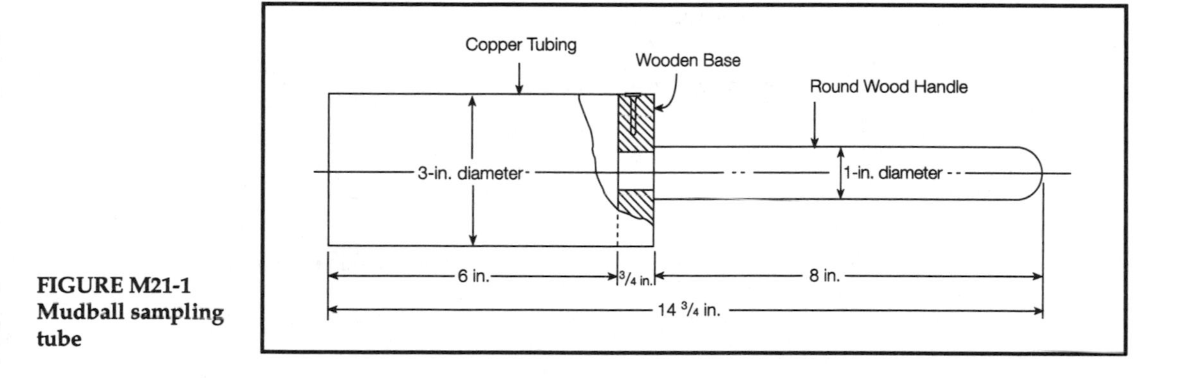

FIGURE M21-1 Mudball sampling tube

in volume in the graduated cylinder. The percent volume of mudballs can then be determined.[1] Mathematically, this is

$$\text{percent volume of mudballs} = \frac{\text{volume of mudballs}}{\text{total volume of sample collected}} \times 100$$

The following example problems illustrate the calculation of percent mudballs.

Example 1

A sample of sand taken from the filter for mudball determination has a volume of 3,500 mL. If the volume of mudballs in the sample is found to be 110 mL, what is the percent volume of mudballs in the sample?

$$\text{percent volume of mudballs} = \frac{\text{volume of mudballs}}{\text{total volume of sample}} \times 100$$

$$= \frac{110 \text{ mL}}{3{,}500 \text{ mL}} \times 100$$

$$= 0.0314 \times 100$$

$$= 3.14\% \text{ mudballs}$$

Example 2

A sample of sand taken from the filter for mudball determination has a volume of 5,600 mL. If the volume of mudballs in the sample is found to be 30 mL, what is the percent volume of mudballs in the sample?

$$\text{percent volume of mudballs} = \frac{\text{volume of mudballs}}{\text{total volume of sample}} \times 100$$

$$= \frac{30 \text{ mL}}{5{,}600 \text{ mL}} \times 100$$

$$= 0.0054 \times 100$$

$$= 0.54\% \text{ mudballs}$$

---

[1]Mathematics 7, Percent.

MATHEMATICS 22

Detention Time

The concept of detention time is used in conjunction with many treatment plant processes, including flash mixing, coagulation–flocculation, and sedimentation. Detention time refers to the length of time a drop of water or a suspended particle remains in a tank or chamber. For example, if water entering a 50-ft long tank has a flow velocity of 1 ft/s, an average of 50 s would elapse from the time a drop entered the tank until it left the tank.

Detention time may also be thought of as the number of minutes or hours required for each tank to empty. You may find it helpful to form a mental image of the flow from the time water enters the tank until it leaves the tank completely ("plug flow"), as shown in Figure M22-1.

Typical ranges of detention times for various treatment processes are

Tank	Detention Time Range
Flash-mixing basin	30–60 s
Flocculation basin	20–60 min
Sedimentation basin	1–12 h

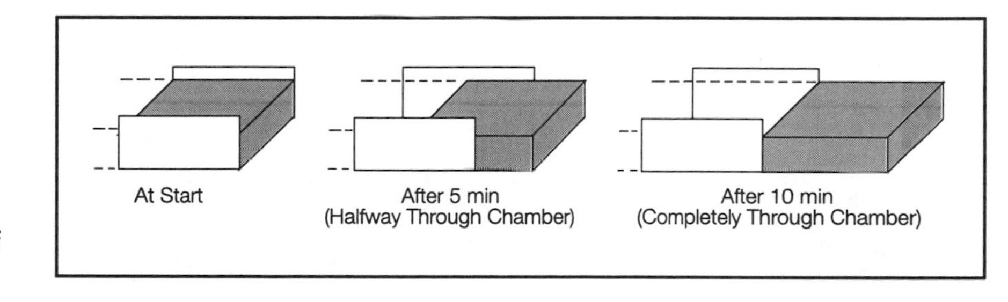

FIGURE M22-1 Simplified schematic for a detention time of 10 min

The equation used to calculate detention time is

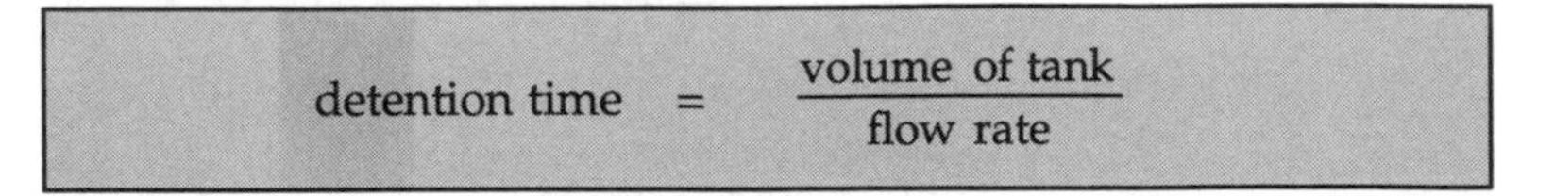

$$\text{detention time} = \frac{\text{volume of tank}}{\text{flow rate}}$$

If the flow rate used in the equation is gallons *per day,* then the detention time calculated will be expressed in *days*. If the flow rate used in the equation is gallons *per minute,* the detention time calculated will be expressed in *minutes*. The calculation method is illustrated in the following examples.

Example 1

A sedimentation tank has a capacity of 60,000 gal. If the hourly flow to the clarifier is 20,800 gph, what is the detention time?

Since the flow rate is expressed in hours, the detention time calculated is also in hours:

$$\text{detention time} = \frac{\text{volume of tank}}{\text{flow rate}}$$

$$= \frac{60{,}000 \text{ gal}}{20{,}800 \text{ gph}}$$

$$= 2.88 \text{ h}$$

Example 2

A flocculation basin is 20 ft long, 10 ft wide, and has a side water depth (SWD) of 10 ft. If the flow to the basin is 520 gpm, what is the detention time?

Since the flow rate is expressed in minutes, the detention time calculated is also in minutes:

$$\text{detention time} = \frac{\text{volume of tank}}{\text{flow rate}}$$

$$= \frac{(20 \text{ ft})\ (10 \text{ ft})\ (10 \text{ ft})\ (7.48 \text{ gal/ft}^3)}{520 \text{ gpm}}$$

$$= \frac{14{,}960 \text{ gal}}{520 \text{ gpm}}$$

$$= 28.77 \text{ min}$$

Example 3

A flash mixing basin has a capacity of 6,800 L. If the flow to the mixing basin is 140 L/s, what is the detention time?

Since the flow rate is expressed in seconds, the detention time is expressed in seconds:

$$\text{detention time} = \frac{\text{volume of tank}}{\text{flow rate}}$$

$$= \frac{6{,}800 \text{ L}}{140 \text{ L/s}}$$

$$= 48.57 \text{ s}$$

Example 4

The flow rate through a circular clarifier is 4,752,000 gpd. If the clarifier is 65 ft in diameter and 12 ft deep, what is the clarifier detention time in hours?

To calculate a detention time in hours, first express the gallons-per-day flow rate as gallons per hour.

$$\frac{4{,}752{,}000 \text{ gpd}}{24 \text{ h/d}} = 198{,}000 \text{ gph}$$

Then calculate the detention time (being certain to express tank volume in gallons):

$$\text{detention time} = \frac{\text{volume of tank}}{\text{flow rate}}$$

$$= \frac{(0.785)\ (65 \text{ ft})\ (65 \text{ ft})\ (12 \text{ ft})\ (7.48 \text{ gal/ft}^3)}{198{,}000 \text{ gph}}$$

$$= \frac{297{,}700 \text{ gal}}{198{,}000 \text{ gph}}$$

$$= 1.5 \text{ h}$$

It should be noted that the calculations for detention time illustrated in this chapter are only theoretical. For many situations, these calculations are sufficient for use in normal water system operations. The flow of water through a pipeline, for instance, is for the most part plug flow, that is to say, with essentially every drop of water entering a section of pipe and leaving the other end in the same amount of time.

Tanks and basins, on the other hand, will not actually be plug flow. Some of the reasons why some particles of water are detained, while others speed from the inlet to the outlet (called short-circuiting) are poor distribution of water at the inlet, dead spaces in the corners of the basin, stratification due to temperature differentials, inadequate baffling, and surface effects from wind.

The contact time (*CT*) of disinfectants and water as they pass through treatment before the water is delivered to the first customer must now be considered under the federal Surface Water Treatment Rule (SWTR). Contact time is the average time that a disinfectant is in contact with the water as it passes through a basin. For a well-designed basin, *CT* may be relatively close to the theoretical detention time. For a poorly designed or unbaffled basin, it may be only a fraction of the theoretical time.

Additional information on methods of determining *CT* through basins may be found in AWWA publications describing details of the SWTR.

MATHEMATICS 23

Well Problems

The three basic calculations related to water well performance are well yield, drawdown, and specific capacity. These calculations provide information for selecting appropriate pumping equipment and for identifying any changes in the productive capacity of the well.

Well Yield

The well yield is the volume of water that is discharged from a well during a specified time period. This discharge may be a result of pumping, as in most cases, or of free flow in the base of a flowing artesian well. Well yield is normally measured as the pumping rate in gallons per minute (gpm). This can be expressed mathematically as

$$\text{well yield} = \frac{\text{gallons}}{\text{minutes}}$$

The easiest method of measuring well yield is to place a flowmeter on the downstream, or discharge, side of the pump. However, to determine relatively small well yields, another method often used is measuring the time required to fill a container of known volume. The following examples illustrate the second method of calculating well yield. A similar calculation is discussed elsewhere in this handbook.[1]

---

[1]Hydraulics 6, Pumping Problems.

Example 1

During a test for well yield, the time required to fill a 55-gal barrel was 25 s. Based on this pumping rate, what was the well yield in gallons per minute?

$$\text{well yield} = \frac{\text{gallons}}{\text{minutes}}$$

The well yield is to be reported in gallons per *minute*, so first express the 25 s given in the problem as minutes:

$$\frac{25 \text{ s}}{60 \text{ s/min}} = 0.42 \text{ min}$$

Now fill in the information given and complete the calculation:

$$\text{well yield} = \frac{55 \text{ gal}}{0.42 \text{ min}}$$

$$= 130.95 \text{ gpm}$$

Example 2

If it takes a well pump 41 s to fill a 210-L barrel, what is the well yield in liters per second?

The equation used in calculating this well yield is

$$\text{well yield} = \frac{\text{liters}}{\text{seconds}}$$

You can solve the well yield problem by filling in the given information and completing the division indicated:

$$\text{well yield} = \frac{210 \text{ L}}{41 \text{ s}}$$

$$= 5.12 \text{ L/s}$$

Example 3

What is the well yield in gallons per minute if it takes the pump 3.5 min to fill a 5-ft^2 tank to a depth of 3 ft?

The equation used in calculating well yield is

$$\text{well yield} = \frac{\text{gallons}}{\text{minutes}}$$

Before information can be filled in the equation, the gallon volume of the tank must be calculated:[2]

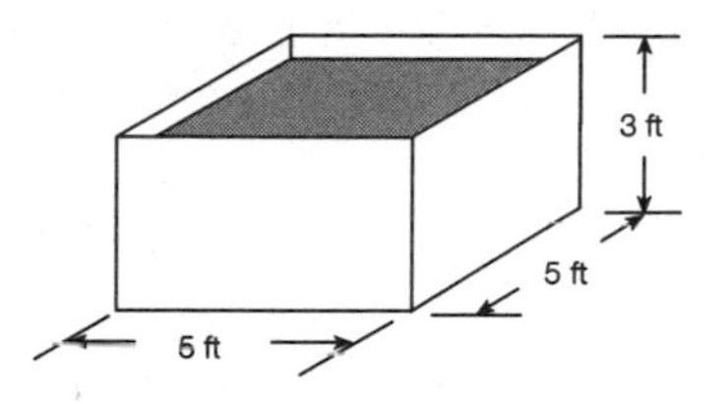

$$\text{volume} = (\text{area})\ (\text{third dimension})$$

$$= (5\ \text{ft})\ (5\ \text{ft})\ (3\ \text{ft})$$

$$= 75\ \text{ft}^3$$

Convert volume from cubic feet to gallons:[3]

$$(75\ \text{ft}^3)\ (7.48\ \text{gal/ft}^3) = 561\ \text{gal}$$

Now you can solve this well yield problem by filling in the given information and completing the calculation:

$$\text{well yield} = \frac{561\ \text{gal}}{3.5\ \text{min}}$$

$$= 160.29\ \text{gpm}$$

Drawdown

The *drawdown* of a well is the amount the water level *drops* once pumping begins. As illustrated in Figure M23-1, drawdown is the difference between the *static water level* (SWL, the level when no water is being taken from the aquifer, either by pumping or by free flow) and the *pumping water level* (PWL, the level when the pump is in operation).

---

[2]Mathematics 10, Volume Measurements.

[3]Mathematics 11, Conversions.

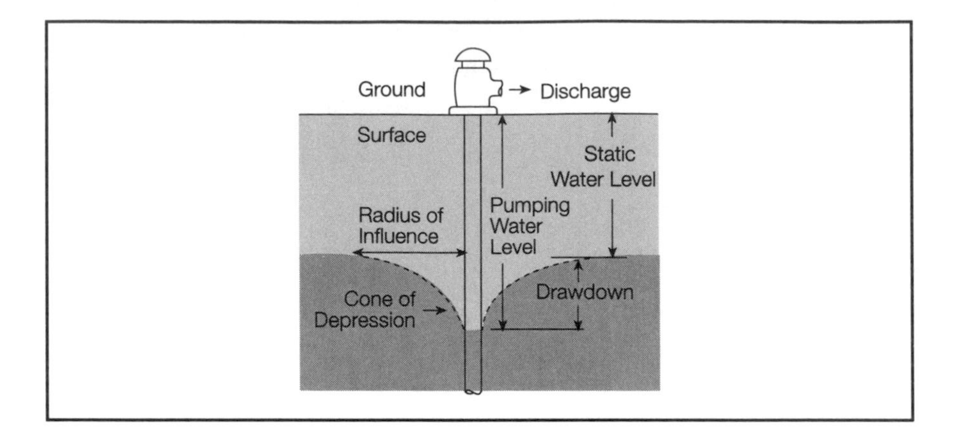

FIGURE M23-1 Diagram depicting drawdown of a well

This is expressed mathematically as

$$\text{drawdown} = \text{pumping water level} - \text{static water level}$$

The following two examples illustrate how to calculate well drawdown.

Example 4

The water level in a well is 25 ft below the ground surface when the pump is not in operation. If the water level is 48 ft below the ground surface when the pump is in operation, what is the drawdown in feet?

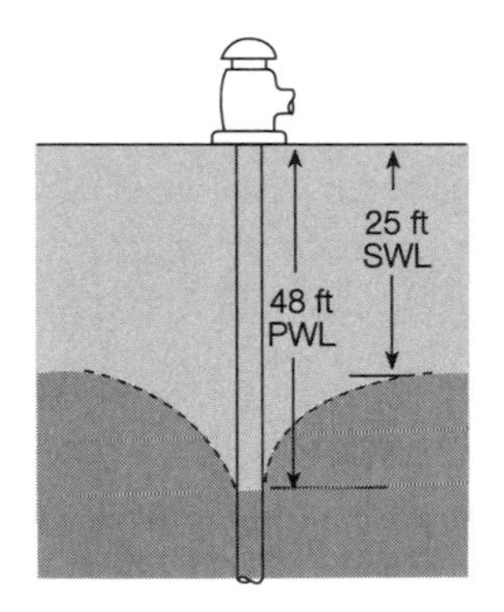

Drawdown is the measure of water level *drop* once the pump has been turned on. In this problem, the drawdown is

$$\text{drawdown} = \text{pumping water level} - \text{static water level}$$

$$= 48 \text{ ft} - 25 \text{ ft}$$

$$= 23 \text{ ft}$$

Example 5

When the pump is not in operation, the water level in a well is 39 ft below ground surface. The water level drops to 57 ft when the pump is in operation. What is the drawdown in feet?

$$\text{drawdown} = \text{pumping water level} - \text{static water level}$$

$$= 57 \text{ ft} - 39 \text{ ft}$$

$$= 18 \text{ ft}$$

Specific Capacity

The specific capacity of a well is a measure of the well yield per unit of drawdown. It is usually expressed in terms of gallons-per-minute well yield per foot of drawdown. Therefore, the equation used to calculate specific capacity is

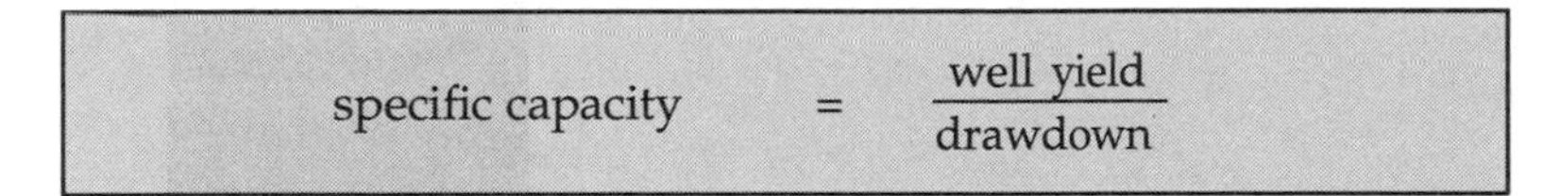

$$\text{specific capacity} = \frac{\text{well yield}}{\text{drawdown}}$$

The following examples illustrate the calculation of specific capacity.

Example 6

It takes a well pump 0.6 min to fill a 55-gal barrel. If the drawdown while the pump is in operation is 11 ft, what is the specific capacity of the well?

To calculate the specific capacity of the well, you must know the gallons-per-minute well yield and the feet of drawdown.

$$\text{well yield} = \frac{55 \text{ gal}}{0.6 \text{ min}}$$

$$= 91.67 \text{ gpm}$$

Using the gallons-per-minute well yield and drawdown information, calculate the specific capacity of the well:

$$\text{specific capacity} = \frac{\text{well yield in gpm}}{\text{drawdown in ft}}$$

$$= \frac{91.67 \text{ gpm}}{11 \text{ ft}}$$

$$= 8.33 \text{ gpm/ft}$$

Example 7

If the well yield is 10.7 L/s when the drawdown for the well is 5.5 m, what is the specific capacity of the well?

The equation used in calculating specific capacity is

$$\text{specific capacity} = \frac{\text{well yield in L/s}}{\text{drawdown in m}}$$

To solve the problem, fill in the information given in the equation and complete the calculation:

$$\text{specific capacity} = \frac{10.7 \text{ L/s}}{5.5 \text{ m}}$$

$$= 1.95 \text{ (L/s)/m}$$

Basic Science Concepts and Applications

Hydraulics

HYDRAULICS 1

Density and Specific Gravity

When we say that one substance is heavier than another, we mean that any given volume of the substance is heavier than the same volume of the other substance. For example, any given volume of steel is heavier than the same volume of aluminum, so we say that steel is heavier than — or has greater density than — aluminum.

Density

For scientific and technical purposes, the density of a body or material is precisely defined as *the weight per unit of volume*. In the water supply field, perhaps the most common measures of density are pounds per cubic foot (lb/ft^3) and pounds per gallon (lb/gal). The density of a dry material, such as sand, activated carbon, lime, and soda ash, is usually expressed in pounds per cubic foot. The density of a liquid, such as water, liquid alum, or liquid chlorine, can be expressed either as pounds per cubic foot or as pounds per gallon. The density of a gas, such as air, chlorine gas, methane, or carbon dioxide, is normally expressed in pounds per cubic foot.

The density of a substance changes slightly as the temperature of the substance changes. This happens because substances usually increase in size (volume) as they become warmer, as illustrated in Figure H1-1. Because of expansion with warming, the same weight is spread over a larger volume, so the density is lower when a substance is warm than when it is cold.

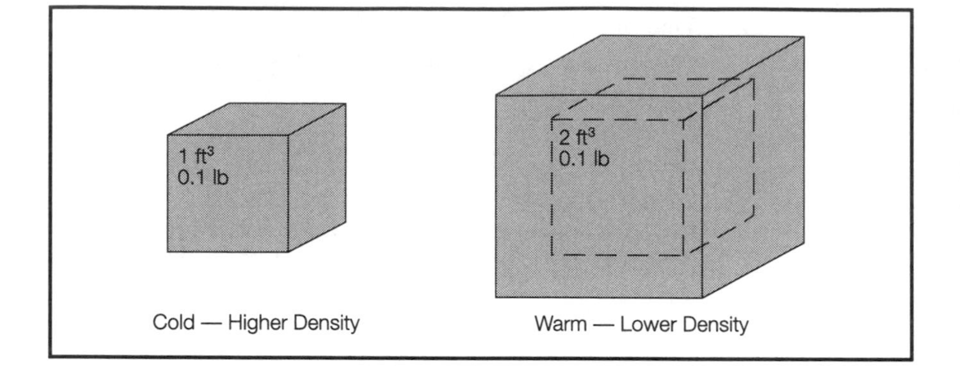

FIGURE H1-1 Changes in density based on temperature

The effects of pressure and temperature on solids and liquids are very small and are usually ignored. However, temperature and pressure have a significant effect on the density of gases. Whenever the density of a gas is given, the temperature and pressure at that density are usually also given.

Table H1-1 indicates how the density of water (usually specified as a constant 62.4 lb/ft^3) varies only slightly with temperature, especially within the water temperature ranges in effect at most water supply operations.

TABLE H1-1 Density of water

Temperature, °*F*	°*C*	Density, *lb/ft*^{3}
32.0	0	62.416
35.0	1.7	62.421
39.2	4.0	62.424
40.0	4.4	62.423
50.0	10.0	62.408
60.0	15.6	62.366
70.0	21.1	62.300
80.0	26.7	62.217
90.0	32.2	62.118
100.0	37.8	61.998
120.0	48.9	61.719
140.0	60.0	61.386
160.0	71.1	61.006
180.0	82.2	60.586
200.0	93.3	60.135
212.0	100.0	59.843

Water is unusual in that it is most dense at 39.2°F (4.0°C), and becomes less dense when the temperature rises or falls.

Table H1-2 shows some densities of typical solid, liquid, and gaseous substances. You'll notice that some of the solids given in the table have density reported as *bulk density*. Bulk density is defined as the weight of a cubic foot of material as it would be shipped from the supplier to the treatment plant. Bulk density is much less than laboratory density because its calculation includes the volume of the air mixed in with the material; the amount of air (and therefore the bulk density) varies according to whether the material comes in rock, crystal, pellet, granular, or powder form. For example, for laboratory purposes the density of pure sodium chloride (table salt) is about 135 lb/ft^3. However, the bulk density of sodium chloride as it is shipped in rock form (rock salt) is only 50–60 lb/ft^3. This means that over half the volume of a bulk container of rock salt is occupied by air between the individual pieces.

Specific Gravity

Since density can be expressed as pounds per cubic foot, pounds per gallon, pounds per cubic inch, or even grams per cubic centimeter, it is sometimes difficult to compare the density of one substance with that of another. Specific gravity is one way around this problem. Although there may be many numbers that express the density of the same substance (depending on the units used), there is only one specific gravity associated with each substance (for one particular temperature and pressure). The specific gravity of a substance is the density of that substance relative to a "standard" density.

Specific Gravity of Solids and Liquids

The standard density used for solids and liquids is that of water, 62.4 lb/ft^3, or 8.34 lb/gal. Therefore, the specific gravity of a solid or liquid is the density of that solid or liquid *relative to the density of water*. It is the ratio[1] of the density of that substance to the density of water. In a calculation of specific gravity, it is *essential* that the densities be expressed in the *same units*. Otherwise the calculation will be wrong.

---

[1]Mathematics 5, Ratios and Proportions.

TABLE H1-2 Densities of various substances

Substance	Density, lb/ft^3	Density, lb/gal
Solids		
Activated carbon*†	8–28 (Avg. 12)	
Lime*†	20–50	
Dry alum*†	60–75	
Aluminum (at 20°C)	168.5	
Steel (at 20°C)	486.7	
Copper (at 20°C)	555.4	
Liquids		
Propane (–44.5°C)	36.5	4.88
Gasoline†	43.7	5.84
Water (4°C)	62.4	8.34
Fluorosilicic acid (30%, –8.1°C)	77.8–79.2	10.4–10.6
Liquid alum (36°Bé, 15.6°C)	83.0	11.09
Liquid chlorine (–33.6°C)	97.3	13.01
Sulfuric acid (18°C)	114.2	15.3
Gases		
Methane (0°C, 14.7 psia)	0.0344	
Air (20°C, 14.7 psia)	0.075	
Oxygen (0°C, 14.7 psia)	0.089	
Hydrogen sulfide†	0.089	
Carbon dioxide†	0.115	
Chlorine gas (0°C, 14.7 psia)	0.187	

*Bulk density of substance.

†Temperature and/or pressure not given.

For example, the density of granite rock is about 162 lb/ft^3, and the density of water is 62.4 lb/ft^3. The specific gravity of granite is found by this ratio:

$$\text{specific gravity} = \frac{\text{density of granite}}{\text{density of water}} = \frac{162 \text{ lb/ft}^3}{62.4 \text{ lb/ft}^3}$$

$$= 2.60$$

In this case, the specific gravity (the ratio of the density of granite to the density of water) indicates that a cubic foot of granite weighs about 2½ times as much as a cubic foot of water.

Let's look at another example. The density of SAE 30 motor oil is about 56 lb/ft^3. Its specific gravity is therefore

$$\text{specific gravity} = \frac{56\ lb/ft^3}{62.4\ lb/ft^3}$$

$$= 0.90$$

In other words, specific gravity in this example tells you that oil is only 9⁄10 as dense as water. Because a cubic foot of oil weighs less than a cubic foot of water, oil floats on the surface of water.

Table H1-3 lists specific gravities for various liquids and solids.

Example 1

Aluminum weighs approximately 168 lb/ft^3. What is the specific gravity of aluminum?

To calculate specific gravity, compare the weight of a cubic foot of aluminum with the weight of a cubic foot of water:

$$\text{specific gravity of aluminum} = \frac{\text{density of aluminum}}{\text{density of water}}$$

$$= \frac{168\ lb/ft^3}{62.4\ lb/ft^3}$$

$$= 2.69$$

Example 2

If the specific gravity of a certain oil is 0.92, what is the density (in pounds per cubic foot) of that oil?

Approach this problem in the same general way as the previous example. First fill in the given information:

$$\text{specific gravity of oil} = \frac{\text{density of oil}}{\text{density of water}}$$

$$0.92 = \frac{x\ lb/ft^3}{62.4\ lb/ft^3}$$

TABLE H1-3 Specific gravities of various solids and liquids

Substance	Specific Gravity
Solids	
Aluminum (20°C)	2.7
Steel (20°C)	7.8
Copper (20°C)	8.9
Activated carbon*†	0.13–0.45 (avg. 0.19)
Lime*†	0.32–0.80
Dry alum*†	0.96–1.2
Soda ash*†	0.48–1.04
Coagulant aids (polyelectrolytes)*†	0.43–0.56
Table salt*†	0.77–1.12
Liquids	
Liquid alum (36°Bé, 15.6°C)	1.33
Water (4°C)	1.00
Fluorosilicic acid (30%, –8.1°C)	1.25–1.27
Sulfuric acid (18°C)	1.83
Ferric chloride (30%, 30°C)	1.34

*Bulk density used to determine specific gravity.
†Temperature and/or pressure not given.

Then solve for the unknown value:[2]

$$(62.4)(0.92) = x$$

$$57.41 \text{ lb/ft}^3 = x$$

Therefore the density of the oil is 57.41 lb/ft^3.

The most common use of specific gravity in water treatment operations is in gallons-to-pounds conversions. In many such cases, the liquids being handled have a specific gravity of 1.00 or very nearly 1.00 (between 0.98 and 1.02), so 1.00 may be used in the calculations without introducing more than a 2 percent error. However, in calculations involving a liquid with a specific gravity less than 0.98 or greater than 1.02 (such as liquid alum), the conversions from gallons to pounds must take specific gravity into account. The technique is illustrated in the following two examples.

---

[2]Mathematics 4, Solving for the Unknown Value.

Example 3

Suppose you wish to pump a certain liquid at the rate of 25 gpm. How many pounds per day will you be pumping if the liquid weighs 74.9 lb/ft^3?

If the liquid being pumped were water, then you would make the following calculations:[3]

$$(25 \text{ gpm})(8.34 \text{ lb/gal})(1{,}440 \text{ min/d}) = 300{,}240 \text{ lb/d}$$

However, the liquid being pumped has a greater density than water; therefore, you will have to adjust the factor of 8.34 lb/gal. First, determine the specific gravity of the liquid:

$$\text{specific gravity of the liquid} = \frac{\text{density of the liquid}}{\text{density of water}}$$

$$= \frac{74.9 \text{ lb/ft}^3}{62.4 \text{ lb/ft}^3}$$

$$= 1.20$$

Next, calculate the corrected factor:

$$(8.34 \text{ lb/gal})(1.20) = 10.01 \text{ lb/gal}$$

Then convert the gallon per minute flow rate to pounds per day:

$$(25 \text{ gpm})(10.01 \text{ lb/gal})(1{,}440 \text{ min/d}) = 360{,}360 \text{ lb/d pumped}$$

Example 4

There are 1,240 gal of a certain liquid in a tank. If the specific gravity of the liquid is 0.93, how many pounds of liquid are in the tank?

Normally, for a conversion from gallons to pounds, the factor 8.34 lb/gal (the density of water) would be used if the substance's specific gravity were between 0.98 and 1.02.[4] However, in this example the substance has a specific gravity outside this range, so

---

[3]Mathematics 11, Conversions.

[4]Mathematics 11, Conversions.

the 8.34 factor must be adjusted. Multiply 8.34 lb/gal by the specific gravity to obtain the adjusted factor:

$$(8.34 \text{ lb/gal})(0.93) = 7.76 \text{ lb/gal}$$

Then convert 1,240 gal to pounds using the corrected factor:

$$(1{,}240 \text{ gal})(7.76 \text{ lb/gal}) = 9{,}622 \text{ lb}$$

Specific Gravity of Gases

The specific gravity of a gas is usually determined by comparing the density of the gas with the density of air, which is 0.075 lb/ft^3 at a temperature of 20°C and a pressure of 14.7 psia (pounds per square inch absolute) — the pressure of the atmosphere at sea level.[5] For example, the density of chlorine gas is 0.187 lb/ft^3. Its specific gravity would be calculated as follows:

$$\text{specific gravity of } Cl_2 \text{ gas} = \frac{\text{density of } Cl_2 \text{ gas}}{\text{density of air}}$$

$$= \frac{0.187 \text{ lb/ft}^3}{0.075 \text{ lb/ft}^3}$$

$$= 2.49$$

[5] Hydraulics 2, Pressure and Force.

This tells you that chlorine gas is about 2½ times as dense as air. Therefore, when chlorine gas is introduced into a room, it will concentrate at the bottom of the room. This is important to know because chlorine is a deadly poisonous gas. Table H1-4 lists specific gravities for various gases.

TABLE H1-4 Specific gravities of various gases

Gas	Specific Gravity	
Hydrogen (0°C; 14.7 psia)	0.07	*When released in a room,*
Methane (0°C; 14.7 psia)	0.46	*these gases will first*
Carbon monoxide*	0.97	*rise to the ceiling area.*
Air (20°C; 14.7 psia)	1.00	
Nitrogen (0°C; 14.7 psia)	1.04	
Oxygen (0°C; 14.7 psia)	1.19	*When released in a room,*
Hydrogen sulfide*	1.19	*these gases will first settle*
Carbon dioxide*	1.53	*to the floor area.*
Chlorine gas (0°C; 14.7 psia)	2.49	
Gasoline vapor*	3.0	

*Temperature and pressure not given.

HYDRAULICS 2

Pressure and Force

The flow of water in a system is dependent on the amount of *force* causing the water to move. *Pressure* is defined as the amount of force acting (pushing) on a unit area.

Pressure

As shown in Figure H2-1, pressure may be expressed in different ways depending on the unit area selected. Normally, however, the unit area of a square inch and the expression of pressures in pounds per square inch (psi) are preferred. In metric units, pressure is generally expressed in kilopascals (kPa).

In the operation of a water treatment system, you will be primarily concerned with the pressures exerted by water. Water pressures are directly related to the height (depth) of water. Suppose, for example, you have a container 1 ft by 1 ft by 1 ft (a cubic-foot container) that is filled with water. What is the pressure on the square-foot bottom of the container?

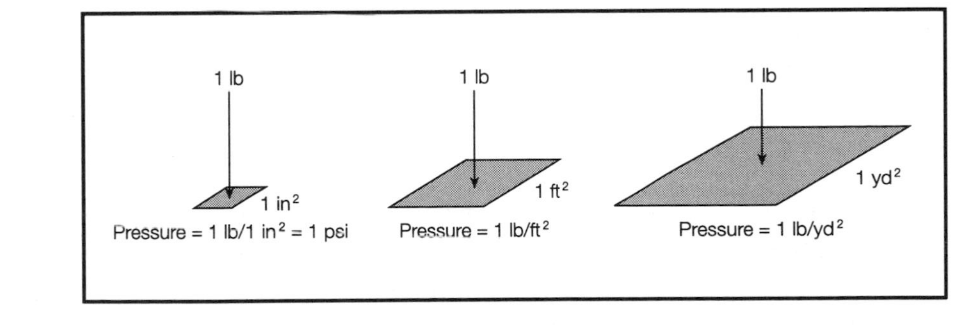

FIGURE H2-1 Forces acting on areas

Pressure in this case is expressed in pounds per unit area. In this case, since the density of water is 62.4 lb/ft^2, the force of the water pushing down on the square foot surface area is 62.4 lb (Figure H2-2).

From this information, the pressure in pounds per square inch (psi) can also be determined. Convert pounds per square foot to pounds per square inch (psi):

$$\frac{62.4 \text{ lb}}{\text{ft}^2} = \frac{62.4 \text{ lb}}{(1 \text{ ft})(1 \text{ ft})}$$

$$= \frac{62.4 \text{ lb}}{(12 \text{ in.})(12 \text{ in.})}$$

$$= \frac{62.4 \text{ lb}}{144 \text{ in.}^2}$$

$$= 0.433 \frac{\text{lb}}{\text{in.}^2}$$

$$= 0.433 \text{ psi}$$

This means that a foot-high column of water over a square-inch surface area weighs 0.433 lb, resulting in a pressure of 0.433 psi (Figure H2-3). The factor 0.433 allows you to convert from pressure measured in feet of water to pressure measured in pounds per square inch, as shown later in this section. A conversion factor can also be developed for converting from pounds per square inch to feet.

Since 1 ft is equivalent to 0.433 psi, set up a ratio[1] to determine how many feet of water are equivalent to 1 psi (that is, how many feet high a water column must be to create a pressure of 1 psi):

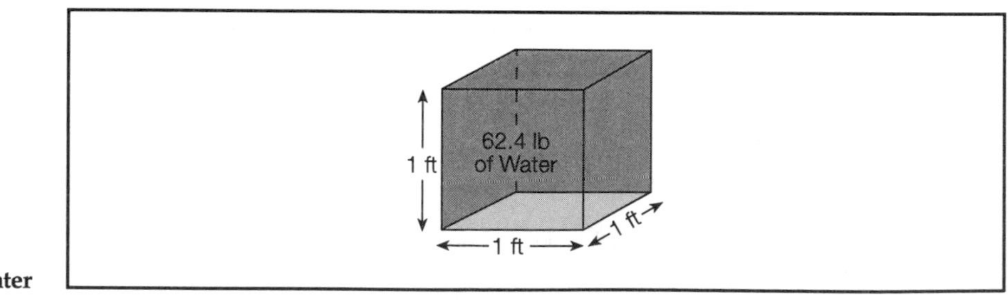

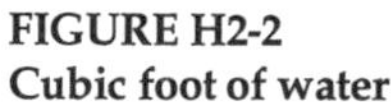
FIGURE H2-2
Cubic foot of water

[1]Mathematics 5, Ratios and Proportions.

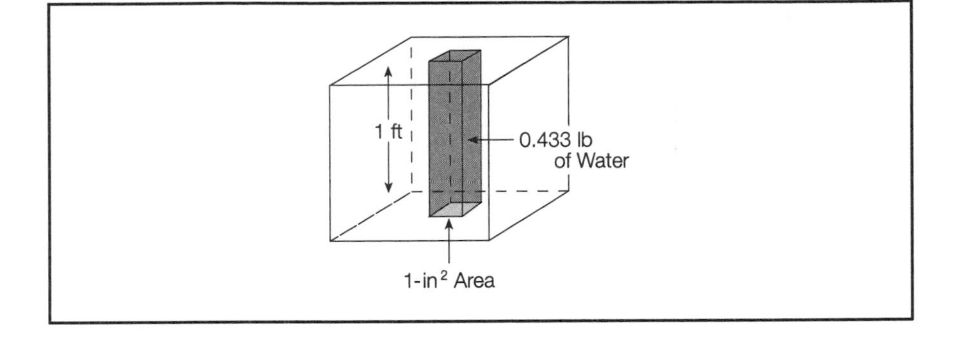

FIGURE H2-3
One-foot column of water over a 1-in.² area

$$\frac{1 \text{ ft}}{0.433 \text{ psi}} = \frac{x \text{ ft}}{1 \text{ psi}}$$

Then solve for the unknown value:[2]

$$\frac{(1)(1)}{0.433} = x$$

$$2.31 \text{ ft} = x$$

Therefore, 1 psi is equivalent to the pressure created by a column of water 2.31 ft high (Figure H2-4).

Since the density of water is assumed to be a constant 62.4 lb/ft^3 in most hydraulic calculations, the height of the water is the most important factor in determining pressures. It is the height of the water that determines the pressure over the square-inch area. Pressure measured in terms of the height of water (in meters or feet) is referred to as *head*. As long as the height of the water stays the same, changing the shape of the container does not change the pressure at the bottom or any other level in the container. For example, see Figure H2-5. Each of the containers is filled with water to the same height. Therefore, the pressures against the bottoms of the containers are the same. And the pressure at any depth in one container is the same as the pressure at the same depth in either of the other containers.

In water system operation, the shape of the container can help to maintain a usable volume of water at higher pressures. For example, suppose you have an elevated storage tank and a standpipe that contain equal

---

[2]Mathematics 4, Solving for the Unknown Value.

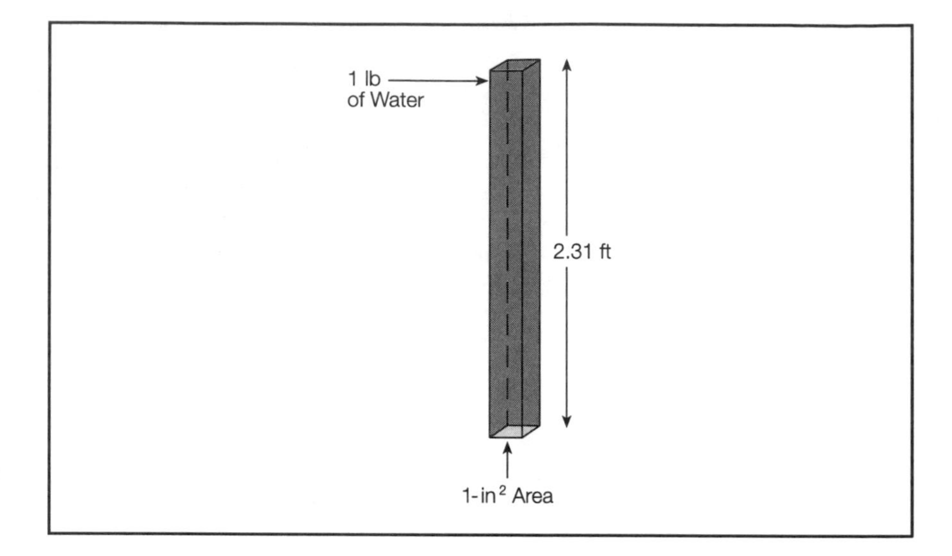

FIGURE H2-4
A 2.31-ft water column (which creates a pressure of 1 psi at the bottom surface)

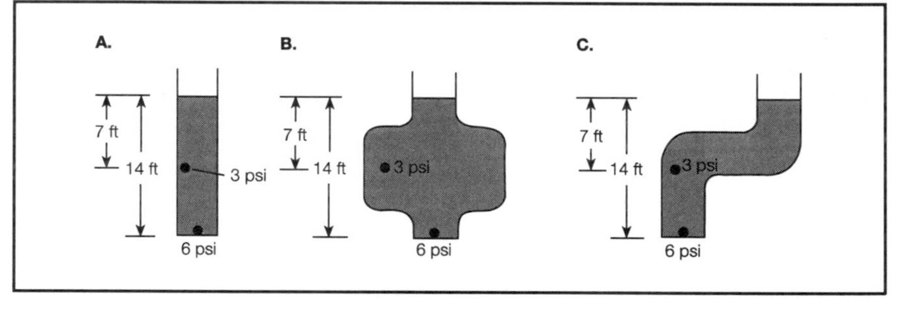

FIGURE H2-5
Pressure depends on weight of water, not shape of container

amounts of water. When the water levels are the same, the pressures at the bottom of the tanks are the same (Figure H2-6A).

However, if half of the water is withdrawn from each tank, the pressure at the bottom of the elevated tank will be greater than the pressure at the bottom of the standpipe (Figure H2-6B).

Because of the direct relationship between the pressure in *pounds per square inch* at any point in water and the *height in feet* of water above that point, pressure can be measured either in pounds per square inch or in feet (of water), called "head."

When the water pressure in a main or in a container is measured by a gauge, the pressure is referred to as *gauge pressure*. If measured in pounds per square inch, the gauge pressure is expressed as *pounds per square inch gauge*

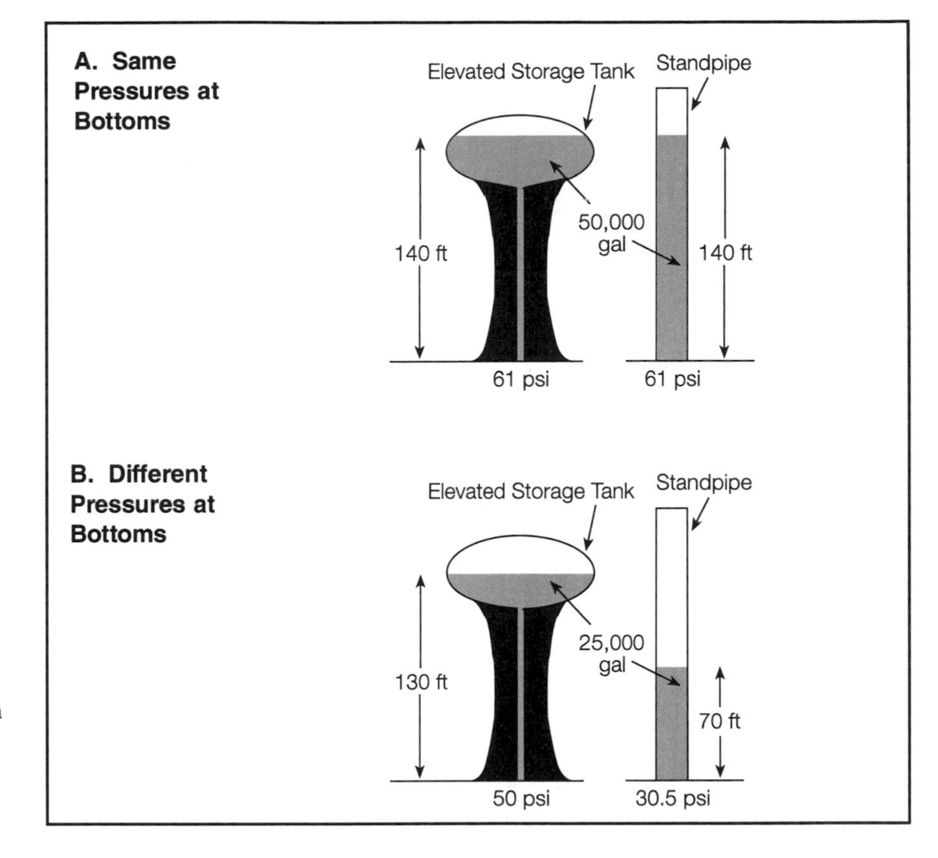

FIGURE H2-6 Comparison of maximum pressure values in containers with different shapes but same volume of water

(psig). The gauge pressure is not the total pressure within the main. Gauge pressure does not show the pressure of the atmosphere, which is equal to approximately 14.7 psi at sea level. Because atmospheric pressure is exerted everywhere (against the outside of the main as well as the inside, for instance), it can generally be neglected in water system calculations. However, for certain calculations, the total (or absolute) pressure must be known. The *absolute pressure* (expressed as pounds per square inch absolute [psia]) is obtained by adding the gauge and atmospheric pressures. For example, in a main under 50-psi gauge pressure, the absolute pressure would be 50 psig + 14.7 psi = 64.7 psia. A line under a partial vacuum, with a gauge pressure of –2, would have an absolute pressure of (14.7 psi) + (–2 psig) = 12.7 psia (Figure H2-7).

Pressure gauges can also be calibrated in feet of head. A pressure gauge reading of 14 ft of head, for example, means that the pressure is equivalent to

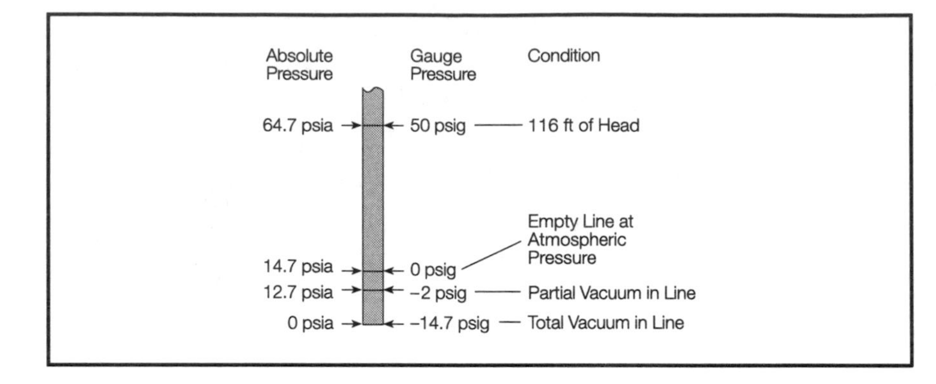

FIGURE H2-7
Gauge versus absolute pressure

the pressure exerted by a column of water 14 ft high. The equations that relate gauge pressure in pounds per square inch to pressure in feet of head are given below. In this handbook, conversions from one term to another will use the first equation only.

1 psig	=	2.31 ft head
1 ft head	=	0.433 psig

The following example problems illustrate conversions from feet of head to pounds per square inch gauge and from pounds per square inch gauge to feet of head. The method used is similar to the conversion approach (box method) discussed in the Mathematics section.[3]

Example 1

Convert a gauge pressure of 14 ft to pounds per square inch gauge.

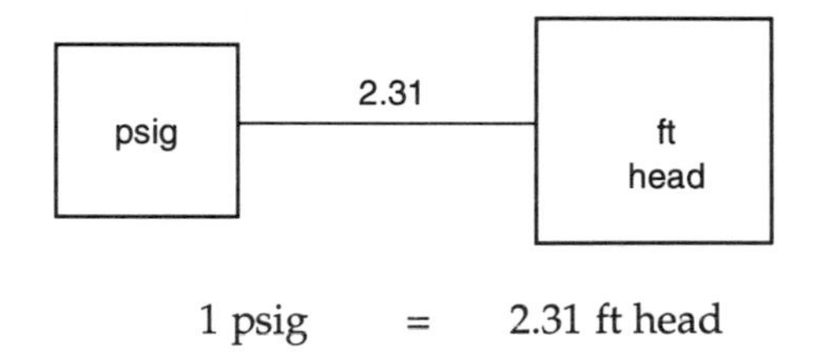

1 psig = 2.31 ft head

[3]Mathematics 11, Conversions.

Using the preceding diagram, in moving from feet of head to pounds per square inch gauge, you are moving from a larger box to a smaller box. Therefore, you should divide by 2.31:

$$\frac{14 \text{ ft}}{2.31 \text{ ft/psig}} = 6.06 \text{ psig}$$

Example 2

A head of 250 ft of water is equivalent to what pressure in pounds per square inch gauge?

$$1 \text{ psig} = 2.31 \text{ ft head}$$

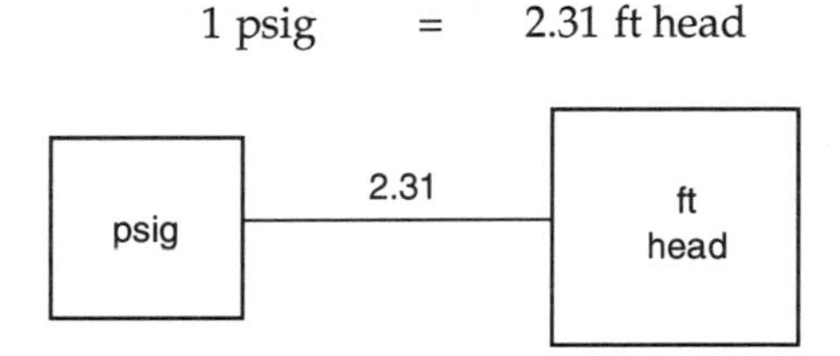

When you move from a larger box to a smaller box, division by 2.31 is indicated:

$$\frac{250 \text{ ft}}{2.31 \text{ ft/psig}} = 108.23 \text{ psig}$$

Example 3

A pressure of 210 kPa (gauge) is equivalent to how many meters of head?

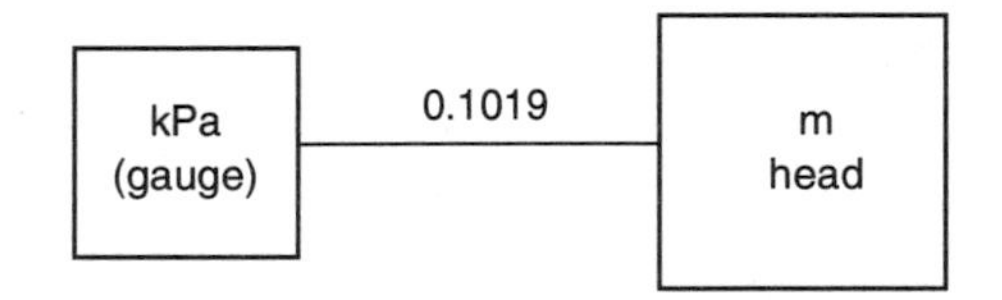

When you move from a smaller box to a larger box, multiplication by 0.1019 is indicated:

$$[210 \text{ kPa(gauge)}] (0.1019 \text{ m/kPa}) = 21.4 \text{ m}$$

Example 4

What would the pounds-per-square-inch gauge pressure readings be at points A and B in the diagram in Figure H2-8?

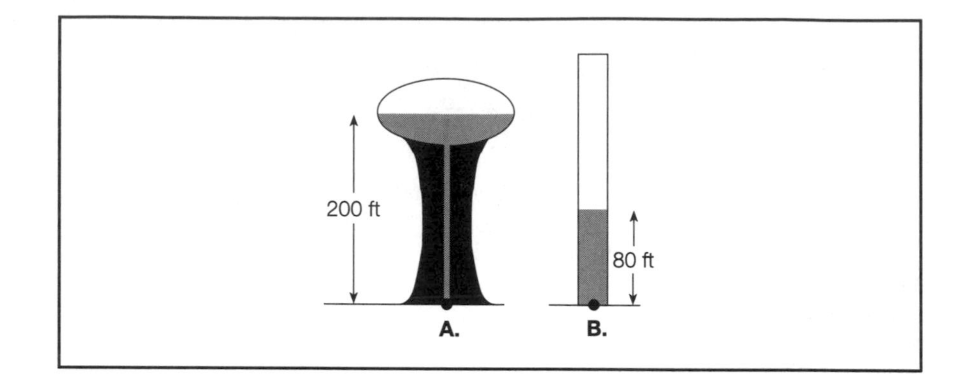

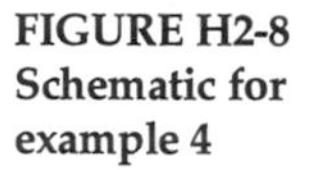
FIGURE H2-8 Schematic for example 4

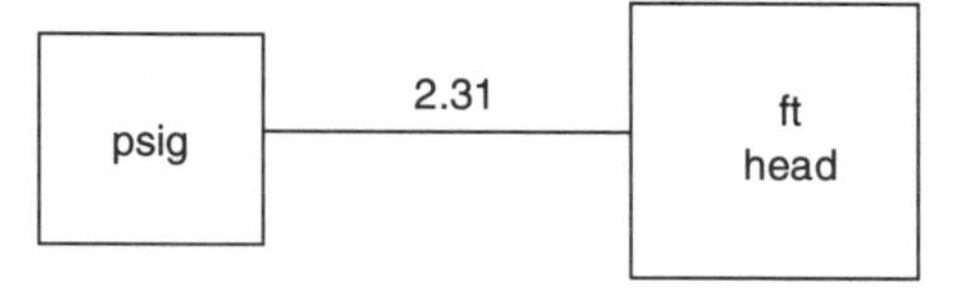

In each case, the conversion is from feet of head to pressure in pounds per square inch gauge. Therefore, when you move from a larger box to a smaller box, division by 2.31 is indicated:

$$\text{pressure at A} = \frac{200 \text{ ft}}{2.31 \text{ ft/psig}}$$

$$= 86.58 \text{ psig}$$

$$\text{pressure at B} = \frac{80 \text{ ft}}{2.31 \text{ ft/psig}}$$

$$= 34.63 \text{ psig}$$

Example 5

Gauges are being used in the water system in Figure H2-9 to measure dynamic pressure at various points. The readings are in units of pounds per square inch gauge. What is the equivalent pressure at each point in feet of head?

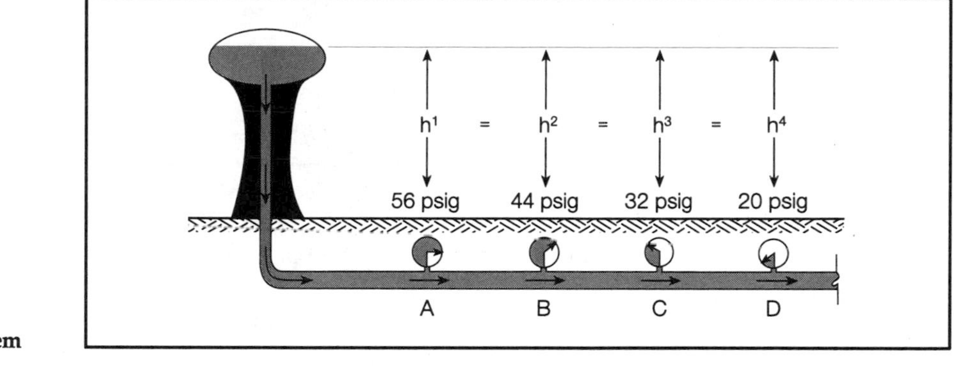

FIGURE H2-9 Schematic for example 5—dynamic system

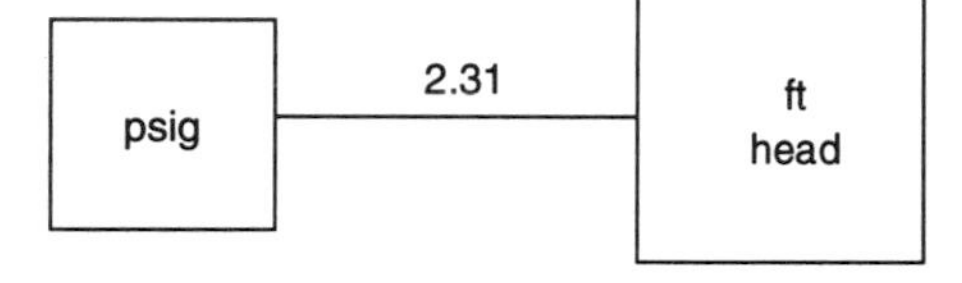

In each case, the conversion is from pounds per square inch gauge to feet of head. Because you are moving from the smaller box to the larger box, multiplication by 2.31 is indicated:

$$\text{pressure at A} = (56 \text{ psig})(2.31 \text{ ft/psig})$$
$$= 129.36 \text{ ft of head}$$

$$\text{pressure at B} = (44 \text{ psig})(2.31 \text{ ft/psig})$$
$$= 101.64 \text{ ft of head}$$

$$\text{pressure at C} = (32 \text{ psig})(2.31 \text{ ft/psig})$$
$$= 73.92 \text{ ft of head}$$

$$\text{pressure at D} = (20 \text{ psig})(2.31 \text{ ft/psig})$$
$$= 46.2 \text{ ft of head}$$

It should be noted for the previous example that when there is no water flowing, the system will be static. Therefore, all of the static pressure measurements at points A, B, C, and D would be the same for static conditions.

Force

The pressure exerted on a surface corresponds to the force applied to that surface.

The general equation that is used in calculating force due to pressure is

$$\text{force} = \text{pressure} \times \text{area}$$

For example, if a pressure of 5 psig is exerted on a surface 2 in. by 3 in. then a force of 5 lb is pressing down on each square inch of surface area. The total force exerted on the surface would be 30 lb (Figure H2-10). In this example, the force would have been calculated as[4]

$$\begin{aligned} \text{force} &= (5 \text{ psig})(2 \text{ in.})(3 \text{ in.}) \\ &= (5 \text{ psig})(6 \text{ in.}^2) \\ &= 30 \text{ lb force} \end{aligned}$$

It is a basic principle of hydraulics that, if two containers of static (nonmoving) fluid are connected with a pipeline, then a force applied to the fluid will exert the same pressure everywhere within the system. This principle and the force/pressure equation just given can be used to explain the operation of a hydraulic jack.

The jack in Figure H2-11 has an operating piston with a surface area of 5 in.^{2} and a lifting piston with a surface area of a 100 in.^{2} A force of 150 lb is applied to the operating piston. What pressure is created within the hydraulic system of the jack, and what is the resulting force exerted by the lifting piston?

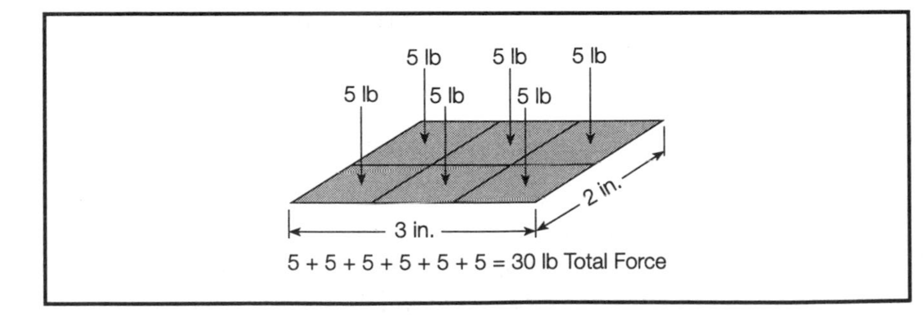

FIGURE H2-10 Force of 5 psig applied to a 6-in.^{2} area

[4]Mathematics 9, Area Measurement.

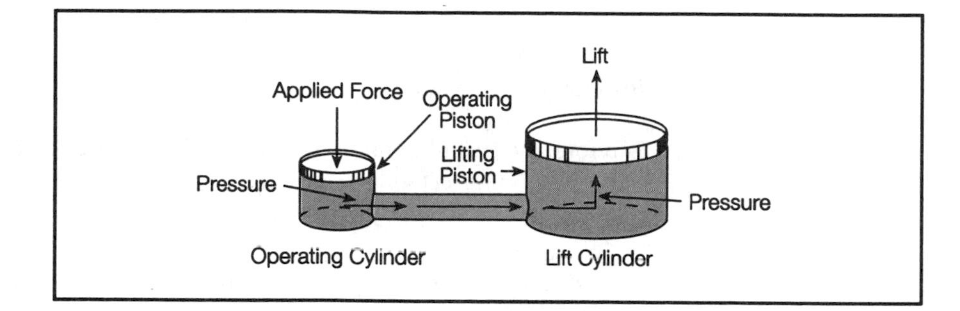

FIGURE H2-11
Hydraulic jack

The pressure created within the jack by the operating piston is calculated as follows, based on the force/pressure equation:

$$\text{force} = \text{pressure} \times \text{area}$$

$$\text{force on operating piston} = \left(\text{pressure within jack}\right)\left(\text{area of operating piston}\right)$$

Fill in the known information:

$$150 \text{ lb} = (x \text{ psig})(5 \text{ in.}^2)$$

Then solve for the unknown value:

$$150 = (x)(5)$$

$$\frac{150}{5} = x$$

$$30 \text{ psig} = x$$

Therefore the pressure exerted by the operating piston is 30 psig. And because the pressure is the same everywhere within the closed hydraulic system of the jack (at least, until the water starts moving), the pressure on the lift piston is the same value. Use the force–pressure equation again to determine the total lifting force exerted by the lift piston:

$$\text{force} = \text{pressure} \times \text{area}$$

$$\text{force on lifting cylinder} = \left(\text{pressure within jack}\right)\left(\text{area of lifting cylinder}\right)$$

$$= (30 \text{ psig}) (100 \text{ in.}^2)$$

$$\text{force on lifting cylinder} = 3{,}000 \text{ lb}$$

Therefore, the jack illustrated increases the force applied to the operating cylinder about 20 times (150-lb force compared to 3,000-lb force).

Example 6

The pressure on a particular surface is 12 psig. If the surface area is 120 in.^{2}, what is the force exerted on that surface?

$$\text{force} = \text{pressure} \times \text{area}$$

$$= (12 \text{ psig}) (120 \text{ in.}^2)$$

$$= 1{,}440 \text{ lb total force}$$

Example 7

If there is a pressure of 40 psig against a surface 2 ft by 1 ft, what is the force against the surface?

The area of the surface is

$$(2 \text{ ft}) (1 \text{ ft}) = 2 \text{ ft}^2$$

However, since the area must be expressed in square inches in order to use the equation, square feet must be converted to square inches:[5]

$$(2 \text{ ft}^2) (144 \text{ in.}^2/\text{ft}^2) = 288 \text{ in.}^2$$

Now calculate force:

$$\text{force} = \text{pressure} \times \text{area}$$

$$= (40.0 \text{ psig}) (288 \text{ in.}^2)$$

$$= 11{,}520 \text{ lb}$$

[5]Mathematics 11, Conversions.

HYDRAULICS 3

Piezometric Surface and Hydraulic Grade Line

Many important hydraulic measurements are based on the difference in height between the *free water surface* and some point in the water system. The free water surface is the surface of water that is in contact with the atmosphere. The *piezometric surface* can be used to locate the free water surface in a container, where it cannot be observed directly.

Piezometric Surface

If you connect an open-ended tube (similar to a straw) to the side of a tank or pipeline, the water will rise in the tube to indicate the level of the water in the tank. Such a tube, shown in Figure H3-1, is called a *piezometer*, and the level of the top of the water in the tube is called the piezometric surface. If the water-containing vessel is not under pressure (as is the case in Figure H3-1), the piezometric surface will be the same as the free water surface in the vessel, just as it would be if a soda straw (the piezometer) were left standing in a glass of water.

If the tank and pipeline are under pressure, as they often are, the pressure will cause the piezometric surface to rise above the level of the water in the tank. The greater the pressure, the higher the piezometric surface (Figure H3-2).

Notice in Figure H3-2A that the free water surface shown by the piezometer is the same level as the water surface in the tank; but once a pressure is applied, as in Figures H3-2B and H3-2C, the free water surface rises above the level of the tank water surface.

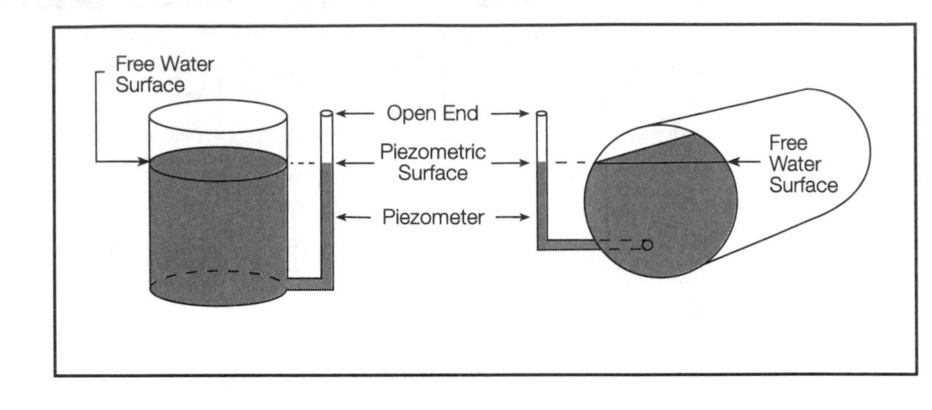

FIGURE H3-1
Piezometer and piezometric surface

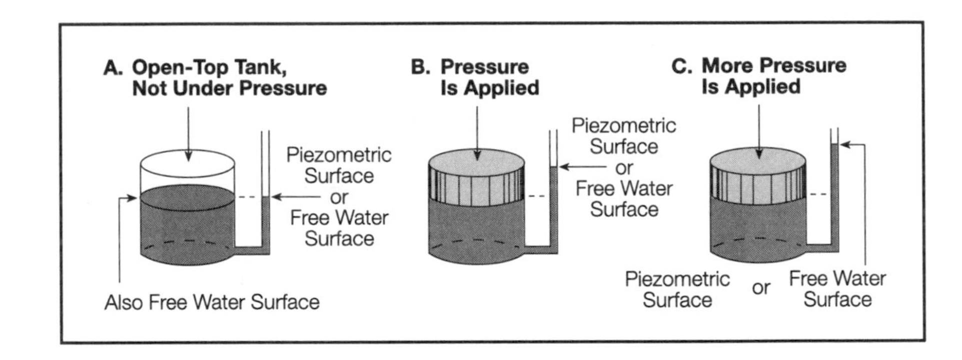

FIGURE H3-2
Piezometric surface varying with pressure

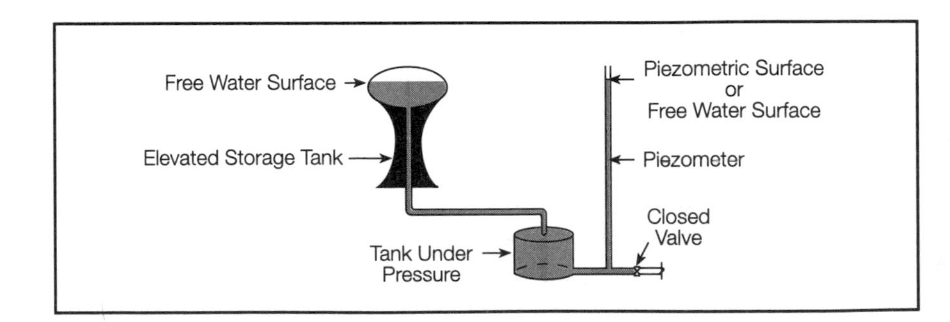

FIGURE H3-3
Piezometric surface caused by elevated tank

The applied pressure caused by a piston in Figures H3-2B and H3-2C can also be caused by water standing in a connected tank at a higher elevation. For example, Figure H3-3 shows that the pressure caused by water in an elevated storage tank is transmitted down the standpipe, through the pipeline, into a closed, low-level tank, and into the piezometer. The pressure causes the water to rise in the piezometer to the height of the water surface in the storage tank. Notice that the relationship between the tank under pressure and the piezometric surface shown in Figure H3-3 is very similar to the relationship between the pressurized tank and the piezometric surface in Figure H3-2C. In both cases, the piezometric surface is higher than the surface of the water in the pressurized tank to which the piezometer is connected. Note, however, that Figure H3-3 is also similar to Figure H3-2A, in that the piezometric surface in both figures is ultimately at the same level as the free water surface in the open-top tank.

Example 1

Locate the piezometric surface (free water surface) in Figures H3-4A and H3-5A.

The answers are shown in Figures H3-4B and H3-5B. In each case, the piezometric surface is the same as the water surface in the main body of water. This is true no matter where the piezometer is connected and no matter what slope or shape the piezometer takes.

So far only the piezometric surface for a body of standing water (static water) has been considered. The piezometers have shown that the water always rises to the water level of the main body of water, *but only when the water is standing still.*

Changes in the piezometric surface occur when water is flowing. Figure H3-6 shows an elevated storage tank feeding a distribution system pipeline. When the valve is closed (Figure H3-6A), all the piezometric surfaces are the same height as the free water surface in storage. When the valve opens and water begins to flow (Figure H3-6B), the piezometric surfaces *drop*. The farther along the pipeline, the lower the piezometric surface, because some of the pressure is used up keeping the water moving over the rough interior surface of the pipe. The pressure that is lost (called head loss) is no longer available to push water up in the piezometer. As water continues down the pipeline, less and less pressure is left.

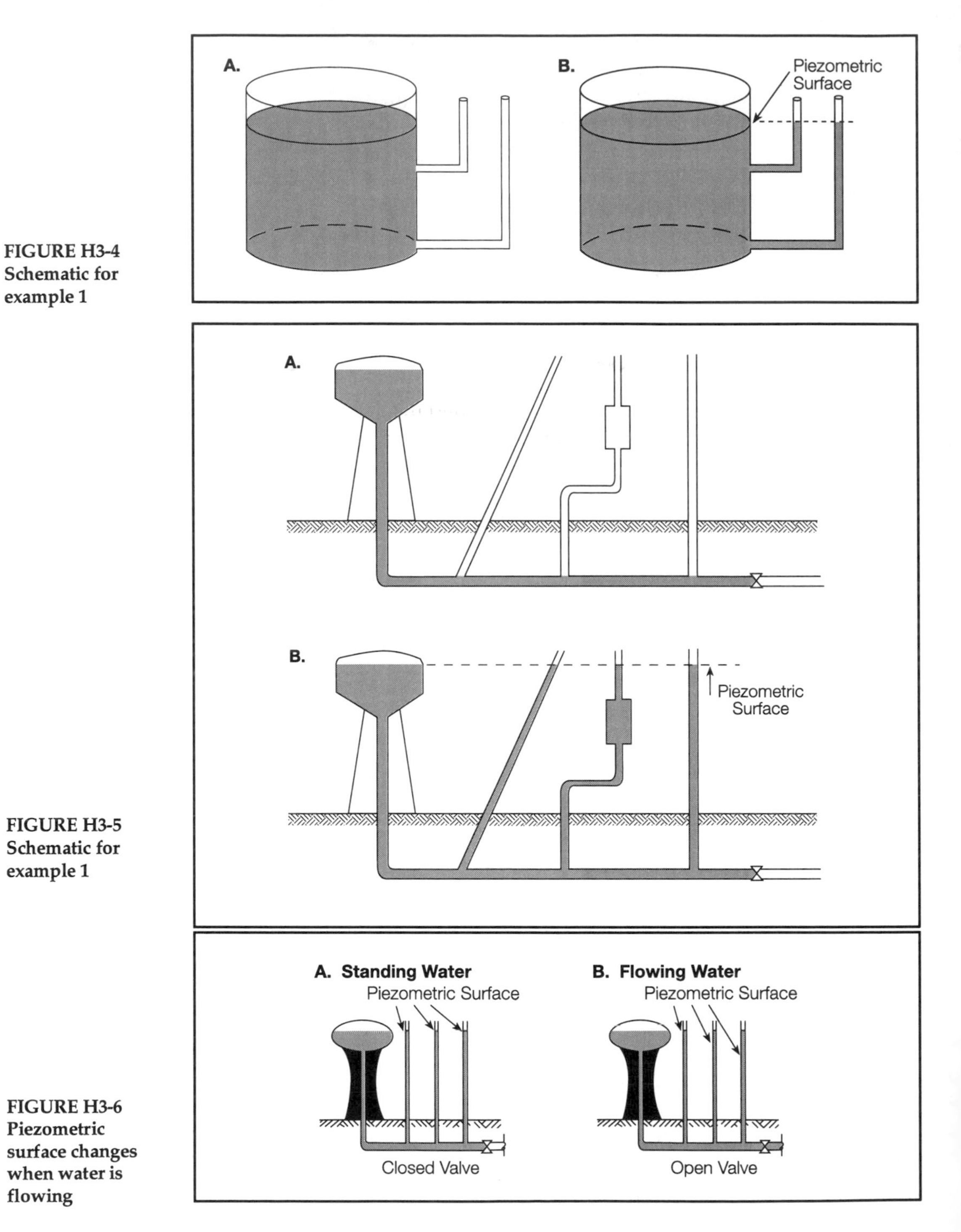

FIGURE H3-4 Schematic for example 1

FIGURE H3-5 Schematic for example 1

FIGURE H3-6 Piezometric surface changes when water is flowing

Hydraulic Grade Line

The hydraulic grade line (HGL) is the line that connects all the piezometric surfaces along the pipeline. It is important from an operating standpoint because it can be used to determine the pressure at any point in a water system.

To better understand HGL, you should know how the HGL is located and drawn. In this section, two techniques are discussed:

- locating HGLs from piezometric surface information
- locating HGLs from pressure gauge information

Locating HGLs From Piezometric Surface Information

First let's look at how to find the HGL of a *static water system* (a system in which the water is not moving). The pipeline in Figure H3-7 is fitted with four piezometers. With the valve closed, water rises in three of them, up to the free water surface. To find the HGL, draw a horizontal line from the free water surface through the piezometric surfaces, as shown.

This demonstrates two important facts about HGLs for *static* water systems:

- The static HGL is always horizontal.
- The static HGL is always at the same height as the free water surface.

Example 2

Locate and draw the HGL for the two static water conditions shown in Figures H3-8A and H3-8C.

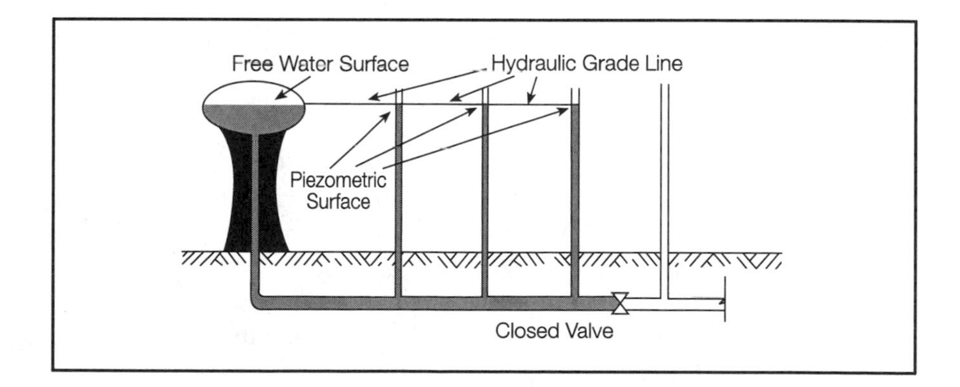

FIGURE H3-7 Hydraulic grade line of static water system

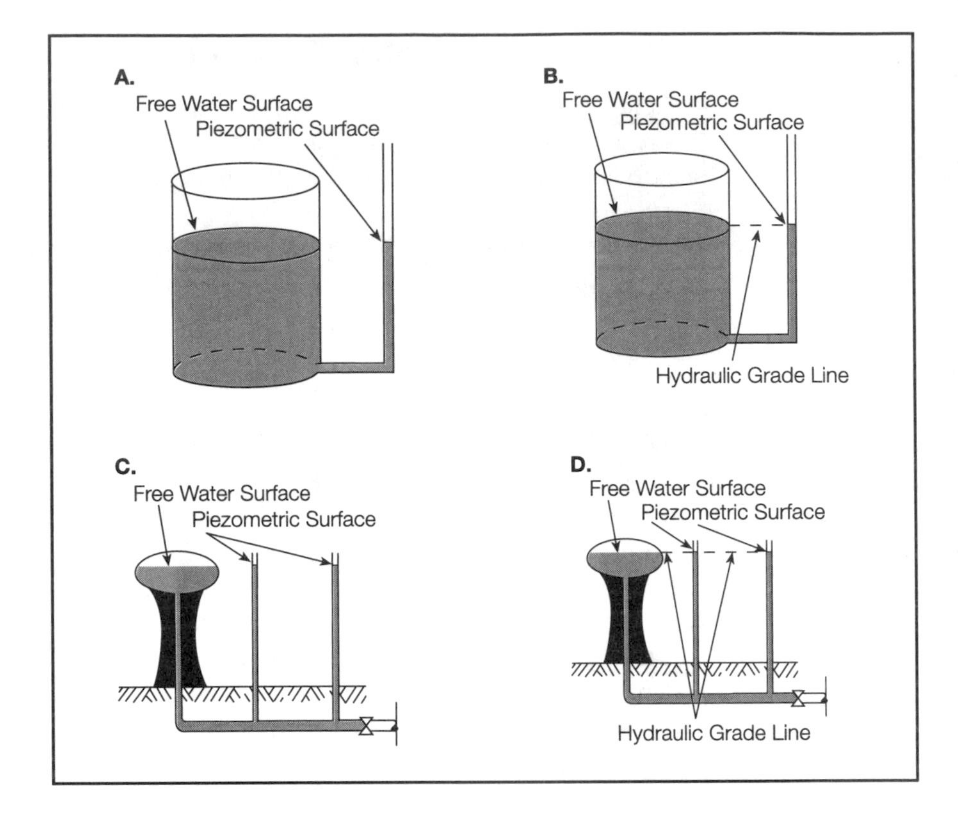

FIGURE H3-8
Schematics for example 2

First, the HGL will pass through the piezometric surfaces. Also, since the water is static, the following principles apply:

- The HGL will be horizontal.
- The HGL will be at the free water surface.

Using this information, you can draw the HGL by connecting the piezometric surfaces to the free water surface of the reservoir with a horizontal line. The resulting HGLs are shown in Figures H3-8B and H3-8D.

Now consider how to find the HGL of a *dynamic water system* (a system in which the water is in motion). Figure H3-9 shows water moving from an elevated storage tank into the distribution system. As before, you draw the HGL by connecting the free water surface and all the piezometric surfaces. The result can be one straight line, as shown in Figure H3-9, or it can be a series of connected straight lines at different angles, as shown in Figure H3-10.

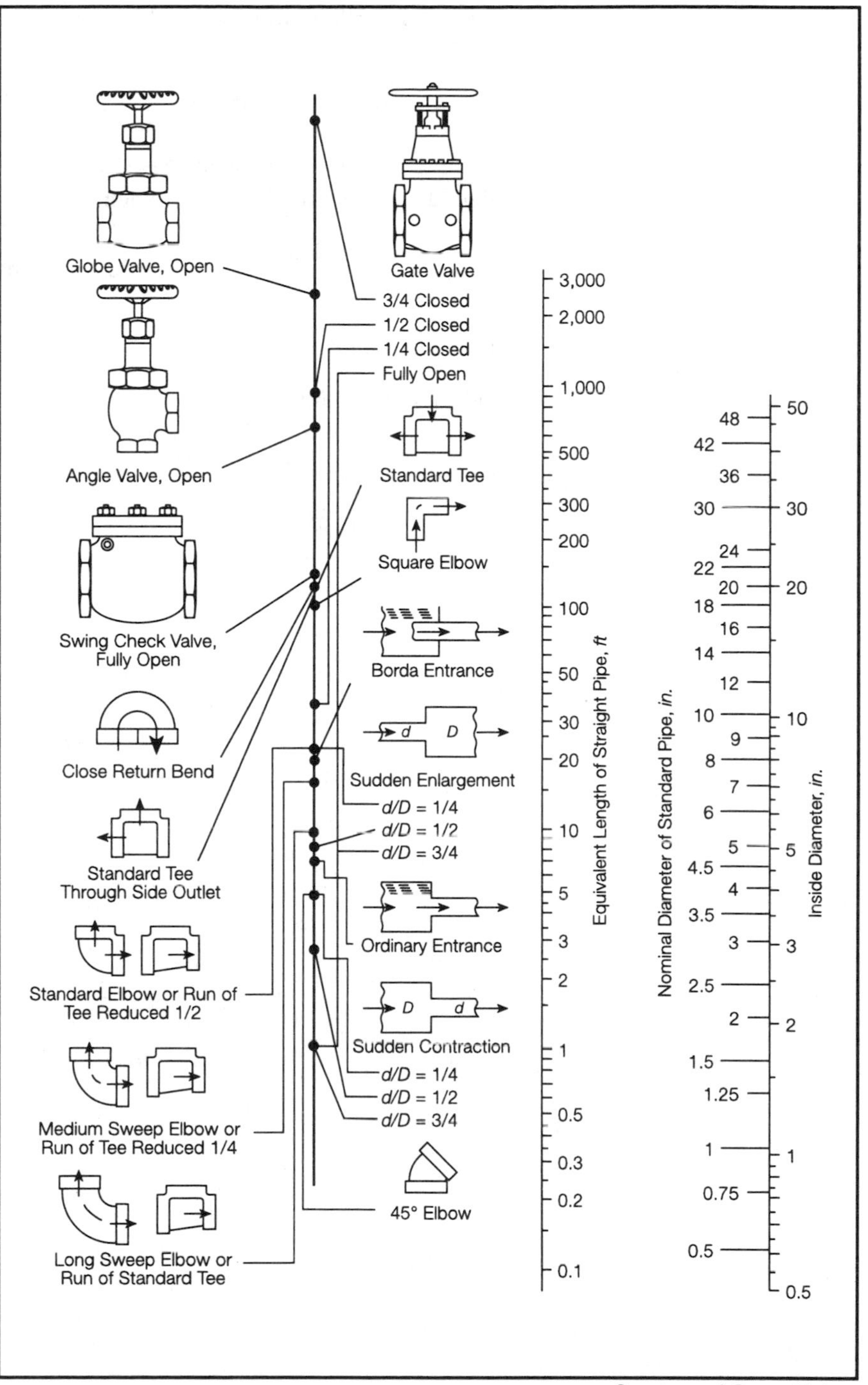

Courtesy of Crane Valves

FIGURE H5-2 Resistance of valves and fittings to flow of fluids

From Table H5-3, find the head loss per 100 ft for 6-in. pipe at a flow of 700 gpm.

$$\text{loss per 100 ft} = 6.21 \text{ ft head}$$

Now determine total head loss (friction + minor losses). First, calculate the number of 100-ft increments:

$$\frac{223 \text{ ft}}{100 \text{ ft}} = 2.23 \text{ increments}$$

Then calculate the total head loss:

$$\begin{aligned}\text{total head loss} &= (\text{head loss per 100-ft increment})\ (\text{no. of increments})\\ &= (6.21 \text{ ft})\ (2.23)\\ &= 13.85 \text{ ft}\end{aligned}$$

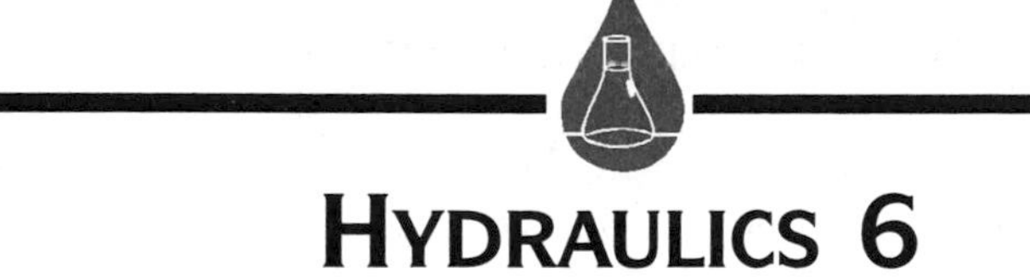

HYDRAULICS 6

Pumping Problems

The rate of flow produced by a pump is expressed as the volume of water pumped during a given period of time.

Pumping Rates

The mathematical equation used in pumping rate problems can usually be determined from the verbal statement of the problem:

VERBAL: What is the pumping rate in "gallons *per* minute"?

MATH: $$\text{pumping rate} = \frac{\text{gallons}}{\text{minutes}}$$

VERBAL: What is the pumping rate in "gallons *per* hour"?

MATH: $$\text{pumping rate} = \frac{\text{gallons}}{\text{hours}}$$

The number of gallons pumped during a period can be determined either by a flowmeter or by measuring the number of gallons pumped into or out of a tank.

Example 1

The totalizer of the meter on the discharge side of your pump reads in hundreds of gallons. If the totalizer shows a reading of 108 at 1:00 p.m. and 312 at 1:30 p.m., what is the pumping rate expressed in gallons per minute?

The problem asks for pumping rate in *gallons per minute* (gpm), so the mathematical setup is

$$\text{pumping rate} = \frac{\text{gallons}}{\text{minutes}}$$

To solve the problem, fill in the blanks (number of gallons and number of minutes) in the equation. The total gallons pumped is determined from the totalizer readings:

$$\begin{array}{rl} 31{,}200 & \text{gal} \\ -\ 10{,}800 & \text{gal} \\ \hline 20{,}400 & \text{gal} \end{array}$$

The volume was pumped between 1:00 p.m. and 1:30 p.m., for a total of 30 min. From this information calculate the gallons-per-minute pumping rate:

$$\text{pumping rate} = \frac{20{,}400 \text{ gal}}{30 \text{ min}}$$

$$= 680 \text{ gpm pumping rate}$$

Instead of using totalizer readings to calculate the average pumping rate for a period of a few minutes or hours (as in example 1), you can read the *instantaneous* pumping rate or flow rate — the flow rate or pumping at *one particular moment* — directly from many flowmeters.[1] Other flowmeters require that you perform calculations to determine the instantaneous flow rate, as illustrated in the following example.

Example 2

The venturi meter on the discharge side of your pump has a throat diameter of 4 in. The head differential between the high-pressure tap and the low-pressure tap is 1.5 ft. Use the nomograph shown in Figure H6-1 to determine the pumping rate in gallons per minute.

To determine the gallons-per-minute pumping rate, first determine the cubic-feet-per-second pumping rate. Then convert to gallons per minute.

---

[1]Hydraulics 7, Flow Rate Problems.

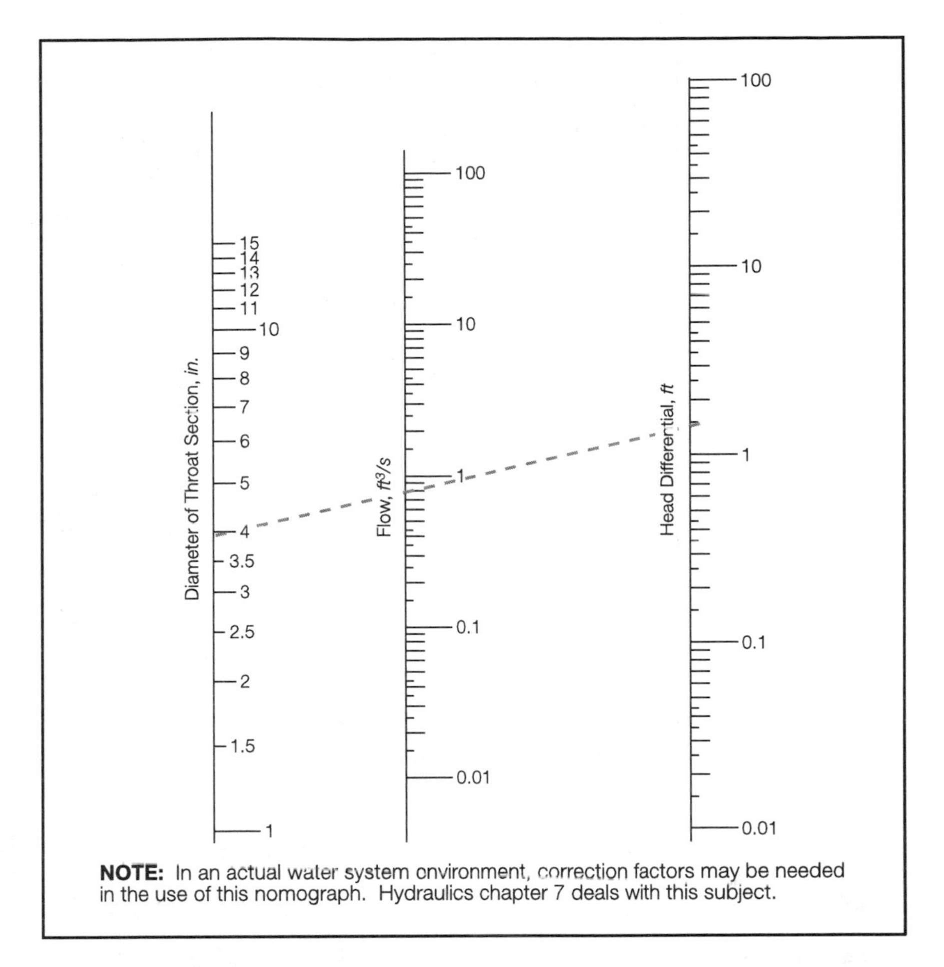

FIGURE H6-1 Flow rate nomograph for venturi meter

Draw a line on the nomograph from 4 in. on the *diameter of throat section* scale to 1.5 ft on the *head differential* scale. The point where the line crosses the middle scale indicates the approximate flow of 0.85 ft^3/s. Now convert the cubic-feet-per-second pumping rate to gallons per minute:[2]

$$(0.85\ ft^3/s)\ (7.48\ gal/ft^3)\ (60\ s/min) \quad = \quad 381.48\text{-gpm pumping rate}$$

[2]Mathematics 11, Conversions.

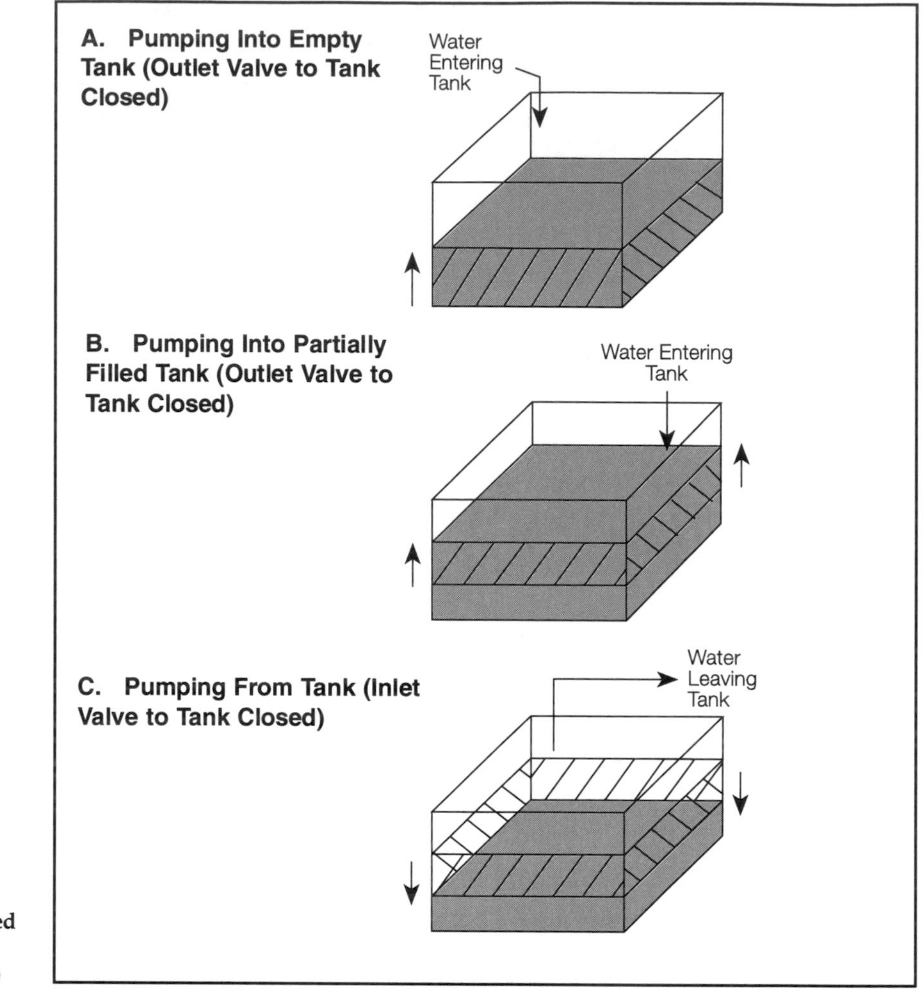

FIGURE H6-2 Determining pumping rate based on tank flow (crosshatched areas indicate volume pumped)

When there is no meter on the discharge side of the pump, you can determine the pumping rate by measuring the number of gallons pumped into or out of a tank during a given time period. Let's look at three examples of determining total gallons pumped, illustrated by Figures H6-2.

In the first example (Figure H6-2A), the pump is discharging into an empty tank with the outlet valve to the tank closed. The total gallons pumped during the given time is the number of gallons in the tank at the end of the pumping test.

Since it is not always possible or practical to pump into an empty tank, the pumping test is sometimes conducted by pumping into a tank that already contains water (Figure H6-2B). The outlet valve is closed and the total gallons (crosshatched area) pumped during the given time is determined from the *rise in water level*.

In Figure H6-2C, the pump is located on the outlet side of the tank. To conduct a pumping test, the inlet valve of the tank is shut off. When the pump is turned on, the total gallons (crosshatched area) pumped during the given time is determined from the *fall in water level*.

Example 3

During a 15-min pumping test, 15,820 gal were pumped into an empty rectangular tank (Figure H6-3). What is the pumping rate in gallons per minute?

The problem asks for the pumping rate in gallons per minute, so the mathematical setup is

$$\text{pumping rate} = \frac{\text{gallons}}{\text{minutes}}$$

Fill in the information given and perform the calculations to complete the problem:

$$\text{pumping rate} = \frac{15{,}820 \text{ gal}}{15 \text{ min}}$$

$$= 1{,}055 \text{ gpm pumping rate}$$

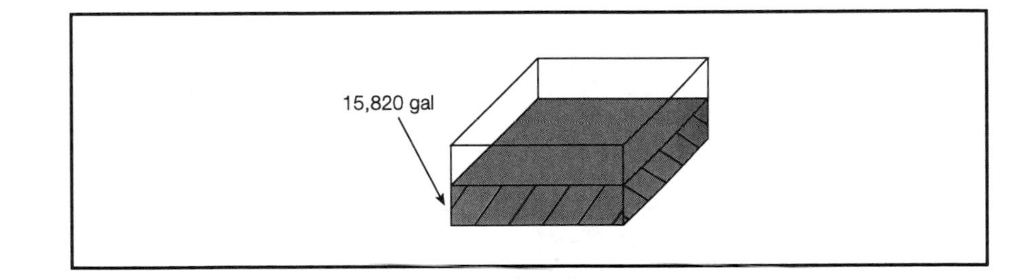

FIGURE H6-3 Schematic for example 3

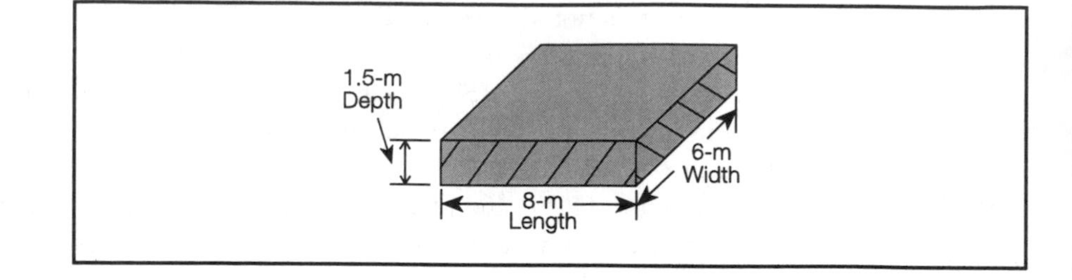

FIGURE H6-4 Schematic for example 4

Example 4

An empty rectangular tank 8 m long and 6 m wide (Figure H6-4) can hold water to a depth of 1.5 m. If this tank is filled by a pump in 55 min, what is the pumping rate in liters per second?

In this example, the entire tank was filled during the 55-min pumping test. Therefore the total volume pumped is equal to the capacity of the tank in liters.[3]

$$\text{volume of tank} = (\text{area of rectangle})(\text{depth})$$

$$= (8\text{ m})(6\text{ m})(1.5\text{ m})$$

$$= 72\text{ m}^3$$

Convert this volume to liters:[4]

$$(72\text{ m}^3)(1{,}000\text{ L/m}^3) = 72{,}000\text{ L}$$

Then use the total volume pumped and the time period of the pumping test [55 min, which is equivalent to (55) (60) = 3,300 s] to calculate the pumping rate:

$$\text{pumping rate} = \frac{\text{liters}}{\text{seconds}}$$

$$= \frac{72{,}000\text{ L}}{3{,}300\text{ s}}$$

$$= 21.82\text{ L/s}$$

[3]Mathematics 10, Volume Measurements.

[4]Mathematics 11, Conversions.

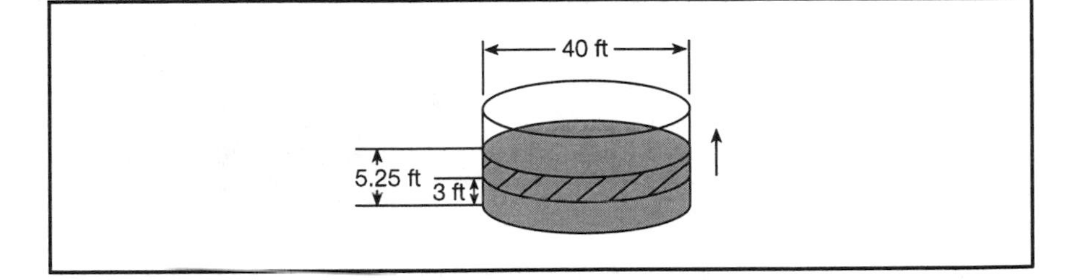

FIGURE H6-5
Schematic for example 5

Example 5

A tank 40 ft in diameter is filled with water to a depth of 3 ft (Figure H6-5). To conduct a pumping test, the outlet valve to the tank is closed and the pump is allowed to discharge into the tank. After 75 min the water level is 5.25 ft. What is the pumping rate in gallons per minute?

In this problem, the total volume pumped is represented by the crosshatched area on the diagram (tank diameter = 40 ft and water depth = 2.25 ft). Calculate the volume pumped in cubic feet:

$$\text{volume pumped} = (\text{area of circle})(\text{depth})$$

$$= (0.785)\,(40\text{ ft})\,(40\text{ ft})\,(2.25\text{ ft})$$

$$= 2{,}826\text{ ft}^3$$

And convert the cubic-feet volume to gallons:

$$(2{,}826\text{ ft}^3)\,(7.48\text{ gal/ft}^3) = 21{,}138\text{ gal}$$

The pumping test was conducted over a period of 75 min. Using the volume pumped (in gallons) and the time (in minutes) it took to pump the volume, calculate the pumping rate in gallons per minute:

$$\text{pumping rate} = \frac{\text{gallons}}{\text{minutes}}$$

$$= \frac{21{,}138\text{ gal}}{75\text{ min}}$$

$$= 281.84\text{ gpm}$$

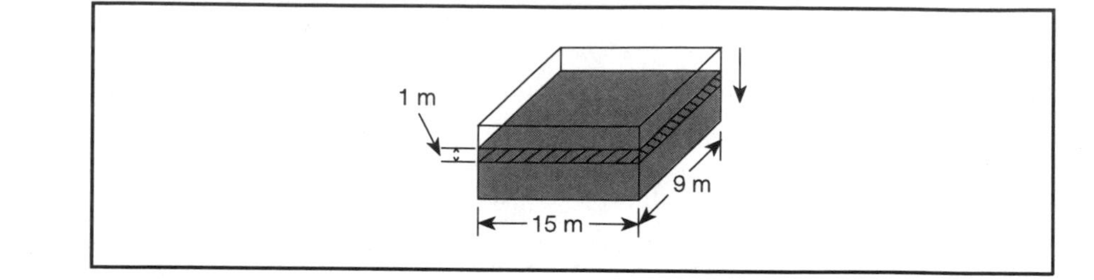

FIGURE H6-6 Schematic for example 6

Example 6

A 1-hour pumping test is run on a pump located on the outlet side of a tank (Figure H6-6). The inlet valve is closed and the pump is started. At the end of the test, the water level in the tank has dropped 1 m. If the tank water depth before pumping was 2 m, what is the pumping rate in cubic metres per second?

The total volume pumped is represented by the crosshatched area on the diagram. (Notice that information pertaining to the water depth before pumping is not needed in solving this problem. The drop in the water level is the essential depth information.) Calculate the total volume pumped:

$$\text{volume pumped} = (\text{area of rectangle})(\text{depth})$$

$$= (15\text{ m})(9\text{ m})(1\text{ m})$$

$$= 135\text{ m}^3$$

Convert 1 hour to seconds:

$$(1\text{ h})(3{,}600\text{ s/h}) = 3{,}600\text{ s}$$

Now determine the pumping rate:

$$\text{pumping rate} = \frac{\text{cubic meters}}{\text{seconds}}$$

$$= \frac{135\text{ m}^3}{3{,}600\text{ s}}$$

$$= 0.0375\text{ m}^3/\text{s}$$

Pump Heads

Pump head measurements are used to determine the amount of energy a pump can or must impart to the water. These heads, measured in feet, are a special application of the pressure and velocity heads discussed in the Hydraulics Section, chapter 4. Specific terms are used to describe heads measured at various points and under different conditions of the pump system.

Suction and Discharge

In pump systems, the words *suction* and *discharge* identify the inlet and outlet sides of the pump. As shown in Figure H6-7, the *suction side* of the pump is the *inlet*, or *low-pressure*, side. The *discharge side* is the *outlet*, or *high-pressure*, side. Any time a pump term includes the words *suction* or *discharge*, you can recognize immediately which side of the pump system is being discussed.

The *pump center line* is represented by a horizontal line drawn through the center of the pump. This line is important because it is the reference line from which pump head measurements are made.

The terms *suction* and *discharge* help establish the location of the particular pump head measurement. The terms *static* and *dynamic*, however, describe the condition of the system when the measurement is taken. Static heads are measured when the pump is off, and therefore the water is not moving. Dynamic heads are measured with the pump running and water flowing through the system.

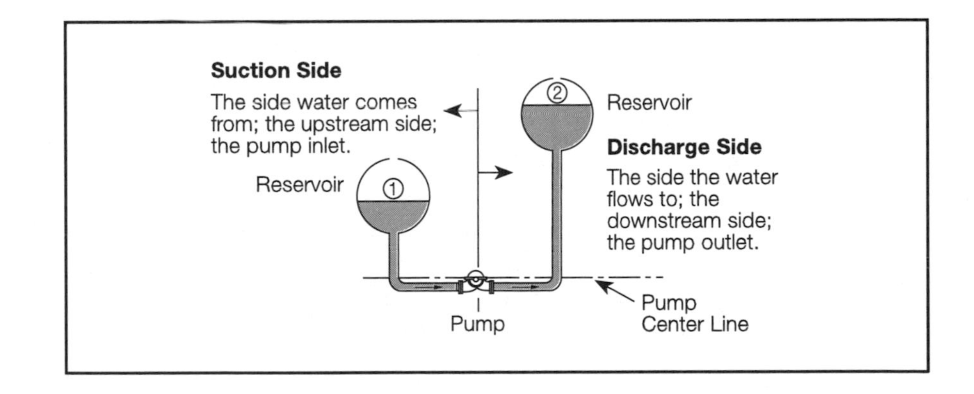

FIGURE H6-7 Suction and discharge sides of a pump

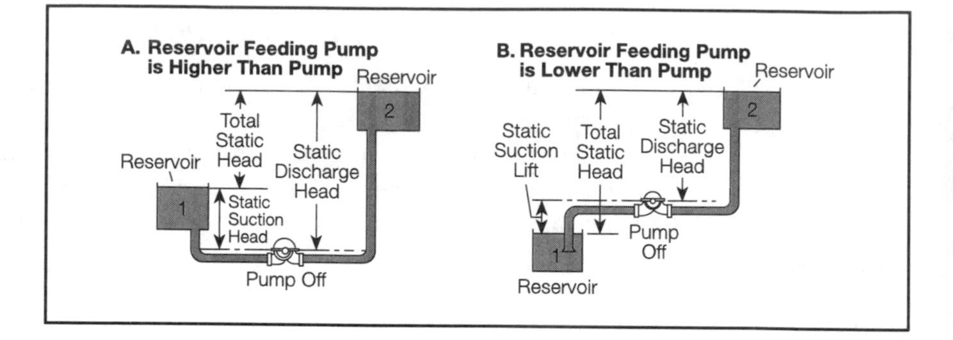

FIGURE H6-8
Static heads

Static Heads

Figure H6-8 illustrates the two basic pumping configurations found in water systems. The labeled vertical distances show the four types of static pump head:

- static suction head
- static suction lift
- static discharge head
- total static head

In a system where the reservoir feeding the pump is higher than the pump (Figure H6-8A), the difference in elevation (height) between the pump center line and the free water surface of the reservoir feeding the pump is termed *static suction head*. But in a system where the reservoir feeding the pump is lower than the pump (as in Figure H6-8B), the difference in elevation between the center line and the free water surface of the reservoir feeding the pump is termed *static suction lift*. Notice that a single system will have either a static head measurement or a static suction lift measurement, but not both.

Static discharge head is defined as the difference in height between the pump center line and the level of the discharge free water surface.

Total static head is the total height that the pump must lift the water when moving it from *reservoir 1* to *reservoir 2*. It is defined precisely as the vertical distance from the suction free water surface to the discharge free water surface. In Figure H6-8A, total static head is found by subtracting static suction head from static discharge head, whereas in Figure H6-8B, it is the sum of the static suction lift and static discharge head.

Static heads can be calculated from measurements of reservoir and pump elevations, or from pressure gauge readings taken when the pump is not running.

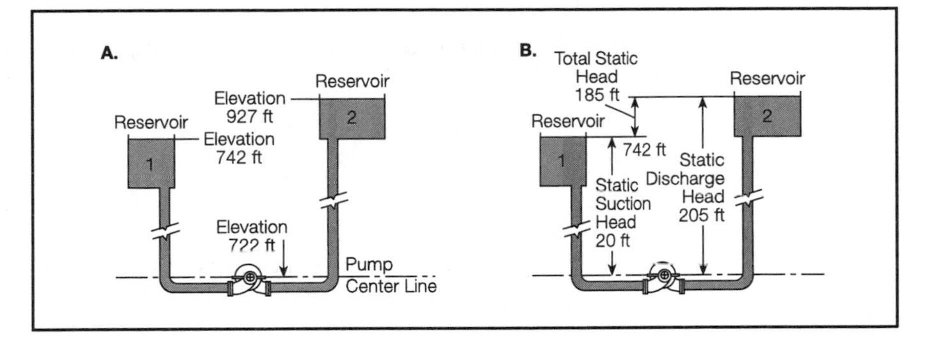

FIGURE H6-9
Schematic for example 7

Example 7

For Figure H6-9A, locate, label, and calculate the following (in feet):

(a) static suction head

(b) static discharge head

(c) total static head

First, locate and label the three types of static head, as shown in Figure H6-9B. Then, using the elevation information given for the problem, calculate each head as follows:

$$\text{static suction head} = 742 \text{ ft} - 722 \text{ ft}$$

$$= 20 \text{ ft}$$

$$\text{static discharge head} = 927 \text{ ft} - 722 \text{ ft}$$

$$= 205 \text{ ft}$$

Total static head can be found in two ways:

$$\text{total static head} = \text{elevation of reservoir 1} - \text{elevation of reservoir 2}$$

$$= 927 \text{ ft} - 742 \text{ ft}$$

$$= 185 \text{ ft}$$

or

total static head = static discharge head – static suction head

= 205 ft – 20 ft

= 185 ft

Example 8

For Figure H6-10A, locate, label, and calculate the following (in meters):

(a) static suction lift

(b) static discharge head

(c) total static head

First, locate and label the three heads as shown in Figure H6-10B. Then, using the elevation information given, calculate the heads as follows:

static suction lift = 296 m – 293 m

= 3 m

static discharge head = 481 m – 296 m

= 185 m

total static head = 185 m + 3 m

= 188 m

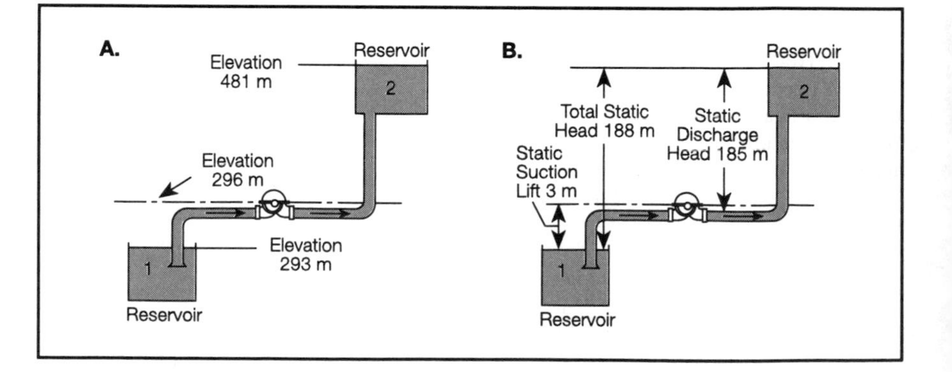

FIGURE H6-10 Schematic for example 8

or

$$\text{total static head} = 481\text{ m} - 293\text{ m}$$

$$= 188\text{ m}$$

If static heads were the only heads involved in pump operations, then any force exerted by the pump greater than the static head would cause water to be pumped. However, this doesn't happen. As soon as the pump is turned on, the pressures and energies within the system change. Measurements made while the pump is running, called *dynamic heads*, are used to describe the total energy that the pump must develop for pumping to take place.

Dynamic Heads

When water flows through a pipe, the water rubs against the walls of the pipe, and energy is lost to friction. In addition, there is a certain amount of resistance to flow as the water passes through valves, fittings, inlets, and outlets of a piping system. These additional energy losses that the pump must overcome are called *friction head losses* and *minor head losses*.[5]

Let's look at the effect of these additional energy losses on each side of the pump. Figure H6-11 shows a comparison of the head that exists on the

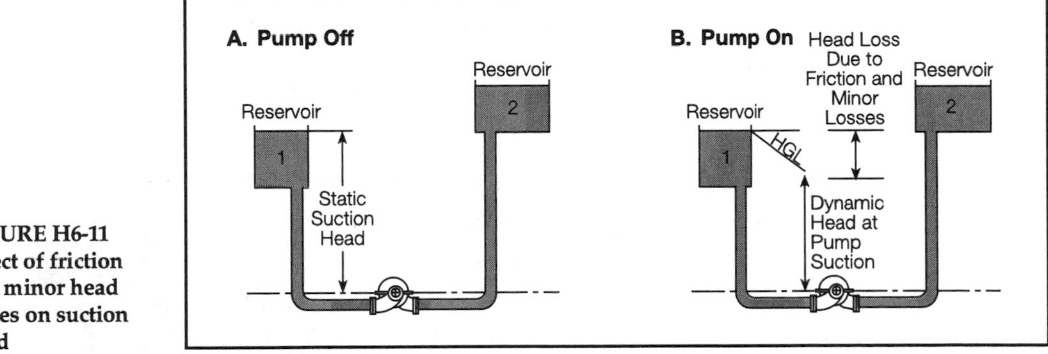

FIGURE H6-11 Effect of friction and minor head losses on suction head

[5]Hydraulics 5, Head Loss.

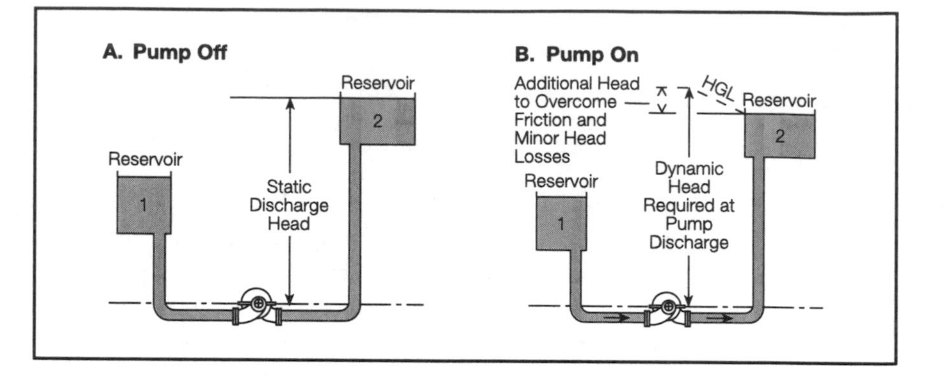

FIGURE H6-12 Effect of friction and minor head losses on discharge head

suction side of the pump before and after the pump is turned on. Notice that loss of head is caused by friction losses when the pump is turned on.

Part of the static suction head (which could otherwise aid the pump in pumping the water up to reservoir 2) is lost because of friction and minor losses as the water moves from reservoir 1 to the pump.

Similarly, friction and minor head losses develop on the discharge side of the pump as the water moves from the discharge side of the pump to reservoir 2. This creates an additional load or head against which the pump must operate (Figure H6-12).

If the pump does not add enough energy to overcome the friction and minor head losses that occur on both the suction and discharge sides of the pump, the pump will not be able to move the water.

Now that you have a basic understanding of the reasons for a difference between static and dynamic conditions, consider the four measurements of dynamic head (shown in Figure H6-13):

- dynamic suction head
- dynamic suction lift
- dynamic discharge head
- total dynamic head

In a system where the hydraulic grade line (HGL) on the suction side is higher than the pump (as in Figure H6-13A), the *dynamic suction head* is measured from the pump center line at the *suction* of the pump to the point on the HGL directly *above* it.

In a system where the HGL on the suction side is lower than the pump (as in Figure H6-13B), the *dynamic suction lift* is measured from the pump center line at the *suction* of the pump to the point on the HGL directly *below* it.

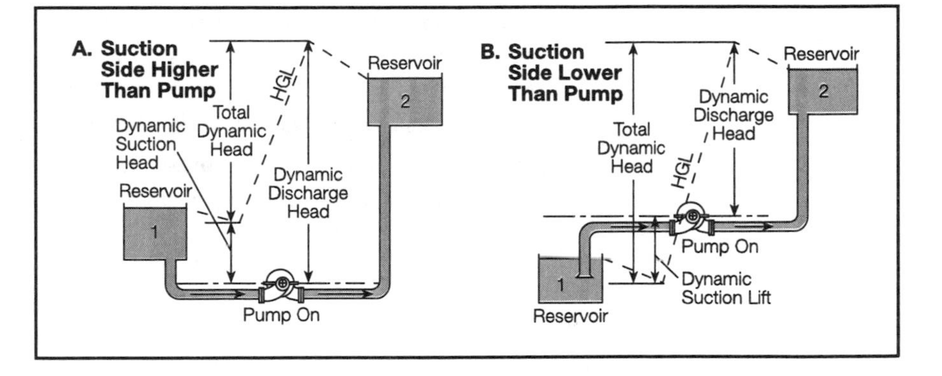

FIGURE H6-13
Dynamic heads

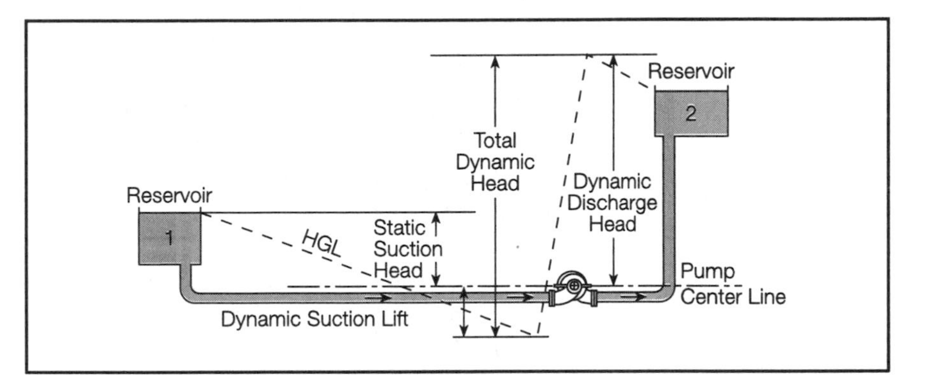

FIGURE H6-14
Dynamic suction lift for reservoir above pump

A single system will have a dynamic suction head or a dynamic suction lift, but not both. The location of the HGL, not of the reservoir, determines which condition exists. As shown in Figure H6-14, it is possible to have a dynamic suction lift condition even though the reservoir feeding the pump is above the pump center line; in such cases, the HGL on the suction side is below the pump center line when the pump is running.

The *dynamic discharge head* is measured from the pump center line at the *discharge* of the pump to the point on the HGL directly *above* it.

The *total dynamic head* is the difference in height between the HGL on the discharge side of the pump and the HGL on the suction side of the pump. This head is a measure of the total energy that the pump must impart to the water to move it from reservoir 1 to reservoir 2.

Dynamic pump heads can be calculated in either of two ways:

- by adding friction and minor losses to static head. (Static heads can be determined using existing elevation information such as shown in examples 7 and 8 in this chapter; and friction and minor losses can be found by using tables such as those shown in the head loss section.)
- by direct measurement using pressure gauges. (Pressure gauges cannot measure the velocity head component. As discussed in the Hydraulics Section, chapter 4, though, velocity head can often be ignored without significant error.)

Examples 9 and 10 illustrate how dynamic pump heads can be measured using pressure gauge information. As noted before, this method of determining heads can also be used to determine static heads by simply turning the pump off before taking the gauge readings.

Example 9

Using the pressure gauge information taken when the pump was running, determine the dynamic heads for the system in Figure H6-15A. Gauge readings are shown in Figure H6-15A in pounds per square inch gauge (psig).

Find the dynamic suction and dynamic discharge heads by converting the pressure gauge readings to feet of head.[6]

Moving from a smaller box to a larger box, you should multiply by 2.31.

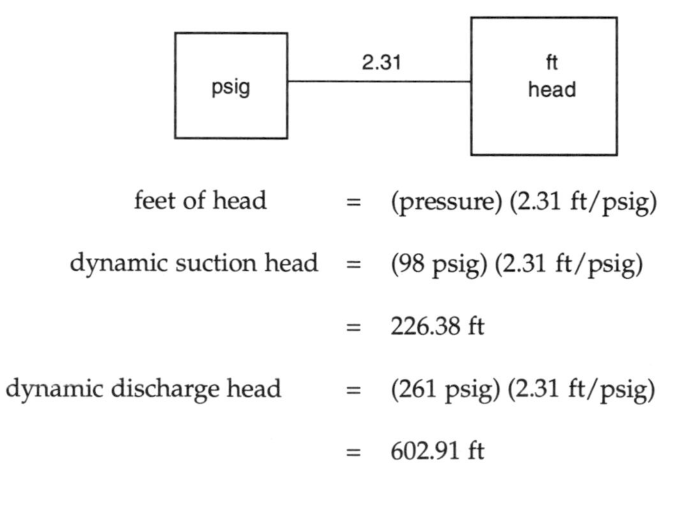

feet of head = (pressure) (2.31 ft/psig)

dynamic suction head = (98 psig) (2.31 ft/psig)

= 226.38 ft

dynamic discharge head = (261 psig) (2.31 ft/psig)

= 602.91 ft

[6]Hydraulics 2, Pressure and Force.

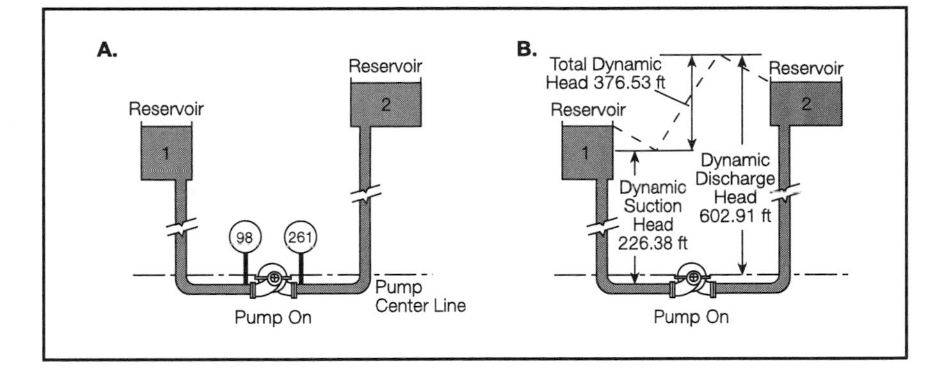

FIGURE H6-15 Schematic for example 9

The total dynamic head in this problem is the *difference* between the head on the discharge side of the pump and the head on the suction side of the pump (Figure H6-15B).

$$\text{total dynamic head} = 602.91 \text{ ft} - 226.38 \text{ ft}$$

$$= 376.53 \text{ ft}$$

Example 10

Using the two gauge pressure readings given in Figure H6-16A, calculate the dynamic heads. Gauge readings are in kilopascals and were taken when the pump was operating.

To calculate the dynamic heads shown, first convert the pressure gauge readings to meters of head.

$$\text{dynamic suction lift} = (41 \text{ kPa})\,(0.1019 \text{ m/kPa})$$

$$= 4.18 \text{ m}$$

The minus sign associated with the 41-kPa reading in the diagram is just a reminder that the dynamic suction lift is a vertical distance *below* the pump center line.

$$\text{dynamic discharge head} = (331 \text{ kPa})\,(0.1019 \text{ m/kPa})$$

$$= 33.73 \text{ m}$$

In a situation involving a suction lift, the total dynamic head against which the pump must operate is the *sum* of the heads on the suction and discharge sides of the pump (Figure H6-16B).

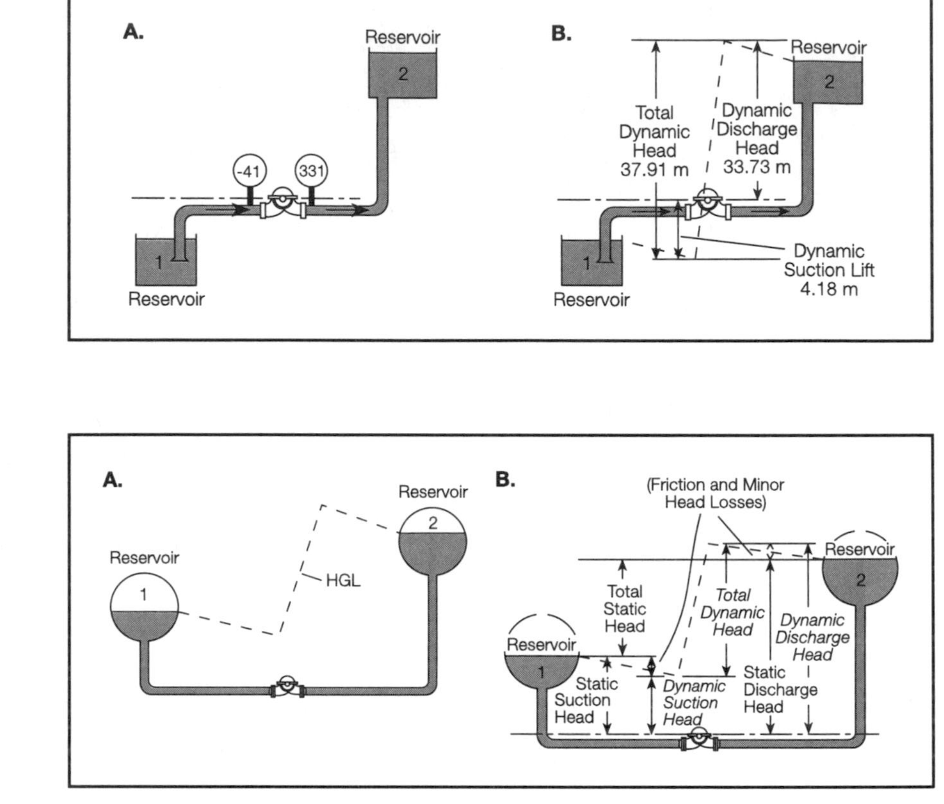

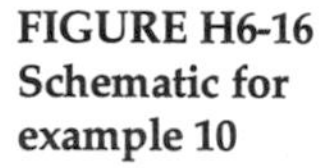
FIGURE H6-16
Schematic for example 10

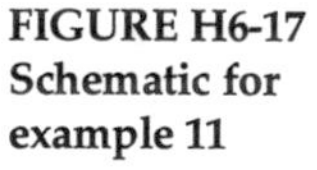
FIGURE H6-17
Schematic for example 11

$$\text{total dynamic head} = 33.73 \text{ m} + 4.18 \text{ m}$$

$$= 37.91 \text{ m}$$

The following example 11 combines most of the pump head terms.

Example 11

Using Figure H6-17A, draw and label the static and dynamic heads. Also label the two distances that represent the combined effects of friction head loss and minor head loss.

First, draw the pump center line. Next, extend the two free water surfaces, shown by the solid lines in Figure H6-17B. Then draw the vertical lines representing the three static and three dynamic heads. Finally, find the friction and minor head losses by measuring the difference between the dynamic suction head and the static suction head (this is the friction and minor loss on the suction side), and by measuring the difference between the static discharge head and the dynamic discharge head (this is the friction and minor loss on the discharge side).

The completed diagram is shown in Figure H6-17B.

Horsepower and Efficiency

Calculations of pump horsepower and efficiency are made in conjunction with many water transmission, treatment, and distribution operations. The selection of a pump or combination of pumps to provide an adequate pumping capacity will depend on the flow rate required and the effective height through which the flow must be pumped. The *effective height* is defined as the total feet of head against which the pump must work.[7]

Horsepower

To understand horsepower, you must first understand the technical meaning of the term *work*. Work is defined as the operation of a force over a specific distance; for example, lifting a *one-pound* object *one foot*. Thus, the *amount of work* done is measured in foot-pounds (ft-lb):

(feet) (pounds) = foot-pounds

Because it always requires the same amount of work to lift a 1-lb object 1 ft straight up, that amount of work is used as a standard measure: 1 ft-lb. However, work performed on an object of any weight can be measured in foot-pounds (Figure H6-18). Work can be performed in any direction (Figure

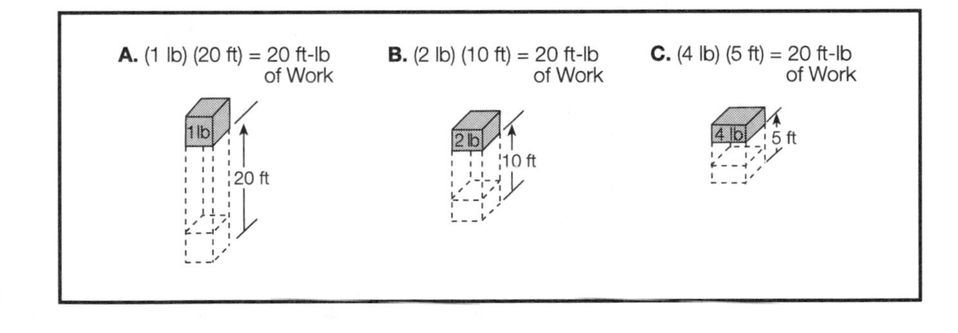

FIGURE H6-18 Equal amounts of work applied to objects of different weight

[7]Hydraulics 4, Head.

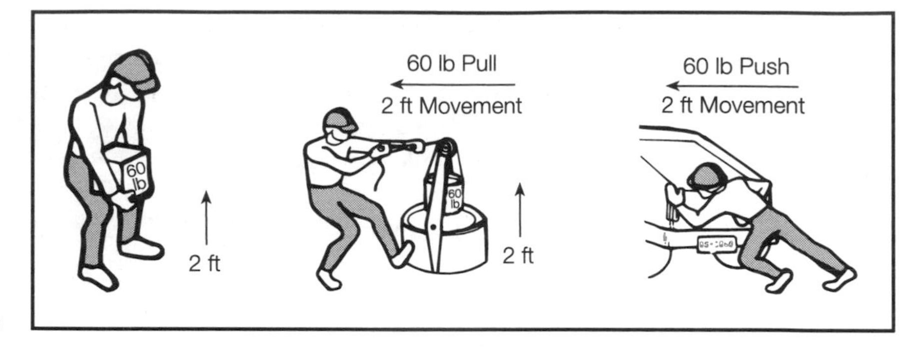

FIGURE H6-19 Work being performed in different directions

H6-19). And, as shown in Figure H6-19C, work doesn't need to involve lifting — an engine pushing a car along a level highway and a pump pushing water through a level pipeline are both examples of work.

The rate of doing work — that is, the measure of how much work is done in a given time, — is called *power*. Therefore, to make power calculations, you must know the time required to perform the work. The basic unit for power measurement is foot-pounds per minute (ft-lb/min). Note that this is work performed *per* time. One equation for calculating power in foot-pounds per minute is

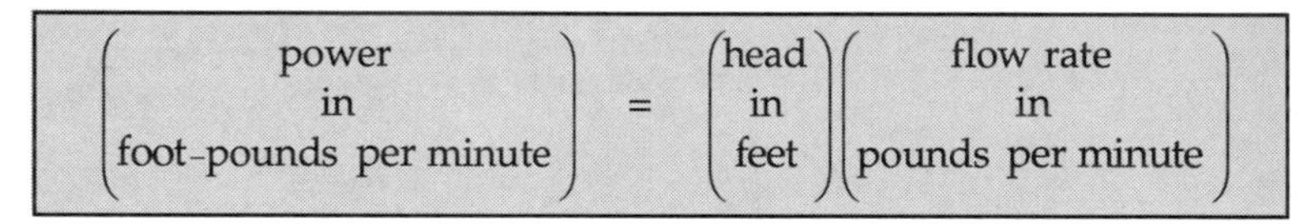

$$\left(\begin{array}{c}\text{power}\\ \text{in}\\ \text{foot-pounds per minute}\end{array}\right) = \left(\begin{array}{c}\text{head}\\ \text{in}\\ \text{feet}\end{array}\right)\left(\begin{array}{c}\text{flow rate}\\ \text{in}\\ \text{pounds per minute}\end{array}\right)$$

You will often work with measurements of power expressed in horsepower (hp), which is related to foot-pounds per minute by the conversion equation

$$1 \text{ hp} = 33{,}000 \text{ ft-lb/min}$$

When doing horsepower problems using the basic power equation, you must first convert the measurements of head, flow rate, and power into the proper units:

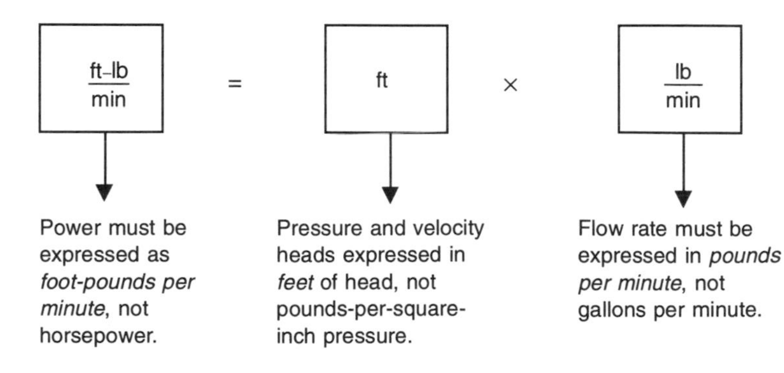

The power calculated in foot-pounds per minute can be converted to horsepower by dividing by 33,000. (The conversion equation is 1 hp = 33,000 ft-lb/min.) This is called *water horsepower* (WHP), since it is the amount of horsepower required to lift the water. Another equation that can be used to calculate water horsepower is

$$\text{whp} = \frac{(\text{flow rate in gallons per minute})(\text{total head in feet})}{3{,}960}$$

However, only the equation first shown is used in this section since it is more flexible and more closely related to the concept of power.

Example 12

A pump must pump 2,000 gpm against a total head of 20 ft. What horsepower (water horsepower) will be required to do the work?

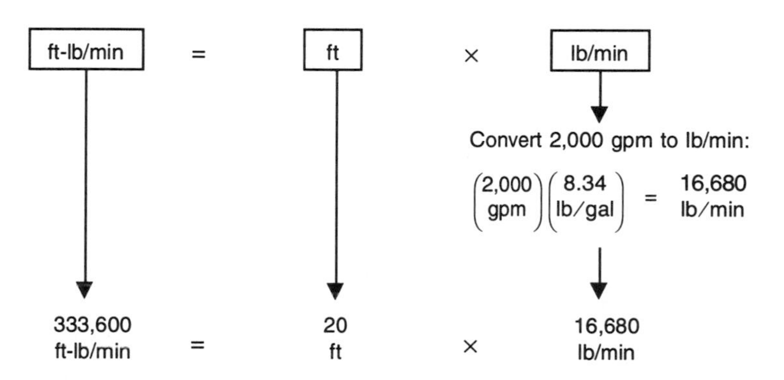

Now convert foot-pounds per minute to horsepower. You know that 1 hp = 33,000 ft-lb/min; in asking, "How much horsepower does this represent?" you are asking, "How many 33,000's are there in 333,600?" Mathematically, this is written as

$$\frac{333{,}600 \text{ ft-lb/min}}{33{,}000 \text{ ft-lb/min/hp}} = 10.11 \text{ hp}$$

As an alternative, you can use the box diagram method[8] to convert foot-pounds per minute to horsepower.

[8] Mathematics 11, Conversions.

You are moving from a larger box to a smaller box. Therefore, division by 33,000 is indicated:

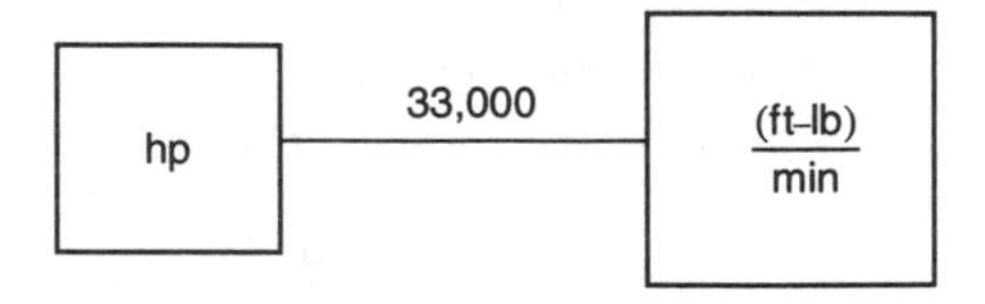

$$\frac{333{,}600 \text{ ft-lb/min}}{33{,}000 \text{ ft-lb/min/hp}} = 10.11 \text{ hp}$$

Example 13

A flow of 450 gpm must be pumped against a head of 50 ft. What is the water horsepower required?

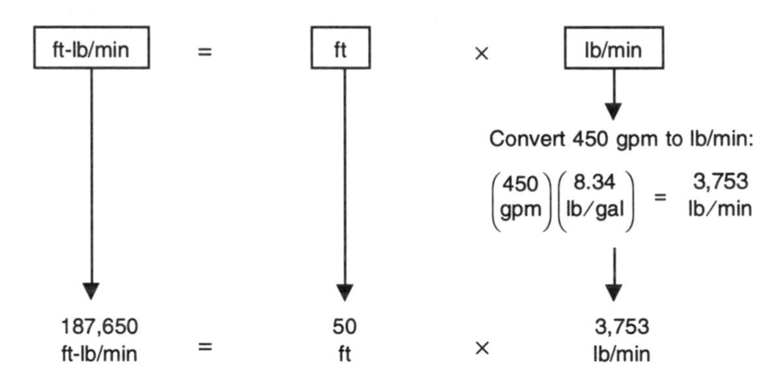

Convert 187,650 ft-lb/min to horsepower (1 hp = 33,000 ft-lb/min) by asking, "How many 33,000's are there in 187,650?" Mathematically, this is written

$$\frac{187{,}650}{33{,}000} = 5.69 \text{ hp}$$

Examples 12 and 13 illustrate the calculation of horsepower when *water* is being pumped. In almost all horsepower calculations pertaining to water supply, treatment, and distribution, water will be the liquid being pumped. However, you should know what happens to the horsepower calculation when a liquid being pumped has a specific gravity[9] different than that of water.

---

[9]Hydraulics 1, Density and Specific Gravity.

Example 14

Given the same flow rate and feet of head as described in example 13, what horsepower would be required if gasoline (specific gravity = 0.75) were being pumped rather than water?

For this problem, use the same equation to calculate horsepower. The principal difference between this problem and examples 12 and 13 is the conversion of gallons per minute to pounds per minute. Because gasoline has a different specific gravity than water, one gallon does not weigh 8.34 lb. Instead, one gallon of gasoline weighs

$$(8.34 \text{ lb/gal})(0.75) = 6.26 \text{ lb/gal}$$

Therefore, in making the conversion from gallons per minute to pounds per minute, use 6.26 lb/gal rather than 8.34 lb/gal. The calculation of required horsepower is as follows:

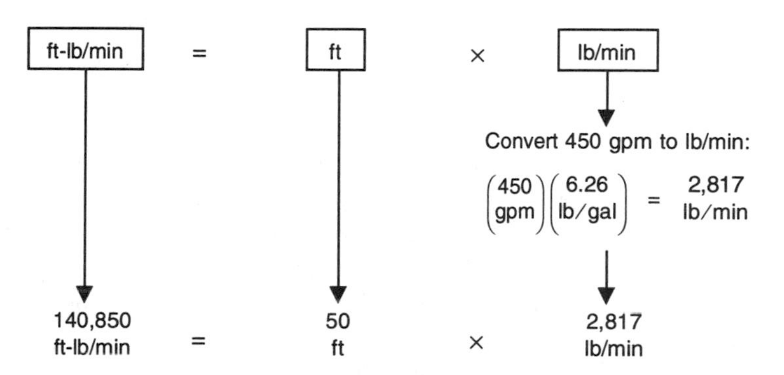

Then convert foot-pounds per minute to horsepower.

$$\frac{140{,}850 \text{ ft-lb/min}}{33{,}000 \text{ ft-lb/min/hp}} = 4.27 \text{ hp}$$

Because gasoline is not as dense as water, the horsepower required to pump gasoline is less than the horsepower required to pump water.

Most of the liquids being pumped in water supply and treatment have a specific gravity of 1 or very nearly 1 (0.98 to 1.2), so in the remaining example problems, a specific gravity of 1 will be assumed.

Example 15

Suppose a pump is pumping against a total head of 46.2 ft. If 800 gpm are to be pumped, what is the water horsepower requirement?

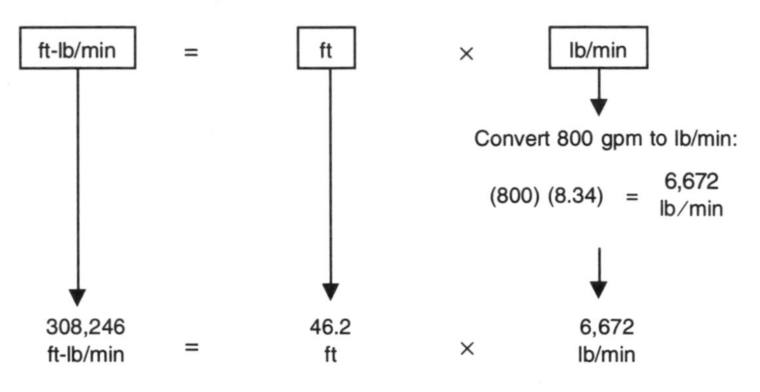

Then convert foot-pounds per minute to horsepower.

$$\frac{308{,}246 \text{ ft-lb/min}}{33{,}000 \text{ ft-lb/min/hp}} = 9.34 \text{ hp}$$

In all example problems thus far, the unknown value has been horsepower (and therefore, indirectly, foot-pounds per minute). However, either of the other two factors in the equation (feet or pounds per minute) might be the unknown value in a particular problem. The basic approach to the problem will remain unchanged regardless of which of the factors is unknown. The following two examples illustrate this concept.

Example 16

A pump is putting out 5 whp and delivering a flow of 430 gpm. What is the total feet of head against which the pump is operating?

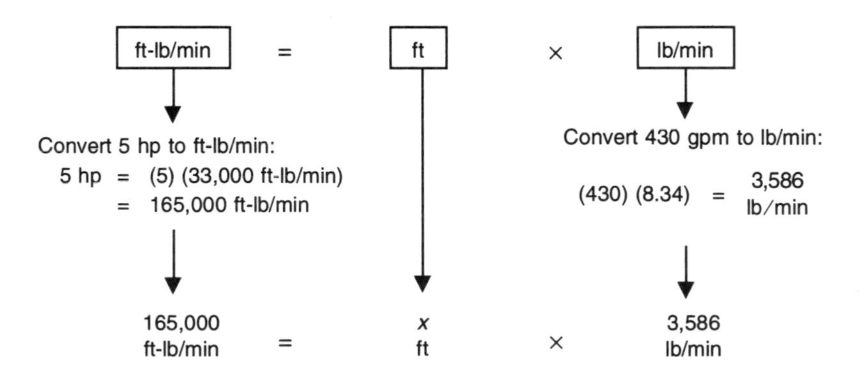

Now solve for the unknown value in the equation:

$$(x \text{ ft})\ (3{,}586) = 165{,}000$$

$$x = \frac{165{,}000}{3{,}586}$$

$$x = 46.01 \text{ ft total head}$$

Example 17

What is the maximum pumping rate (in gallons per minute) of a pump that is producing 15 whp against a head of 65 ft?

First calculate the maximum flow rate in pounds per minute:

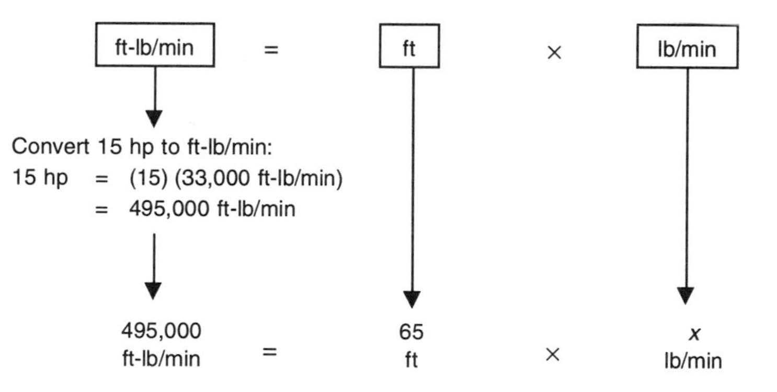

Solve for the unknown value:

$$(65)\ (x \text{ lb/min}) = 495{,}000$$

$$x = \frac{495{,}000}{65}$$

$$x = 7{,}615 \text{ lb/min}$$

Now express this maximum pumping rate in gallons per minute:

$$\frac{7{,}615 \text{ lb/min}}{8.34 \text{ lb/gal}} = 913.70 \text{ gpm}$$

Efficiency

In the preceding examples, you learned how to calculate water horsepower, the amount of power that must be applied directly to water to move it at a given rate against a given head. To apply the power to the water requires a pump, which in turn must be driven by a motor, which is powered by electric current.

Neither the pump nor the motor will ever be 100 percent efficient (Figure H6-20). This means that not all of the power supplied by the motor to the pump (called *brake horsepower*, bhp) will be used to lift the water (*water horsepower*) — some of the power is used to overcome friction within the pump. Similarly, not all of the power of the electric current driving the motor (called *motor horsepower*, mhp) will be used to drive the pump — some of the current is used to overcome friction within the motor, and some current is lost in the conversion of electrical energy to mechanical power.

Depending on size and type, pumps are usually 50–85 percent efficient, and motors are usually 80–95 percent efficient. The efficiency of a particular motor or pump is given in the manufacturer's information accompanying the unit.

In some installations, you will know only the combined efficiency of the pump and motor. This is called the *wire-to-water efficiency* and is obtained by multiplying the motor and pump efficiencies together. For example, if a motor is 82 percent efficient and a pump is 67 percent efficient, the overall, or wire-to-water, efficiency is 55 percent ($0.82 \times 0.67 = 0.55$).

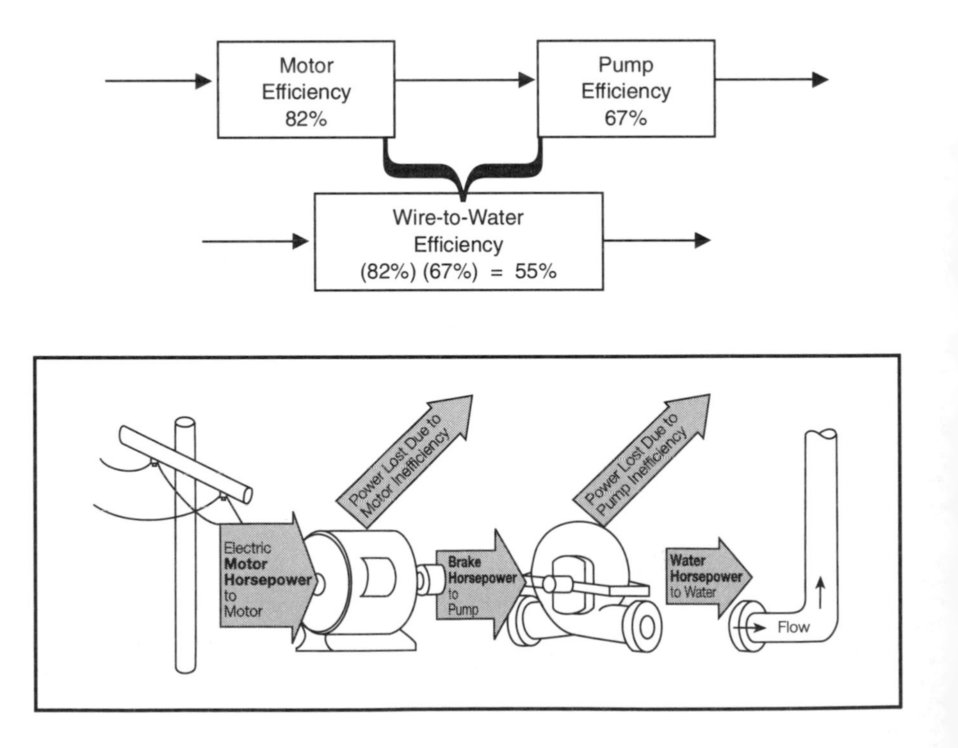

FIGURE H6-20 Power loss due to motor and pump inefficiency

In practical horsepower calculations, you must take into account pump and motor efficiencies so that you can determine what size pump and motor and how much power will be necessary to move the water at the desired flow rate.

To show how horsepower and efficiency calculations are used, consider the system used for example 15, in which the pump was pumping 800 gpm against a total head of 46.2 ft. The water horsepower required to perform this work was calculated to be 9.34 whp. Suppose that the pump was only 85 percent efficient and the motor 95 percent efficient. To account for these inefficiencies, more than 9.34 whp would have to be supplied to the pump, and even more horsepower would have to be supplied to the motor.

Techniques for calculating the required horsepower are discussed in the next series of examples. To use the techniques, however, you must know that power input to the motor is usually expressed in terms of electrical power rather than horsepower.

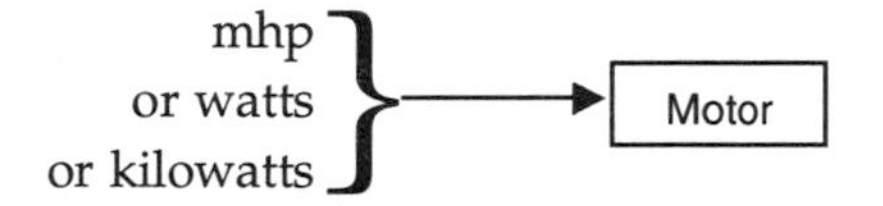

The relationship between horsepower and watts or kilowatts is given below. With these equations, you can convert from one term to the other.

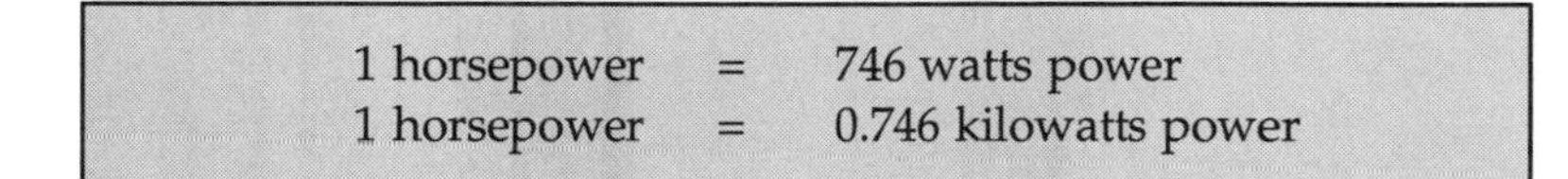

Example 18

If a pump is to deliver 460 gpm of water against a total head of 95 ft, and the pump has an efficiency of 75 percent, what horsepower must be supplied to the pump?

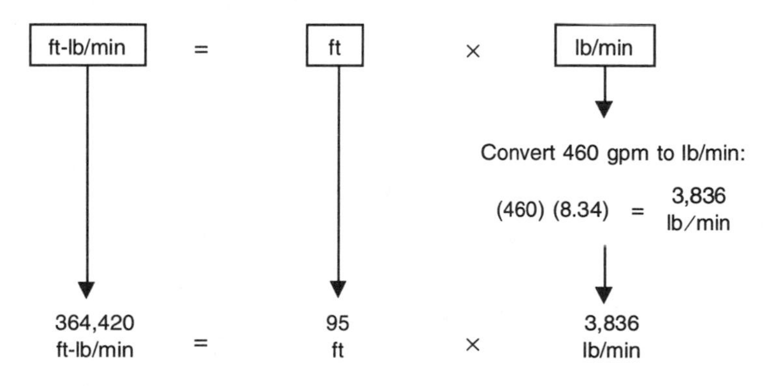

In this problem, brake horsepower is to be calculated. First calculate the water horsepower, based on the work to be accomplished. Then determine the brake horsepower required.

Convert foot-pounds per minute to water horsepower:

$$\frac{364{,}420 \text{ ft-lb/min}}{33{,}000 \text{ ft-lb/min/hp}} = 11.04 \text{ whp}$$

This calculation shows that 11.04 whp (water horsepower) is required.

Now determine the amount of horsepower that must be supplied to the pump (brake horsepower):

As stated in the problem, the pump is 75 percent efficient. Therefore, the brake horsepower will have to be *more than 11.04 hp* in order to accomplish the 11.04 hp work. Since the usable brake horsepower (75 percent) must equal 11.04 hp, the mathematical equation is[10]

$$(75\%)\ (\text{bhp}) = 11.04 \text{ whp}$$

Restate the percent as a decimal number:

$$(0.75)\ (x \text{ bhp}) = 11.04 \text{ whp}$$

And then solve for the unknown value:

$$(0.75)\ (x \text{ bhp}) = 11.04$$

$$x = \frac{11.04}{0.75}$$

$$x = 14.72 \text{ bhp}$$

Therefore, the 75 percent efficient pump must be supplied with 14.72 hp in order to accomplish the 11.04 hp of work:

[10]Mathematics 7, Percent.

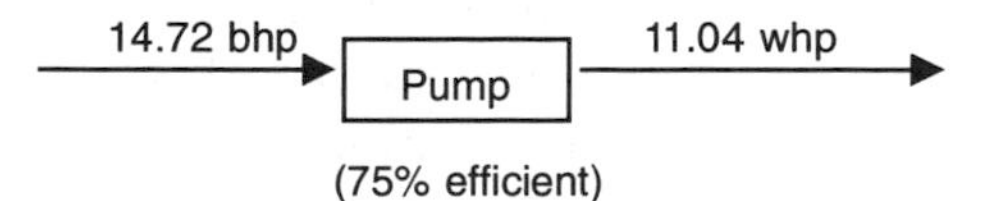

Example 19

The motor nameplate indicates that the output of a certain motor is 15 hp. How much horsepower must be supplied to the motor if the motor is 90 percent efficient?

In this problem, the brake horsepower is known and the motor horsepower must be calculated:

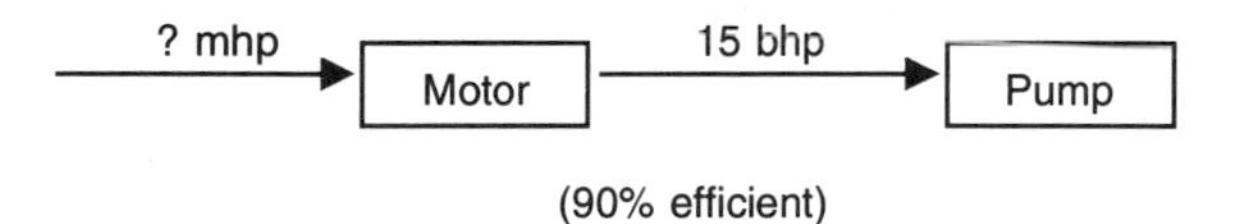

Since only 90 percent of the horsepower supplied to the motor is usable, you want that 90 percent to equal 15 hp. Mathematically, this is stated as

$$(90\%)\ (\text{mhp}) = 15\ \text{bhp}$$

Express the percent as a decimal number:

$$(0.90)\ (x\ \text{mhp}) = 15\ \text{bhp}$$

And then solve for the unknown value:

$$(0.90)\ (x\ \text{mhp}) = 15$$

$$x = \frac{15}{0.90}$$

$$x = 16.67\ \text{mhp}$$

Example 20

You have calculated that a certain pumping job will require 6 whp. If the pump is 80 percent efficient and the motor is 92 percent efficient, what motor horsepower will be required?

Often, when you are given both the motor and pump efficiencies, the information is combined and treated as one overall efficiency, as follows:

0.92	×	0.80	=	0.74
motor efficiency		pump efficiency		overall efficiency

This problem, then, becomes very similar to the two previous examples:

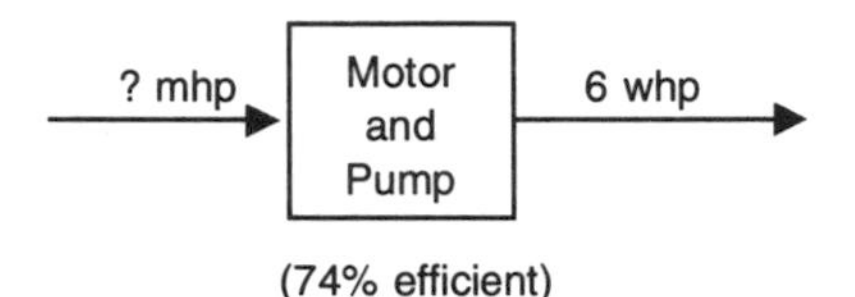

Since 74 percent of the motor horsepower must equal 6 hp, the mathematical setup is

$$(74\%)\,(\text{mhp}) = 6 \text{ whp}$$

Restate the percent as a decimal number:

$$(0.74)\,(x \text{ mhp}) = 6 \text{ whp}$$

And then solve for the unknown value:

$$(0.74)\,(x \text{ mhp}) = 6$$

$$x = \frac{6}{0.74}$$

$$x = 8.11 \text{ hp}$$

In the preceding problems, the pump and motor efficiencies were used to calculate the required brake and motor horsepower. Occasionally, however, the motor, brake, and water horsepowers will be known and the *efficiencies* unknown.

A calculation of motor or pump efficiency is essentially a percent calculation. The general equation for percent calculations is[11]

$$\text{percent} = \frac{\text{part}}{\text{whole}} \times 100$$

---

[11]Mathematics 7, Percent.

In calculations of efficiency, the "whole" is the total horsepower supplied to the unit, and the "part" is the horsepower output of the unit:

$$\text{percent efficiency} = \frac{\text{hp output}}{\text{hp supplied}} \times 100$$

From this general equation, the specific equations to be used for motor, pump, and overall efficiency calculations are

$$\text{percent motor efficiency} = \frac{\text{bhp}}{\text{mhp}} \times 100$$

$$\text{percent pump efficiency} = \frac{\text{whp}}{\text{bhp}} \times 100$$

$$\text{percent overall efficiency} = \frac{\text{whp}}{\text{mhp}} \times 100$$

Let's look at three examples of efficiency calculations.

Example 21

Based on the flow rate to be pumped and the total head the pump must pump against, the water horsepower requirement was calculated to be 8.5 whp. If the motor supplies the pump with 11 hp, what is the efficiency of the pump?

A diagram will help sort out the information given in the problem:

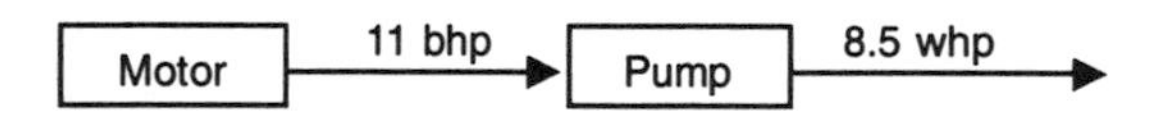

The diagram indicates the horsepower going into the pump and the horsepower coming out of the pump. From this information, calculate the pump efficiency:

$$\begin{aligned}
\text{percent pump efficiency} &= \frac{\text{hp output}}{\text{hp supplied}} \times 100 \\
&= \frac{8.5 \text{ whp}}{11 \text{ bhp}} \times 100 \\
&= 0.77 \times 100 \\
&= 77\%
\end{aligned}$$

Example 22

What is the wire-to-water efficiency if electric power equivalent to 20 hp is supplied to the motor and 13.5 hp of work is accomplished by the pump?

In this problem, overall efficiency must be determined. The approach to solving the problem is similar to the previous example. Diagram the information given in the problem:

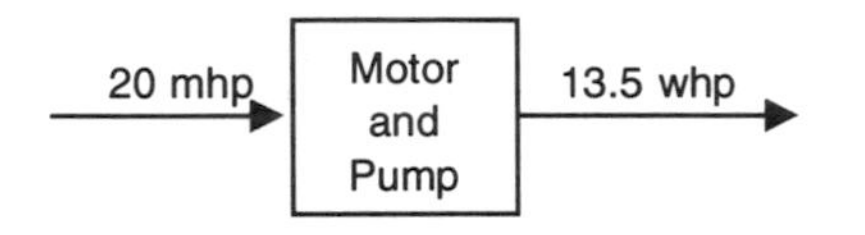

Calculate the percent overall efficiency:

$$\text{percent overall efficiency} = \frac{\text{hp output}}{\text{hp supplied}} \times 100$$

$$= \frac{13.5 \text{ whp}}{20 \text{ mhp}} \times 100$$

$$= 0.68 \times 100$$

$$= 68\% \text{ wire-to-water efficiency}$$

Example 23

Suppose that 11 kW (kilowatts) power is supplied to the motor. If the brake horsepower is 13 bhp, what is the efficiency of the motor?

First, diagram the information given in the problem:

11 kW → Motor → 13 bhp → Pump

To calculate the percent efficiency of the motor, the kilowatts power is converted to horsepower.

Based on the fact that 1 hp = 0.746 kW, the box diagram method can be used to perform the conversion:

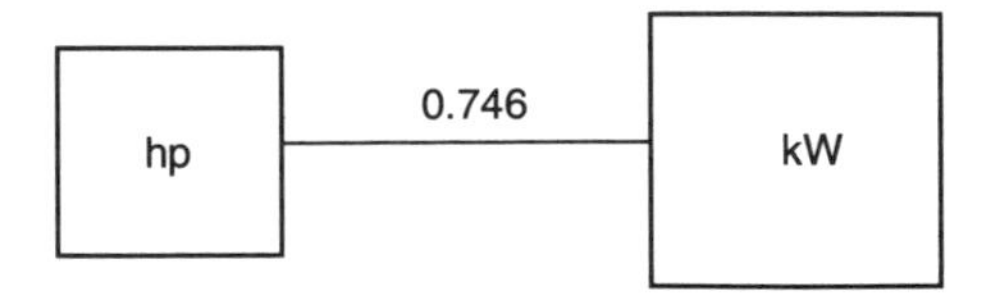

In converting from kilowatts to horsepower, you are moving from a larger box to a smaller box. Therefore *division* by 0.746 is indicated:

$$\frac{11 \text{ kW}}{0.746 \text{ kW/hp}} = 14.75 \text{ hp}$$

Now calculate the percent efficiency of the motor:

$$\text{percent efficiency} = \frac{\text{hp output}}{\text{hp supplied}} \times 100$$

$$= \frac{13 \text{ bhp}}{14.75 \text{ mhp}} \times 100$$

$$= 0.88 \times 100$$

$$= 88\%$$

Pumping Costs

Power is generally sold in units of kilowatt-hours (kW•h). If a motor draws 1 kW of power and runs for 1 hour, then the electric company will charge for 1 kW•h. Therefore, to calculate pumping costs, you will need to know the power requirements (power demand) of the motor and the length of time the motor runs. In most situations, you will know, or be able to determine, the total head and flow rate against which the pump is working, so the water horsepower can be calculated. Then, using the wire-to-water efficiency of the motor and pump, you can determine the motor horsepower and the kilowatts of power demand. (Power demand can also be calculated from a meter reading, but that calculation is not covered in this text.) The following three examples illustrate how pumping costs can be calculated.

Example 24

The motor horsepower required for a particular pumping job is 25 hp. If your power cost is \$0.03/kW•h, what is the cost of operating the motor for 1 hour?

First, convert the motor horsepower demand to kilowatt power demand:

$$1 \text{ hp} = 0.746 \text{ kW}$$

Therefore,

$$25 \text{ hp} = (25)(0.746 \text{ kW})$$
$$= 18.65 \text{ kW power demand}$$

Since the kilowatt demand is 18.65 kW, a value of 18.65 kW•h is required in 1 hour. Therefore, the cost of operating the pump for an hour is

$$(18.65 \text{ kW·h})(\$0.03/\text{kW·h}) = \$0.56$$

Example 25

You have calculated that the minimum motor horsepower requirement for a particular pumping problem is 15 mhp. If the cost of power is $0.025/kW•h, what is the power cost in operating the pump for 12 hours?

To determine power costs, as in the previous example, first express the motor horsepower requirement or demand in terms of kilowatts demand:

$$15 \text{ hp} = (15)(0.746 \text{ kW})$$
$$= 11.19 \text{ kW power demand}$$

In 12 hours, the number of kilowatt-hours required to operate the pump is

$$(11.19 \text{ kW})(12 \text{ h}) = 134.28 \text{ kW·h}$$

Thus, the power cost associated with operating the pump for 12 hours is

$$(134.28 \text{ kW·h})(\$0.025/\text{kW·h}) = \$3.36$$

Example 26

A pump is discharging 1,200 gpm against a head of 65 ft. The wire-to-water efficiency is 68 percent. If the cost of power is $0.022/kW•h, what is the cost of the power consumed during a week in which the pump runs 78 hours?

To determine the power cost, first calculate the water horsepower required and then the motor horsepower demand of the motor:

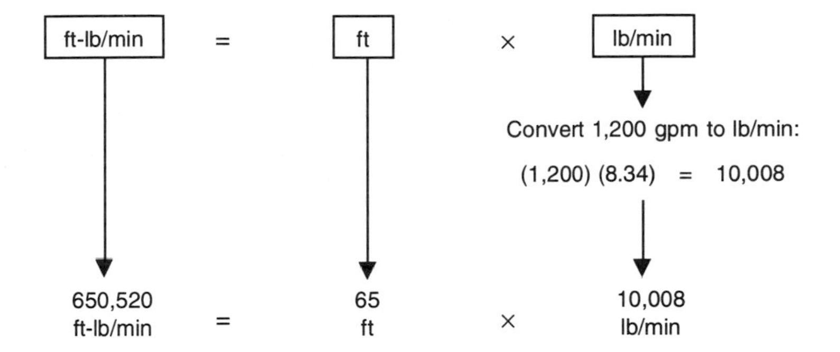

Express foot-pounds per minute as horsepower:

$$\frac{650{,}520 \text{ ft-lb/min}}{33{,}000 \text{ ft-lb/min/hp}} = 19.71 \text{ whp}$$

Now the water horsepower requirement has been determined. And from the statement of the problem, you know wire-to-water (overall) efficiency. A diagram illustrates what information is now known:

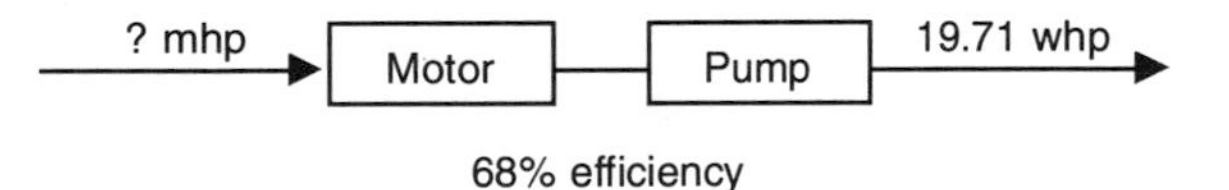

Since only 68 percent of the horsepower supplied to the motor is usable as water horsepower, 68 percent mhp must equal 19.71 hp. Mathematically, this is stated as

$$(68\%)\ (x \text{ mhp}) = 19.71 \text{ whp}$$

Change the percent to a decimal number:

$$(0.68)\ (x \text{ mhp}) = 19.71 \text{ whp}$$

Solve for the unknown value:

$$(0.68)\ (x \text{ mhp}) = 19.71$$

$$x = \frac{19.71}{0.68}$$

$$x = 28.99 \text{ mhp}$$

If this was merely a horsepower problem, the problem would end right here, as in similar problems of the previous section. However, this is where a cost problem begins.

To calculate power costs, the motor horsepower demand will have to be expressed in terms of kilowatts power demand. Recall that

$$1 \text{ hp} = 0.746 \text{ kW}$$

Therefore,

$$28.99 \text{ hp} = (28.99)(0.746 \text{ kW})$$

$$= 21.63 \text{ kW power demand}$$

For 78 hours, the number of kilowatt-hours required to operate the pump is

$$(21.63 \text{ kW})\,(78 \text{ h}) = 1{,}687 \text{ kW·h}$$

And the power cost associated with operating the pump for that week is

$$(1{,}687 \text{ kW·h})\,(\$0.022/\text{kW·h}) = \$37.11$$

Demand Charges

All water utilities pay an electrical demand charge for the pumping that is done in the treatment plants or in the distribution system. A demand charge is the charge for the maximum instantaneous power demand in kilowatts (kW). When added to the energy charge, the customer is able to calculate the power bill for the billing period. An average demand charge is $8–$10/kW, which can be significant.

A 1,000-hp pump will have a 746-kW demand when run. If the demand charge is $10/kW, to turn the pump on will cost $7,460 and the energy cost will be approximately $37.30/hour of operation, if energy costs $0.05/kW·h. The system demand is usually established on a monthly basis. Many utilities run a computerized demand model to minimize the power demand. Usually a rate is set, and the operators try to stay within the rate set. Turning on a number of large pumps at the same time will greatly increase demand charges. Because of the cost factor, it is most important to not oversize motors and pumps.

Reading Pump Curves

There is a great deal of information written about pumps, pump performance, and pump curves. The following section covers just those basic topics needed to understand and use pump curves.

Description of Pump Curves

A pump curve is a graph showing the *characteristics* of a particular pump. For this reason, these graphs are commonly called "pump characteristic curves." The characteristics commonly shown on a pump curve are

- capacity (flow rate)
- total head
- power (brake horsepower)
- efficiency

If a pump is designed to be driven by a variable-speed motor, then "speed" is also a characteristic shown on the graph. For the purposes of this discussion, it will be assumed that the pump is driven by a constant-speed motor, so the graphs used will have only four curves.

The four pump characteristics (capacity, head, power, and efficiency) are related to each other. This is extremely important, for it is this interrelationship that enables the four pump curves to be plotted on the same graph.

Experience has shown that the *capacity* (flow rate) of a pump changes as the *head* against which the pump is working changes. Pump capacity also changes as the *power* supplied to the pump changes. Finally, pump capacity changes as *efficiency* changes. Consequently, head, power, and efficiency can all be graphed "as a function of" pump capacity. This simply means that *capacity* is shown along the horizontal (bottom) scale of the graph, and *head, power,* and *efficiency* (any one or a combination of them) are shown along the vertical (side) scales of the graph. The following material discusses each of these relationships individually.

The H–Q Curve

The H–Q curve shows the relationship between total head H against which the pump must operate and pump capacity Q. A typical H–Q curve is shown in Figure H6-21. The curve indicates what flow rate the pump will produce at any given total head.

For example, if the total head is 100 ft, the pump will produce a flow of about 1,910 gpm. The H–Q curve also indicates that the capacity of the pump

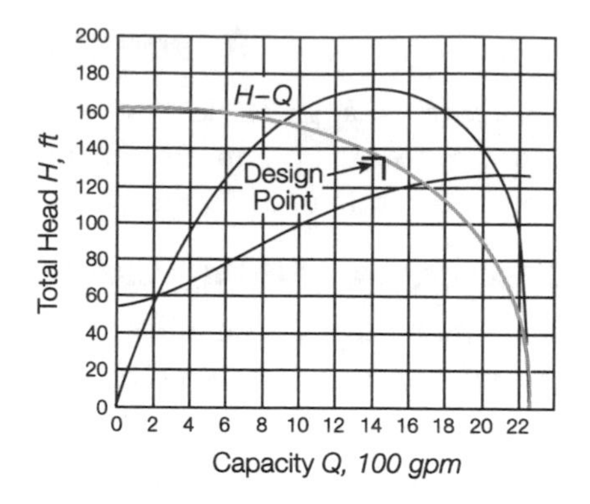

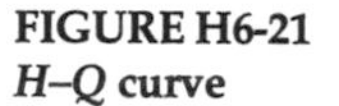
FIGURE H6-21
***H–Q* curve**

decreases as the total head *increases*. (Generally speaking, when the force against which the pump must work increases, the flow rate decreases.) The way total head controls the capacity is a characteristic of a particular pump.

The *H–Q* curve also identifies two very important operational facts. Normally on the curve you will find a mark like this:⌝. The mark defines the *design point*, the head and capacity at which the pump is intended to operate for best efficiency in a particular installation.

The *P–Q* Curve

The *P–Q* curve (Figure H6-22) shows the relationship between power *P* and capacity *Q*. In this figure, pump capacity is measured as gallons per minute, and power is measured as brake horsepower. For example, if you pump at a rate of 1,480 gpm, then the power used to drive the pump is about 58 bhp. This is valuable information. As explained in the preceding section on horsepower calculations, if you know the brake horsepower and the

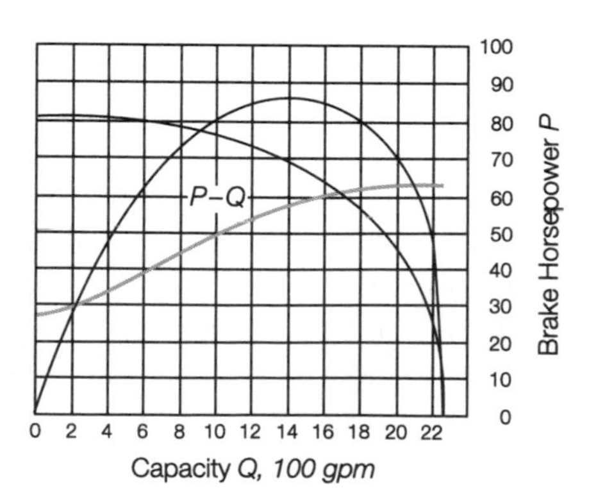

FIGURE H6-22
***P–Q* curve**

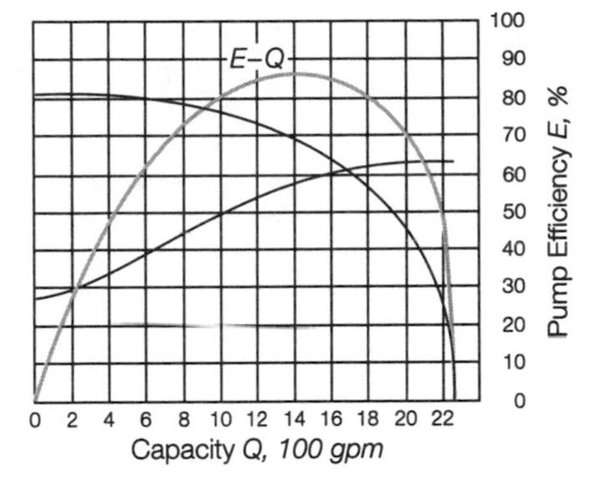

FIGURE H6-23
***E–Q* curve**

motor efficiency, you can then determine the motor horsepower (the power required to drive the motor).

Knowledge of what power the pump requires is valuable for checking the adequacy of an existing pump and motor system. The knowledge is also important when you find yourself scrambling to rig a temporary system from spare or surplus parts.

The *E–Q* Curve

The *E–Q* curve shows the relationship between pump efficiency *E* and capacity *Q*, as shown in Figure H6-23. (Sometimes the Greek letter eta — η — is used to stand for efficiency, and the efficiency-capacity curve is designated the η–*Q* curve.) In sizing a pump system, the engineer attempts to select a pump that will produce the desired flow rate at or near peak pump efficiency.

The more efficient your pump is, the less costly it is to operate. Stated another way, the more efficient your pump is, the more water you can pump for each dollar's worth of power. Knowledge of pump efficiency allows you to compute the cost of pumping water. Therefore, from an operational point of view, pump efficiency is a "must know" item.

Reading the Curve

Figure H6-24 is the complete pump curve. By using the pump curve, you can determine *any three* of the four pump characteristics (capacity, head, efficiency, and horsepower) when given information about the fourth

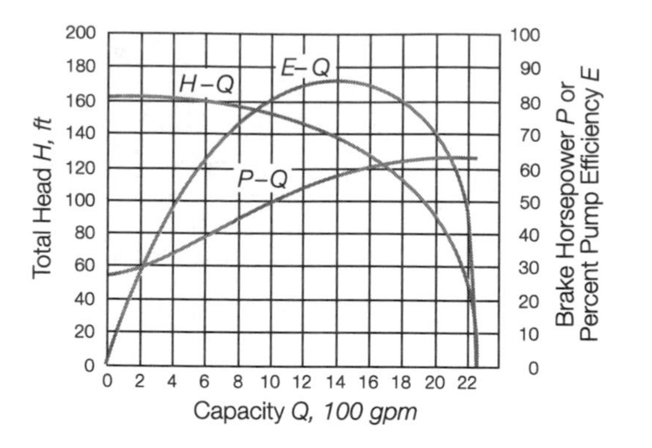

FIGURE H6-24 Complete pump curve

characteristic. (Normally the total head or the capacity is the known characteristic.) The following examples illustrate how to read pump curves.[12]

Example 27

Using the pump curve given in Figure H6-24, determine the pump head, power, and efficiency when operating at a flow rate of 1,600 gpm.

The key to reading pump curves is to find a certain vertical line and to note where the line intersects all three pump characteristic curves.

In this problem, the known characteristic is *capacity* (1,600 gpm). To determine head, power, and efficiency, you should first draw a vertical line upward from the number 16 on the horizontal scale (Figure H6-25).

Note that the vertical line intersects each of the three curves, as indicated by the three circled dots. Following the line upward, you first reach the intersection with the *P–Q* (power) curve. To determine what the power is at this point, move *horizontally to the right* toward the bhp scale; you should hit the scale at about 60 bhp.

Now return to the vertical line (1,600 gpm) and move to the intersection with the *H–Q* (head) curve. To determine the head at this point, move *horizontally to the left,* toward the total head scale; you should hit the scale at about 128 ft.

Finally, look at the point where the vertical line (1,600 gpm) intersects the *E–Q* (efficiency) curve. To determine the efficiency at

[12]Mathematics 12, Graphs and Tables.

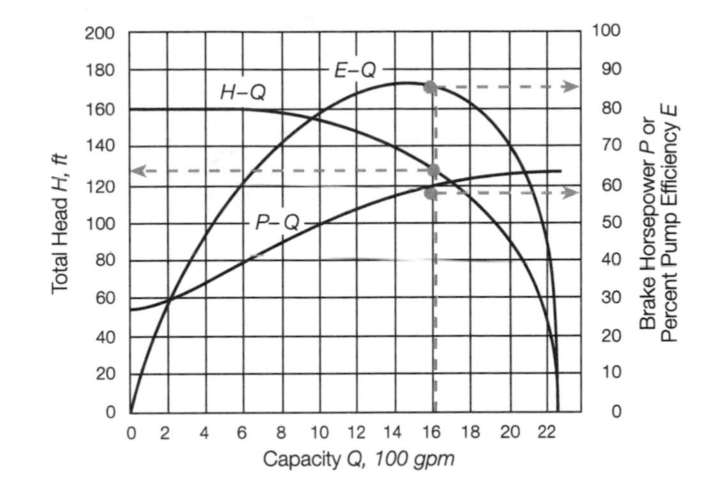

FIGURE H6-25 Pump curve for example 27

this point, move *horizontally to the right* toward the efficiency scale. The reading is about 85 percent.

From the known pump capacity of 1,600 gpm, then, the following characteristics were determined:

brake horsepower P = 60 bhp

pump head H = 128 ft

pump efficiency E = 85%

Example 28

The total head against which a pump must operate is 35 ft. Using the pump curve shown in Figure H6-26, determine pump capacity, power, and efficiency at this head.

As mentioned in the previous example, the key in pump curve problems is to find a certain vertical line and to note where it intersects all three characteristic curves. Since the given characteristic is head (35 ft), locate the value 35 ft on the total head scale and draw a horizontal line to intersect with the head curve (*H–Q*). Draw a vertical line through the point where the horizontal line intersects the curve.

The vertical line enables you to determine the three unknown characteristics.

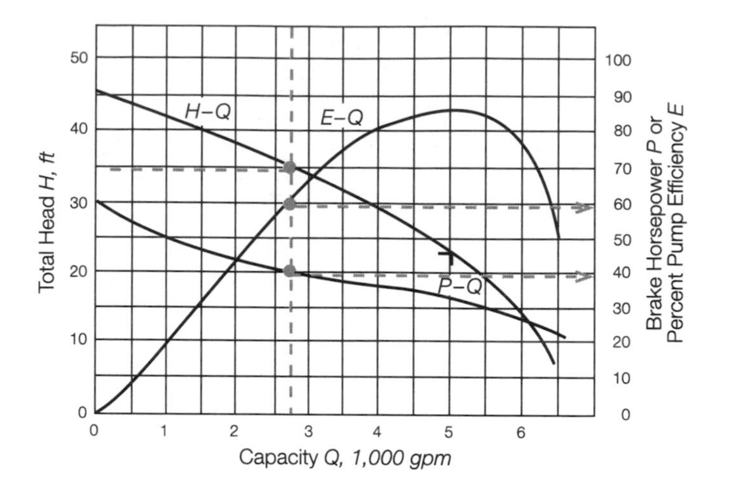

FIGURE H6-26 Pump curve for example 28

First, the line falls halfway between 2.5 and 3 on the *capacity* scale. Therefore the reading on the scale is 2.75, which indicates a capacity of 2,750 gpm (the scale is given in 1,000 gpm).

Following the vertical line upward, you will first come to the intersection on the *P–Q* curve. As before, to determine the power at this point move *horizontally to the right* toward the bhp scale. About 40 bhp is indicated.

The next intersection point up on the vertical line is the *E–Q* curve. To determine pump efficiency at this point, move *horizontally to the right* toward the efficiency scale. The reading is about 60 percent.

The last intersection with the vertical line is at the *H–Q* curve. You already know that characteristic to be 35 ft.

Given a head of 35 ft, then, the following characteristics were determined:

pump capacity = 2,750 gpm

brake horsepower = 40 bhp

pump efficiency = 60%

Example 29

Your pump is producing 5,500 gpm, and your motor is 85 percent efficient. Use the pump curve in Figure H6-27 to determine brake horsepower, and then calculate the motor horsepower.

First draw the vertical line at the point corresponding with 5,500 gpm. Where this line intersects the *P–Q* curve, brake horsepower can be determined.

On this graph, brake horsepower is given on the right scale. Therefore, the brake horsepower that corresponds with a flow rate of 5,500 gpm is about 129 bhp.

So far, the brake horsepower and motor efficiency are known:

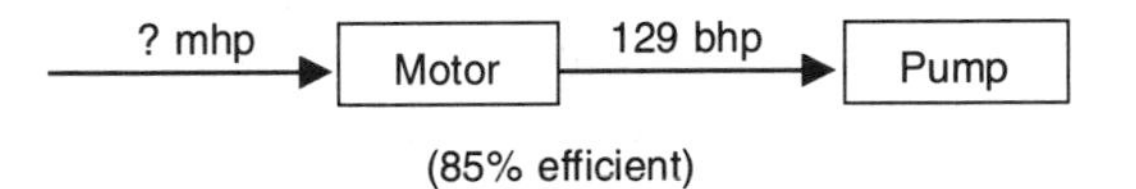

Use this information to calculate the motor horsepower:

$$(85\%)\ (x \text{ mhp}) \quad = \quad 129 \text{ bhp}$$

Express the percent as a decimal number:

$$(0.85)\ (x \text{ mhp}) \quad = \quad 129 \text{ bhp}$$

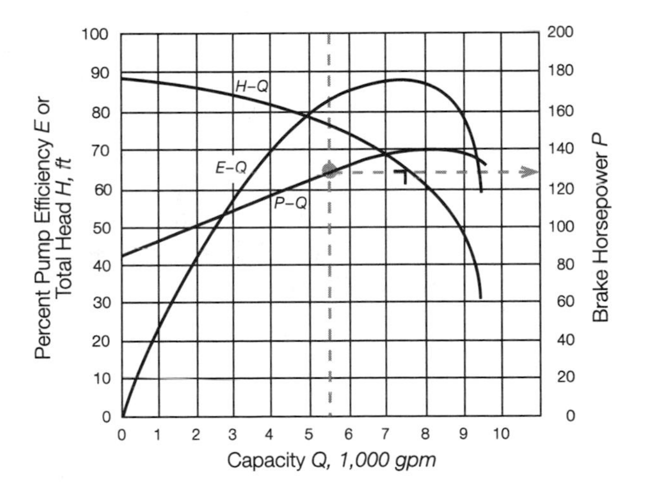

FIGURE H6-27
Pump curve for example 29

And then solve for the unknown value:

$$x = \frac{129}{0.85}$$

$$x = 151.76 \text{ mhp}$$

Example 30

A pump operates against a total head of 32 ft. The motor is 78 percent efficient. Power costs \$0.03/kW•h.

Use the pump curve in Figure H6-28 to determine brake horsepower required under the given conditions. Then calculate the cost of operating the pump for 100 hours.

To locate the brake horsepower that corresponds with a total head of 32 ft, first draw a horizontal line from the value 32 ft on the total head scale until the line intersects the *H–Q* curve. Draw a vertical line through the intersection.

The vertical line intersects the *P–Q* curve at a point corresponding to about 75 bhp. Using this figure and the motor efficiency given in the problem, calculate the motor horsepower required.

$$(78\%)\,(x \text{ mhp}) = 75 \text{ bhp}$$

Express the percent as a decimal number:

$$(0.78)\,(x \text{ mhp}) = 75 \text{ bhp}$$

And then solve for the unknown value:

$$x = \frac{75}{0.78}$$

$$x = 96.15 \text{ mhp}$$

Once the motor horsepower requirement has been determined, power demand can be expressed in terms of kilowatts (recall that 1 hp = 0.746 kW).

$$96.15 \text{ hp} = (96.15)\,(0.746 \text{ kW})$$

$$= 71.73 \text{ kW demand}$$

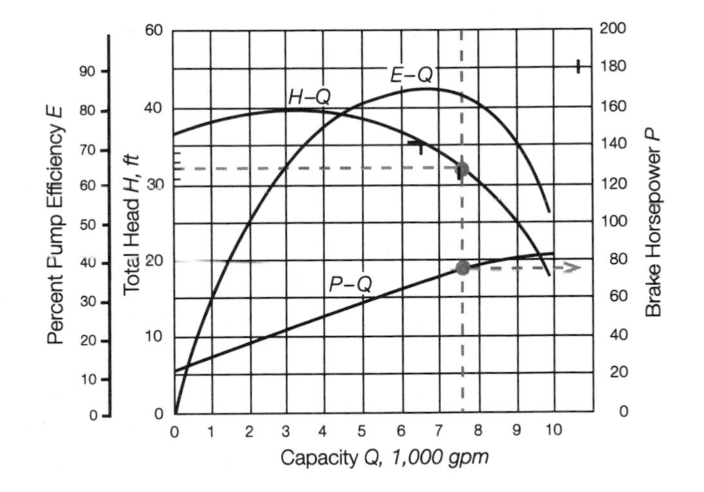

FIGURE H6-28 Pump curve for example 30

In 100 hours, then, the number of kilowatt-hours used is

$$(71.73 \text{ kW}) (100 \text{ h}) \quad = \quad 7{,}173 \text{ kW·h}$$

At a cost of \$0.03/kW·h, the cost of operating the pump for one day is

$$(7{,}173 \text{ kW·h}) (\$0.03/\text{kW·h}) \quad = \quad \$215.19$$

HYDRAULICS 7

Flow Rate Problems

The measurement of water flow rates is an important part of efficient water supply operations. Flow rate information is used in determining chemical dosages, assessing the cost of treatment, evaluating the plant efficiency, determining the amounts of water used by various commercial and industrial sources, and planning for future expansion of the treatment facility and its distribution system.

Two types of flow rates are important in water supply operations: (1) *instantaneous flow rates* (flow rates at a particular moment) and (2) *average flow rates* (the average of the instantaneous flow rates over a given period of time, such as a day). Instantaneous flow rates are calculated based on the cross-sectional area and velocity of water in a channel or pipe. Such calculations are discussed in the first part of this chapter. Average flow rates are calculated from records of instantaneous flow rates or from records of time and total volume, as discussed in the Mathematics portion of this book.[1]

Flow rates in water supply operations are measured by various metering devices, such as weirs or Parshall flumes for open-channel flow, and venturi or orifice meters for closed conduit flow. The second part of this chapter illustrates how graphs and tables are used in determining flow rates through these devices.

[1]Mathematics 16, Average Daily Flow.

Instantaneous Flow Rate Calculations

The flow rate of water through a channel or pipe at a particular moment depends on the cross-sectional area[2] and the velocity of the water moving through it. This is stated mathematically as follows:

$$Q = AV$$

Where:

Q = flow rate
A = area
V = velocity

Figure H7-1A illustrates the $Q = AV$ equation as it pertains to flow in an open channel. Since flow rate and velocity must be expressed for the same unit of time, the flow rate in the open channel is expressed as

Q		A	V
flow rate	=	(width) (depth)	(velocity)
ft^3/time		ft ft	ft/time

In using the $Q = AV$ equation, the time units given for the velocity must match the time units for cubic-feet flow rate. For example, if the velocity is expressed as feet per second (ft/s), then the resulting flow rate must be expressed as cubic feet per second (ft^3/s). If the velocity is expressed as feet per minute (ft/min), then the resulting flow rate must be expressed as cubic feet per minute (ft^3/min). And if the velocity is expressed as feet per day (ft/d), then the resulting flow rate must be expressed as cubic feet per day (ft^3/d). Figures H7-1B through H7-1D illustrate this concept.

Using the $Q = AV$ equation, if you know the cross-sectional area of the water in the channel (the width and depth of the rectangle), and you know the velocity, then you will be able to calculate the flow rate. This approach to flow rate calculations can be used if the problem involves an open-channel flow with a rectangular cross section, as shown previously, or if the problem involves a pipe flow with a circular cross section, as shown in Figure H7-1E. Only the calculation of the cross-sectional area will differ.

---

[2]Mathematics 9, Area Measurements.

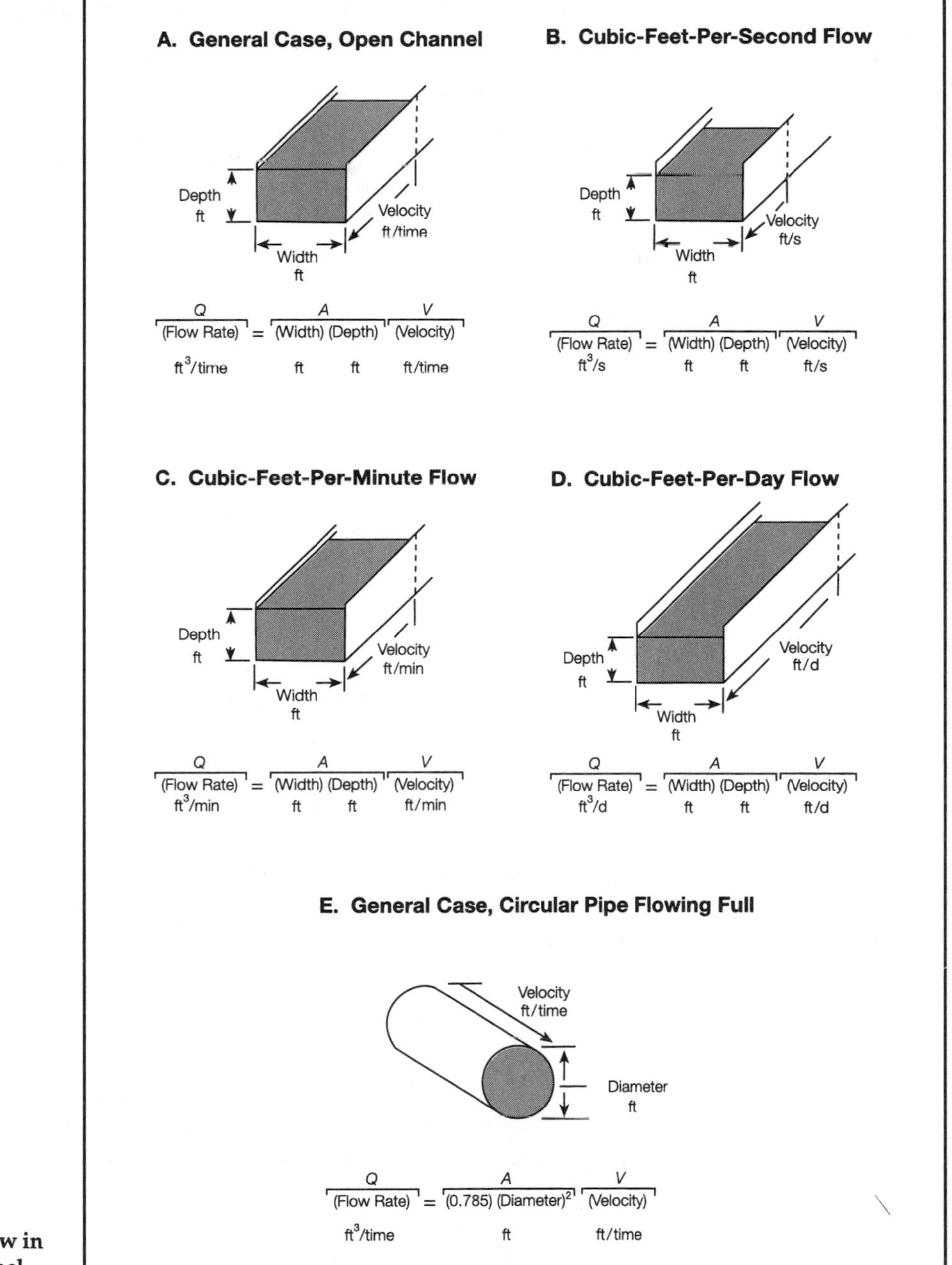

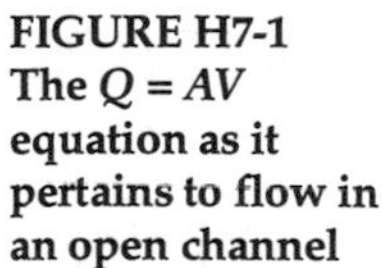
FIGURE H7-1
The $Q = AV$ equation as it pertains to flow in an open channel

If a circular pipe is flowing full (as it will be in most situations in water supply and treatment), the resulting flow rate is expressed as

Q		A	V
(flow rate)	=	(0.785) (diameter^2)	(velocity)
ft^3/time		ft^2	ft/time

As in the case of the rectangular channel, if the velocity is expressed as feet per second, then the resulting flow rate must be expressed as cubic feet per second. If the velocity is expressed as feet per minute, then the resulting flow rate must be expressed as cubic feet per minute. If the velocity is expressed as feet per day, then the resulting flow rate must be expressed as cubic feet per day.

Example 1

A 15-in.-diameter pipe is flowing full. What is the gallons-per-minute flow rate in the pipe if the velocity is 110 ft/min?

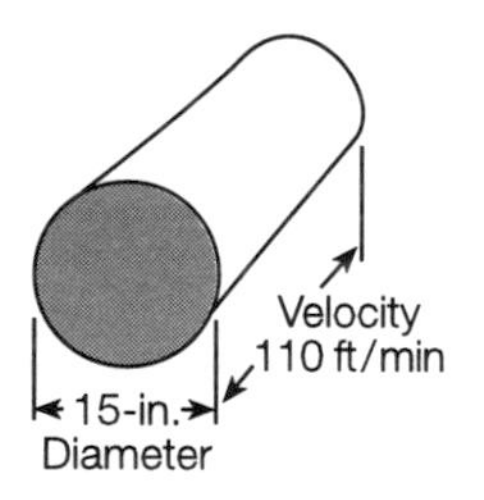

To use the $Q = AV$ equation, the diameter and velocity terms should be expressed using feet. Therefore, convert the 15-in. diameter to feet:[3]

$$\frac{15 \text{ in.}}{12 \text{ in./ft}} = 1.25 \text{ ft}$$

Now use the $Q = AV$ equation to calculate the flow rate. Since the velocity is expressed in *feet per minute*, first calculate the cubic-feet-per-minute flow rate, then convert to gallons per minute:

Q		A	V
(flow rate)	=	(0.785) (diameter)^2	(velocity)

[3]Mathematics 11, Conversions.

$$= (0.785)\ (1.25\ \text{ft})\ (1.25\ \text{ft})\ (110\ \text{ft/min})$$

$$= 134.92\ \text{ft}^3/\text{min}$$

Convert cubic feet per minute to gallons per minute:[4]

$$(134.92\ \text{ft}^3/\text{min})\ (7.48\ \text{gal/ft}^3) = 1{,}009\ \text{gpm}$$

Example 2

What is the million-gallons-per-day flow rate through a channel that is 3 ft wide with the water flowing to a depth of 16 in. at a velocity of 2 ft/s?

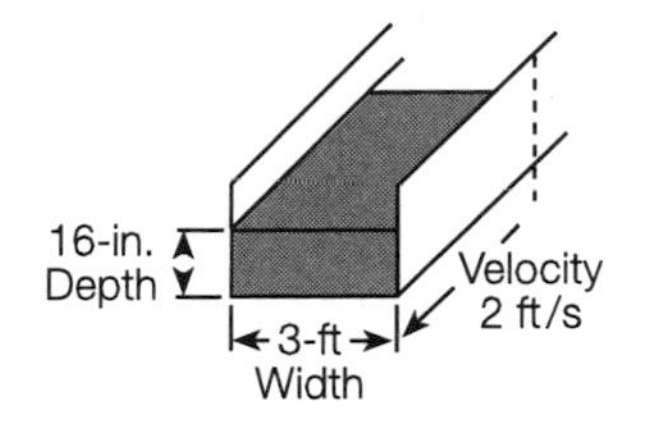

Before beginning the calculation, convert the 16-in. depth to feet:

$$\frac{16\ \text{in.}}{12\ \text{in./ft}} = 1.33\ \text{ft}$$

Now use the $Q = AV$ equation to calculate flow rate through the channel. Since the velocity is given in feet per second, first calculate the cubic-feet-per-second flow rate, then convert to gallons per day:

$$\underset{\text{flow rate}}{Q} = \underset{\text{(width) (depth)}}{A} \quad \underset{\text{(velocity)}}{V}$$

$$= (3\ \text{ft})\ (1.33\ \text{ft})\ (2\ \text{ft/s})$$

$$= 7.98\ \text{ft}^3/\text{s}$$

Convert cubic feet per second to gallons per day:

$$(7.98\ \text{ft}^3/\text{s})\ (60\ \text{s/min})\ (7.48\ \text{gal/ft}^3)\ (1{,}440\ \text{min/d}) = 5{,}157{,}251\ \text{gpd}$$

[4]Mathematics 11, Conversions.

The "millions comma" in a gallons-per-day flow rate always indicates the million-gallons-per-day flow rate. In this example,

5,157,251 gpd = 5.16 mgd

↑

millions comma

In examples 1 through 3 in this chapter, Q is the unknown value in the formula. However, in some problems, V or A might be the unknown value. Such problems would be set up in the same basic manner as examples 1 through 3; then the equations would be solved for the unknown value.[5]

Example 3

A channel is 1 m wide with water flowing to a depth of 0.6 m. The velocity in the channel is found to be 0.55 m/s. What is the flow rate in the channel in cubic meters per second?

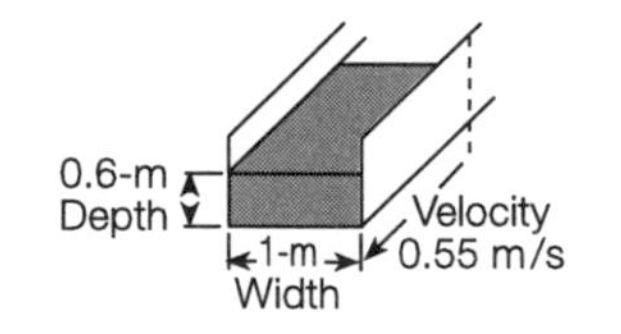

$$
\begin{aligned}
\underset{\text{(flow rate)}}{Q} &= \underset{\text{(width) (depth)}}{A} \quad \underset{\text{(velocity)}}{V} \\
&= (1 \text{ m})\ (0.6 \text{ m})\ (0.55 \text{ m/s}) \\
&= 0.33 \text{ m}^3\text{/s}
\end{aligned}
$$

Example 4

A channel is 3 ft wide. If the flow in the channel is 7.5 mgd and the velocity of the flow is 185 ft/min, what is the depth (in feet) of water in the channel?

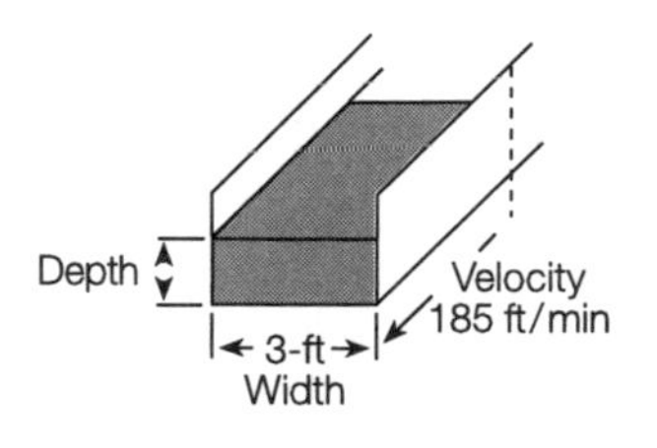

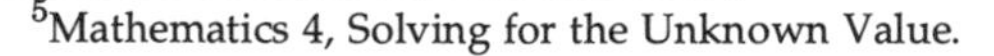

[5]Mathematics 4, Solving for the Unknown Value.

The velocity is given in *feet per minute*. The corresponding flow rate is therefore *cubic feet per minute*. So before continuing with the $Q = AV$ calculation, convert the million-gallons-per-day flow rate given in the problem to gallons-per-day flow rate and then to cubic-feet-per-minute flow rate:

$$7.5 \text{ mgd} = 7{,}500{,}000 \text{ gpd}$$

↑ *millions comma*

Convert gallons-per-day flow rate to cubic feet per minute:

$$\frac{7{,}500{,}000 \text{ gpd}}{(1{,}440 \text{ min/d})\ (7.48 \text{ gal/ft}^3)} = 696.3 \text{ ft}^3\text{/min}$$

Now that the flow rate is expressed in the correct units, use the $Q = AV$ equation to determine the depth of water in the channel:

$$\underset{\text{flow rate}}{Q} = \underset{\text{(width) (depth)}}{A} \quad \underset{\text{(velocity)}}{V}$$

$$696.3 \text{ ft}^3\text{/min} = (3 \text{ ft})\ (x)\ (185 \text{ ft/min})$$

Then solve for the unknown value:

$$\frac{696.3}{(3)\ (185)} = x$$

$$1.25 \text{ ft} = x$$

In estimating flow in an open channel, you will usually be able to determine the approximate cross-sectional area A without much difficulty. In addition, you will often have to estimate the velocity V of the flow in the channel so that flow rate Q may be determined.

You may estimate flow velocity by placing a float (a stick or a cork) in the water and recording the time it takes to travel a measured distance. Usually the flow on the surface is slightly greater or less than the average flow velocity, so the estimated velocity will be fairly inaccurate.

Example 5

A float is placed in a channel. It takes 2.5 min to travel 300 ft. What is the flow velocity in feet per minute in the channel? (Assume the float is traveling at the average velocity of the water.)

This problem asks for the velocity in feet per minute:

$$\text{velocity} = \frac{\text{feet}}{\text{minute}}$$

Fill in the information given:

$$\text{velocity} = \frac{300 \text{ ft}}{2.5 \text{ min}}$$

$$= 120 \text{ ft/min}$$

Example 6

A 305-mm-diameter pipe flowing full is carrying 35 L/s. What is the velocity of the water (in meters per second) through the pipe?

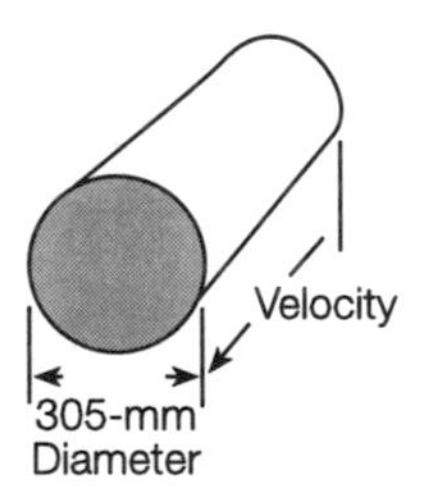

You are asked to determine the velocity V given the flow rate Q and diameter. Note first that the diameter of 305 mm is equivalent to 0.305 m. To use the $Q = AV$ formula to calculate velocity in a *pipe*, you should use the following mathematical setup:

$$\underset{\text{(flow rate)}}{Q} = \underset{(0.785)\ (\text{diameter})^2}{A} \ \underset{\text{(velocity)}}{V}$$

Since you want to know velocity in meters per second, the flow rate must also be expressed in terms of meters and seconds (cubic meters per second). The information given in the problem expresses the flow rate as *liters* per second. Therefore, the flow rate must be converted to cubic meters per second before you begin the $Q = AV$ calculation:

$$35 \text{ L/s} \times 0.001 \text{ m}^3\text{/L} = 0.035 \text{ m}^3\text{/s}$$

Now determine the velocity using the $Q = AV$ equation:

$$\begin{array}{ccc} Q & A & V \\ 0.035\ \text{m}^3/\text{s} = & (0.785)\ (0.305\ \text{m})^2 & (x) \end{array}$$

And then solve for the unknown value:

$$\frac{0.035\ \text{m}^3/\text{s}}{(0.785)\ (0.305\ \text{m})^2} = x$$

$$0.48\ \text{m/s} = x$$

Example 7

A cork is placed in a channel and travels 10 m in 20 s. What is the velocity of flow in meters per second?

In this problem, the velocity is requested in meters per second:

$$\text{velocity} = \frac{\text{meters}}{\text{second}}$$

Fill in the given information:

$$\text{velocity} = \frac{10\ \text{m}}{20\ \text{s}}$$

$$= 0.5\ \text{m/s}$$

Example 8

A channel is 4 ft wide with water flowing to a depth of 2.3 ft. If a float placed in the channel takes 3 min to travel a distance of 500 ft, what is the cubic-feet-per-minute flow rate in the channel?

Before using the $Q = AV$ equation to determine the flow rate in the channel, calculate the estimated velocity:

$$\text{velocity} = \frac{\text{feet}}{\text{min}}$$

$$= \frac{500\ \text{ft}}{3\ \text{min}}$$

$$= 166.67\ \text{ft/min}$$

Now use this information, along with channel width and depth, to determine the cubic-feet-per-minute flow rate in the channel:

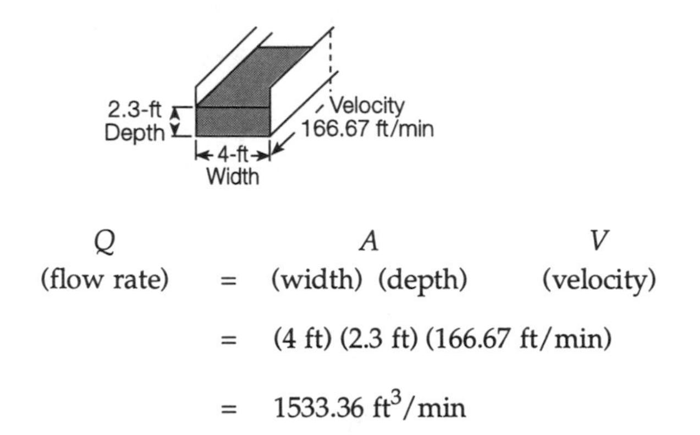

Q		A	V
(flow rate)	=	(width) (depth)	(velocity)
	=	(4 ft) (2.3 ft) (166.67 ft/min)	
	=	1533.36 ft^3/min	

Once you have learned how to solve $Q = AV$ problems, you can solve more complex problems dealing with flow rate. The *rule of continuity* states that the flow Q that enters a system must also be the flow that leaves the system. Mathematically, this can be stated as follows:

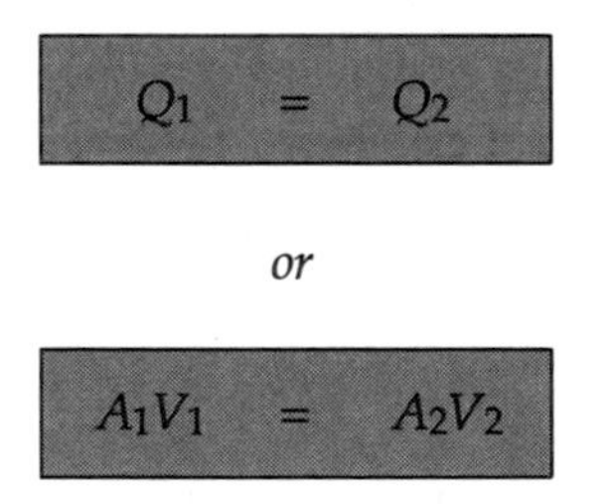

$$Q_1 = Q_2$$

or

$$A_1V_1 = A_2V_2$$

The following examples illustrate the rule of continuity.

Example 9

If the velocity in the 10-in.-diameter section of pipe shown in the following diagram is 3.5 ft/s, what is the feet-per-second velocity in the 8-in.-diameter section?

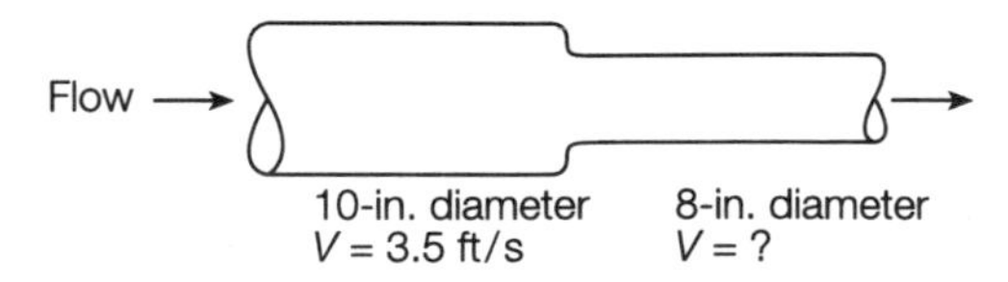

According to the rule of continuity, the flow rate Q in the 10-in. section must equal the flow rate Q in the 8-in. section:

$$Q_1 = Q_2 \text{ and } A_1V_1 = A_2V_2$$

To calculate the velocity V_2 in the 8-in. pipe, fill in the information given in the problem and solve for the unknown value. But first, as in the previous examples, express the pipe diameters in feet:

$$\frac{10\text{ in.}}{12\text{ in./ft}} = 0.83\text{ ft} \qquad \frac{8\text{ in.}}{12\text{ in./ft}} = 0.67\text{ ft}$$

Now fill in the information given in the problem:

$$\underbrace{(0.785)(0.83\text{ ft})(0.83\text{ ft})}_{A_1}\;\underbrace{(3.5\text{ ft/s})}_{V_1} = \underbrace{(0.785)(0.67\text{ ft})(0.67\text{ ft})}_{A_2}\;\underbrace{(x\text{ ft/s})}_{V_2}$$

Solve for the unknown value:

$$\frac{(0.785)(0.83)(0.83)(3.5)}{(0.785)(0.67)(0.67)} = x$$

$$\frac{1.89}{0.35} = x$$

$$5.4\text{ ft/s} = x$$

Example 10

The flow entering the leg of a tee connection is 0.25 m^3/s, as shown in the diagram below. If the flow through one branch of the tee is 0.14 m^3/s, what is the flow through the other branch?

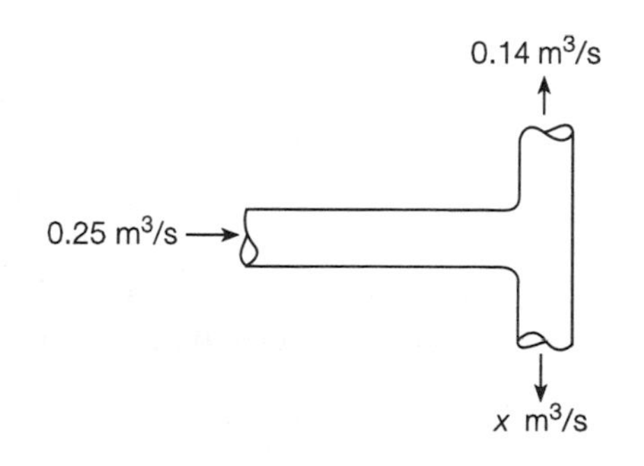

According to the rule of continuity, the flow rate Q going into the tee connection will be the same as the flow rate Q coming out of the tee connection. Mathematically, this is expressed as follows:

$$0.25 \text{ m}^3/\text{s} = 0.14 \text{ m}^3/\text{s} + x \text{ m}^3/\text{s}$$

Therefore, $x = 0.25 \text{ m}^3/\text{s} - 0.14 \text{ m}^3/\text{s} = 0.11 \text{ m}^3/\text{s}$.

Example 11

Given the following diagram and information, determine the velocity in feet per second at points A, B, and C.

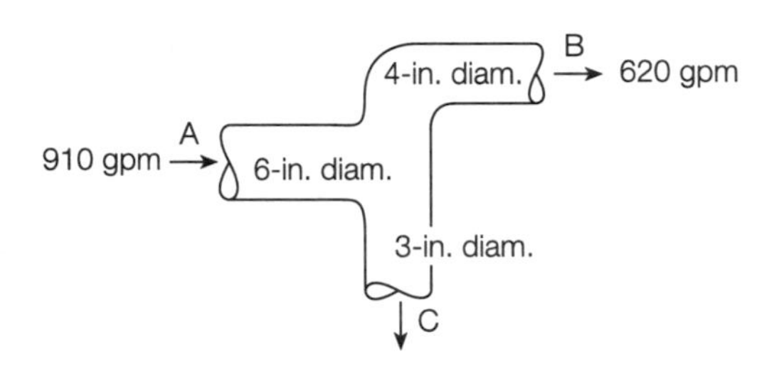

To determine the velocity at points A, B, and C, you will need to know the values Q and A at each point. First, determine the flow rate at point C. The flow rate into the system is 910 gpm. This must also be the flow rate out of the system, so

$$\underset{\text{(point A)}}{910 \text{ gpm}} = \underset{\text{(point B)}}{620 \text{ gpm}} + \underset{\text{(point C)}}{x \text{ gpm}}$$

The flow at point C is:

$$\begin{array}{rl} 910 & \text{gpm} \\ -\ 620 & \text{gpm} \\ \hline 290 & \text{gpm} \end{array}$$

You now know all the information needed to determine the velocity at points A, B, and C. Therefore, the $Q = AV$ equation can be used to solve for the velocity at each point.

Since the velocity in this problem must be expressed as feet per second, each of the flow rates should be converted to the corresponding units (cubic feet per second). In addition, the pipe diameters given in inches should be converted to feet.[6]

[6]Mathematics 11, Conversions.

gallons per minute to cubic feet per second conversions

$$\frac{910 \text{ gpm}}{(7.48 \text{ gal/ft}^3)(60 \text{ s/min})} = 2.03 \text{ ft}^3\text{/s}$$

$$\frac{620 \text{ gpm}}{(7.48 \text{ gal/ft}^3)(60 \text{ s/min})} = 1.38 \text{ ft}^3\text{/s}$$

$$\frac{290 \text{ gpm}}{(7.48 \text{ gal/ft}^3)(60 \text{ s/min})} = 0.65 \text{ ft}^3\text{/s}$$

inches to feet conversions

$$\frac{6 \text{ in.}}{12 \text{ in./ft}} = 0.5\text{-ft diameter}$$

$$\frac{4 \text{ in.}}{12 \text{ in./ft}} = 0.33\text{-ft diameter}$$

$$\frac{3 \text{ in.}}{12 \text{ in./ft}} = 0.25\text{-ft diameter}$$

Based on these calculations, the revised diagram is as follows:

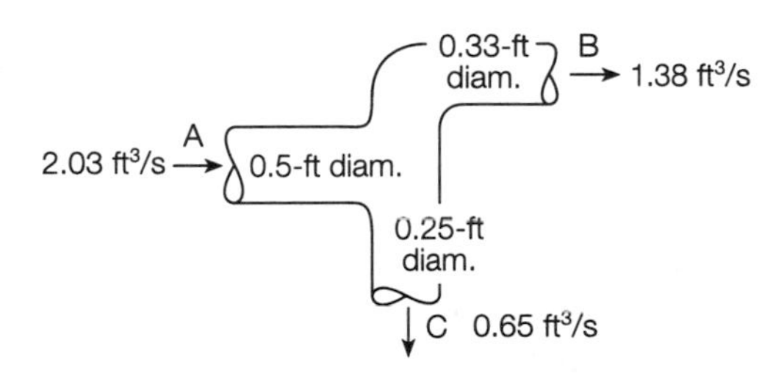

Now calculate the velocity at each of the three points by filling in the given information for each point:

At point A:

$$Q_A = A_A V_A$$

$$2.03 \text{ ft}^3\text{/s} = (0.785)(0.5 \text{ ft})(0.5 \text{ ft})(V_A \text{ ft/s})$$

$$\frac{2.03}{(0.785)(0.5)(0.5)} = V_A \text{ ft/s}$$

$$10.34 \text{ ft/s} = V_A \text{ ft/s}$$

At point B:

$$Q_B = A_B V_B$$

$$1.38 \text{ ft}^3/\text{s} = (0.785)\,(0.33 \text{ ft})\,(0.33 \text{ ft})\,(V_B \text{ ft/s})$$

$$\frac{1.38}{(0.785)\,(0.33)\,(0.33)} = V_B \text{ ft/s}$$

$$16.14 \text{ ft/s} = V_B \text{ ft/s}$$

At point C:

$$Q_C = A_C V_C$$

$$0.65 \text{ ft}^3/\text{s} = (0.785)\,(0.25 \text{ ft})\,(0.25 \text{ ft})\,(V_C \text{ ft/s})$$

$$\frac{0.65}{(0.785)\,(0.25)\,(0.25)} = V_C \text{ ft/s}$$

$$13.25 \text{ ft/s} = V_C \text{ ft/s}$$

Flow-Measuring Devices

Most devices for measuring flow rate can be equipped with instrumentation that will give a continuous recording of flow rates entering or leaving the treatment plant. To calibrate the instrumentation, and in some situations where there is no automatic continuous monitoring, you must be able to calculate the flow rates through the measuring devices. Although the most accurate interpretation of readings from these devices usually involves calculations beyond the scope of this handbook, graphs and tables are available for day-to-day use in determining flow rates. The following discussion provides examples and illustrates the use of a few such graphs and tables.[7]

[7]Mathematics 12, Graphs and Tables.

Weirs

The two most commonly used weirs are the V-notch and rectangular weirs, illustrated in Figure H7-2. To read a flow rate graph or table pertaining to a weir, you must know two measurements: (1) the height H of the water above the weir crest; and (2) the *angle* of the weir (V-notch weir) *or* the *length* of the crest (rectangular weir).

Example 12

Given the discharge curve in Figure H7-3 for a 60° V-notch weir, what is the flow rate (in gallons per day) when the height of the water above the weir crest is 3 in.?

Since the height, or head above the weir crest, is 3 in., enter the *head* scale of Figure H7-3 at 3. This head indicates a flow of 30,000 gpd.

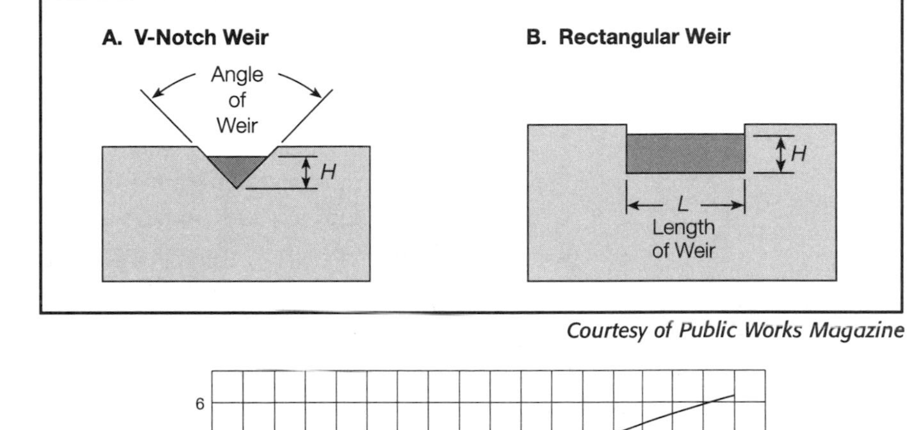

FIGURE H7-2
Types of weirs

Courtesy of Public Works Magazine

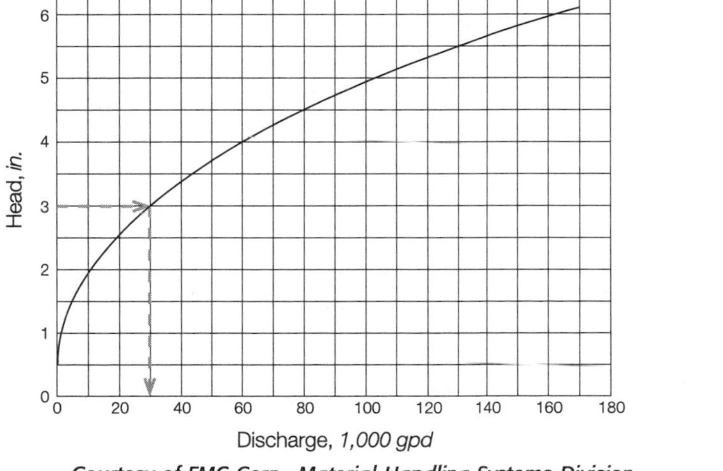

FIGURE H7-3
Discharge curve, 60° V-notch weir

Courtesy of FMC Corp., Material Handling Systems Division

Example 13

Given the discharge curves shown in Figure H7-4 for 60° and 90° V-notch weirs, (a) what is the flow rate (in million gallons per day) when the height (head of the water above the 60° V-notch weir crest) is 6 in.? (b) What is the flow rate (in gallons per day) when the height of water above the 90° V-notch weir crest is 3.5 in.?

First, notice that the part of the graph for 60° V-notch weirs presents information similar to the graph used in example 12. The principal difference between the two graphs is that the graph in example 12 uses arithmetic scales, whereas the graph in this problem uses logarithmic scales.

(a) Enter the *head* scale at 6 in.; this head indicates a million-gallons-per-day flow rate somewhere between 0.15 and 0.2 mgd for the 60° V-notch weir.

(b) The flow rate that corresponds with a head of 3.5 in. is 0.078 mgd for the 90° V-notch weir. This value is equivalent to 78,000 gpd.

Example 14

A nomograph for 60° and 90° V-notch weirs is given in Figure H7-5. Using this nomograph, determine (a) the flow rate in gallons per minute if the height of water above the 60° V-notch weir crest is 12 in.; (b) the gallons-per-day flow rate over a 90° V-notch weir when the height of the water is 12 in. over the crest.

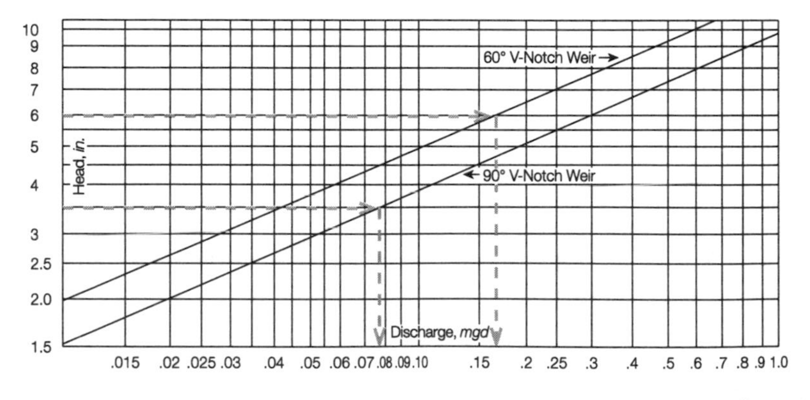

FIGURE H7-4 Discharge curves, 60° and 90° V-notch weirs

Courtesy of Leeds & Northrup

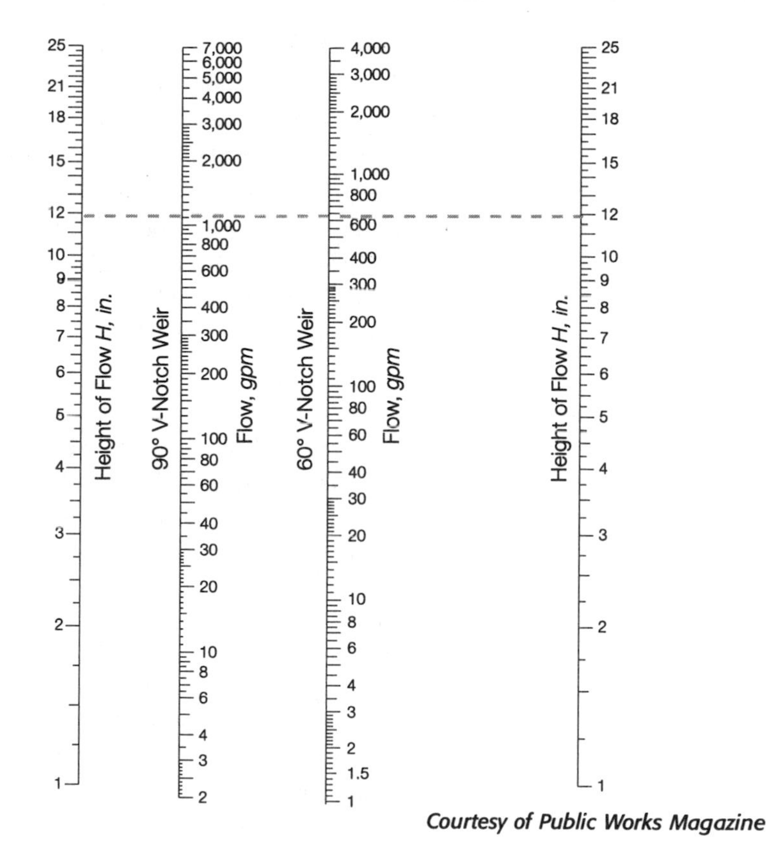

FIGURE H7-5 Flow rate nomograph for 60° and 90° V-notch weirs

Courtesy of Public Works Magazine

(a) As in the previous example, the scales used on this graph are logarithmic. This information is important because it determines how interpolation should be performed when the indicated flow falls between two known values.

First, draw a horizontal line from 12 on the *height* scale on the left to 12 on the *height* scale on the right. Then on the scale for a 60° V-notch, read the flow rate indicated by 12 in. head. The flow rate falls between 600 and 700 gpm at approximately 650 gpm.

(b) On the scale for a 90° V-notch, the indicated flow rate is between 1,000 and 2,000 gpm. More precisely, it falls between 1,100 and 1,200 gpm at a reading of about 1,150 gpm. Convert gallons-per-minute rate to gallons per day:

$$(1{,}150 \text{ gpm})\ (1{,}440 \text{ min/d}) \quad = \quad 1{,}656{,}000 \text{ gpd}$$

Example 15

Table H7-1 pertains to the discharge of 45° V-notch weirs. Use the table to determine (a) flow rate in cubic feet per second when the head above the crest is 0.75 ft; (b) the gallons-per-day flow rate when the head is 1.5 ft.

(a) In the table, part of the head (0.7) is given on the vertical scale, and the remainder (0.05) is given on the horizontal scale (0.7 + 0.05 = 0.75). The cubic-feet-per-second flow rate indicated by a head of 0.75 ft is 0.504 ft^3/s.

(b) A head of 1.5 ft is read as 1.5 on the vertical scale and 0.00 on the horizontal scale (1.5 + 0.00 = 1.50). The million-gallons-per-day flow rate indicated by this head is 1.84 mgd. This is equal to a flow rate of 1,840,000 gpd. (The *mgd* column was read in this problem because it is easier to convert to gallons per day from million gallons per day than from cubic feet per second.)

Example 16

A discharge curve for a 15-in. contracted rectangular weir is given in Figure H7-6. Using this curve, determine the gallons-per-day flow rate if the head above the weir crest is 2.75 in.

The scales for the discharge curve are arithmetic, so it is much easier to determine the indicated flow rate when the arrow falls between two known values. In this case, a head of 3.75 ft indicates a flow rate between 250,000 and 300,000 gpd, about 285,000 gpd.

Example 17

A nomograph for rectangular weirs (contracted and suppressed) is given in Figure H7-7. Using this nomograph, determine (a) the flow in gallons per minute over a suppressed rectangular weir if the length of the weir is 3 ft and the height of the water over the weir is 4 in.; (b) the flow in gallons per minute over a contracted rectangular weir, for the same weir length and head as in (a).

To use the nomograph, you must know the difference between a contracted rectangular weir (one *with* end contractions) and a suppressed rectangular weir (one *without* end contractions). As shown in Figure H7-8, a contracted rectangular weir comes in somewhat from the side of the channel before the crest cutout begins. On a suppressed rectangular weir, however, the crest cutout stretches from one side of the channel to the other.

TABLE H7-1 Discharge of 45° V-notch weirs

	.00		.01		.02		.03		.04		.05		.06		.07		.08		.09	
Head, *ft*	ft^3/s	*mgd*	ft^3/s	*mgd*	ft^3/s	*mgd*	ft^3/s	*mgd*	ft^3/s	*mgd*	ft^3/s	*mgd*	ft^3/s	*mgd*	ft^3/s	*mgd*	ft^3/s	*mgd*	ft^3/s	*mgd*
0.1	.003	.002	.004	.003	.005	.003	.006	.004	.008	.005	.009	.006	.011	.007	.012	.008	.014	.009	.016	.011
0.2	.019	.012	.021	.014	.024	.015	.026	.017	.029	.019	.032	.021	.036	.023	.039	.025	.043	.028	.047	.030
0.3	.051	.033	.055	.036	.060	.039	.065	.042	.070	.045	.075	.048	.081	.052	.086	.056	.092	.060	.098	.064
0.4	.105	.068	.111	.072	.118	.077	.126	.081	.133	.086	.141	.091	.149	.096	.157	.101	.165	.107	.174	.112
0.5	.183	.118	.192	.124	.202	.130	.212	.137	.222	.143	.232	.150	.243	.157	.254	.164	.265	.171	.277	.179
0.6	.289	.187	.301	.194	.313	.203	.326	.211	.339	.219	.353	.228	.366	.237	.380	.246	.395	.255	.410	.265
0.7	.425	.274	.440	.284	.455	.294	.471	.305	.488	.315	.504	.326	.521	.337	.539	.348	.556	.360	.574	.371
0.8	.593	.383	.611	.395	.630	.407	.650	.420	.670	.433	.690	.446	.710	.459	.731	.472	.752	.486	.774	.500
0.9	.796	.514	.818	.529	.841	.543	.864	.558	.887	.573	.911	.589	.935	.604	.960	.620	.985	.636	1.01	.653
1.0	1.04	.669	1.06	.686	1.09	.703	1.11	.721	1.14	.738	1.17	.756	1.20	.774	1.23	.793	1.25	.811	1.28	.830
1.1	1.31	.849	1.34	.869	1.37	.888	1.41	.908	1.44	.929	1.47	.949	1.50	.970	1.53	.991	1.57	1.01	1.60	1.03
1.2	1.63	1.06	1.67	1.08	1.70	1.10	1.74	1.12	1.77	1.15	1.81	1.17	1.84	1.19	1.88	1.22	1.92	1.24	1.96	1.26
1.3	1.99	1.29	2.03	1.31	2.07	1.34	2.11	1.36	2.15	1.39	2.19	1.42	2.23	1.44	2.27	1.47	2.32	1.50	2.36	1.52
1.4	2.40	1.55	2.44	1.58	2.49	1.61	2.53	1.64	2.58	1.66	2.62	1.69	2.67	1.72	2.71	1.75	2.76	1.78	2.81	1.81
1.5	2.85	1.84	2.90	1.87	2.95	1.91	3.00	1.94	3.05	1.97	3.10	2.00	3.15	2.03	3.20	2.07	3.25	2.10	3.30	2.13
1.6	3.35	2.17	3.41	2.20	3.46	2.23	3.51	2.27	3.57	2.30	3.62	2.34	3.68	2.38	3.73	2.41	3.79	2.45	3.84	2.48
1.7	3.90	2.52	3.96	2.56	4.02	2.60	4.08	2.63	4.13	2.67	4.19	2.71	4.25	2.75	4.32	2.79	4.38	2.83	4.44	2.87
1.8	4.50	2.91	4.56	2.95	4.63	2.99	4.69	3.03	4.75	3.07	4.82	3.11	4.89	3.16	4.95	3.20	5.02	3.24	5.08	3.29
1.9	5.15	3.33	5.22	3.37	5.29	3.42	5.36	3.46	5.43	3.51	5.50	3.55	5.57	3.60	5.64	3.64	5.71	3.69	5.78	3.74
2.0	5.86	3.79	5.93	3.83	6.00	3.88	6.08	3.93	6.15	3.98	6.23	4.03	6.31	4.08	6.38	4.13	6.46	4.18	6.54	4.23

Source: Adapted from Leupold & Stevens, Inc., P.O. Box 688, Beaverton, Oregon 97005, from Stevens Water Resources Data Book. *Formula:* $ft^3/s = 1.035\ H^{5/2}$; $mgd = ft^3/s \times 0.646$.

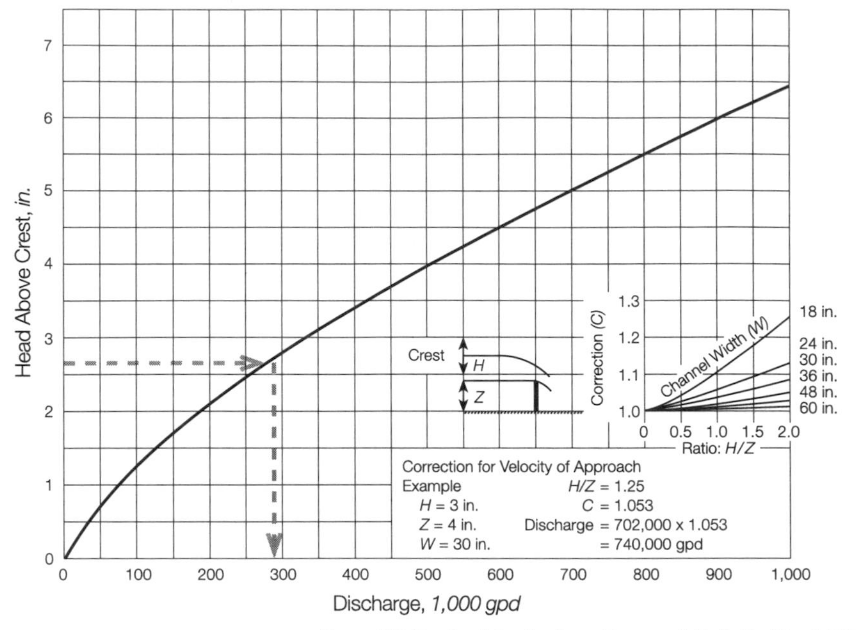

FIGURE H7-6
Discharge curve, 15-in. contracted rectangular weir

From US Dept. of Agriculture Farmers' Bulletin No. 1683

(a) To determine the flow over the suppressed weir, draw a line on Figure H7-7 from L = 3 ft on the left-hand scale, through H = 4 in. (right side of the middle scale). A flow rate of 850 gpm is indicated where the line crosses the right-hand scale. This is the flow over the suppressed rectangular weir.

(b) To determine the flow rate over a contracted rectangular weir using the nomograph, first determine the flow rate over a suppressed weir given the weir length and head, as in part (a). Then subtract the flow indicated on the middle scale.

In this example, the flow rate over a 3-ft-long suppressed weir with a head of 4 in. is 850 gpm. To determine the flow rate over a contracted weir 3 ft long with a head of 4 in., a correction factor must be subtracted from the 850 gpm. As indicated by the middle scale, the correction factor is 20 gpm.

850	gpm	suppressed rectangular weir
– 20	gpm	
830	gpm	contracted rectangular weir

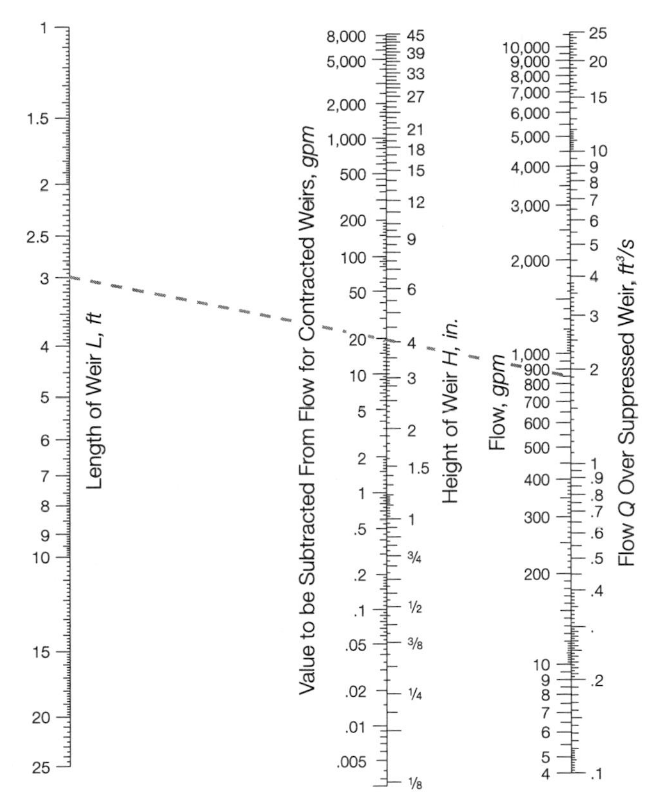

Courtesy of Public Works Magazine

FIGURE H7-7 Flow rate nomograph for rectangular weirs

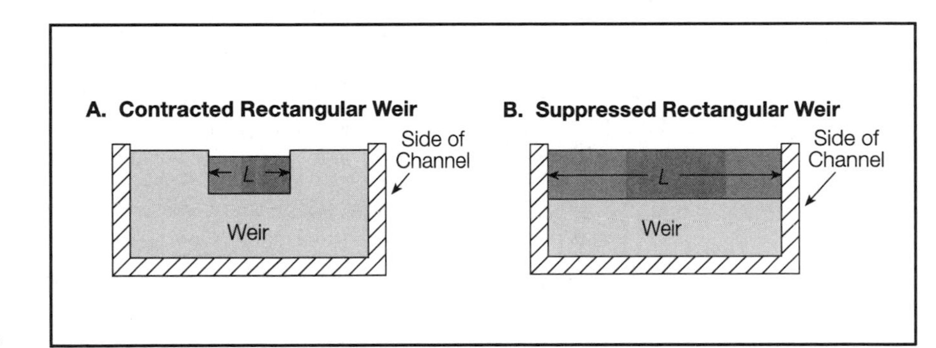

FIGURE H7-8 Two types of rectangular weirs

Example 18

Use Table H7-2 to determine the flow rate (in million gallons per day) over a contracted rectangular weir if the length of the weir crest is 3 ft and the head is 0.58 ft.

TABLE H7-2 Flow through contracted rectangular weirs

	Length of Weir Crest, *ft*											
	1		1 ½		2		3		4		5	
Head, *ft*	ft^3/s	*mgd*	ft^3/s	*mgd*	ft^3/s	*mgd*	ft^3/s	*mgd*	ft^3/s	*mgd*	ft^3/s	*mgd*
.36	.667	.431	1.026	.663	1.386	.895	2.105	1.360	2.824	1.824	3.543	2.289
.37	.695	.448	1.070	.690	1.445	.932	2.195	1.416	2.945	1.900	3.695	2.384
.38	.721	.465	1.111	.717	1.501	.969	2.281	1.473	3.061	1.976	3.841	2.480
.39	.748	.483	1.153	.745	1.559	1.006	2.370	1.530	3.181	2.054	3.992	2.577
.40	.775	.500	1.196	.772	1.617	1.044	2.459	1.588	3.301	2.132	4.143	2.676
.41	.802	.518	1.239	.800	1.676	1.083	2.550	1.647	3.424	2.216	4.298	2.776
.42	.830	.536	1.283	.829	1.736	1.121	2.642	1.707	3.548	2.292	4.454	2.877
.43	.858	.554	1.327	.857	1.797	1.160	2.736	1.767	3.675	2.373	4.614	2.979
.44	.886	.572	1.372	.886	1.858	1.200	2.830	1.827	3.802	2.455	4.774	3.082
.45	.915	.591	1.417	.915	1.920	1.240	2.925	1.889	3.930	2.538	4.935	3.187
.46	.943	.609	1.462	.945	1.982	1.280	3.021	1.951	4.060	2.621	5.099	3.292
.47	.972	.628	1.508	.974	2.045	1.320	3.118	2.013	4.191	2.706	5.264	3.399
.48	1.001	.646	1.554	1.004	2.108	1.361	3.215	2.076	4.322	2.791	5.429	3.506
.49	1.030	.665	1.601	1.034	2.172	1.403	3.314	2.140	4.456	2.878	5.598	3.615
.50	1.059	.684	1.647	1.064	2.236	1.444	3.413	2.204	4.590	2.965	5.767	3.725
.51	1.089	.703	1.695	1.095	2.302	1.486	3.515	2.269	4.728	3.052	5.941	3.835
.52	1.119	.722	1.743	1.126	2.368	1.529	3.617	2.335	4.866	3.141	6.115	3.947
.53	1.149	.742	1.791	1.156	2.434	1.571	3.719	2.401	5.004	3.230	6.289	4.060
.54	1.178	.761	1.838	1.188	2.499	1.614	3.820	2.467	5.141	3.321	6.462	4.174
.55	1.209	.781	1.888	1.219	2.567	1.658	3.925	2.534	5.283	3.411	6.641	4.288
.56	1.240	.800	1.938	1.251	2.636	1.701	4.032	2.602	5.428	3.503	6.824	4.404
.57	1.270	.820	1.986	1.282	2.703	1.745	4.136	2.670	5.569	3.595	7.002	4.520
.58	1.300	.840	2.035	1.314	2.771	1.790	4.242	2.739	5.713	3.689	7.184	4.638
.59	1.331	.859	2.085	1.347	2.840	1.830	4.349	2.808	5.858	3.783	7.367	4.757
.60	1.362	.879	2.136	1.380	2.910	1.879	4.458	2.879	6.006	3.877	7.554	4.876
.61	1.393	.899	2.186	1.412	2.980	1.924	4.567	2.948	6.154	3.972	7.741	4.997
.62	1.424	.920	2.237	1.444	3.050	1.970	4.676	3.019	6.302	4.068	7.928	5.118
.63	1.455	.940	2.287	1.477	3.120	2.015	4.785	3.090	6.450	4.165	8.115	5.240
.64	1.487	.960	2.339	1.510	3.192	2.061	4.897	3.162	6.602	4.262	8.307	5.363
.65	1.518	.980	2.390	1.544	3.263	2.107	5.008	3.234	6.753	4.360	8.498	5.487
.66	1.550	1.001	2.443	1.577	3.336	2.153	5.122	3.306	6.908	4.459	8.694	5.612
.67	1.581	1.021	2.494	1.611	3.407	2.200	5.233	3.379	7.059	4.558	8.885	5.738
.68	1.613	1.042	2.546	1.644	3.480	2.247	5.347	3.453	7.214	4.658	9.081	5.864
.69	1.646	1.062	2.600	1.680	3.555	2.295	5.464	3.527	7.373	4.759	9.282	5.991
.70	1.677	1.083	2.652	1.713	3.627	2.342	5.577	3.601	7.527	4.860	9.477	6.120
.71	1.709	1.104	2.705	1.747	3.701	2.390	5.693	3.676	7.685	4.962	9.677	6.249
.72	1.741	1.124	2.758	1.781	3.775	2.438	5.809	3.751	7.843	5.065	9.877	6.379
.73	1.774	1.145	2.812	1.816	3.851	2.486	5.928	3.827	8.005	5.168	10.08	6.510

Source: Adapted from Leupold & Stevens, Inc., P.O. Box 688, Beaverton, Oregon 97005, from Stevens Water Resources Data Book.

Enter the table under the *head* column at 0.58; move right until you come under the 3 heading for *length of weir crest*. The indicated flow rate is 2.739 mgd.

A few of the charts and graphs available for determining flows in Parshall flumes, venturi meters, and orifice meters are shown in the remaining examples of this section. Most such graphs and tables are very similar to the ones discussed in previous examples.

Parshall Flumes

The Parshall flume is another type of flow-metering device designed to measure flows in open channels. Top and side views of a flume are shown in Figure H7-9.

To use most nomographs pertaining to flow in a Parshall flume, you must know the width of the throat section of the flume (W), the upstream depth of water (depth H_a in the tube of one stilling well), and the depth of water at the foot of the throat section (depth H_b in the tube of the other stilling well).

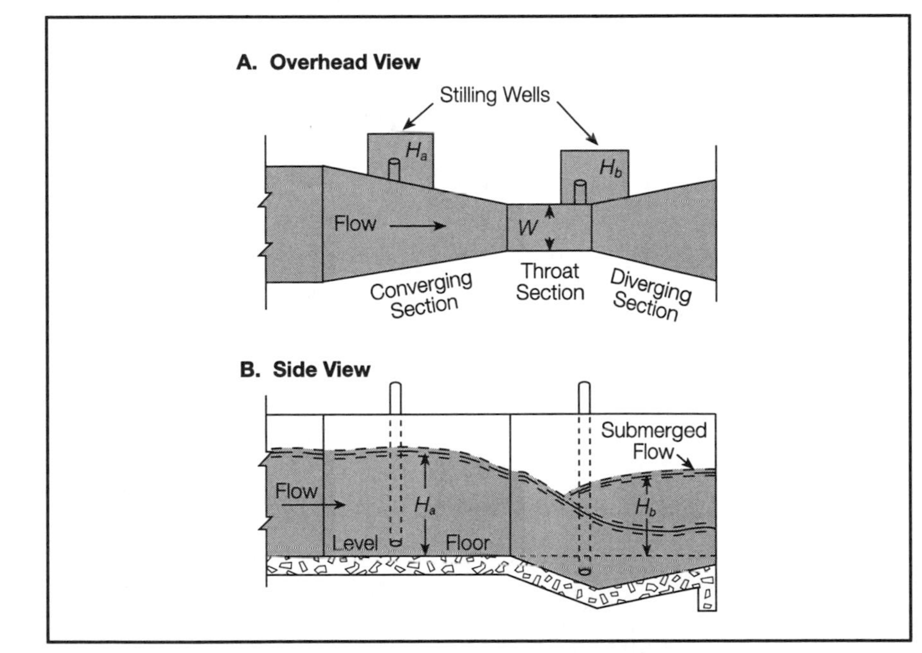

FIGURE H7-9
Parshall flume

Courtesy of Public Works Magazine

At normal flow rates, a nomograph may be used directly to determine the flow through the Parshall flume. At higher flows, however, the throat begins to become submerged (as shown in the side view of Figure H7-9). When this occurs, a correction to the nomograph reading is required. For a flume that has a 1-in. throat width, this correction factor is not required until the percent of submergence $[(H_b/H_a) \times 100]$ is greater than 50 percent. For a flume that has a 1-ft throat width, the correction is not required until the percent of submergence exceeds 70 percent. The following examples use the Parshall flume nomograph and corrections graph shown in Figure H7-10.

Example 19

Suppose that your treatment plant has a Parshall flume with a 4-ft throat width. If the upstream depth of the water (depth H_a) is 1 ft and the depth of water in the throat section (depth H_b) is 0.65 ft, what is the gallons-per-minute flow rate through the flume?

First, determine the approximate flow rate using the flume throat width and the upstream (H_a) depth reading. Then use the correction graph if required.

Line A is drawn on the nomograph from 4 ft on the *throat width* (left) scale through 1 ft on the H_a *depth* scale (right side of the middle scale); the flow rate indicated where the line crosses the right scale is about 7,200 gpm.

To determine whether a correction to this flow rate is required, first calculate the percent submergence of the throat:[8]

$$\text{percent submergence} = \frac{H_b}{H_a} \times 100$$

$$= \frac{0.65 \text{ ft}}{1 \text{ ft}} \times 100$$

$$= 0.65 \times 100$$

$$= 65\%$$

The correction line corresponding to a 4-ft throat width does not even begin until the percent submergence reaches about 78 percent.

[8] Mathematics 7, Percent.

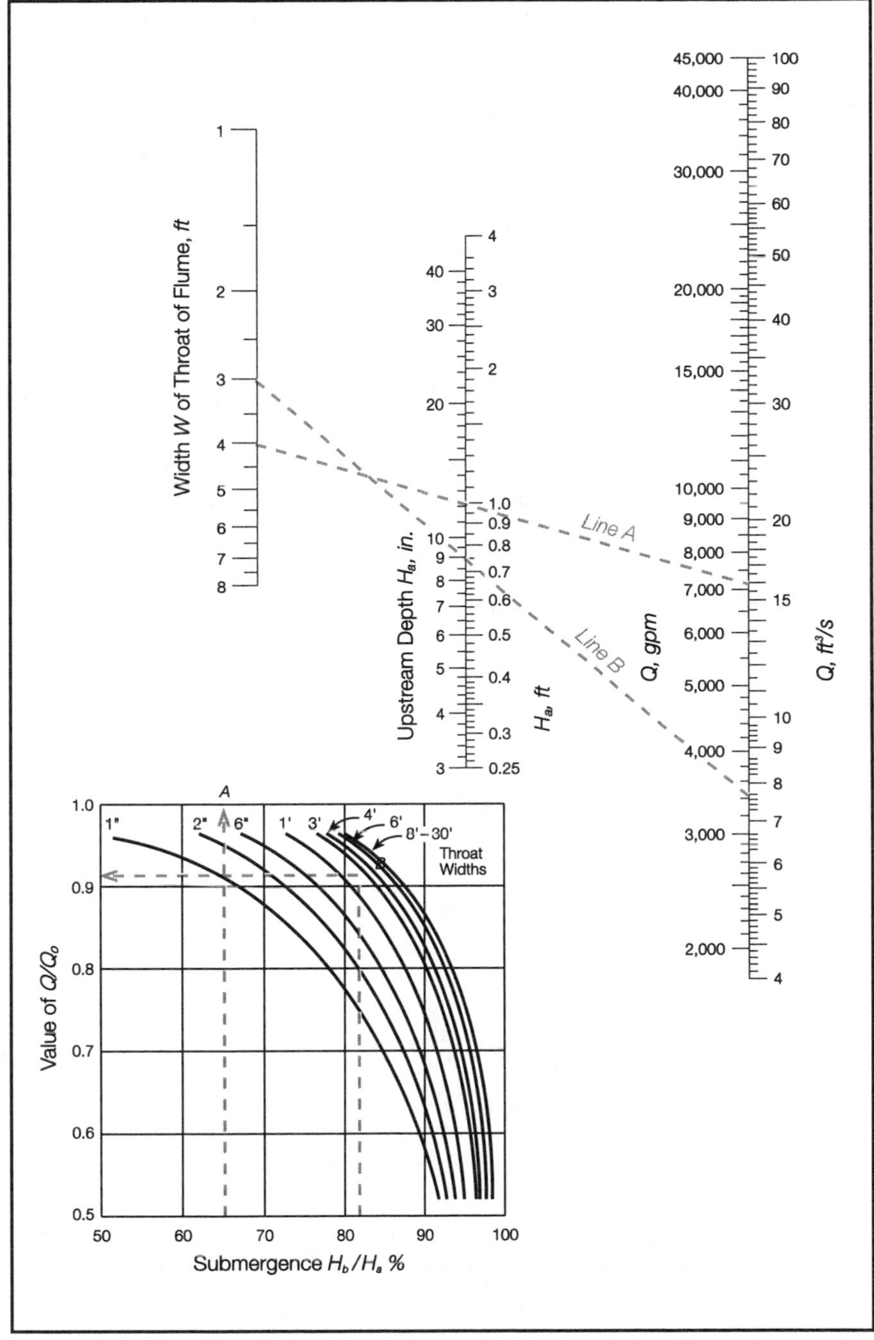

FIGURE H7-10 Parshall flume flow rate nomograph and corrections graph

Courtesy of Public Works Magazine

Therefore, in this problem, no correction to the 7,200-gpm flow rate is required.

Example 20

A flume has a throat width of 3 ft. If the upstream depth of the water (depth H_a) is 9 in., and the depth of water in the throat section (depth H_b) is 7.5 in., what is the gallons-per-day flow rate through the flume?

As for example 19, first determine the approximate flow rate in the flume using the flume throat width and the upstream (H_a) depth reading. Then use the correction graph if needed.

Line B is drawn on the nomograph from 3 ft on the *throat width* (left) scale through 9 in. on the H_a *depth* scale (left side of the middle scale); the flow rate indicated where the line crosses the right scale is 3,400 gpm. Convert this flow rate to gallons per day:

$$(3{,}400 \text{ gpm})(1{,}440 \text{ min/d}) = 4{,}896{,}000 \text{ gpd}$$

To determine whether a correction to this flow rate is needed, first calculate the percent submergence of the throat:

$$\text{percent submergence} = \frac{H_b}{H_a} \times 100$$

$$= \frac{7.5 \text{ in.}}{9 \text{ in.}} \times 100$$

$$= 0.83 \times 100$$

$$= 83\%$$

Looking at the correction line corresponding to a 3-ft throat width, you can see that a correction is needed once the submergence exceeds about 77 percent. Since the submergence calculated is 83 percent, a correction is required. Draw a vertical line up from the indicated 83 percent until it crosses the 3-ft correction line (point B on the correction graph). Then, move directly to the left until you reach the Q/Q_0 scale. Calculate the corrected flow through the flume by multiplying the flow read from the nomograph by the Q/Q_0 value.

In this case, the Q/Q_0 value is about 0.92. Multiply this by the flow rate obtained from the nomograph:

$$(4{,}896{,}000 \text{ gpd})(0.92) = 4{,}504{,}320 \text{ gpd corrected flow}$$

Venturi Meters

Venturi meters are designed to measure flow in closed-conduit or pressure pipeline systems. As shown in Figure H7-11, flow measurement in this type of meter is based on differences in pressure between the upstream section (diameter D_1) and throat section (diameter D_2) of the meter. One type of nomograph available for use with venturi meters is shown in Figure H7-12. The graph applies to meters for which the diameter at D_1 divided by the diameter at D_2 is equal to 0.5. It may be used for meters having D_1/D_2 values as high as 0.75, but the results will be about 1 percent high. For high flow rates, the nomograph may be used with no correction factors necessary. For low flow rates, however, the correction factors graph shown in Figure H7-12 must be used to find a C' value. The value of Q obtained from the nomograph is then multiplied by the C' value to obtain the final answer.

Example 21

Based on the nomograph in Figure H7-12, what is the flow rate in cubic feet per second in a venturi meter with a 1-in. throat, if the difference in head between D_1 and D_2 is 4 ft?

First, determine the cubic-feet-per-second flow rate using the nomograph, then use a correction factor if needed.

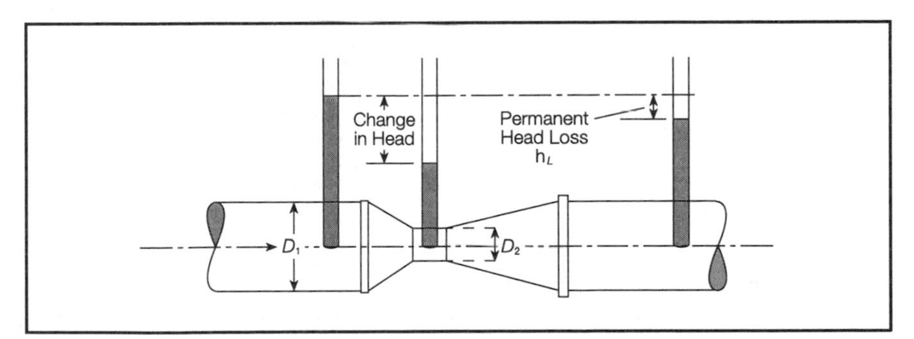

FIGURE H7-11
Venturi meter

Courtesy of Public Works Magazine

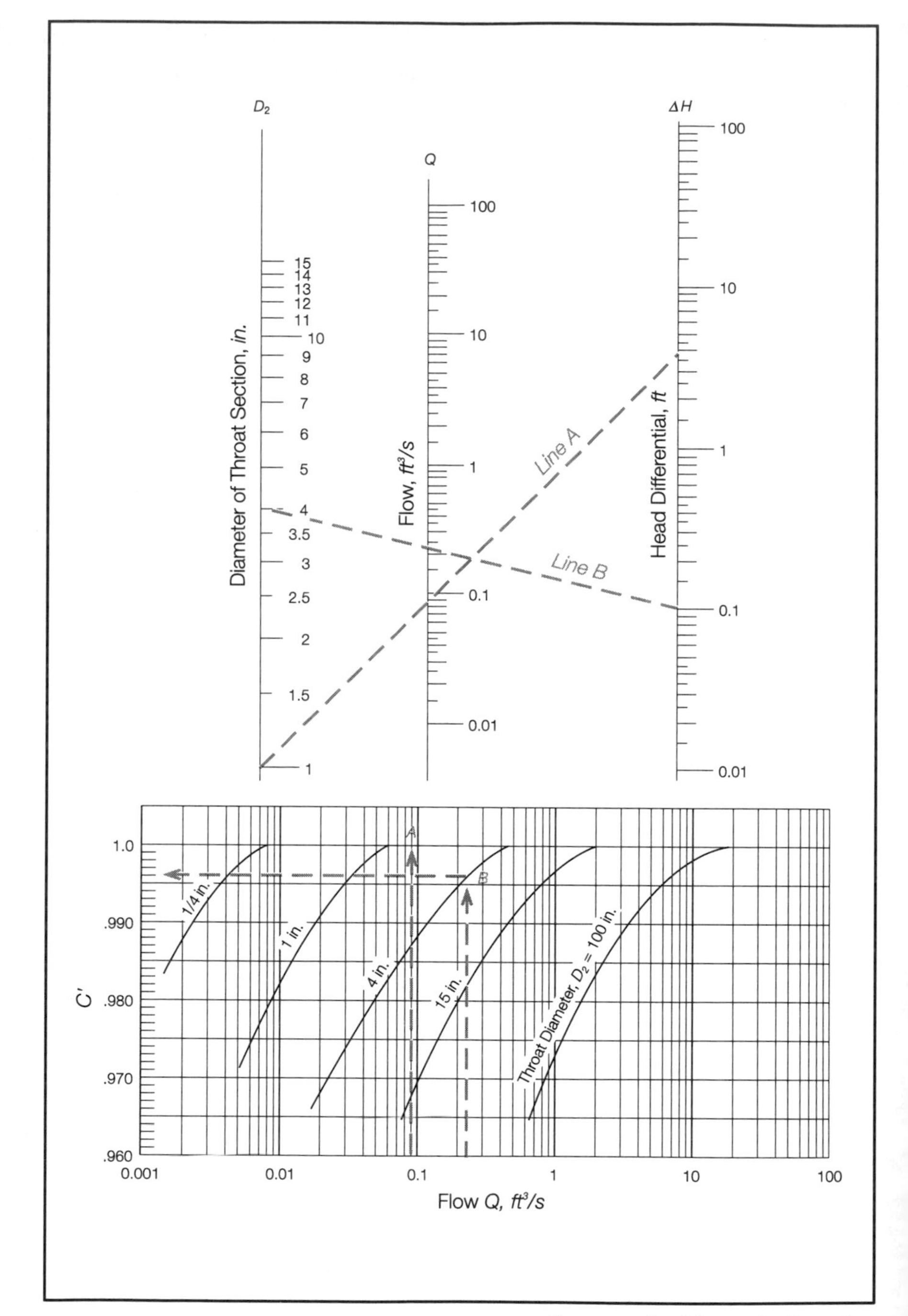

FIGURE H7-12 Venturi meter flow rate nomograph and corrections graph

Courtesy of Public Works Magazine

Line A is drawn from 1 in. on the *throat diameter* (left) scale to 4 ft on the *head differential* (right) scale. The flow rate indicated where the line crosses the middle scale is about 0.09 ft^3/s.

To determine if a flow correction is required, first locate 0.09 on the *flow* (bottom) scale of the correction graph. Then move upward until you cross the 1-in. throat diameter line (point A). The flow rate corresponds with a C' value of 1.0. Therefore, the flow obtained from the nomograph requires no correction, because

$$\text{corrected flow} = (Q \text{ from nomograph}) (C' \text{ from correction graph})$$

$$= (0.09\ ft^3/s)\ (1.0)$$

$$= 0.09\ ft^3/s$$

The corrected flow is the same as the flow obtained from the nomograph.

Examine the correction graph closely, and notice that any flow *at or above* 0.06 ft^3/s for a 1-in. throat diameter would need no correction, because the C' values at these points equal 1.0. Any flow less than 0.06 ft^3/s would require correction.

Example 22

The venturi meter at your treatment plant has a throat diameter of 4 in. If the pressure differential between D_1 and D_2 is 0.1 ft, what is the flow rate in cubic feet per second?

As in the previous example, first determine the cubic-feet-per-second flow rate using the nomograph. Then use a correction factor if necessary.

On the nomograph in Figure H7-12, line B is drawn from 4 in. on the *throat diameter* (left) scale to 0.1 ft on the *head differential* (right) scale; the flow rate indicated on the middle scale is about 0.22 ft^3/s.

Now, to determine if a correction to this flow rate is required, locate 0.22 ft^3/s on the *flow* scale of the correction graph, and move upward until you intersect the 4-in. throat diameter line at point B. The flow corresponds to a C' value of 0.996. Therefore, the corrected flow rate is

$$\text{corrected flow} = (Q \text{ from nomograph}) (C' \text{ from correction graph})$$

$$= (0.22\ ft^3/s)\ (0.996)$$

$$= \quad 0.219 \text{ ft}^3/\text{s}$$

The corrected flow is essentially the same as the flow indicated on the nomograph (0.219 ft^3/s compared with 0.22 ft^3/s).

Orifice Meters

Orifice meters are another type of head differential meter designed for use in closed conduits. As shown in Figure H7-13, there is a pressure tap on each side of the orifice plate, and flow rate calculations are based on the difference in pressure between the upstream (high-pressure) tap (point 1) and the downstream (low-pressure) tap (point 2). The nomograph shown in Figure H7-14 can be used to estimate the flow in orifice meters.

Example 23

If the head differential between pressure taps 1 and 2 is 10.2 ft, the diameter of the pipe is 6 in., and the diameter of the orifice is 3 in., what is the flow rate in the pipeline in cubic feet per second?

First calculate the ratio of the pipe diameter to the orifice diameter and locate the result on the right side of the middle scale of the nomograph in Figure H7-14.

$$\frac{D_1}{D_2} = \frac{6 \text{ in.}}{3 \text{ in.}}$$

$$= 2.0$$

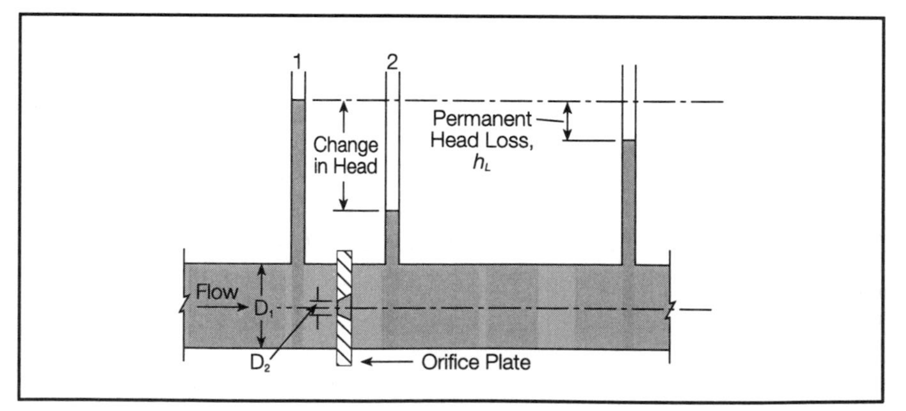

FIGURE H7-13
Orifice meter

Courtesy of Public Works Magazine

Now draw a line from 10.2 on the right-hand scale, through 2.0 on the right side of the middle scale, and across to the velocity scale. The indicated velocity is 4 ft/s.

The diameter of the pipeline is known (so that area can be calculated), and the velocity of the flow is known, so use the $Q = AV$ equation to determine the flow rate in the pipe:

$$Q = AV$$

$$= (0.785)\ (0.5\ \text{ft})\ (0.5\ \text{ft})\ (4\ \text{ft/s})$$

$$= 0.79\ \text{ft}^3/\text{s}$$

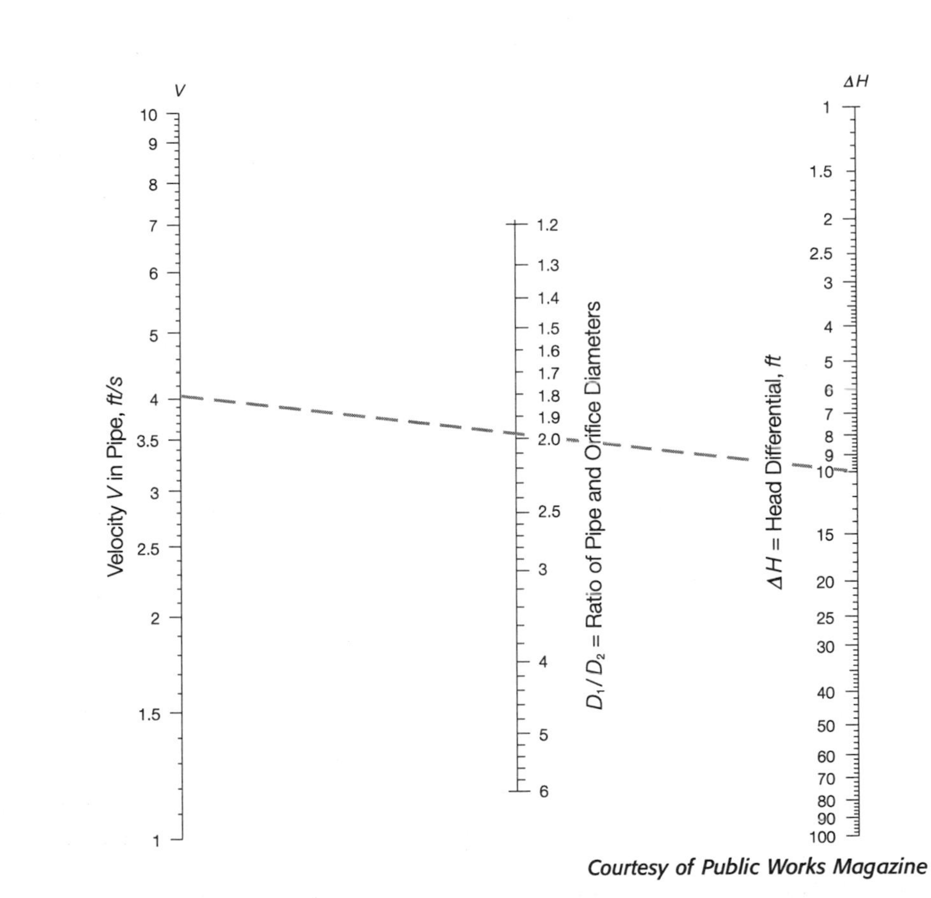

FIGURE H7-14 Flow rate nomograph for orifice meter in example 23

Courtesy of Public Works Magazine

HYDRAULICS 8

Thrust Control

Water under pressure and water in motion can exert tremendous forces inside a pipeline. One of these forces, *thrust*, pushes against fittings, valves, and hydrants, causing couplings to leak or to pull apart entirely. Although it is not possible to eliminate thrust, it is possible and absolutely necessary to control it.

Thrust can be caused by any factor that produces a pressure or force on a fitting, but it is primarily caused by water pressure and *water hammer* (the force that results in a pipe from a sudden change in water velocity). As shown in Figure H8-1, thrust usually acts perpendicular (at 90°) to the inside surface it pushes against. Note in the figure how thrust acts against the outer curve of the fitting, tending to push the fitting away from both sections of pipeline. You can picture the force as being caused by water as it "rounds the bend." If the flow direction were reversed, the direction of thrust would be unchanged. Uncontrolled, the thrust can cause movement in the fitting or

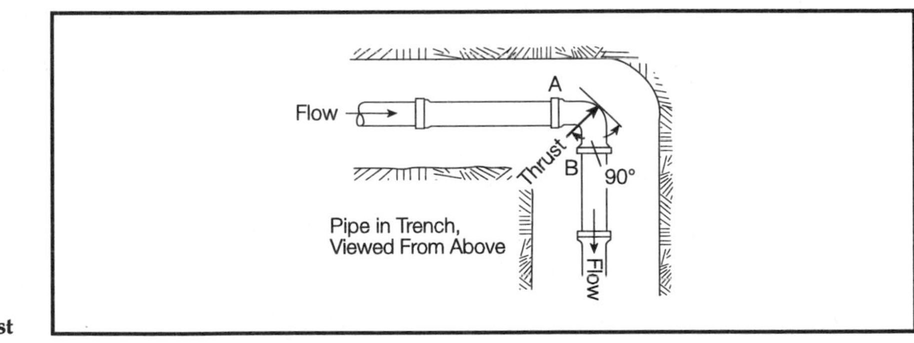

FIGURE H8-1
Direction of thrust

pipeline, which will lead to leakage or complete separation at coupling A or B, or at some other nearby coupling upstream or downstream of the fitting.

There are two types of devices commonly used to control thrust: thrust blocks and thrust anchors. A *thrust block* is a mass of concrete, cast in place between the fitting being restrained and the undisturbed soil at the side or bottom of the pipe trench. An example is shown in Figure H8-2. This type of thrust control can be used either to control thrust forces that act horizontally or to control thrusts that act downward.

A *thrust anchor* is a massive block of concrete, often a cube, cast in place below the fitting to be anchored. As shown in Figure H8-3, imbedded steel shackle rods anchor the fitting to the concrete block, effectively resisting upward thrusts.

The size and shape of a thrust control device depend primarily on five factors:

- type of fitting
- diameter of fitting
- water pressure

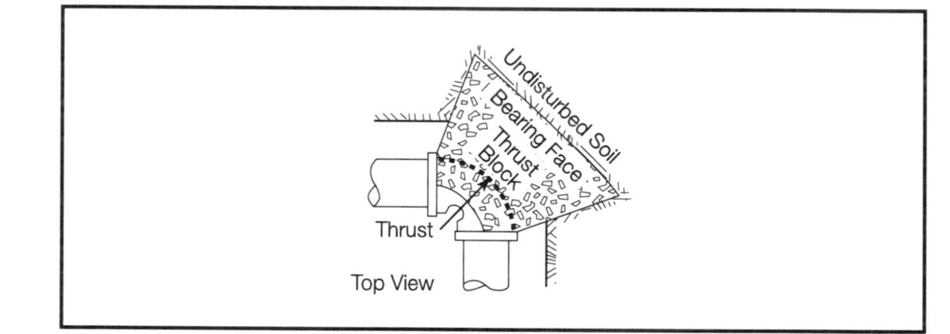

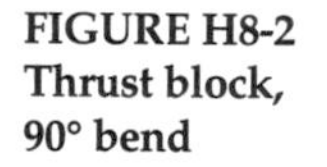

FIGURE H8-2
Thrust block, 90° bend

Courtesy of J-M Manufacturing Co., Inc.

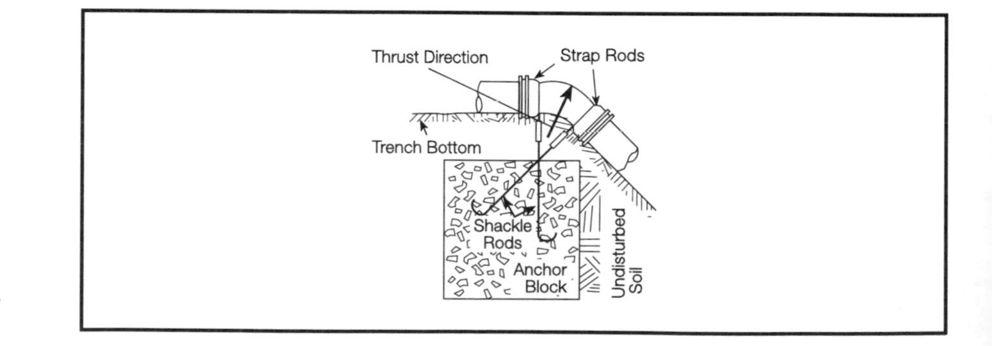

FIGURE H8-3
Thrust anchor, 45° bend

- water hammer
- soil type

If you are given information on these five factors, then you can solve many thrust control sizing problems by reading and interpreting tables, such as the ones shown in this chapter. Be cautious in applying the information given in this chapter in the field. Although the following procedures will work quite adequately in many situations, they should not be used as a substitute for professional engineering design, particularly in situations involving large diameter pipelines (greater than 12 in.), high-velocity situations (greater than 10 ft/s), or soils where soil type or stability may be questionable. (One of the more common causes of thrust block failure is the installation of thrust blocks in unstable soil, or in locations too close to trenches for other pipelines.)

Thrust Block Calculations

To determine the proper size thrust block for a particular fitting, you will first have to determine the total thrust in pounds against the fitting. You can then determine the area of the bearing face based on the type of soil and how much pressure it can withstand. The following two examples illustrate how to size thrust blocks using this method.

Example 1

Thrust is developed at the tee fitting shown in Figure H8-4. The pipeline has an 8-in. diameter and will operate at a maximum pressure of 200 psig.

The undisturbed soil type in the trench is alluvial soil. Determine (a) thrust on the tee, in pounds, and (b) area and dimensions of the necessary bearing face for the thrust block.

(a) First determine the thrust on the 8-in. tee using Table H8-1. Move across the row beginning with 8-in. diameter and read the answer from the last column (tees and dead ends) as 5,000 lb. Notice that the table gives thrust values *per 100 psig of pressure*. The maximum thrust against the fitting depends on the maximum pressure in the pipeline. The maximum water pressure that is expected to occur is the sum of the maximum operating pressure plus an allowance for pressure resulting from water hammer. In this problem, the maximum operating pressure is stated as 200 psig. From Table H8-2, determine that the water

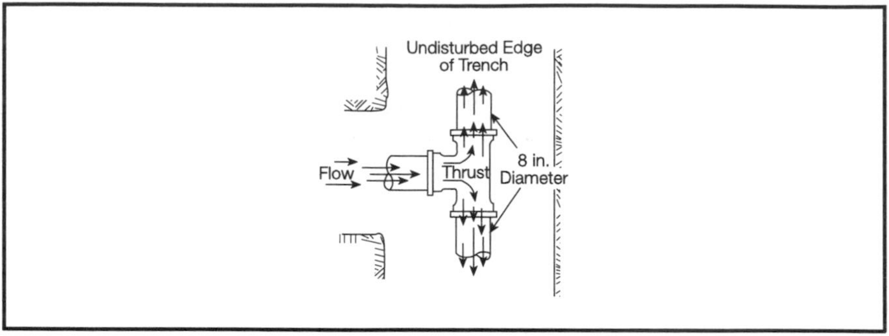

FIGURE H8-4 Schematic for example 1

Courtesy of J-M Manufacturing Co., Inc.

TABLE H8-1 Fitting thrust at 100-psig water pressure

	Fitting Thrust, *lb*				
Diameter, *in.*	11 ¼° Bend	22 ½° Bend	45° Bend	90° Bend	Tees and Dead Ends
3	140	280	540	1,000	710
4	250	490	960	1,800	1,300
6	550	1,100	2,200	4,000	2,800
8	990	2,000	3,800	7,100	5,000
10	1,500	3,100	6,000	11,100	7,900
12	2,200	4,400	8,700	16,000	11,300
14	3,000	6,000	11,800	21,800	15,400
16	3,900	7,800	15,400	28,400	20,100
18	5,000	9,900	19,500	36,000	25,400
20	6,200	12,300	24,000	44,400	31,400
24	7,500	14,800	29,100	53,800	38,000
30	13,900	27,600	54,100	100,000	70,700
36	20,000	40,000	77,900	144,000	102,000
42	27,200	54,100	106,000	196,000	139,000
48	35,500	70,600	138,000	256,000	181,000
54	44,900	89,400	175,000	324,000	229,000
60	55,400	110,000	216,000	400,000	283,000

TABLE H8-2 Allowances for water hammer

Pipeline Diameter, *in.*	Water Hammer Pressure Allowance, *psig*
3–10	120
12–14	110
16–18	100
20	90
24	85
30	80
36	75
42–60	70

hammer allowance for an 8-in. pipeline is 120 psig. Adding these two pressures, you know that the total water pressure that may occur is

maximum operating pressure	200	psig
water hammer allowance	+ 120	psig
maximum total water pressure	320	psig

Using this water pressure information, the maximum expected thrust against the fitting can be calculated. As determined from Table H8-1, the thrust against the fitting is 5,000 lb for every 100 psig of pressure. Since 320/100 = 3.2, there are 3.2 units of 100 in the number 320. Therefore, the maximum thrust is

$$(5{,}000 \text{ lb per } 100 \text{ psig})\ (3.2 \text{ units of } 100 \text{ psig}) = 16{,}000 \text{ lb}$$

(b) To find the area and dimensions of the thrust block bearing face, first find out how much bearing pressure the soil can support (the allowable soil bearing pressure) from Table H8-3 for alluvial soil:

$$\text{allowable soil bearing pressure} = 1{,}000 \text{ lb/ft}^2$$

This means that each square foot of soil should have no more than 1,000 lb of force against it. For a maximum thrust of 16,000 lb, the area needed for the bearing face is

$$\frac{16{,}000 \text{ lb}}{1{,}000 \text{ lb/ft}^2} = 16 \text{ ft}^2$$

TABLE H8-3 Allowable bearing pressure for soil types

Soil Type	Bearing Pressure, *lb/ft*^{2}
Peat or muck	0
Alluvial soil	1,000
Soft clay	2,000
Sand	4,000
Sand and gravel	6,000
Sand and gravel with clay	8,000
Shale	12,000
Rock	20,000

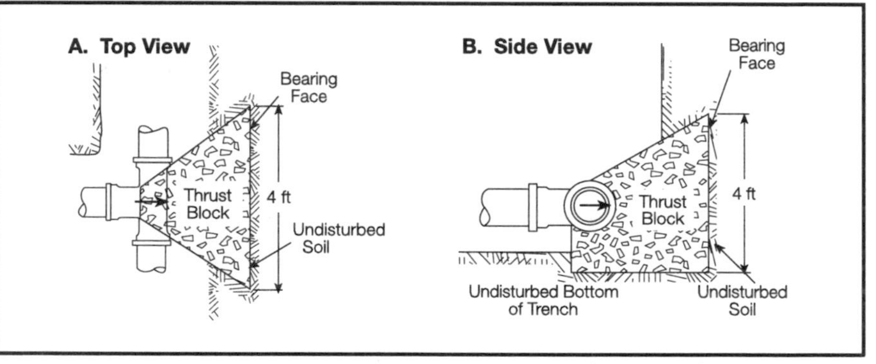

FIGURE H8-5 Correctly sized thrust block for example 1

Courtesy of J-M Manufacturing Co., Inc.

It is desirable to have square bearing faces. You can find the dimensions for such faces by taking the square root of the area (square root tables or calculators with the square root function are available for finding the square root). In this case,

$$\sqrt{16 \text{ ft}^2} = 4 \text{ ft on a side}$$

The correctly sized thrust block is shown in Figure H8-5.

Often pipelines are pressure-tested after construction is complete. The test pressure is always greater than the pipeline would experience in normal day-to-day operation. Since such tests normally last for two hours or so, the fittings need to be designed to withstand the test pressures. Therefore, in such cases, you should use the test pressure instead of the operating pressure in determining the size of thrust blocks. The following example will illustrate the procedure.

Example 2

The 10-in. pipeline shown in Figure H8-6 is designed to operate at 75 psig. After construction is completed, the pipeline will be tested at twice the design operating pressure. The soil type is sand. Determine the size of the thrust block required.

Follow the same procedure as in example 1. First find the thrust against the 10-in.-diameter 90° bend using Table H8-1. The thrust listed is 11,100 lb for each 100 psig. To determine the maximum thrust against the fitting, you will have to know the maximum water pressure that could occur in the pipeline:

maximum test pressure (2 × 75)	150	psig
water hammer allowance (from Table H8-2)	+ 120	psig
maximum total water pressure	270	psig

Using this pressure information, you can calculate the total thrust against the fitting. As determined from Table H8-1, the thrust against the fitting is 11,100 lb for every 100 psig of pressure. Since 270/100 = 2.7, there are 2.7 units of 100 in the number 270. Therefore, the maximum thrust is

$$(11{,}100 \text{ lb}/100 \text{ psig})\ (2.7 \text{ units of } 100 \text{ psig}) = 29{,}970 \text{ lb}$$

From Table H8-3, determine that the allowable soil bearing pressure for sand is 4,000 lb/ft^2. This means that each square foot of soil should have no more than 4,000 lb of force against it. For a total thrust of 29,970 lb, the area of the thrust block bearing face should be

$$\frac{29{,}970 \text{ lb}}{4{,}000 \text{ lb/ft}^2} = 7.49 \text{ ft}^2$$

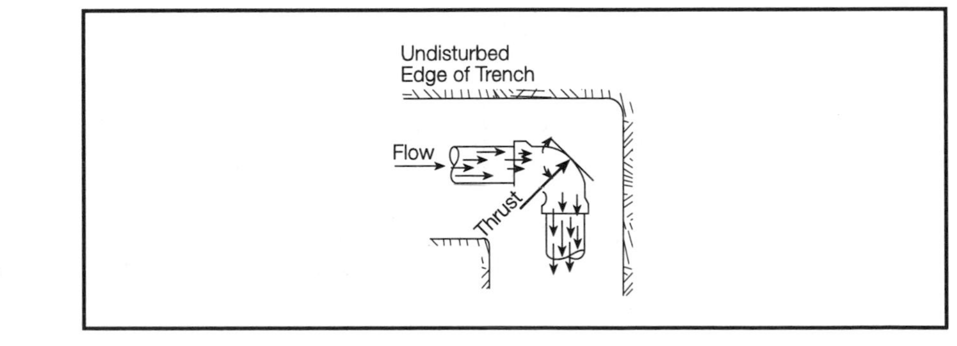

FIGURE H8-6 Schematic for example 2

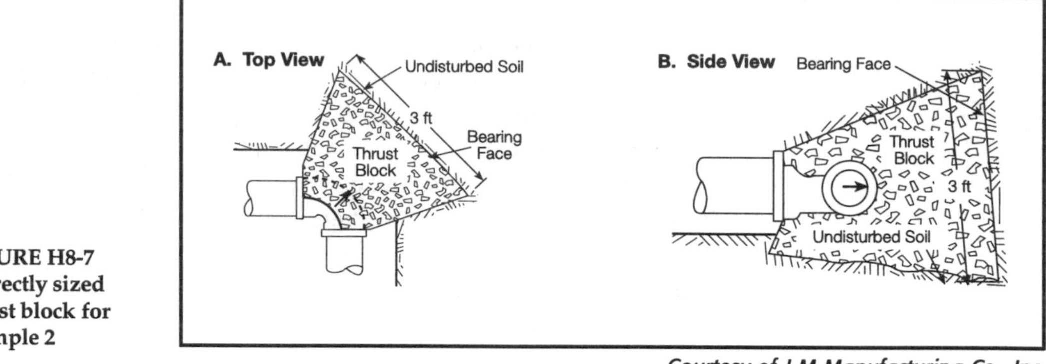

FIGURE H8-7 Correctly sized thrust block for example 2

Courtesy of J-M Manufacturing Co., Inc.

The bearing face should be square, so you can find the dimensions by taking the square root of the area:

$$\sqrt{7.49 \text{ ft}^2} \quad = \quad 2.74 \text{ ft}$$

Rounding up to the nearest foot, you find that the correctly sized thrust block is 3 ft on a side (Figure H8-7).

Thrust Anchor Calculations

Sometimes standard engineering plans and specifications are prepared for a variety of routinely used devices such as thrust anchors. Tables like Table H8-4 allow the nonengineer to size thrust anchors for routine situations with a minimum of effort. Note that such standard tables are prepared for a particular set of trench conditions, and can be used only when those conditions exist. (The set of trench conditions applicable for Table H8-4 have arbitrarily been denoted as "type X" conditions.) Other standard tables would have to be used for differing trench conditions. There is no single tabular procedure that applies to all thrust anchor problems, and there is no set of tables that will replace a competent design engineer. The following two examples illustrate how to size thrust anchors using standard tables.

Example 3

Thrust acts on a 22½° vertical bend, as shown in Figure H8-8. Assuming type X trench conditions, use Table H8-4 to determine how the fitting can be properly anchored to resist upward thrust. The fitting is 8 in. in diameter, and the pipeline will be tested after construction at a pressure of 300 psig.

TABLE H8-4 Thrust anchoring for 11¼°, 22½°, 30°, and 45° vertical bends — applicable only to "type X" trench conditions

Pipe Size Nominal Diameter, *in.*	Test Pressure, *psi*	Vertical Bend, *degrees*	Volume of Concrete Blocking, ft^3	Side of Cube, *ft*	Diameter of Shackle Rods, *in.**	Depth of Rods in Concrete, *ft*
4	300	11¼	8	2	¾	1.5
		22½	11	2.2	¾	2.0
		30	17	2.6	¾	2.0
		45	30	3.1	¾	2.0
6	300	11¼	11	2.2	¾	2.0
		22½	25	2.9	¾	2.0
		30	41	3.5	¾	2.0
		45	68	4.1	¾	2.0
8	300	11¼	16	2.5	¾	2.0
		22½	47	3.6	¾	2.0
		30	70	4.1	¾	2.5
		45	123	5.0	¾	2.0
12	250	11¼	32	3.2	¾	2.0
		22½	88	4.5	⅞	3.0
		30	132	5.1	⅞	3.0
		45	232	6.1	¾	2.5
16	225	11¼	70	4.1	⅞	3.0
		22½	184	5.7	1⅛	4.0
		30	275	6.5	1¼	4.0
		45	478	7.8	1⅛	4.0
20	200	11¼	91	4.5	⅞	3.0
		22½	225	6.1	1¼	4.0
		30	330	6.9	1⅜	4.5
		45	560	8.2	1¼	4.0
24	200	11¼	128	5.0	1	3.5
		22½	320	6.8	1⅜	4.5
		30	480	7.9	1⅝	5.5
		45	820	9.4	1⅜	4.5

*Four rods for 45° vertical bends, two rods for all others.

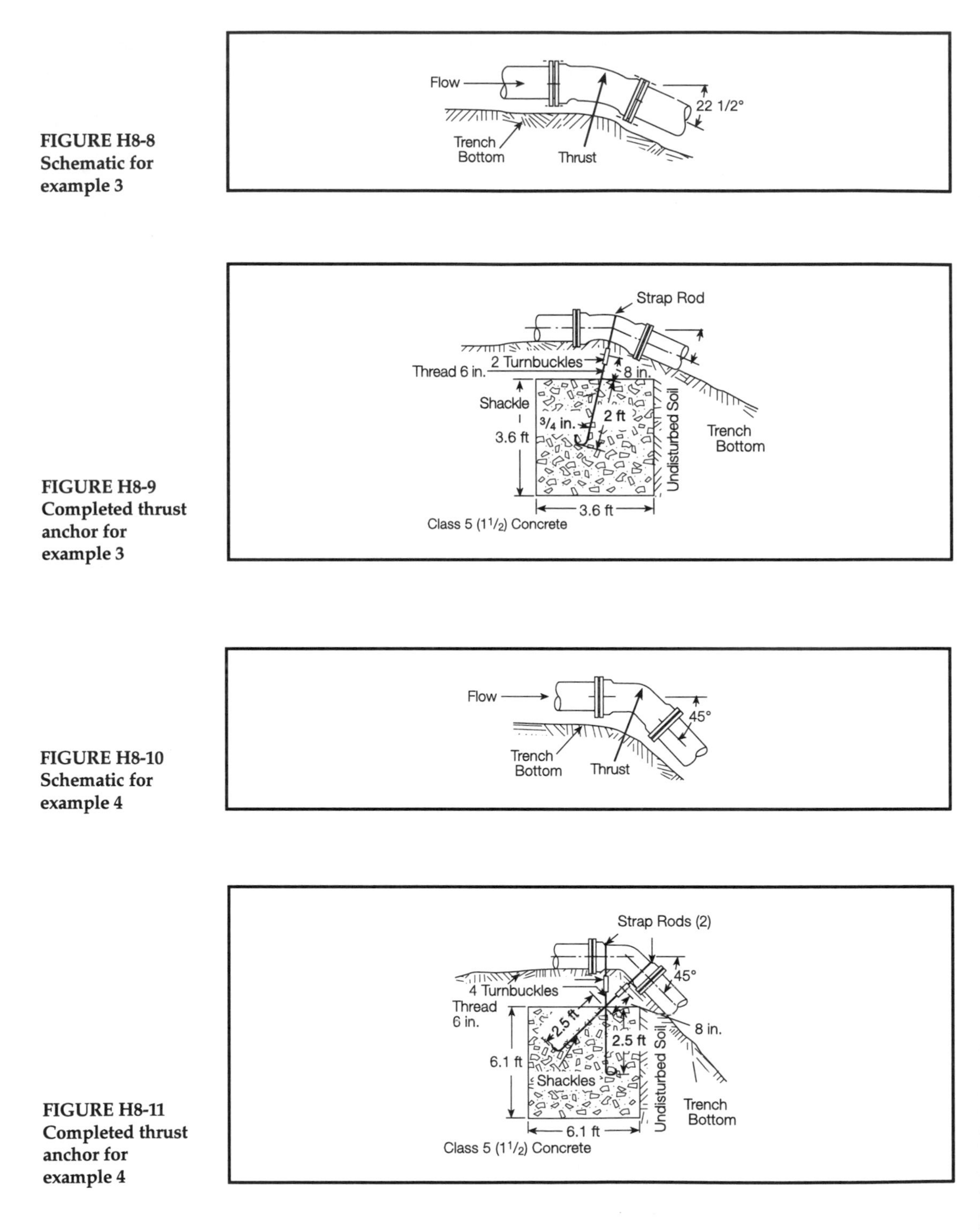

FIGURE H8-8 Schematic for example 3

FIGURE H8-9 Completed thrust anchor for example 3

FIGURE H8-10 Schematic for example 4

FIGURE H8-11 Completed thrust anchor for example 4

Using Table H8-4, you will find that thrust anchor sizing is a simple task. Enter the table at 8 in. in the pipe size column; then read the sizing needs from the last four columns corresponding to 22½° as follows:

volume of concrete block	47 ft^3
dimension of cubic concrete block	3.6 ft on a side
diameter of shackle rods (two required)	¾ in.
imbedded depth of rods in concrete	2 ft

This information is now used to draw the thrust anchor shown in Figure H8-9. Notice that two shackles are required (one is directly behind the one shown). One shackle is connected to each threaded end of the strap rod by a turnbuckle.

Example 4

Thrust acts on a 45° vertical bend as shown in Figure H8-10. Based on thrust anchor data shown in Table H8-4, make a drawing of the correct thrust anchor for type X trench conditions and label the dimensions. The fitting is 12 in. in diameter and will be tested at a pressure of 250 psig.

Enter Table H8-4 at a diameter of 12 in. and read the sizing information from the last four columns as follows:

volume of concrete block	232 ft^3
dimension of cubic concrete block	6.1 ft on a side
diameter of shackle rods (four required)	¾ in.
imbedded depth of rods in concrete	2.5 ft

Now make a drawing of the information, as shown in Figure H8-11. Note that two of the four required shackles are hidden directly behind the two shown.

Basic Science Concepts and Applications

Chemistry

CHEMISTRY 1

The Structure of Matter

If you could take a sample of an element and divide it into smaller and smaller pieces, you would eventually come down to a tiny particle that, if subdivided any more, would no longer show the characteristics of the original element. The smallest particle that still retains the characteristics of the element is called an *atom,* from the Greek word *atomos* meaning "uncut" or "indivisible."

Although an atom is the smallest particle that still retains the characteristics of the element it is taken from, and is so small that it can't be seen with today's most powerful microscopes, the atom itself can be broken down into even smaller particles called *subatomic* particles. The different number and arrangement of subatomic particles distinguish the atoms of one element from those of another, and give each element specific qualities. A general understanding of the structure of atoms is basic to an understanding of chemical reactions and equations.

Atomic Structure

A great many subatomic particles have been identified, many of which exist for only a fraction of a second. In the study of chemistry and chemical reactions, however, the structure of the atom is adequately explained on the basis of three fundamental particles: the *proton*, the *neutron*, and the *electron*.

As shown in Figure C1-1, the center of the atom, called the *nucleus* (plural: *nuclei*), is made up of positively charged particles called protons and uncharged particles called neutrons. Negatively charged particles called electrons occupy the space surrounding the nucleus and make up most of the volume of the atom. The electrons are said to occupy the space around the nucleus in "shells."

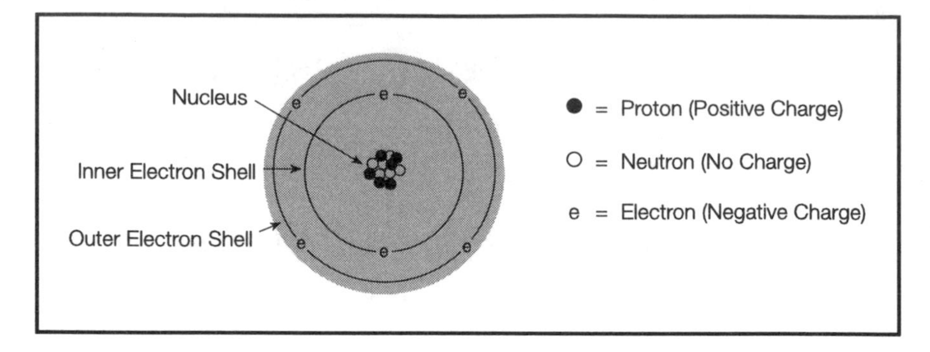

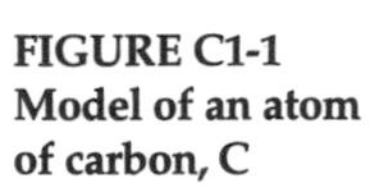
FIGURE C1-1 Model of an atom of carbon, C

The basic defining characteristic of the atoms of any one element is the number of protons in the nucleus. An atom of carbon, for example, always has six protons in the nucleus; and any atom with exactly six protons in the nucleus must be a carbon atom. Boron atoms have five protons in the nucleus, whereas nitrogen atoms have seven. The number of protons in the nucleus of an atom is called the *atomic number*. Therefore, the atomic number of carbon is six; boron, five; and nitrogen, seven.

The nucleus is extremely small in comparison with the total size of the atom. If the atom were the size of a football stadium, the nucleus would be no larger than a small insect flying in the middle. Nonetheless, because neutrons and protons (which have nearly identical weights) are much heavier than electrons, the nucleus contains most of the mass (weight) of the atom. The *atomic weight* of an atom is defined as the sum of the number of protons and the number of neutrons in the nucleus. In the nucleus of a carbon atom there are six protons and six neutrons; therefore, the atomic weight of the atom is 12.

Atomic weights are not "weights" in the usual sense of the word. They do not indicate the number of pounds or grams an atom weighs, but merely how the weight of one atom *compares* with the weight of another. For example, the atomic weight of hydrogen is 1 (the hydrogen nucleus has one proton and no neutrons) and the atomic weight of carbon is 12. Therefore, an atom of carbon weighs 12 times as much as an atom of hydrogen.

Isotopes

All atoms of a given element have the same number of protons in the nucleus, but the number of neutrons may vary. Atoms that are of the same element but contain varying numbers of neutrons in the nucleus are called *isotopes* of that element. The atoms of each element that have the most common numbers of neutrons are called the *principal isotopes* of the element.

The atomic weight of an element is generally given in tables as a whole number with decimals, the result of averaging together the atomic weights of the principal (most common) isotopes. The average also takes into account how often each isotope occurs. The element chlorine, for example, has a listed atomic weight of 35.45 — an average of the principal isotopes of chlorine. (About 75 percent of all chlorine atoms have atomic weight 35, and about 25 percent have atomic weight 37.)

Ions

As shown in Figure C1-1, there are the same number of electrons (negative charges) surrounding the carbon atom as there are protons (positive charges) in the nucleus. The atom is said to be electrically stable.

Now consider the effect on the atom's charge if an electron is removed from the outer electron shell, or if an electron is added to the outer shell (Figure C1-2). In either case, the charges on the atom are no longer balanced. With one electron removed from the outer shell, there are six protons (positive charges) counterbalanced by only five electrons (negative charges), resulting in a net charge on the atom of *plus 1*. On the other hand, with one electron added to the outer shell, there are six protons (positive charges) counterbalanced by seven electrons (negative charges), resulting in a net charge on the atom of *minus 1*.

When the charges on the atom are *not* balanced, the atom is no longer stable (that is, it has a plus or minus charge). In this unstable condition, the atom is called an *ion*. When the net charge on an atom is positive (more protons than electrons), the ion is called a *cation*. When the net charge on an atom is negative (more electrons than protons), the ion is called an *anion*.

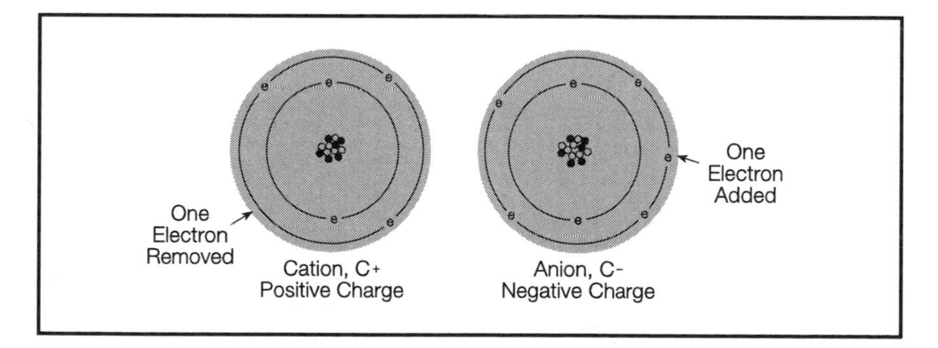

FIGURE C1-2 Models of ionized carbon atoms

Of the three fundamental particles — protons, neutrons, and electrons — the electron is the most important particle in understanding basic chemistry. During chemical reactions, the nucleus of an atom remains unchanged; only the electrons of atoms interact, and only those in the outermost shell.

The Periodic Table

The elements can be arranged according to the number of electron shells they have, and according to similarities of chemical properties. When so arranged, a table is formed called the *periodic table* (see appendix B for the complete periodic table). In the periodic table, the horizontal rows are called *periods*, and the vertical columns are called *groups*. Elements of the same period have the same number of electron shells; elements in the same group have similar chemical properties.

As an example of members of the same period, hydrogen and helium are members of the first horizontal row (period) in the periodic table. Each has only one electron shell. Lithium, beryllium, boron, carbon, nitrogen, oxygen, fluorine, and neon are members of the second period, and all have two electron shells.

The vertical columns (groups) are important because the elements within groups tend to have similar chemical properties. For example, notice that the two elements chlorine and iodine, both of which can be used for disinfecting water, are in the same chemical group.

Although the individual boxes of a large periodic table may contain as many as nine or more kinds of information about each element (including element name, symbol, electron structure, atomic number and weight, oxidation states, boiling points, melting points, and density), the four basic kinds of information included in almost all periodic tables are (1) atomic weight, (2) element symbol, (3) element name, and (4) atomic number.

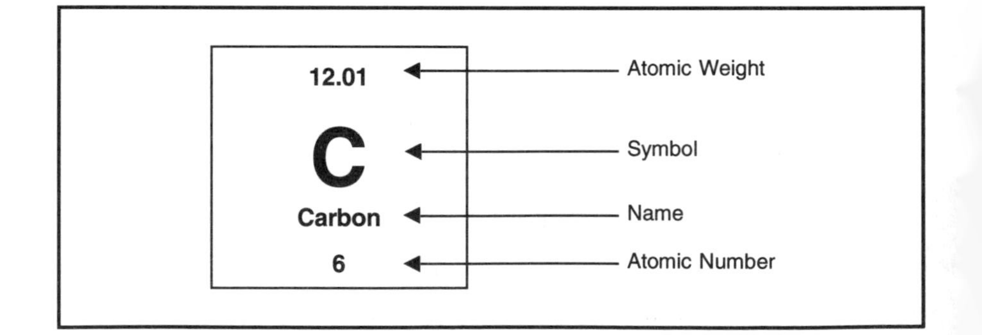

FIGURE C1-3 Carbon as it appears in the periodic table

Consider the carbon atom model again, this time as it appears in the periodic table (Figure C1-3).

The number at the top of the box is the atomic weight of the element. As discussed before, this number represents the average weight of the common isotopes of the element. The letter in the box is the standard abbreviation, or symbol, that has been assigned to the element. Often, two letters are used because a single letter has already been assigned to a specific element. For example, *H* stands for hydrogen; *Hg* for mercury. The bottom number in the box is the atomic number — the number of protons in the nucleus of an atom of that element. (It also indicates the number of electrons in a stable atom of the element, since the number of protons is the same as the number of electrons in a stable atom.)

CHEMISTRY 2

The Classification of Matter

Matter is anything that occupies space and has weight (mass). Matter includes subatomic particles — protons and electrons — as well as the atoms that such particles form. Matter also includes everything formed by atoms — nearly everything you encounter in the world.

Matter exists in three forms: (1) solids, (2) liquids, and (3) gases. Solids, liquids, and gases may exist in pure form, may combine chemically with other elements to form compounds, or may be mixed together without chemically combining to form mixtures.

Pure Elements

As discussed previously, elements are matter built up from subatomic particles, with properties determined by the element's nucleus (protons and neutrons) and electron shells. All of the atoms of an element have the same number of protons in their nuclei. Elements do not break down into simpler elements. There are more than 100 elements known; 92 occur naturally, and others have been produced in the laboratory. Elements important in water chemistry are listed in Table C2-1.

A few elements exist in pure form. Carbon is an example — a diamond is pure carbon in a particular arrangement. Oxygen in the air is another example of an element in its pure form. However, most elements are unstable and are usually found combined with other elements in the form of compounds.

TABLE C2-1 Elements important in water treatment

Element	Symbol	Element	Symbol
Antimony*	Sb	Lead*	Pb
Aluminum	Al	Magnesium	Mg
Arsenic*	As	Manganese	Mn
Barium*	Ba	Mercury*	Hg
Beryllium*	Be	Nickel*	Ni
Boron	B	Nitrogen	N
Bromine	Br	Oxygen	O
Cadmium*	Cd	Phosphorus	P
Calcium	Ca	Potassium	K
Carbon	C	Radium*	Ra
Chlorine	Cl	Selenium*	Se
Chromium*	Cr	Silicon	Si
Copper*	Cu	Silver*	Ag
Fluorine†	F	Sodium	Na
Hydrogen	H	Strontium*	Sr
Iodine	I	Sulfur	S
Iron	Fe	Thallium*	Tl

*This element must be monitored according to the requirements of the Safe Drinking Water Act.

†Fluoride, an anion of the element fluorine, must be monitored according to the requirements of the Safe Drinking Water Act.

Compounds

Compounds are two or more elements that are "stuck" (bonded) together by a chemical reaction (explained in the Chemistry section, chapter 3). A compound can be broken down into its original elements only by a reversal of the chemical reaction that formed it. The weight of the atoms of any one element in a compound is always a definite fraction (or proportion) of the weight of the entire compound. For example, in any given weight of water, $2/18$ of the weight comes from atoms of hydrogen.

When atoms of two or more elements are bonded together to form a compound, the resulting particle is called a *molecule*. A molecule may be only two atoms of one or more elements bonded together; or it may be dozens of atoms bonded together, and may consist of several elements. For example, when two atoms of hydrogen and one of oxygen combine, a molecule of water is formed. When one atom of carbon and two of oxygen combine, a molecule of carbon dioxide is formed. When two atoms of oxygen combine, a molecule of oxygen is formed.

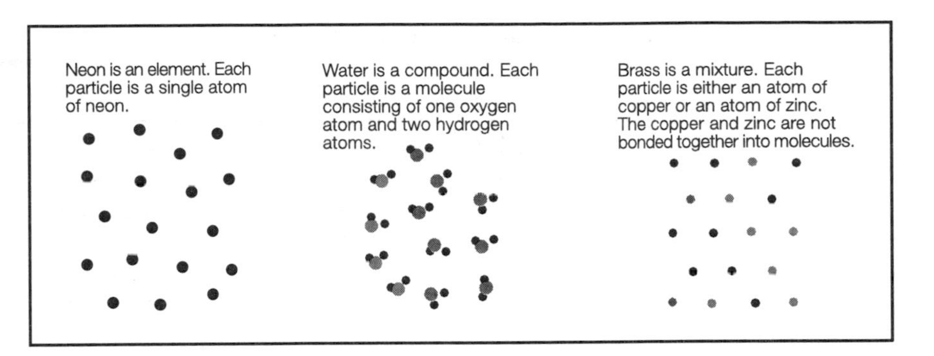

FIGURE C2-1 Models of an element, a compound, and a mixture

Other examples of compounds are

- salt (sodium and chlorine)
- sulfuric acid (hydrogen, sulfur, and oxygen)
- ammonia (nitrogen and hydrogen)
- rust (iron and oxygen)
- lime (calcium, oxygen, and hydrogen)
- sand (silicon and oxygen)

The number of compounds that can be formed by chemical reaction between elements is enormous. Well over two million compounds have been identified by chemists, and the number is still increasing.

Mixtures

When two or more elements, compounds, or both, are mixed together and no chemical reaction (bonding between individual particles) occurs, then the result is a *mixture*. No new compounds are formed, and the elements or compounds may be mixed in any proportion. Any mixture can be separated into its original elements or compounds by "physical" means, such as filtering, settling, or distillation. For example, a mixture of salt water can be separated into its compounds, salt and water, by the process of distillation — heating the mixture causes the water to evaporate, leaving the salt behind.

Other examples of mixtures are

- air (mostly oxygen, carbon dioxide, and nitrogen)
- glass (sand, various metals, and borax)
- steel (primarily iron and carbon)

It is helpful to think of the differences between elements, compounds, and mixtures by considering Figure C2-1.

CHEMISTRY 3

Valence, Chemical Formulas, and Chemical Equations

As explained in the previous chapter, atoms of elements can combine to form molecules of compounds. Experience has shown that only certain combinations of atoms will react (bond) together. For example, two atoms of iron will bond to three atoms of oxygen to form a molecule of ferric oxide (rust), but atoms of iron will not bond to atoms of magnesium. Iron and magnesium can be blended into a mixture — an alloy — but not combined into a compound.

It has also been found through experiments that the number of atoms of each element in a molecule is very definite. A water molecule is formed of two hydrogen atoms and one oxygen atom — no other combination of hydrogen and oxygen makes water, and other compounds of hydrogen and oxygen can be formed only under special circumstances. Similarly, exactly one atom of hydrogen is required to combine with one atom of chlorine to form a molecule of hydrogen chloride.

The following paragraphs contain a brief discussion of why only certain molecules occur, and of how chemists describe the characteristics of an atom that determine which chemical combinations it can enter into.

Valence

In nature, atoms of elements tend to form molecules whenever the molecule is more chemically stable than the individual element. The number of electrons on the very outside of each atom (in the outermost shell) is the most important factor in determining which atoms will combine with other atoms to achieve greater stability. The electrons in the outermost shell are called the *valence electrons*.

Based on experience, chemists have assigned to every element in the periodic table one or more numbers, indicating the ability of the element to react with other elements. The numbers, which depend on the number of valence electrons, are called the *valences* of the element.

In the formation of chemical compounds from the elements, the valence electrons are transferred from the outer shell of one atom to the outer shell of another atom, or they are shared among the outer shells of the combining atoms. When electrons are transferred, the process is called *ionic bonding*. If electrons are shared, it is called *covalent bonding*. This rearrangement of electrons produces *chemical bonds*. The actual number of electrons that an atom gains, loses, or shares in bonding with one or more other atoms is the valence of the atom. For example, if an atom gives away one electron in a reaction, then it has a valence of +1. Similarly, if an atom must gain one electron to complete a reaction, then it has a valence of –1.

The following example of ionic bonding (transfer of electrons) shows how valence works. A diagram of the sodium and chlorine atoms and how they react to form sodium chloride (NaCl) is shown in Figure C3-1. Sodium has an atomic number of 11, indicating that it has 11 protons in a nucleus surrounded by 11 electrons. As illustrated, there is only one electron in the outermost shell, or ring. Chlorine, with an atomic number of 17, has 17 protons in the nucleus, surrounded by 17 electrons; 7 of the chlorine electrons are in the outermost shell. As the diagram indicates, to form a molecule of sodium chloride, the sodium atom transfers one electron to the chlorine atom. The molecule has greater chemical stability than the separate elements, so when sodium and chlorine are mixed together (under conditions that make the electron transfer possible), sodium chloride molecules will form. *Note that mixing pure sodium with chlorine will cause a violent explosion.*

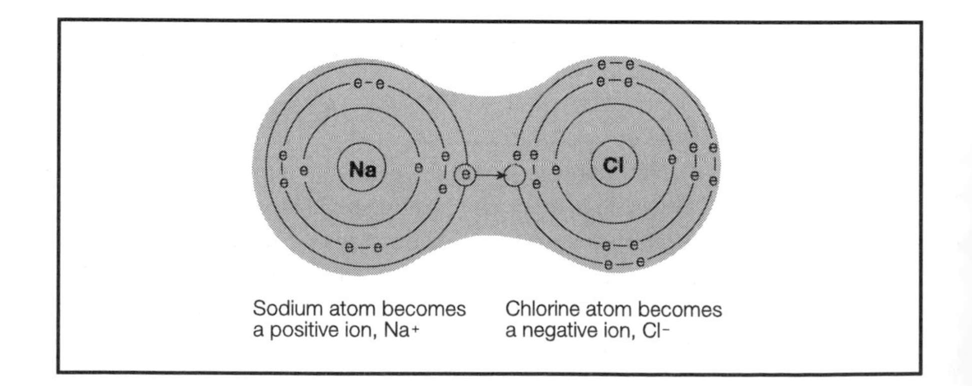

FIGURE C3-1 Ionic bonding, illustrated by sodium chloride (NaCl)

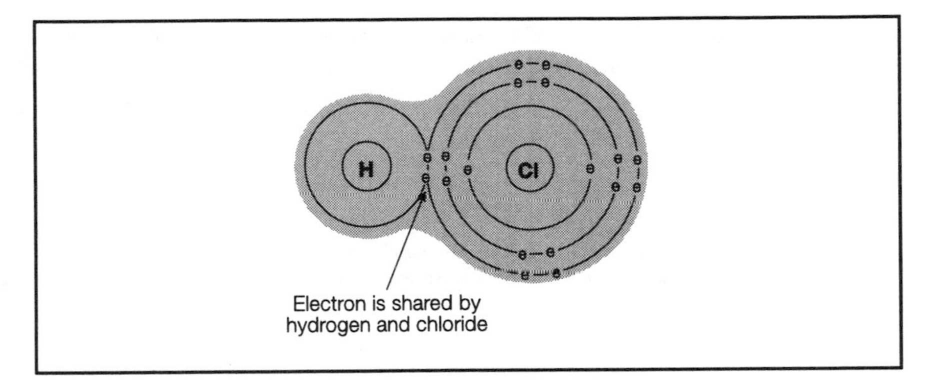

FIGURE C3-2 Covalent bonding, illustrated by hydrogen chloride (HCl)

Covalent bonding is a process similar to ionic bonding, but the electrons are shared rather than transferred, as illustrated for hydrogen chloride in Figure C3-2. Since electrons are not lost or gained, the valence of an atom involved in a covalent bond is expressed without a + or – sign. In the example, the valence of both hydrogen and chlorine is 1, since the bond is formed by sharing a single electron.

Many elements (iron and copper, for example) have more than one valence. The number of electrons involved in a reaction — the valence number — depends on several factors, such as the conditions under which the reaction occurs and the other elements involved. Iron has a valence of +2 when it forms $FeSO_4$, ferrous sulfate; but it has a valence of +3 when it forms Fe_2O_3, common red rust.

Some groups of elements bond together and act like single atoms or ions in forming compounds. Such groups of elements are called *radicals*. For example, the sulfate ion (SO_4^{-2}) and the nitrate ion (NO_3^{-1}) are important radicals in drinking water quality. The valences of some common elements and radicals are listed in Tables C3-1 and C3-2.

Chemical Formulas and Equations

A group of chemically bonded atoms forms a particle called a molecule. The simplest molecules contain only one type of atom, such as when two atoms of oxygen combine (O_2) or when two atoms of chlorine combine (Cl_2). Molecules of compounds are made up of the atoms of at least two different elements; for example, one oxygen atom and two hydrogen atoms form a molecule of the compound water (H_2O). "H_2O" is called the *formula* of water. The formula is a shorthand way of writing *what* elements are present in a molecule of a compound, and *how many* atoms of each element are present in each molecule.

TABLE C3-1 Oxidation numbers of various elements

Element	Common Valences	Element	Common Valences
Aluminum (Al)	+3	Lead (Pb)	+2, +4
Arsenic (As)	+3, +5	Magnesium (Mg)	+2
Barium (Ba)	+2	Manganese (Mn)	+2, +4
Boron (B)	+3	Mercury (Hg)	+1, +2
Bromine (Br)	–1	Nitrogen (N)	+3, –3, +5
Cadmium (Cd)	+2	Oxygen (O)	–2
Calcium (Ca)	+2	Phosphorus (P)	–3
Carbon (C)	+4, –4	Potassium (K)	+1
Chlorine (Cl)	–1	Radium (Ra)	+2
Copper (Cu)	+1, +2	Selenium (Se)	–2, +4
Chromium (Cr)	+3	Silicon (Si)	+4
Fluorine (F)	–1	Silver (Ag)	+1
Hydrogen (H)	+1	Sodium (Na)	+1
Iodine (I)	–1	Strontium (Sr)	+2
Iron (Fe)	+2, +3	Sulfur (S)	–2, +4, +6

TABLE C3-2 Oxidation numbers of common radicals

Radical	Common Valences
Ammonium (NH_4)	+1
Bicarbonate (HCO_3)	–1
Hydroxide (OH)	–1
Nitrate (NO_3)	–1
Nitrite (NO_2)	–1
Carbonate (CO_3)	–2
Sulfate (SO_4)	–2
Sulfite (SO_3)	–2
Phosphate (PO_4)	–3

Reading Chemical Formulas

The following are examples of chemical formulas and what they indicate.

Example 1

The chemical formula for calcium carbonate is

$$CaCO_3$$

According to the formula, what is the chemical makeup of the compound?

First, the letter symbols given in the formula indicate the three elements that make up the calcium carbonate compound:

Ca = calcium

C = carbon

O = oxygen

Second, the subscripts (numbers) in the formula indicate how many atoms of each element are present in a single molecule of the compound. There is no number just to the right of the Ca or C symbols; this indicates that only one atom of each is present in the molecule. The subscript 3 to the right of the O symbolizing oxygen indicates that there are three oxygen atoms in each molecule.

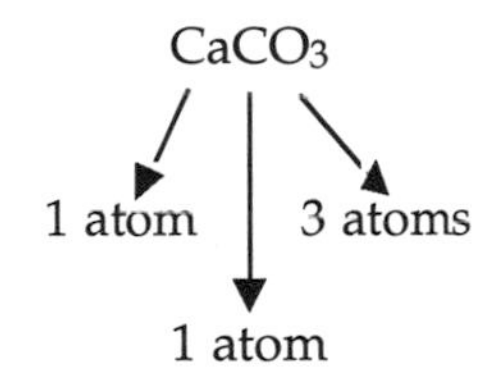

Example 2

The chemical formula for sulfuric acid is H_2SO_4. On the basis of this formula, what can be said about the chemical makeup of the compound?

The chemical symbols in the formula indicate that sulfuric acid contains three elements:

H = hydrogen

S = sulfur

O = oxygen

The formula also indicates the number of atoms of each element present:

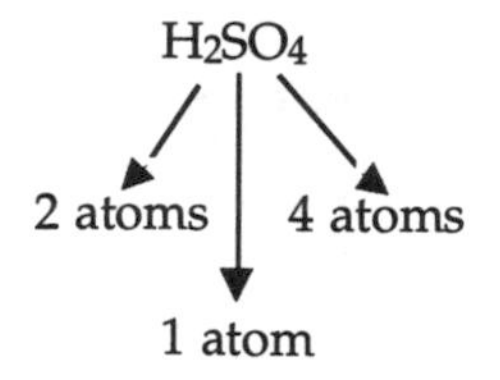

In some chemical formulas, you will see a subscript just outside a parentheses, as shown in the two following examples:

$$Ca(HCO_3)_2 \qquad Al_2(SO_4)_3$$

Notice that in both examples, *radicals* are inside the parentheses. As explained previously, radicals are a group of atoms bonded together into a unit and acting as a single atom (ion). The parentheses and subscripts are used to indicate how many of these units are involved in the reaction. In the case of calcium bicarbonate, $Ca(HCO_3)_2$, one calcium atom reacts with two bicarbonate units (ions):

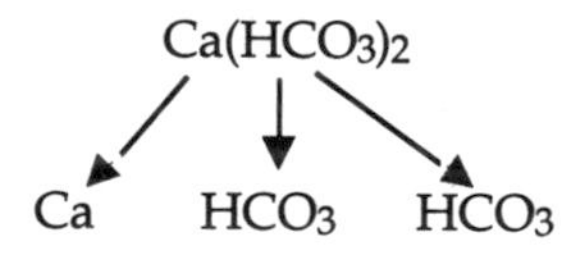

Therefore, the number of atoms of each element in a molecule of the compound are as follows:

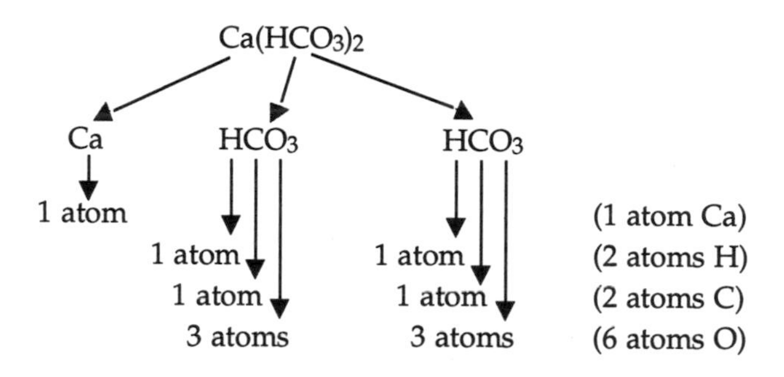

Notice that there are 6 atoms of oxygen, because there are 3 atoms of oxygen present in each bicarbonate radical (HCO_3), and there are 2 bicarbonate radicals present in each molecule of $Ca(HCO_3)_2$. Multiplying the subscripts, $3 \times 2 = 6$.

In the case of aluminum sulfate, $Al_2(SO_4)_3$, two aluminum atoms react with three sulfate ions:

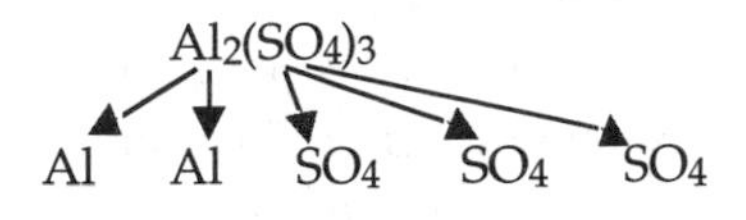

Therefore, the number of atoms of each element in a molecule of the compound is

$Al_2(SO_4)_3$

2 atoms

12 atoms [3 radicals × 4 atoms per radical]

3 atoms [3 radicals × 1 atom per radical]

Example 3

The chemical formula for calcium hydroxide (lime) is $Ca(OH)_2$. Determine the number of atoms of each element in a molecule of the compound.

The formula indicates that one atom of calcium reacts with two hydroxyl ions:

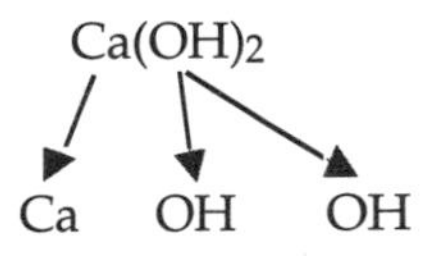

The number of atoms of each element is

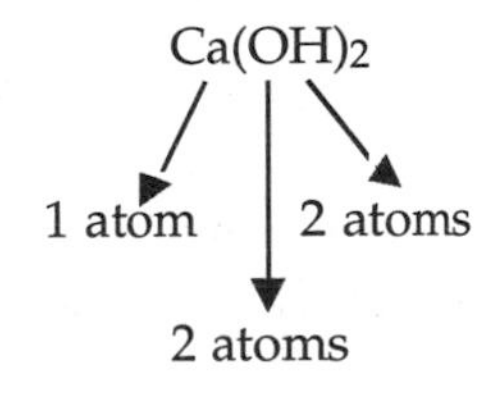

Determining Percent by Weight of Elements in a Compound

If 100 lb of sodium chloride (NaCl) were separated into the elements that make up the compound, there would be 39.3 lb of pure sodium (Na) and 60.7 lb of pure chlorine (Cl). We say that sodium chloride is 39.3 *percent* sodium *by weight* and that it is 60.7 *percent* chlorine *by weight*.[1] The *percent by weight* of each element in a compound can be calculated using the compound's chemical formula and atomic weights from the periodic table.

The first step in calculating percent by weight of an element in a compound is to determine the *molecular weight* (sometimes called *formula weight*) of the compound. The molecular weight of a compound is defined as the sum of the atomic weights of all the atoms in the compound.

For example, to determine the molecular weight of sodium chloride, first count how many atoms of each element a single molecule contains:

Na	Cl
1 atom	1 atom

Next, find the atomic weight of each atom, using the periodic table:

$$\text{atomic weight of Na} = 22.99$$

$$\text{atomic weight of Cl} = 35.45$$

Finally, multiply each atomic weight by the number of atoms of that element in the molecule, and total the weights:

	Number of Atoms		Atomic Weight		Total Weight
sodium (Na)	1	×	22.99	=	22.99
chlorine (Cl)	1	×	35.45	=	35.45
molecular weight of NaCl				=	58.44

Once the molecular weight of a compound is determined, the percent by weight of each element in the compound can be found with the following formula:

$$\text{percent element by weight} = \frac{\text{weight of element in compound}}{\text{molecular weight of compound}} \times 100$$

[1]Mathematics 7, Percent.

Using the formula, first calculate the percent by weight of sodium in the compound:

$$\text{percent Na by weight} = \frac{\text{weight of Na in compound}}{\text{molecular weight of compound}} \times 100$$

$$= \frac{22.99}{58.44} \times 100$$

$$= 0.393 \times 100$$

$$= 39.3\% \text{ sodium by weight}$$

Then, calculate percent by weight of chlorine in the compound:

$$\text{percent Cl by weight} = \frac{\text{weight of Cl in compound}}{\text{molecular weight of compound}} \times 100$$

$$= \frac{35.45}{58.44} \times 100$$

$$= 0.607 \times 100$$

$$= 60.7\% \text{ chlorine by weight}$$

To check the calculations, add the percents — the total should be 100:

39.3%	Na
+ 60.7%	Cl
100.0%	NaCl

Example 4

The formula for calcium carbonate is $CaCO_3$. What percent by weight of the compound is calcium, what percent by weight is carbon, and what percent by weight is oxygen?

To calculate the percent by weight of each of the elements, first find the atomic weight of each element in the compound, then calculate the molecular weight of $CaCO_3$.

	Number of Atoms		Atomic Weight		Total Weight
calcium (Ca)	1	×	40.08	=	40.08
carbon (C)	1	×	12.01	=	12.01
oxygen (O)	3	×	16.00	=	48.00
molecular weight of $CaCO_3$				=	100.09

Now calculate the percent by weight for each element:

$$\text{percent Ca by weight} = \frac{\text{weight of Ca in compound}}{\text{molecular weight of compound}} \times 100$$

$$= \frac{40.08}{100.09} \times 100$$

$$= 0.40 \times 100$$

$$= 40\% \text{ Ca by weight}$$

$$\text{percent C by weight} = \frac{\text{weight of C in compound}}{\text{molecular weight of compound}} \times 100$$

$$= \frac{12.01}{100.09} \times 100$$

$$= 0.12 \times 100$$

$$= 12\% \text{ C by weight}$$

$$\text{percent O by weight} = \frac{\text{weight of O in compound}}{\text{molecular weight of compound}} \times 100$$

$$= \frac{48.00}{100.09} \times 100$$

$$= 0.48 \times 100$$

$$= 48\% \text{ O by weight}$$

Check the calculations by making sure the sum of the percents is 100:

40%	Ca
12%	C
+ 48%	O
100%	$CaCO_3$

Example 5

The formula for calcium bicarbonate is $Ca(HCO_3)_2$. What percent of the weight of this compound is calcium? What are the percents of hydrogen, carbon, and oxygen?

To calculate the percent of the weight represented by each element, first determine the molecular weight of the compound:

	Number of Atoms		Atomic Weight		Total Weight
calcium (Ca)	1	×	40.08	=	40.08
hydrogen (H)	2	×	1.01	=	2.02
carbon (C)	2	×	12.01	=	24.02
oxygen (O)	6	×	16.00	=	96.00
molecular weight of $Ca(HCO_3)_2$				=	162.12

Now calculate the percent weight represented by each of the elements:

$$\text{percent Ca by weight} = \frac{\text{weight of Ca in compound}}{\text{molecular weight of compound}} \times 100$$

$$= \frac{40.08}{162.12} \times 100$$

$$= 0.247 \times 100$$

$$= 24.7\% \text{ Ca by weight}$$

$$\text{percent H by weight} = \frac{\text{weight of H in compound}}{\text{molecular weight of compound}} \times 100$$

$$= \frac{2.02}{162.12} \times 100$$

$$= 0.012 \times 100$$

$$= 1.2\% \text{ H by weight}$$

$$\text{percent C by weight} = \frac{\text{weight of C in compound}}{\text{molecular weight of compound}} \times 100$$

$$= \frac{24.02}{162.12} \times 100$$

$$= 0.148 \times 100$$

$$= 14.8\% \text{ C by weight}$$

$$\text{percent O by weight} = \frac{\text{weight of O in compound}}{\text{molecular weight of compound}} \times 100$$

$$= \frac{96.00}{162.12} \times 100$$

$$= 0.592 \times 100$$

$$= 59.2\% \text{ O by weight}$$

Checking these calculations shows the sum of the percents equal to 99.9% instead of 100%. This slight difference is due to rounding of the answers.

Once you have calculated the percent composition by weight, if you know how many pounds of the chemical you have, then you can calculate the actual weight of any element present. The following examples illustrate the procedure.

Example 6

If you have 50 lb of $Ca(HCO_3)_2$, then how many pounds of each element in the compound do you have? (Use the percent information determined in example 5: Ca = 24.7%; H = 1.2%; C = 14.8%; and O = 59.3%.)

The percents given indicate what percent of the total weight each element makes up. For example, 24.7% of the total weight of the compound is calcium. This means that 24.7% of the 50 lb of $Ca(HCO_3)_2$ is calcium:

$$[0.247][50 \text{ lb } Ca(HCO_3)_2] = 12.4 \text{ lb calcium}$$

Similar calculations may be made for the other elements:

$$[0.012][50 \text{ lb } Ca(HCO_3)_2] = 0.6 \text{ lb hydrogen}$$

$$[0.148][50 \text{ lb } Ca(HCO_3)_2] = 7.4 \text{ lb carbon}$$

$$[0.593][50 \text{ lb } Ca(HCO_3)_2] = 29.7 \text{ lb oxygen}$$

The total of these numbers should be 50 lb. In this case the total is 50.1 lb because of rounding.

Example 7

Suppose you have 75 kg of sodium carbonate (Na_2CO_3). How many kilograms of sodium, carbon, and oxygen are in this much sodium carbonate?

To determine the weight of each element, you will first have to determine *what percent* of the total weight is represented by each element. First find the molecular weight:

	Number of Atoms		Atomic Weight		Total Weight
sodium (Na)	2	×	22.99	=	45.98
carbon (C)	1	×	12.01	=	12.01
oxygen (O)	3	×	16.00	=	48.00
molecular weight of Na_2CO_3				=	105.99

Next calculate the percent composition by weight for each element:

$$\text{percent Na by weight} = \frac{\text{weight of Na in compound}}{\text{molecular weight of compound}} \times 100$$

$$= \frac{45.98}{105.99} \times 100$$

$$= 0.434 \times 100$$

$$= 43.4\% \text{ Na by weight}$$

$$\text{percent C by weight} = \frac{\text{weight of C in compound}}{\text{molecular weight of compound}} \times 100$$

$$= \frac{12.01}{105.99} \times 100$$

$$= 0.113 \times 100$$

$$= 11.3\% \text{ C by weight}$$

$$\text{percent O by weight} = \frac{\text{weight of O in compound}}{\text{molecular weight of compound}} \times 100$$

$$= \frac{48.00}{105.99} \times 100$$

$$= 0.453 \times 100$$

$$= 45.3\% \text{ O by weight}$$

Now that the percent of the weight represented by each element has been calculated, determine how much of the 75 kg is made up by each element:

43.4% sodium: (0.434) (75 kg Na_2CO_3) = 32.6 kg sodium

11.3% carbon: (0.113) (75 kg Na_2CO_3) = 8.5 kg carbon

45.3% oxygen: (0.453) (75 kg Na_2CO_3) = 34.0 kg oxygen

Check the calculation by adding together the weights of the individual elements — the sum is 75.1 kg, differing from the 75 kg total because of rounding.

Chemical Equations

A *chemical equation* is a shorthand way, through the use of chemical formulas, to write the reaction that takes place when certain chemicals are brought together. As shown in the following example, the left side of the equation indicates the *reactants*, or chemicals that will be brought together; the arrow indicates which direction the reaction occurs; and the right side of the equation indicates the *products*, or results, of the chemical reaction.

calcium bicarbonate	*plus*	calcium hydroxide	*react to form*	calcium carbonate	*plus*	water
$Ca(HCO_3)_2$	+	$Ca(OH)_2$	$\rightarrow$	$2CaCO_3$	+	$2H_2O$
		Reactants			*Products*	

The 2 in front of $CaCO_3$ is called a *coefficient*. A coefficient indicates the relative number of molecules of the compound that are involved in the chemical reaction. If no coefficient is shown, then only one molecule of the compound is involved. For example, in the equation above, one molecule of calcium bicarbonate reacts with one molecule of calcium hydroxide to form two molecules of calcium carbonate and two molecules of water. Without the coefficients, the equation could be written

$$Ca(HCO_3)_2 + Ca(OH)_2 \rightarrow CaCO_3 + CaCO_3 + H_2O + H_2O$$

If you count the atoms of calcium (Ca) on the left side of the equation, and then count the ones on the right side, you will find that the numbers are the same. In fact, for each element in the equation, as many atoms are shown on the left side as on the right. An equation for which this is true is said to be *balanced*. A balanced equation accurately represents what really happens in a chemical reaction: because matter is neither created nor destroyed, the number of atoms of each element going into the reaction must be the same as the number coming out. Coefficients allow balanced equations to be written compactly.

Coefficients and subscripts can be used to calculate the molecular weight of each term in an equation, as illustrated in the following example.

Example 8

Calculate the molecular weights for each of the four terms in the following equation:

$$Ca(HCO_3)_2 + Ca(OH)_2 \rightarrow 2CaCO_3 + 2H_2O$$

First, calculate the molecular weight of $Ca(HCO_3)_2$:

	Number of Atoms		Atomic Weight		Total Weight
calcium (Ca)	1	×	40.08	=	40.08
hydrogen (H)	2	×	1.01	=	2.02
carbon (C)	2	×	12.01	=	24.02
oxygen (O)	6	×	16.00	=	96.00
molecular weight of $Ca(HCO_3)_2$				=	162.12

The molecular weight for $Ca(OH)_2$ is

	Number of Atoms		Atomic Weight		Total Weight
calcium (Ca)	1	×	40.08	=	40.08
oxygen (O)	2	×	16.00	=	32.00
hydrogen (H)	2	×	1.01	=	2.02
molecular weight of $Ca(OH)_2$				=	74.10

The 2 in front of the next term of the equation ($2CaCO_3$) indicates that two molecules of $CaCO_3$ are involved in the reaction. First find the weight of *one molecule*, then find the weight of *two molecules*:

	Number of Atoms		Atomic Weight		Total Weight
calcium (Ca)	1	×	40.08	=	40.08
carbon (C)	1	×	12.01	=	12.01
oxygen (O)	3	×	16.00	=	48.00
weight of one molecule $CaCO_3$				=	100.09

weight of two molecules $CaCO_3$ = (2)(100.09)

= 200.18

The coefficient in front of the fourth term in the equation ($2H_2O$) also indicates that two molecules are involved in the reaction. As in the last calculation, first determine the weight of one molecule of H_2O, then the weight of two molecules:

	No. of Atoms		Atomic Weight		Total Weight
hydrogen (H)	2	×	1.01	=	2.02
oxygen (O)	1	×	16.00	=	16.00
weight of one molecule H_2O				=	18.02
weight of two molecules H_2O				=	(2)(18.02)
				=	36.04

In summary, the weights that correspond to each term of the equation are

$Ca(HCO_3)_2$	+	$Ca(OH)_2$	→	$2CaCO_3$	+	$2H_2O$
162.12		74.10		200.18		36.04

Notice that the total weight on the left side of the equation (236.22) is equal to the total weight on the right side of the equation (236.22).

The practical importance of the weight of each term of the equation is that the chemicals shown in the equation will always react in the proportions indicated by their weights.

For example, from the calculation above, you know that $Ca(HCO_3)_2$ reacts with $Ca(OH)_2$ in the ratio 162.12:74.10. This means that, given 162.12 lb of $Ca(OH)_2$, you must add 74.10 lb of $Ca(HCO_3)_2$ for a complete reaction. Given twice the amount of $Ca(HCO_3)_2$ (that is, 324.24 lb), you must add twice the amount of $Ca(OH)_2$ (equal to 148.20 lb) to achieve complete reaction. The next two examples illustrate more complicated calculations using the same principle.

Example 9

If 25 g of $Ca(OH)_2$ were added to some $Ca(HCO_3)_2$, how many grams of $Ca(HCO_3)_2$ would react with the $Ca(OH)_2$? The molecular weight of $Ca(HCO_3)_2$ is 162.12; of $Ca(OH)_2$, 74.10.

The molecular weights indicate the weight ratio in which the two compounds will react.

Use this information to set up a proportion[2] in order to determine how many grams of $Ca(HCO_3)_2$ will react with the $Ca(OH)_2$:

known ratio *desired ratio*

$$\frac{74.10 \text{ g } Ca(OH)_2}{162.12 \text{ g } Ca(HCO_3)_2} = \frac{25 \text{ g } Ca(OH)_2}{x \text{ g } Ca(HCO_3)_2}$$

Next solve for the unknown value:[3]

$$\frac{74.10}{162.12} = \frac{25}{x}$$

$$\frac{(x)(74.10)}{162.12} = 25$$

$$x = \frac{(25)(162.12)}{74.10}$$

$$x = 54.7 \text{ g } Ca(HCO_3)_2$$

Given the molecular weights and the chemical equation indicating the ratio by which the two chemicals would combine, we were able to calculate that 54.7 g of $Ca(HCO_3)_2$ would react with 25 g of $Ca(OH)_2$.

Example 10

The equation of the reaction between calcium carbonate ($CaCO_3$) and carbonic acid (H_2CO_3) is shown. If 10 lb of H_2CO_3 are to be used in the reaction, how many pounds of $CaCO_3$ will react with the H_2CO_3?

$$CaCO_3 + H_2CO_3 \rightarrow Ca(HCO_3)_2$$

[2]Mathematics 5, Ratios and Proportions.

[3]Mathematics 4, Solving for the Unknown Value.

To determine how many pounds of $CaCO_3$ will react with the H_2CO_3, first determine the weight *ratios* in the reaction:

$CaCO_3$:

	Number of Atoms		Atomic Weight		Total Weight
calcium (Ca)	1	×	40.08	=	40.08
carbon (C)	1	×	12.01	=	12.01
oxygen (O)	3	×	16.00	=	48.00
molecular weight of $CaCO_3$				=	100.09

H_2CO_3:

	Number of Atoms		Atomic Weight		Total Weight
hydrogen (H)	2	×	1.01	=	2.02
carbon (C)	1	×	12.01	=	12.01
oxygen (O)	3	×	16.00	=	48.00
molecular weight of H_2CO_3				=	62.03

The reacting weight ratios of $CaCO_3$ and H_2CO_3 are

$$\underset{100.09}{CaCO_3} + \underset{62.03}{H_2CO_3} \rightarrow Ca(HCO_3)_2$$

Now set up a proportion to solve for the amount of $CaCO_3$ that will react with 10 lb of H_2CO_3:

known ratio *desired ratio*

$$\frac{100.09 \text{ lb } CaCO_3}{62.03 \text{ lb } H_2CO_3} = \frac{x \text{ lb } CaCO_3}{10 \text{ lb } H_2CO_3}$$

And solve for the unknown value:

$$\frac{100.09}{62.03} = \frac{x}{10}$$

$$\frac{(10)\,(100.09)}{62.03} = x$$

$$16.1 \text{ lb } CaCO_3 = x$$

A list of compounds and chemical equations common to water treatment is given in appendix C.

Definition of Mole

You may sometimes find chemical reactions described in terms of *moles* of a substance. The measurement "mole" (an abbreviation for "gram-mole") is closely related to molecular weight. The molecular weight of water, for example, is 18.02 — and 1 mol of water is defined to be 18.02 g of water. (The abbreviation for mole is *mol*.) The general definition of a mole is as follows:

> A mole of a substance is a number of grams of that substance, where the number equals the substance's molecular weight.

In example 8 you saw that the following equation and molecular weights were correct:

$Ca(HCO_3)_2$	+	$Ca(OH)_2$	→	$2CaCO_3$	+	$2H_2O$
162.12		74.10		2(100.09)		2(18.02)

If 162.12 g of $Ca(HCO_3)_2$ were used in the reaction, then the ratio equations given in example 9 would show that the weights of each of the substances in the reaction were

$Ca(HCO_3)_2$	+	$Ca(OH)_2$	→	$2CaCO_3$	+	$2H_2O$
162.12 g		74.10 g		(2)(100.09) g		(2)(18.02) g

Because of the way a mole is defined, this could also be written in the more compact form:

$Ca(HCO_3)_2$	+	$Ca(OH)_2$	→	$2CaCO_3$	+	$2H_2O$
1 mole		1 mole		2 moles		2 moles

Reading this information, a chemist could state, "One mole of $Ca(HCO_3)_2$ is needed to react with one mole of $Ca(OH)_2$, and the reaction yields two moles of $CaCO_3$ and two moles of water."

When measuring chemicals in moles, always remember that *the weight of a mole of a substance depends on what the substance is.* One mole of water weighs 18.02 g — but one mole of calcium carbonate weighs 100.09 g.

Example 11

A lab procedure calls for 3.0 mol of sodium bicarbonate ($NaHCO_3$) and 0.10 mol of potassium chromate (K_2CrO_4). How many grams of each compound are required?

To find the grams required of $NaHCO_3$, first determine the weight of 1 mol of the compound:

	Number of Atoms		Atomic Weight		Total Weight
sodium (Na)	1	×	22.99	=	22.99
hydrogen (H)	1	×	1.01	=	1.01
carbon (C)	1	×	12.01	=	12.01
oxygen (O)	3	×	16.00	=	48.00
molecular weight of $NaHCO_3$				=	84.01

Therefore, 1 mol of $NaHCO_3$ weights 84.01 g.

Next, multiply the weight of 1 mol by the number of moles required. Three mol $NaHCO_3$ is required. The weight of 3 mol $NaHCO_3$ is (3)(84.01 g) = 252.03 g.

To find grams required of K_2CrO_4, first determine the weight of 1 mol of the compound:

	Number of Atoms		Atomic Weight		Total Weight
potassium (K)	2	×	39.10	=	78.20
chromium (Cr)	1	×	52.00	=	52.00
oxygen (O)	4	×	16.00	=	64.00
molecular weight of K_2CrO_4				=	194.20

Therefore, 1 mol of K_2CrO_4 weighs 194.20 g.

Next, multiply the weight of 1 mol by the number of moles required.

The required amount of K_2CrO_4 is 0.10 mol. This amount weighs (0.10) (194.20 g) = 19.42 g.

CHEMISTRY 4

Solutions

A solution consists of two parts: a *solvent* and a *solute,* which are completely and evenly mixed (they form a *homogenous* mixture[1]). The solute part of the solution is dissolved in the solvent (Figure C4-1).

In a true solution, the solute will remain dissolved and will not settle out. Salt water is a true solution; salt is the solute and water is the solvent. On the other hand, sand mixed into water does not form a solution — the sand will settle out when the water is left undisturbed.

In water treatment, the most common solvent is water. Before it is dissolved, the solute may be solid (such as dry alum), liquid (such as sulfuric acid), or gaseous (such as chlorine).

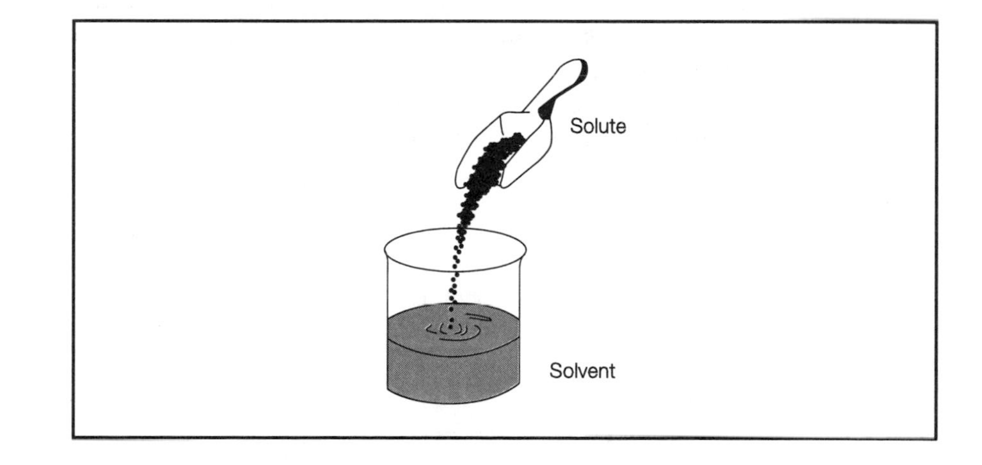

FIGURE C4-1 Solution, composed of a solute and a solvent

[1] Chemistry 2, The Classification of Matter.

The *concentration* of a solution is a measure of the amount of solute dissolved in a given amount of solvent. A concentrated (strong) solution is a solution in which a relatively great amount of solute is dissolved in the solvent. A dilute (or weak) solution is one in which a relatively small amount of solute is dissolved in the solvent.

There are many ways of expressing the concentration of a solution, including

- milligrams per liter
- grains per gallon
- percent strength
- molarity
- normality

Milligrams per Liter and Grains per Gallon

The measurements milligrams per liter (mg/L) and grains per gallon (gpg) each express the *weight* of solute dissolved in a given *volume* of solution. The mathematics needed to deal with grains per gallon[2] and milligrams per liter[3] are covered in other sections of this book, and the measurements will not be discussed further here.

Percent Strength

The percent strength of a solution can be expressed as percent by weight or percent by volume. The percent-by-weight calculation is used more often in water treatment.

Conversions between milligrams per liter and percent are discussed in another section of this book.[4]

[2] Mathematics 11, Conversions.

[3] Chemistry 7, Chemical Dosage Problems.

[4] Chemistry 7, Chemical Dosage Problems.

Percent Strength by Weight

The equation used to calculate percent by weight is as follows:

$$\text{percent strength (by weight)} = \frac{\text{weight of solute}}{\text{weight of solution}} \times 100$$

where

$$\text{weight of solution} = \text{weight of solute} + \text{weight of solvent}$$

Use of the equation is illustrated in the following examples.

Example 1

If 25 lb of chemical is added to 400 lb of water, what is the percent strength of the solution by weight?

$$\text{percent strength (by weight)} = \frac{\text{weight of solute}}{\text{weight of solution}} \times 100$$

The weight of the chemical (the solute) is given in the problem as 25 lb. The weight of the solution, however, is the combined weight of the solute and the solvent:

$$\text{weight of solution} = \text{weight of solute} + \text{weight of solvent}$$

$$= 25 \text{ lb} + 400 \text{ lb}$$

$$= 425 \text{ lb}$$

Using this information, calculate the percent concentration:

$$\text{percent strength (by weight)} = \frac{\text{weight of solute}}{\text{weight of solution}} \times 100$$

$$= \frac{25 \text{ lb chemical}}{425 \text{ lb solution}} \times 100$$

$$= 0.059 \times 100$$

$$= 5.9\% \text{ strength}$$

Example 2

If 40 lb of chemical is added to 120 gal of water, what is the percent strength of the solution by weight?

First, calculate the weight of the solution. The weight of the solution is equal to the weight of the solute plus the weight of solvent. To calculate this, first convert 120 gal of water to pounds of water:[5]

$$(120\text{ gal})(8.34\text{ lb/gal}) = 1{,}001\text{ lb water}$$

Then calculate the weight of solution:

$$\begin{aligned}\text{weight of solution} &= \text{weight of solute} + \text{weight of solvent}\\ &= 40\text{ lb} + 1{,}001\text{ lb}\\ &= 1{,}041\text{ lb}\end{aligned}$$

Now calculate the percent strength of the solution:

$$\begin{aligned}\text{percent strength (by weight)} &= \frac{\text{weight of solute}}{\text{weight of solution}} \times 100\\ &= \frac{40\text{ lb chemical}}{1{,}041\text{ lb solution}} \times 100\\ &= 0.038 \times 100\\ &= 3.8\%\text{ strength}\end{aligned}$$

Example 3

In 300 L of a solution there are 8 kg of chemical. What is the percent strength of the solution? Assume the solution has the same density as water — 1 kg/L.[6]

First express the 300 L of solution in terms of kilograms:

$$(300\text{ L})(1\text{ kg/L}) = 300\text{ kg solution}$$

---

[5] Mathematics 11, Conversions.

[6] Hydraulics 1, Density and Specific Gravity.

Now calculate the percent strength. Notice that in this problem the liters measure the amount of *solution*, not merely the amount of water.

$$\text{percent strength (by weight)} = \frac{\text{weight of solute}}{\text{weight of solution}} \times 100$$

$$= \frac{8 \text{ kg chemical}}{300 \text{ kg solution}} \times 100$$

$$= 0.027 \times 100$$

$$= 2.7\% \text{ strength}$$

In the previous three examples, percent strength was the unknown quantity. However, the same equation can be used to determine the weight of chemical required, provided the percent strength and weight (or percent strength, volume, and density) of the solution are known. The next two examples illustrate the procedure.

Example 4

You need to prepare 25 gal of a 2.5 percent strength solution. Assume the solution will have the same density as water: 8.34 lb/gal. How many pounds of chemical will you need to dissolve in the water?

Use the same equation of percent strength as in the previous problems. First, write into the equation the information given in the problem:

$$\text{percent strength (by weight)} = \frac{\text{weight of solute}}{\text{weight of solution}} \times 100$$

$$2.5\% = \frac{x \text{ lb chemical}}{(25)\,(8.34) \text{ lb solution}} \times 100$$

Then solve for the unknown value:[7]

$$2.5 = \frac{x}{(25)\,(8.34)} \times 100$$

---

[7] Mathematics 4, Solving for the Unknown Value.

This equation may be restated as

$$\frac{2.5}{1} = \frac{(x)\,(100)}{(25)\,(8.34)}$$

Then,

$$\frac{(25)\,(8.34)\,(2.5)}{1} = (x)(100)$$

$$\frac{(25)\,(8.34)\,(2.5)}{100} = x$$

$$5.21 \text{ lb chemical} = x$$

Example 5

You wish to prepare 175 L of a 4 percent strength solution. How much water and chemical should be mixed together? Assume the solution will have the same density as water: 1 kg/L.

First calculate the kilograms of chemical by writing the given information into the equation:

$$\begin{array}{c}\text{percent strength}\\ \text{(by weight)}\end{array} = \frac{\text{weight of solute}}{\text{weight of solution}} \times 100$$

$$4\% = \frac{x \text{ kg chemical}}{(175)\,(1) \text{ kg solution}} \times 100$$

Then solve for the unknown value:

$$4 = \frac{x}{(175)\,(1)} \times 100$$

This can be restated as

$$\frac{4}{1} = \frac{(x)\,(100)}{(175)\,(1)}$$

$$\frac{(175)\,(1)\,(4)}{1} = (x)(100)$$

$$\frac{(175)\,(1)\,(4)}{100} = x$$

$$7 \text{ kg chemical} = x$$

To calculate the kilograms of water, first calculate the total kilograms of solution:

$$(175 \text{ L solution}) (1 \text{ kg/L}) = 175 \text{ kg solution}$$

The 175 kg of solution contains both water and chemical. Since the weight of chemical is known, the weight of water can be determined:

$$175 \text{ kg solution} = x \text{ kg water} + 7 \text{ kg chemical}$$

Therefore,

175	kg solution
− 7	kg chemical
168	kg water

Dilution Calculations

Sometimes a particular strength solution will be made up by diluting a strong solution with a weak solution of the same chemical. The new solution will have a concentration somewhere between the weak and the strong solutions.

Although there are several methods available for determining what amounts of the weak and strong solutions are needed, perhaps the easiest is the *rectangle method* (sometimes called the *dilution rule*), shown in Figure C4-2.

The following three examples illustrate how the rectangle method is used.

Example 6

What volumes of a 3 percent solution and an 8 percent solution must be mixed to make 400 gal of a 5 percent solution?

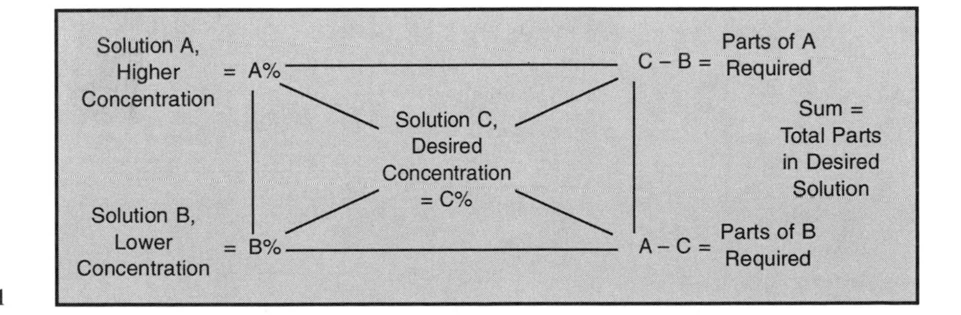

FIGURE C4-2 Schematic for rectangle method

Use the rectangle method to solve the problem. First write the given concentrations into the proper places in the rectangle:

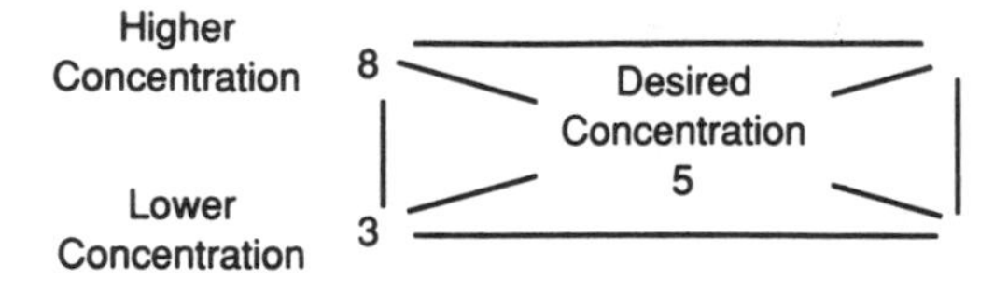

Now complete the rectangle by subtraction:

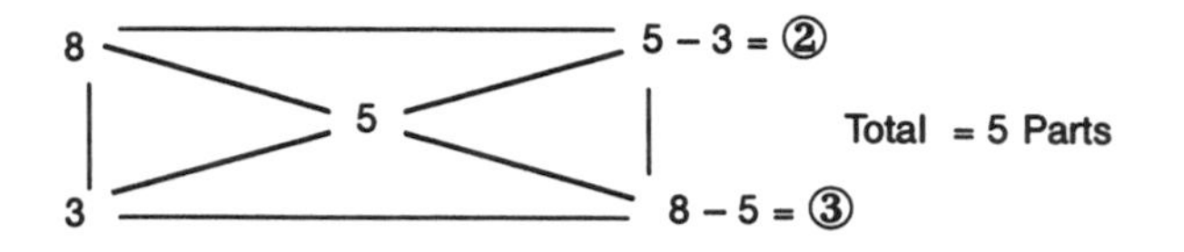

The circled numbers on the right side of the rectangle indicate the volume *ratios* of the solutions to be mixed. The sum of the circled numbers is a total of five parts to be added (2 parts + 3 parts = 5 parts total).

Two parts out of the five parts (2/5) of the new solution should be made up of the 8 percent solution. And three parts out of the five parts (3/5) of the new solution should be made up of the 3 percent solution.

Using the ratios, determine the *number of gallons* of each solution to be mixed:

$$\left(\frac{2}{5}\right)(400 \text{ gal}) = 160 \text{ gal of } 8\% \text{ solution}$$

$$\left(\frac{3}{5}\right)(400 \text{ gal}) = 240 \text{ gal of } 3\% \text{ solution}$$

Mixing these amounts will result in 400 gal of a 5 percent solution.

Example 7

What weights of a 7 percent solution and a 4 percent solution should be added together to obtain 300 lb of a 5 percent solution?

Use the rectangle method. First write in the given concentrations:

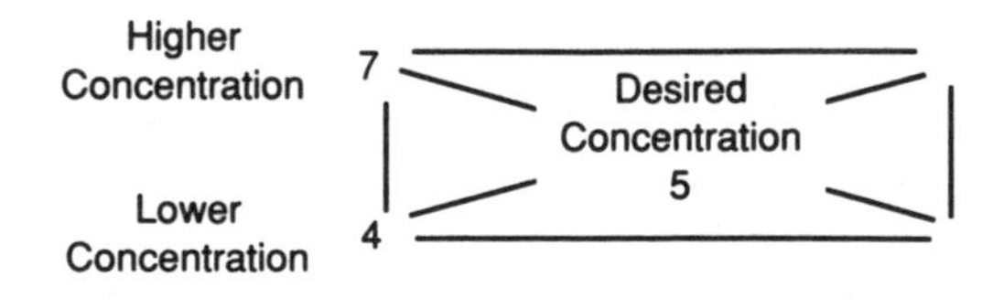

Now complete the right side of the rectangle:

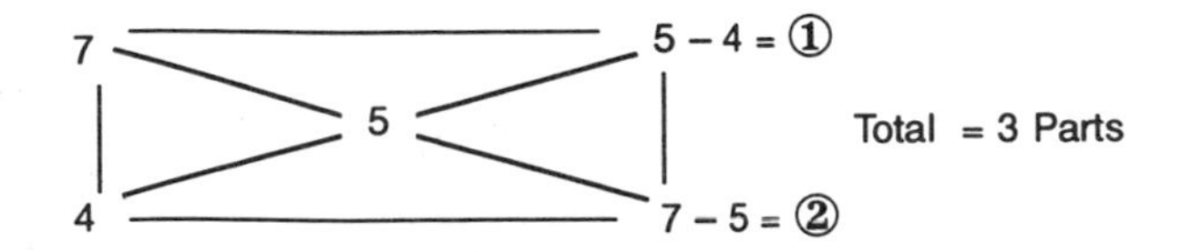

The circled numbers on the right side of the rectangle indicate the weight *ratios* by which the two solutions are to be mixed. There are a total of three parts to the new solution. One part out of three (⅓) comes from the 7 percent solution, and two parts out of three (⅔) come from the 4 percent solution.

The *number of pounds* of each of the two solutions that must be used in making the desired 5 percent solution are calculated as follows:

$$\left(\frac{1}{3}\right)(300 \text{ lb}) \quad = \quad 100 \text{ lb of 7 percent solution}$$

$$\left(\frac{2}{3}\right)(300 \text{ lb}) \quad = \quad 200 \text{ lb of 4 percent solution}$$

When these amounts are mixed, the result will be 300 lb of a 5 percent solution.

Example 8

How many liters of water and a 6 percent solution should be used to make 80 L of a 4 percent solution?

The rectangle method can also be used to solve this problem. Water is considered a 0 percent solution; therefore, the problem can be restated as "How many liters of a 0 percent solution and a 6 percent solution must be mixed to obtain 80 L of a 4 percent solution?"

First write the given information into the rectangle:

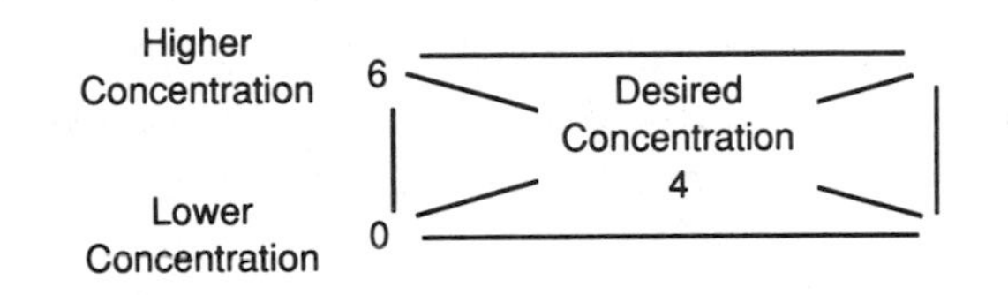

Now, complete the right side of the rectangle:

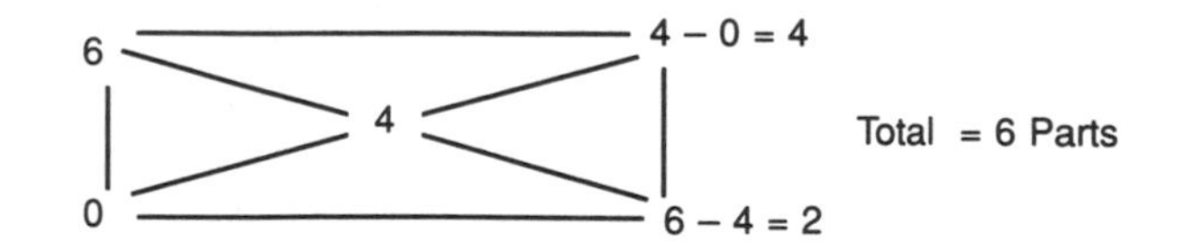

As indicated on the right side of the rectangle, four parts out of six (4/6) is the *ratio* by which the 6 percent solution must be added; and two parts out of six (2/6) is the ratio by which the water must be added.

The *number of liters* of each of the two liquids that must be used in making the 4 percent solution are calculated as follows:

$$\left(\frac{4}{6}\right)(80\text{ L}) = 53.33\text{ L of the 6\% solution}$$

$$\left(\frac{2}{6}\right)(80\text{ L}) = 26.67\text{ L of water}$$

When these amounts are mixed, the resulting solution will be 80 L of a 4 percent solution.

Although the percent strength method of expressing solution concentration is sufficiently accurate in many applications of water treatment, a more precise measurement of solution concentration is required in the laboratory.

Moles and Molarity

A more accurate way of expressing the concentration of a solution than percent strength is *molarity* (usually designated by the letter *M*). Molarity is defined as the number of *moles*[8] of solute per liter of solution. A mole (short for gram-mole and abbreviated as *mol*) is the quantity of a compound that has a weight in grams equal to the compound's molecular weight. For example, the molecular weight of $CaCO_3$ is 100.09. If you had 100.09 g of $CaCO_3$, then you would have 1 mol of $CaCO_3$. If you had 200.18 g of $CaCO_3$, then you would have 2 mol of $CaCO_3$.

[8] Chemistry 3, Valence, Chemical Formulas, and Chemical Equations.

To determine the number of moles you have of a given compound, compare the total weight of the compound in grams with the compound's molecular weight (as a number of grams):

$$\text{number of moles} = \frac{\text{total weight}}{\text{molecular weight}}$$

Example 9

If 150 g of sodium hydroxide (NaOH) is mixed into water to make a solution, how many moles of solute have been used? (Use the periodic table given in appendix B for atomic weights.)

First determine the molecular weight of NaOH:

	Number of Atoms		Atomic Weight		Total Weight
sodium (Na)	1	×	22.99	=	22.99
oxygen (O)	1	×	16.00	=	16.00
hydrogen (H)	1	×	1.01	=	1.01
molecular weight of NaOH				=	40.00

Next, calculate the number of moles of NaOH used in making up the solution:

$$\text{number of moles} = \frac{\text{total weight}}{\text{molecular weight}}$$

$$= \frac{150 \text{ g}}{40.00 \text{ g}}$$

$$= 3.75 \text{ mol of NaOH}$$

Example 10

If 45 g of sulfuric acid (H_2SO_4) were used in making up a solution, how many moles of H_2SO_4 were used?

To determine the number of moles used, first calculate the molecular weight of H_2SO_4:

	Number of Atoms		Atomic Weight		Total Weight
hydrogen (H)	2	×	1.01	=	2.02
sulfur (S)	1	×	32.06	=	32.06
oxygen (O)	4	×	16.00	=	64.00
molecular weight of H_2SO_4				=	98.08

Now determine the number of moles:

$$\text{number of moles} = \frac{\text{total weight}}{\text{molecular weight}}$$

$$= \frac{45 \text{ g}}{98.08 \text{ g}}$$

$$= 0.46 \text{ mol of } H_2SO_4$$

Once the number of moles of solute has been determined, the *molarity* of a solution may be calculated using the following equation:

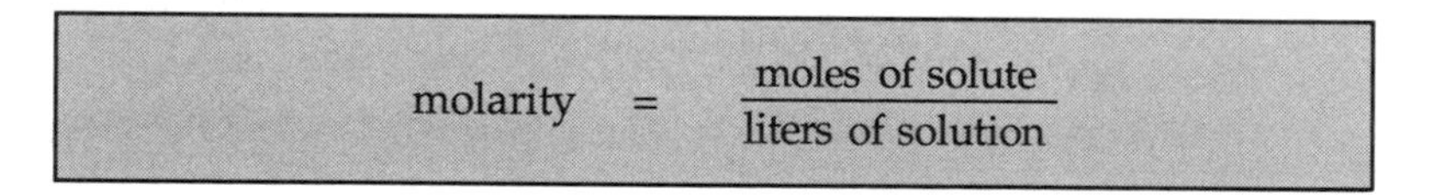

$$\text{molarity} = \frac{\text{moles of solute}}{\text{liters of solution}}$$

The following two examples illustrate the use of the equation.

Example 11

If 0.4 mol of NaOH is dissolved in 2 L of solution, what is the molarity of the solution?

In this problem, the number of moles in the solution has already been calculated. Therefore, the molarity of the solution is

$$\text{molarity} = \frac{\text{moles of solute}}{\text{liters of solution}}$$

$$= \frac{0.4 \text{ mol}}{2 \text{ L solution}}$$

$$= 0.2 \text{ molarity}$$

$$= 0.2M$$

Example 12

In 800 mL of solution, 200 g of NaCl is dissolved. What is the molarity of the solution? (Use the periodic table in appendix B for atomic weights.)

First determine the number of moles of NaCl in the solution:

	Number of Atoms		Atomic Weight		Total Weight
sodium (Na)	1	×	22.99	=	22.99
chlorine (Cl)	1	×	35.45	=	35.45
molecular weight of NaCl				=	58.44

The number of moles used in the solution is therefore

$$\text{number of moles} = \frac{\text{total weight}}{\text{molecular weight}}$$

$$= \frac{200 \text{ g}}{58.44 \text{ g}}$$

$$= 3.42 \text{ mol}$$

The chemical is dissolved in 800 mL of solution. Milliliters must be expressed as liters for the calculation of molarity:[9]

$$800 \text{ mL} = 0.8 \text{ L}$$

Now calculate the molarity of the solution:

$$\text{molarity} = \frac{\text{moles of solute}}{\text{litres of solution}}$$

$$= \frac{3.4 \text{ mol}}{0.8 \text{ L solution}}$$

[9] Mathematics 11, Conversions.

$$= \quad 4.25 \text{ molarity}$$

$$= \quad 4.25M$$

Example 13

You need to mix 2.25 L of a 1.5*M* solution of H_2SO_4.

(a) How many moles of H_2SO_4 must be dissolved in the 2.25 L?

(b) How many grams of H_2SO_4 is this? The molecular weight of H_2SO_4 is 98.08.

(a) Use the same molarity equation as in the previous examples. Write in the given information:

$$\text{molarity} = \frac{\text{moles of solute}}{\text{liters of solution}}$$

$$1.5M = \frac{x \text{ mol } H_2SO_4}{2.25\ L}$$

$$(2.25)(1.5) = x$$

$$3.38 \text{ mol } H_2SO_4 = x$$

(b) Calculate how many grams of H_2SO_4 there are in 3.4 mol. Note that 1 mol H_2SO_4 weighs 98.08 g. Therefore, 3.38 mol of H_2SO_4 weighs

$$(3.38)\ (98.08 \text{ g}) = 331.51 \text{ g } H_2SO_4$$

Equivalent Weights and Normality

Another method of expressing the concentration of a solution is *normality*. Normality depends in part on the valence of an element or compound. An element or compound may have more than one valence, and it is not always clear which valence (and therefore what concentration) a given normality represents. Because of this problem, normality is being replaced by molarity as the expression of concentration used for chemicals in the lab.

In the lab, you will often have detailed, step-by-step instructions for preparing a solution of a needed normality. Nonetheless, it is useful to have a basic idea of what the measurement means. To understand normality, you must first understand equivalent weights.

Equivalent Weights

The equivalent weight of an element or compound is *the weight of that element or compound that in a given chemical reaction, has the same combining capacity as 8 g of oxygen or as 1 g of hydrogen*. The equivalent weight may vary with the reaction being considered. However, one equivalent weight of a reactant will always react with one equivalent weight of the other reactant in a given reaction.

Although you are not expected to know how to determine the equivalent weights of various reactants at this level, it will help if you remember one characteristic: the equivalent weight of a reactant either will be *equal* to the reactant's molecular weight or will be a *simple fraction* of the molecular weight. For example, if the molecular weight of a compound is 60.00 g, then the equivalent weight of the compound in a reaction will be 60.00 g or a simple fraction (usually ½, ⅓, ¼, ⅕, or ⅙) of 60.00 g.

Normality

Normality is defined as the *number of equivalent weights of solute per liter of solution*. Therefore, to determine the normality of a solution, you must first determine how many equivalent weights of solute are contained in the total weight of dissolved solute. Use the following equation:

$$\text{number of equivalent weights} = \frac{\text{total weight}}{\text{equivalent weight}}$$

Example 14

If 90 g of sodium hydroxide (NaOH) were used in making up a solution, how many equivalent weights were used? Use 40.00 g as the equivalent weight for NaOH.

$$\text{number of equivalent weights} = \frac{\text{total weight}}{\text{equivalent weight}}$$

$$= \frac{90\text{ g}}{40\text{ g}}$$

$$= 2.25\text{ equivalent weights}$$

Example 15

If 75 g of sulfuric acid (H_2SO_4) were used in making up a solution, how many equivalent weights of H_2SO_4 were used? Use 49.04 as the equivalent weight for H_2SO_4.

$$\text{number of equivalent weights} = \frac{\text{total weight}}{\text{equivalent weight}}$$

$$= \frac{75\text{ g}}{49.04\text{ g}}$$

$$= 1.53\text{ equivalent weights}$$

When you have determined the number of equivalent weights of dissolved solute, you can determine the normality of the solution using the following equation:

$$\text{normality} = \frac{\text{number of equivalent weights of solute}}{\text{liters of solution}}$$

Example 16

If 2.1 equivalents of NaOH were used in making up 1.75 L of solution, what is the normality of the solution?

$$\text{normality} = \frac{\text{number of equivalent weights of solute}}{\text{liters of solution}}$$

$$= \frac{2.1\text{ equivalents}}{1.75\text{ L}}$$

$$= 1.2\text{ normality}$$

$$= 1.2N$$

Example 17

If 0.5 equivalents of Na_2CO_3 were used in making up 750 mL of solution, what is the normality of the solution?

$$\text{normality} = \frac{\text{number of equivalent weights of solute}}{\text{liters of solution}}$$

$$= \frac{0.5 \text{ equivalents}}{0.750 \text{ L}}$$

$$= 0.67 \text{ normality}$$

$$= 0.67N$$

Hardness

Sometimes in water chemistry, it is not important to know exactly which impurities a sample of water contains, but merely what combined effect those impurities have. A common example of this is hardness.

Hardness is a measurement of the effects that water impurities such as magnesium and calcium have on corrosion, scaling, and soap. Water from one source might contain a great deal of magnesium and little else; or it might contain a great deal of calcium and little else; or it might contain some magnesium, some calcium, and traces of some similarly behaving elements, such as strontium. In each case it would be a "hard water," a water that has noticeable effects associated with hardness. Furthermore, the laboratory test used to determine hardness would show the same result, whether the hardness was the result of calcium, magnesium, or a combination of elements.

Calcium carbonate ($CaCO_3$) is one of the more common causes of hardness, and the total hardness of water is usually expressed in terms of $CaCO_3$. For example, a lab report might read

total hardness: 180 mg/L as $CaCO_3$

This means that the lab has not determined exactly what chemicals are causing the water's hardness, but that their combined effect is the same as if the water contained exactly 180 mg/L of $CaCO_3$ and no other chemicals. By expressing the hardness of every sample in terms of how much calcium carbonate it might contain, the hardness of any two samples can be compared more easily.

In fact, measuring hardness in terms of $CaCO_3$ is so convenient that even when the lab determines exactly what chemicals are in a sample, the result may still be expressed as $CaCO_3$. Thus, a lab report could read

calcium	125 mg/L as $CaCO_3$
total hardness	150 mg/L as $CaCO_3$
bicarbonate alkalinity	122 mg/L as $CaCO_3$

total alkalinity	130 mg/L as $CaCO_3$
carbon dioxide	23 mg/L as $CaCO_3$

In softening and other water treatment operations, you may need to convert hardness expressed in terms of one chemical to hardness expressed in terms of another, usually $CaCO_3$. In a problem where hardness is expressed as the effect of 100 mg/L of chemical A and you want it expressed as the effect of an unknown weight per liter of chemical B, you are interested in what weight of chemical B has the same chemical-combining power as 100 mg of chemical A. The question can be expressed in equivalent weights: you need as many equivalent weights of B as there are equivalent weights of A in 100 mg of A. The following examples illustrate the calculations needed.

Example 18

A water sample contains 136 mg/L Ca as Ca. What is the concentration expressed as $CaCO_3$? For the equivalent weight of Ca, use 20.00 g; for the equivalent weight of $CaCO_3$, use 50.00 g.

First find the number of equivalent weights of Ca in 136 mg of Ca:

$$\begin{array}{c}\text{number of}\\ \text{equivalent weights}\\ \text{of Ca}\end{array} = \frac{\text{total weight of Ca}}{\text{equivalent weight of Ca}}$$

$$= \frac{136 \text{ mg}}{20.00 \text{ g}}$$

$$= \frac{0.136 \text{ mg}}{20.00 \text{ g}}$$

$$= 0.0068 \text{ equivalent weights of Ca}$$

Because the sample behaves as if it has 0.0068 equivalent weights of Ca per liter, it will behave as if it has 0.0068 equivalent weights of $CaCO_3$ per liter. So, calculate the total weight of 0.0068 equivalent weights of $CaCO_3$:

$$\begin{array}{c}\text{total weight}\\ \text{of } CaCO_3\end{array} = \left(\begin{array}{c}\text{number of}\\ \text{equivalent weights}\\ \text{of } CaCO_3\end{array}\right)(\text{equivalent weight of } CaCO_3)$$

$$= (0.0068)\,(50.00 \text{ g})$$

$$= \quad 0.34 \text{ g}$$

$$= \quad 340 \text{ mg}$$

Therefore, the hardness measured as 136 mg/L as Ca can also be expressed as a hardness of 340 mg/L as $CaCO_3$.

The calculations just performed demonstrate the meaning of "equivalent weights." In practice, a simpler calculation is commonly used, combining the two previous equations into a single equation that gives the same result. The simpler calculation is performed as follows: set up the ratio[10] of the equivalent weight of the desired ("new") measure to the equivalent weight of the given ("old") measure; then multiply by the "old" concentration.

$$\left(\frac{\text{equivalent weight of new measure}}{\text{equivalent weight of old measure}}\right)(\text{old concentration}) = \text{new concentration}$$

For the problem in example 18, filling in the values gives

$$\left(\frac{\text{equivalent weight of } CaCO_3}{\text{equivalent weight of Ca}}\right)(\text{concentration as Ca}) = \text{concentration as } CaCO_3$$

$$\left(\frac{50.00 \text{ g}}{20.00 \text{ g}}\right)(136 \text{ mg/L as Ca}) = x \text{ mg/L as } CaCO_3$$

$$= 340 \text{ mg/L as } CaCO_3$$

Example 19

A lab report shows

magnesium	17 mg/L
carbon dioxide	3 mg/L

Express the concentrations as $CaCO_3$. Use 12.16 g for the equivalent weight of magnesium (Mg); use 22.00 g for the equivalent weight of

---

[10] Mathematics 5, Ratios and Proportions.

carbon dioxide (CO_2); and use 50.00 g for the equivalent weight of $CaCO_3$.

First note that the lab report does not state that the concentrations are expressed "as" anything. This means that the concentrations are simply measurements of the chemicals listed — there actually is 17 mg of magnesium in every liter of the water. You could say that the magnesium concentration is expressed "as magnesium," and that the carbon dioxide concentration is expressed "as carbon dioxide."

To convert the measurements to $CaCO_3$, use the equation

$$\left(\frac{\text{equivalent weight of new measure}}{\text{equivalent weight of old measure}}\right)(\text{old concentration}) = \text{new concentration}$$

Fill in the values and solve for the magnesium concentration:

$$\left(\frac{\text{equivalent weight of } CaCO_3}{\text{equivalent weight of Mg}}\right)(\text{concentration as Mg}) = \text{concentration as } CaCO_3$$

$$\left(\frac{50.00 \text{ g}}{12.16 \text{ g}}\right)(17 \text{ mg/L as Mg}) = x \text{ mg/L as } CaCO_3$$

$$= 69.9 \text{ mg/L as } CaCO_3$$

Fill in the values and solve for the carbon dioxide concentration:

$$\left(\frac{\text{equivalent weight of } CaCO_3}{\text{equivalent weight of } CO_2}\right)(\text{concentration as } CO_2) = \text{concentration as } CaCO_3$$

$$\left(\frac{50.00 \text{ g}}{22.00 \text{ g}}\right)(3 \text{ mg/L as } CO_2) = x \text{ mg/L as } CaCO_3$$

$$= 6.82 \text{ mg/L as } CaCO_3$$

Nitrogen Compounds

Nitrogen is one of the more common elements associated with the biochemical processes of life. Chemists sometimes find it convenient to use a special measurement for concentrations of nitrogen compounds. For example, when a chemist measuring nitrate (NO_3^-) writes that a sample contains "8 mg/L NO_3^- as N," the phrase "as N" means that each liter of the sample contains 8 mg of nitrogen atoms; not that it contains 8 mg of NO_3^- ions. Because each nitrogen atom is bonded to three oxygen atoms, the weight of nitrate (NO_3^-) in each liter of solution is equal to the 8 mg of nitrogen atoms *plus* the weight of all the oxygen atoms to which the nitrogen is bonded. In fact, the weight of nitrate in each liter of this solution is 35.4 mg; therefore, the nitrate concentration can also be written "35.4 mg/L NO_3^- as NO_3^-."

To convert between measurements of nitrogen compounds "as N" and measurements of the compounds "as [the compound]," set up a ratio between the molecular weight of the "new" measurement and the molecular weight of the "old" measurement:

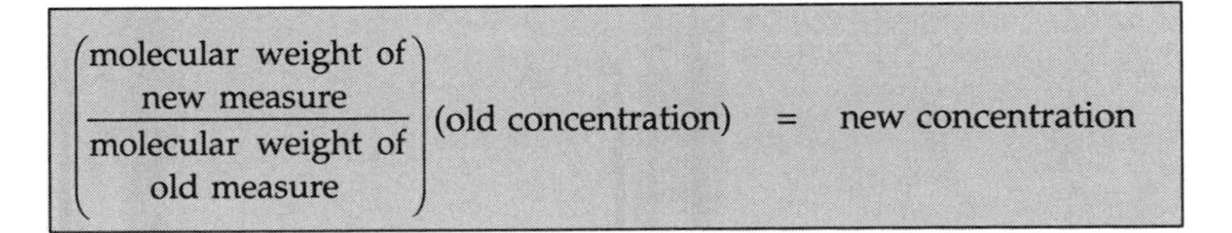

$$\left(\frac{\text{molecular weight of new measure}}{\text{molecular weight of old measure}}\right)(\text{old concentration}) = \text{new concentration}$$

When one of the measurements is "as N," the atomic weight of nitrogen — 14.01 — is used for that measurement's molecular weight. The following example illustrates the use of the equation.

Example 20

Under the Primary Regulations of the Safe Drinking Water Act, the maximum contaminant level (MCL) for nitrate (NO_3^-) is 10 mg/L as N. A lab reports that a water sample contains 42 mg/L NO_3^- as NO_3^-. Does the concentration of nitrate exceed the MCL?

First find the molecular weight of nitrate:

	Number of Atoms		Atomic Weight		Total Weight
nitrogen (N)	1	×	14.01	=	14.01
oxygen (O)	3	×	16.00	=	48.00
molecular weight of NO_3^-				=	62.01

Next set up the ratio, fill in the values, and solve:

$$\left(\frac{\text{molecular weight of new measure}}{\text{molecular weight of old measure}}\right)(\text{old concentration}) = \text{new concentration}$$

$$\left(\frac{\text{molecular weight of N}}{\text{molecular weight of } NO_3^-}\right)(\text{concentration as } NO_3^-) = \text{concentration as N}$$

$$\left(\frac{14.01 \text{ g}}{62.01 \text{ g}}\right)(42 \text{ mg/L as } NO_3^-) = x \text{ mg/L as N}$$

$$= 9.48 \text{ mg/L as N}$$

Therefore, the concentration does not exceed the MCL.

When expressing a nitrogen compound "as N," you can use several ways to write the expression. The following all mean the same:

10 mg/L NO_3^- as N
10 mg/L nitrate as nitrogen
10 mg/L NO_3^-–N
10 mg/L nitrate–nitrogen

Standard Solutions

A standard solution is any solution that has an accurately known concentration. Although there are many uses of standard solutions, they are often used to determine the concentration of substances in other solutions. Standard solutions are generally made up based on one of three characteristics:

- weight per unit volume
- dilution
- reaction

Weight Per Unit Volume

When a standard solution is made up by weight per unit volume, a pure chemical is accurately weighed and then dissolved in some solvent. By the addition of more solvent, the amount of solution is increased to a given volume. The concentration of the standard is then determined in terms of molarity or normality, as discussed previously.

Dilution

When a given volume of an existing standard solution is diluted with a measured amount of solvent, the concentration of the resulting (more dilute) solution can be determined from the following equation:

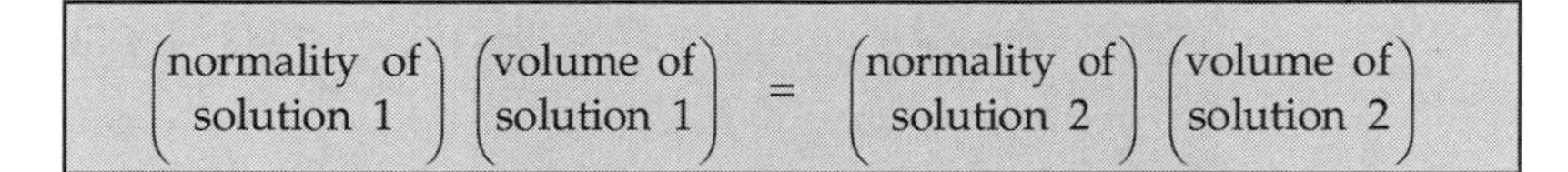

$$\begin{pmatrix}\text{normality of}\\ \text{solution 1}\end{pmatrix}\begin{pmatrix}\text{volume of}\\ \text{solution 1}\end{pmatrix} = \begin{pmatrix}\text{normality of}\\ \text{solution 2}\end{pmatrix}\begin{pmatrix}\text{volume of}\\ \text{solution 2}\end{pmatrix}$$

This equation can be abbreviated as

$$(N_1)(V_1) = (N_2)(V_2)$$

When this equation is being used, it is important to remember that both volumes — V_1 and V_2 — must be expressed in the same units. That is, both must be in liters (L) or both must be in milliliters (mL).

Example 21

You have a standard 1.4N solution of H_2SO_4. How much water must be added to 100 mL of the standard solution to produce a 1.2N solution of H_2SO_4?

First determine the total volume of the new solution by using the relationship between solution concentration and volume:

$$(N_1)(V_1) = (N_2)(V_2)$$

$$(1.4N)(100 \text{ mL}) = (1.2N)(x \text{ mL})$$

Solve for the unknown value:

$$\frac{(1.4)\,(100)}{1.2} = x \text{ mL}$$

$$116.67 = x \text{ mL}$$

Therefore, the total volume of the new solution will be 116.67 mL. Since the volume of the original solution is 100 mL, 16.67 mL of water (that is, 116.67 mL – 100 mL) need to be added to obtain the 1.2*N* solution:

100 mL of 1.4*N* solution	+	16.67 mL of water	=	116.67 mL of 1.2*N* solution

Example 22

How much water should be added to 80 mL of a 2.5*N* solution of H_2CO_3 to obtain a 1.8*N* solution of H_2CO_3?

Use the equation relating solution concentration and volume to calculate the total volume of the new solution:

$$(N_1)\,(V_1) = (N_2)\,(V_2)$$

$$(2.5N)\,(80 \text{ mL}) = (1.8N)\,(x \text{ mL})$$

Solve for the unknown value:

$$\frac{(2.5)\,(80)}{1.8} = x \text{ mL}$$

$$111.11 = x \text{ mL}$$

Therefore, the total volume of the new solution will be 111.11 mL. Since the volume of the original solution is 80 mL, 31.11 mL of water (that is, 111.11 mL – 80 mL) need to be added to obtain the 1.8*N* solution:

80 mL of 2.5*N* solution	+	31.11 mL of water	=	111.11 mL of 1.8*N* solution

Reaction

A similar equation can be used for calculations involving reactions between samples of two solutions, as illustrated in the following example.

Example 23

Thirty-two mL of a 0.1*N* solution of HCl is required to react with (neutralize) 30 mL of a certain base solution. What is the normality of the base solution?

$$(N_1)(V_1) = (N_2)(V_2)$$

$$(0.1N)(32\text{ mL}) = (xN)(30\text{ mL})$$

$$\frac{(0.1)(32)}{30} = x\text{ normality}$$

$$0.11 = x\text{ normality}$$

CHEMISTRY 5

Acids, Bases, and Salts

Inorganic compounds (compounds generally not containing carbon) can be classified into three main groups: (1) acids, (2) bases, and (3) salts. These three terms are commonly used throughout chemistry, and it is important that you understand the basic features that distinguish them. The following definitions are adequate for most water treatment chemistry; however, somewhat different definitions may be used in advanced work.

Acids

An acid is any substance that releases hydrogen ions (H^+) when it is mixed into water. For example, shortly after sulfuric acid (H_2SO_4) is mixed into water, many of the H_2SO_4 molecules *dissociate* (come apart), forming H^+ and $SO_4^=$ ions. The release of H^+ ions indicates that H_2SO_4 is an acid.

Acids that dissociate readily are known as *strong acids*. Most of the molecules of a strong acid dissociate when mixed into water, releasing a large concentration of hydrogen ions. Examples of strong acids are sulfuric (H_2SO_4), hydrochloric (HCl), and nitric (HNO_3).

Acids that dissociate poorly are known as *weak acids*. They release very few hydrogen ions in water. Examples include carbonic acid (H_2CO_3), which is the acid found in soft drinks; and hydrogen sulfide (H_2S), the compound responsible for the rotten-egg odor found naturally in some groundwaters and certain deep surface waters. The following four equations are examples of how acids dissociate when mixed into water:

$HCl \xrightarrow{H_2O} H^+ + Cl$ (a strong acid: generally, more than 99% of the molecules dissociate in water)

$H_2SO_4 \xrightarrow{H_2O} 2H^+ + SO_4^{-2}$ (a strong acid: generally, more than 99% of the molecules dissociate in water)

$H_2S \xrightarrow{H_2O} 2H^+ + SO^{-2}$ (a weak acid: generally, less than 0.1% of the molecules dissociate in water)

$HCO_3 \xrightarrow{H_2O} 2H^+ + CO_3^{-2}$ (a weak acid: generally, less than 0.1% of the molecules dissociate in water)

Solutions that contain significant numbers of H^+ ions are called *acidic*. Three other features that distinguish acids from bases and salts are

1. Acids change the color of chemical color indicators.
   - Acids turn litmus paper red.
   - Acids turn phenolphthalein colorless.
   - Acids turn methyl orange to red.
2. Acids neutralize bases, resulting in the formation of a salt and water.
3. Acids found naturally in foods give the foods a sour taste. The sour flavor of citrus fruits is caused by citric acid. *It is vital to note that tasting the acids found in laboratories and water treatment plants can be dangerous, even fatal.*

Bases

A base is any substance that produces hydroxyl ions (OH^-) when it dissociates in water. Lime [$Ca(OH)_2$], caustic soda [sodium hydroxide, or NaOH], and common household ammonia [NH_4OH] are familiar examples of bases. Strong bases are those that dissociate readily, releasing a large concentration of hydroxyl ions. NaOH is an example of a very strong, caustic base. Weak bases, such as $Ca(OH)_2$ and NH_4OH, dissociate poorly, releasing few OH^- ions. The following equations are examples of how bases dissociate when mixed into water:

$NaOH$	→	Na^+	+	OH^-	(a strong base: generally, more than 99% of the molecules dissociate in water)
KOH	→	K^+	+	OH^-	(a strong base: generally, more than 99% of the molecules dissociate in water)
$Ca(OH)_2$	→	Ca^{+2}	+	$2OH^-$	(a relatively weak base: generally, less than 15% of the molecules dissociate in water)
NH_4OH	→	NH_4^+	+	OH^-	(a weak base: generally, less than 0.5% of the molecules dissociate in water)

Solutions that contain significant numbers of OH^- ions are called *basic solutions* or *alkaline solutions*. The term *alkaline* should not be confused with the term *alkalinity*, which has a special meaning in water treatment.

Three other features that distinguish bases from acids and salts are

1. Bases change the color of chemical color indicators.
   - Bases turn litmus paper blue.
   - Bases turn phenolphthalein red.
   - Bases turn methyl orange to yellow.
2. Bases neutralize acids, resulting in the formation of a salt and water.
3. Bases found naturally in foods give the foods a bitter taste. Baking soda and milk of magnesium both taste bitter because they contain basic compounds. *It is vital to note that tasting the bases found in laboratories and water treatment plants can be dangerous, even fatal.*

Salts

Salts are compounds resulting from an acid–base mixture. The process of mixing an acid with a base to form a salt is called *neutralization*. Calcium sulfate ($CaSO_4$), for example, is a salt formed by the following acid–base neutralization:

$$H_2SO_4 + Ca(OH)_2 \rightarrow CaSO_4 + H_2O$$

Another example is sodium chloride (NaCl):

$$HCl + NaOH \rightarrow NaCl + H_2O$$

Notice that each acid–base reaction results in a salt plus water. Salts generally have no effect on color indicators. When occurring naturally in foods, salts taste salty; however, like all chemicals found in the laboratory or water treatment plants, *salts may be poisonous, and tasting them could be dangerous or fatal.*

pH

Solutions range from very acidic (having a high concentration of H^+ ions) to very basic (having a high concentration of OH^- ions). When there are exactly as many OH^- ions as H^+ ions, the solution is *neutral* — neither acidic nor basic — and each OH^- can combine with an H^+ to form a molecule of H_2O. Pure water is neutral, and most salt solutions are neutral or very nearly so.

The *pH* of a solution is a measurement of how acidic or basic the solution is (Figure C5-1). The pH scale runs from 0 (most acidic) to 14 (most basic). The scale is a log scale, which means that each pH measurement is 10 times greater than the preceding value. For example, if $[OH^-]$ at pH 7 equals 10, at pH 8 $[OH^-]$ equals 100 and at pH 9 $[OH^-]$ equals 1,000. This is also true for $[H^+]$ in acidic solutions moving from pH 7 to pH 1. Pure water has a pH of 7, the center of the range, neither acidic nor basic.

For each treatment process, there is a pH at which the operation is most effective. If the pH of the water is too low (the water is too acidic) for an operation to be effective, then the pH can be increased by the addition of a base, such as lime $[Ca(OH)_2]$. The OH^- ions released by the base will combine with some of the H^+ ions of the acidic water, forming H_2O molecules and lowering the concentration of H^+ ions.

Similarly, if the pH of water is too high (the water is too basic), then the pH can be lowered by the addition of an acid. In water treatment, pH is often lowered by bubbling carbon dioxide (CO_2) gas through the water. The CO_2 reacts with water to form carbonic acid (H_2CO_3), and the acid dissociates into $2H^+ + CO_3^{-2}$. The H^+ ions then combine with some of the OH^- ions of the basic water, forming H_2O molecules and lowering the concentrations of OH^- ions.

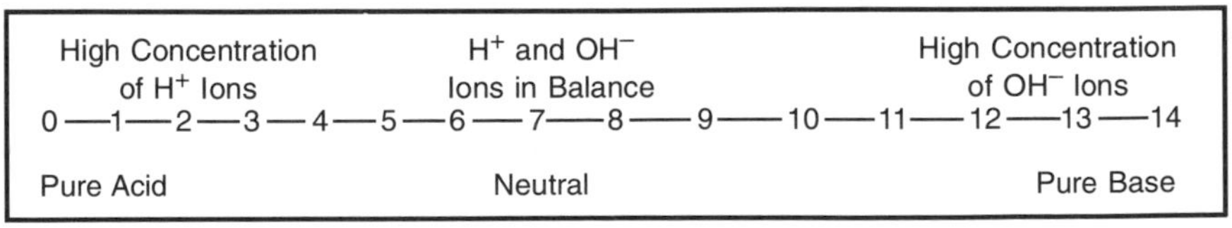

High Concentration of H^+ Ions — H^+ and OH^- Ions in Balance — High Concentration of OH^- Ions

0—1—2—3—4—5—6—7—8—9—10—11—12—13—14

Pure Acid — Neutral — Pure Base

FIGURE C5-1
The pH scale

Alkalinity

When acid is added to water that has a high concentration of OH^- ions (basic water, high in pH), the H^+ ions released by the acid combine with the OH^- ions in the water to form H_2O. As long as the water contains OH^- ions, those ions will *neutralize* the added acid and the water will remain basic. In natural waters, alkalinity is measured by bicarbonate (HCO_3^-) concentration. Bicarbonate maintains pH by taking on H^+:

$$HCO_3^- + H^+ \longrightarrow H_2CO_3$$

or by giving H^+:

$$HCO_3^- \longrightarrow H^+ + CO_3^-$$

These reactions maintain the pH of natural water at around pH 7. After enough acid has been added to combine with all the OH^- ions, addition of more acid will give the water a high concentration of H^+ ions — the water will become acidic.

However, if acid is added to water containing carbonate (CO_3^{-2}) ions in addition to OH^- ions, then some of the H^+ ions released by the acid will combine with the OH^- to form H_2O, and some will combine with the CO_3^{-2} ions to form HCO_3^- (bicarbonate). As more acid is added, the H^+ ions released will combine with the HCO_3^- ions to form H_2CO_3 (carbonic acid). A concentration of H^+ ions will begin to accumulate (the water will become acidic) only after enough acid has been added to convert all the OH^+ to H_2O and to convert all the CO_3^{-2} to HCO_3^- and then to H_2CO_3.

The CO_3^{-2} and HCO_3^- ions in the water increase the capacity of the water to neutralize (or *buffer*) an acid. *Alkalinity* is a measurement of a water's capacity to neutralize an acid, whether the neutralization is the result of OH^-, CO_3^{-2}, HCO_3^-, or other negative ions.

In water treatment, OH^-, CO_3^{-2}, and HCO_3^- are the ions causing most of the alkalinity. Alkalinity caused by OH^- is called *hydroxyl alkalinity;* if caused by CO_3^{-2} it is called *carbonate alkalinity;* and if caused by HCO_3^- it is called *bicarbonate alkalinity*. The combined effect of all three types is reported by a lab as the *total alkalinity*.

The experiment illustrated in Figure C5-2 demonstrates the relationship between alkalinity and pH. In step 1, equal volumes of two solutions are prepared. Solution 1 is made up by mixing a base [lime, $Ca(OH)_2$] into water.

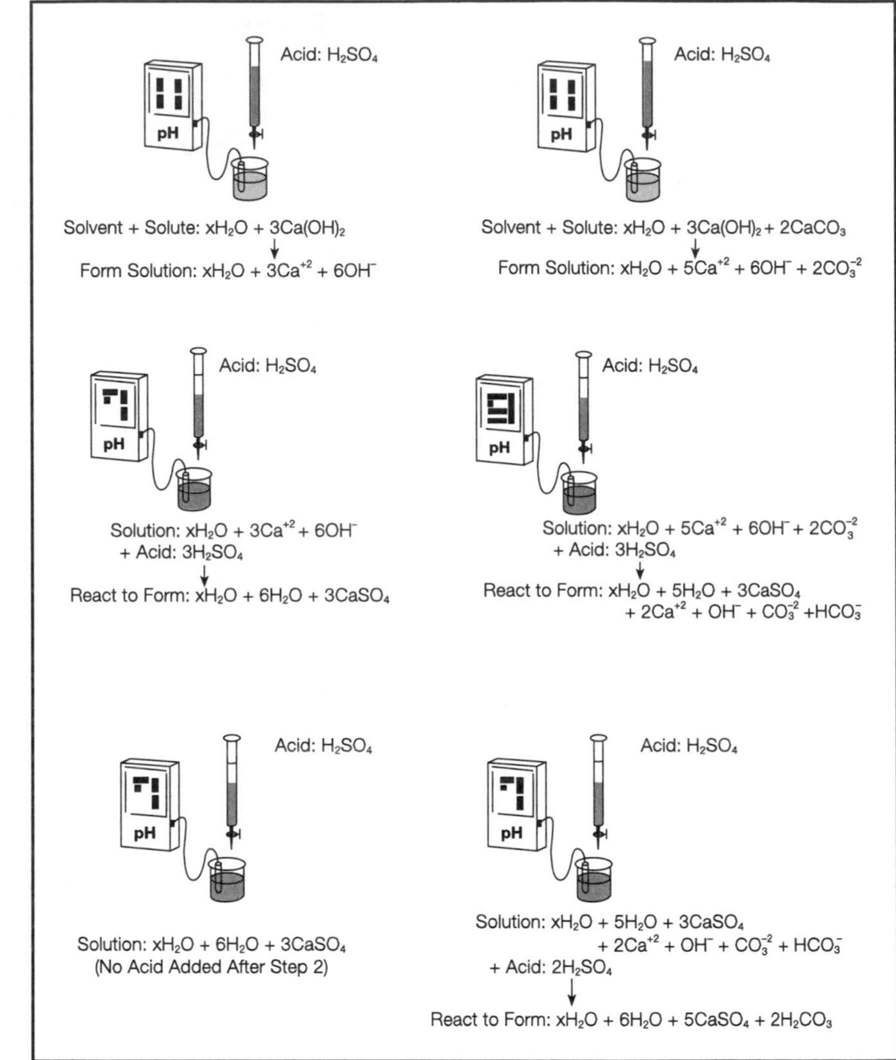

FIGURE C5-2 Alkalinity compared with pH

The pH of solution 1 is 11, and the only alkalinity is provided by the hydroxyl ions (OH^-) released by the base. Solution 2 is made up by mixing the same base [$Ca(OH)_2$] and calcium carbonate [$CaCO_3$] into water. The pH of solution 2 is also 11 — that is, solution 1 and solution 2 have the same concentration of OH^- ions. However, solution 2 has a higher alkalinity than solution 1, because the alkalinity of solution 2 includes carbonate ions (CO_3^{-2}

released by the calcium carbonate) in addition to the hydroxyl ions (OH^- released by the lime).

In step 2, sulfuric acid (H_2SO_4) is slowly added to solution 1 until the pH drops to 7, as indicated by the pH meter. The pH of 7 means that enough H^+ ions have been added (released by the acid) to combine with all the OH^- ions originally in the basic solution. The same volume of acid is added to solution 2, but the pH meter still indicates a basic solution: pH greater than 7. The carbonate ions (CO_3^{-2}) in solution 2 have combined with some of the H^+ ions released by the acid; as a result, there are not yet enough H^+ ions in solution 2 to balance the original concentration of OH^- ions.

To complete the experiment in step 3, more acid is added to solution 2 until the pH drops to a neutral 7. The solutions had the same pH at the beginning of the experiment, and they have the same pH at the end. But solution 2, which had the higher alkalinity, was capable of neutralizing a greater volume of acid than solution 1. (Note that the pH meter readings and equations in Figure C5-2 indicate a slightly simpler behavior of the chemicals than actually would occur in a laboratory.)

CHEMISTRY 6

Chemistry of Treatment Processes

In order to operate a water treatment plant successfully, you will need to understand some of the reactions involved when chemicals are added to water, or when constituents are removed from water. The following chapter discusses the basic chemistry of

- taste-and-odor removal
- coagulation
- iron and manganese removal
- lime–soda ash softening
- recarbonation
- ion exchange softening
- scale and corrosion control (stabilization)
- chlorination

Taste-and-Odor Removal

The removal of tastes and odors may be one of the more difficult problems you will confront. This is due to three important factors:

1. There are many different materials that can cause tastes and odors.
2. No common treatment process can remove all types of tastes and odors.
3. The tests used to measure tastes and odors rely on the keenness of one's sense of taste and smell.

Perhaps the best tools available for measuring taste-and-odor problems are two procedures: the threshold odor number (TON) test and the flavor profile analysis. These procedures are as valuable in maintaining high-quality, odor-free water as they are in identifying and treating an odor problem.

Threshold odor test results indicate if treated water is within the limit of 3 TON, recommended in the US Environmental Protection Agency's secondary drinking water regulations. The results identify specific types of odors and are used with the jar test to find specific treatment methods to remove the odor. Records of threshold odor tests are valuable in identifying those odors that consumers find most offensive. Flavor profile analysis is used to evaluate the sensory characteristics of water and to help water suppliers solve recurring taste-and-odor problems.

Physical removal and chemical oxidation are the two methods normally used to remove tastes and odors. Physical methods for taste-and-odor removal, not discussed in detail in this text, include (1) coagulation followed by filtration, and (2) adsorption. Chemical oxidation to remove tastes and odors can be attained using one of several oxidizing agents:

- oxygen
- chlorine
- potassium permanganate
- chlorine dioxide
- ozone

In chemical oxidation, the undesirable taste-and-odor materials are converted chemically to compounds that do not cause tastes or odors. In some cases, the conversion allows the materials to be removed by precipitation, although such physical removal is not always essential. The effectiveness of each oxidizing agent varies, depending on the type of odor that is present and the way in which the oxidant is applied.

Oxygen

Oxygen, mixed into the water by aeration, can be effective in removing tastes and odors caused by iron, manganese, and hydrogen sulfide. Aeration brings the offending material into direct contact with oxygen in the atmosphere. The materials are then oxidized into stable, nontroublesome forms.

Iron and Manganese

The removal of iron and manganese is discussed later in this chapter.

Hydrogen Sulfide

Aeration removes hydrogen sulfide (H_2S) primarily by the mechanical scrubbing action of air moving through water; however, some oxidation does occur, as shown in Eq C6-1 and Eq C6-2.

$$\underset{\text{hydrogen sulfide}}{2H_2S} + O_2 \xrightarrow{\text{aeration}} 2H_2O + \underset{\text{sulfur}}{2S\downarrow} \qquad \text{(C6-1)}$$

Further

$$\underset{\text{sulfur}}{2S} + 4O_2 \xrightarrow{\text{aeration}} \underset{\text{sulfate ion}}{2SO_4^{-2}} \qquad \text{(C6-2)}$$

Although oxidation decreases the H_2S, oxidation alone does not completely remove or convert all the H_2S. Equation C6-1 shows elemental sulfur (solid sulfur) as the final product, which gives the aerated water a milky-blue color. To remove the sulfur, the water must be chemically coagulated and filtered. (The "↓" after the "S" indicates that sulfur is a precipitate.) Because H_2S is best removed by scrubbing at about pH 4.5, it is common practice to bubble CO_2 through the water prior to aeration in order to reduce the pH.

Notice that in Eq C6-2, further aeration converts the sulfur to sulfate ion. This solves the taste-and-odor problem; however, it leaves the sulfur in the water in sulfate form. Sulfate may be converted back to H_2S if the water becomes anaerobic (without dissolved oxygen), such as it might in a dead end or a slightly used portion of the distribution system.

Chlorine

Chlorination, performed within the proper limits, can be an effective way of oxidizing many substances that cause tastes and odors. For example, algae-caused odors described as fishy, grassy, or septic can be controlled by prechlorination of the water to a free residual of 0.25–5 mg/L. Iron and manganese can effectively be removed by chlorination, as discussed later. Hydrogen sulfide (H_2S) can also be successfully oxidized with chlorine, as discussed later.

Chlorine can also intensify certain tastes and odors. For example, chlorine will combine with phenols in water, producing extremely objectionable medicinal tastes (phenol is an aromatic organic compound that gives water the characteristic odor of creosote or coal tar). Chlorine will also intensify earthy odors caused by certain types of algae and actinomycetes. (Actinomycetes are a group of organisms usually associated with taste-and-odor problems. The group has some characteristics of bacteria and some of fungi.)

Chlorine itself can, under certain circumstances, be the cause of chlorine-like tastes and odors. The level of chlorine residual that will cause tastes and odors will vary with the characteristics of the raw water.

Studies have identified more than 700 different organic chemicals ("organics") in drinking water throughout the United States. Although organics can cause taste-and-odor problems in water, there is a more important consideration. Long-term exposure to certain organics may pose a threat to the public's health, including the risk of cancer. One group of organics, *trihalomethanes*, is formed when chlorine is added to water containing humic acid (a natural organic produced by decaying vegetation). To reduce the amount of trihalomethanes formed in drinking water, prechlorination for taste-and-odor control should be avoided if humic acids are present in the raw-water supply, particularly if other taste-and-odor control methods are available.

Potassium Permanganate

Potassium permanganate ($KMnO_4$), used either alone or in combination with other chemicals, is effective in removing iron and manganese and oxidizing both organic and inorganic materials that cause tastes and odors. When it is added to water containing taste-and-odor compounds, the reaction is

$$2KMnO_4 + H_2O + \text{taste-and-odor compounds} \rightarrow$$

$$\underset{\substack{\text{manganese}\\\text{dioxide}}}{2MnO_2\downarrow} + \underset{\substack{\text{potassium}\\\text{hydroxide}}}{2KOH} + \underset{\text{oxygen}}{3O^{-2}} + \substack{\text{taste-}\\\text{and-odor}\\\text{compounds}} \qquad \text{(C6-3)}$$

This reaction produces oxygen in the water (notice the three Os in the equation above), which oxidizes the organic and inorganic taste-and-odor compounds. The manganese dioxide by-product (insoluble and very finely divided) is removed later, through coagulation, settling, and filtration.

Dosages of $KMnO_4$ will vary from 0.5 to 15 mg/L, although dosages in the range of 0.5–2.5 mg/L are usually adequate to oxidize most taste-and-odor-producing chemicals. Potassium permanganate should be added very early in the treatment process, perhaps at the intake, in order to allow enough time for oxidation to take place. In most cases, detention times of 1–1.5 hours are adequate, but for some particularly stubborn taste-and-odor compounds, detention times of 4–5 hours may be needed.

Water takes on a pink color when potassium permanganate is first added. Then, as it oxidizes to manganese dioxide (MnO_2), the color begins to change, first to yellow, then orange, and finally brown, which is the color of the manganese dioxide precipitate. By the time the water reaches the filters, the pink color should be gone entirely. If not, oxidation will continue after filtration, giving the water an unacceptable yellow to yellow-brown color. This problem of "color breakthrough" can be controlled by either reducing the $KMnO_4$ dosage or by moving the point of chemical addition farther upstream.

Chlorine Dioxide

Chlorine dioxide (ClO_2) is a strong oxidizing agent used primarily to control the odors caused by phenolic compounds. It is about 2–2.5 times as powerful as chlorine. Chlorine dioxide is not purchased as a ready-made chemical like most other water treatment chemicals. Instead, it is made at the treatment plant by adding a solution of sodium chlorite ($NaClO_2$) to a concentrated chlorine solution, as follows:

$$2NaClO_2 + Cl_2 \rightarrow 2ClO_2 + 2NaCl \qquad (C6\text{-}4)$$

The resultant solution should have a pale yellow color. The amount of chlorine and sodium chlorite needed to reach this color usually varies, depending on the strength of the sodium chlorite. The sodium chlorite commonly used is about 80 percent pure. Theoretically, 1.68 lb of sodium chlorite would be mixed with 0.5 lb of chlorine in order to produce 1.0 lb of chlorine dioxide. However, by using more chlorine (at least as much chlorine as sodium chlorite), operators have shown that the reaction occurs faster and the conversion of the chlorite is more complete.

Ozone

Ozone treatment, called *ozonation*, can be an effective method of taste-and-odor control, although experience shows that results are quite variable, depending on the quality of the water being treated. Although ozonation is not common in the United States, it is widely practiced in Europe. Ozone (O_3) is generated at the treatment plant by releasing an electrical spark into a stream of oxygen under controlled circumstances. Although ozone is a powerful oxidizing agent, the cost to produce it is high.

Coagulation

In coagulation, colloidal and finely divided suspended matter gather together as a result of the addition of a floc-forming chemical or biological processes. The actual chemical reactions are very complex and are influenced by a variety of factors, including

- type of coagulant used
- amount of coagulant used
- type and length of flash mixing
- type and length of flocculation
- effectiveness of sedimentation
- other chemicals used
- temperature
- pH
- alkalinity
- zeta potential
- raw-water quality

Alum

Because it is effective, inexpensive, and easy to apply, aluminum sulfate (usually called alum) is presently the most widely used coagulant in the field of water treatment.

Alum is available in two forms. Filter alum is the name given to solid or dry alum [$Al_2(SO_4)_3 \cdot 14H_2O$]. (The portion of the equation showing "$\cdot 14H_2O$" indicates that a certain amount of water is included chemically along with the alum; in this case, 14 molecules of water are included with each molecule of alum. For simplicity, the water portion of the chemical equation is often left off.) Filter alum is ivory-white and is available in lump, ground, rice, or powdered form. Liquid alum [$Al_2(SO_4)_3 \cdot xH_2O$] is alum already in solution. It is available in three strengths — the strongest being less than half the strength of dry filter alum. Liquid alum can vary in color, depending on strength, from a slight, white, iridescent-like color to a yellow-brown.

Depending on the form, filter alum varies in density from 48 to 76 $lb/ft.^3$ Liquid alum has a density of approximately 11 lb/gal, considerably more than that of water.

When alum is added to water, it reacts as follows:

$$\underset{\text{alum}}{Al_2(SO_4)_3} + \underset{\substack{\text{natural}\\ \text{bicarbonate}\\ \text{alkalinity}}}{3Ca(HCO_3)_2} \rightarrow \underset{\substack{\text{aluminum}\\ \text{hydroxide}\\ \text{floc}}}{2Al(OH)_3\downarrow} + \underset{\substack{\text{calcium}\\ \text{sulfate}}}{3CaSO_4\downarrow} + \underset{\substack{\text{carbon}\\ \text{dioxide}}}{6CO_2} \quad \text{(C6-5)}$$

$$\underset{\text{alum}}{Al_2(SO_4)_3} + \underset{\substack{\text{added}\\ \text{carbonate}\\ \text{alkalinity}}}{3Na_2CO_3} + 3H_2O \rightarrow$$

$$\underset{\substack{\text{aluminum}\\ \text{hydroxide}\\ \text{floc}}}{2Al(OH)_3\downarrow} + \underset{\substack{\text{sodium}\\ \text{sulfate}}}{3Na_2SO_4} + \underset{\substack{\text{carbon}\\ \text{dioxide}}}{3CO_2} \quad \text{(C6-6)}$$

$$\underset{\text{alum}}{Al_2(SO_4)_3} + \underset{\substack{\text{added}\\ \text{hydroxide}\\ \text{alkalinity}}}{3Ca(OH)_2} \rightarrow \underset{\substack{\text{aluminum}\\ \text{hydroxide}\\ \text{floc}}}{2Al(OH)_3\downarrow} + \underset{\substack{\text{calcium}\\ \text{sulfate}}}{3CaSO_4\downarrow} \quad \text{(C6-7)}$$

Notice in each case that alkalinity is absolutely necessary for the reaction to take place. It does not matter whether the alkalinity is present naturally, as in Eq C6-5, or whether it is added prior to coagulation, as in Eq C6-6 and Eq C6-7. The important point is that there must be enough alkaline substances to react with the alum to form aluminum hydroxide [$Al(OH)_3$], the sticky floc material. About 0.5 mg/L of alkalinity is required for each milligram of alum added per liter. Without an adequate supply of alkalinity, the alum will not precipitate (i.e., will not form aluminum hydroxide floc) and will pass through the filters. If alkalinity is added later, as it might be for corrosion control, the floc will form then and settle out either in the clearwell or in the distribution system, which can cause serious problems.

When alum is added to an alkaline water, as shown in the equations above, coagulation occurs in three steps. First, the positively charged aluminum ions (Al^{+3}) attract the negatively charged particles that cause color and turbidity, and form tiny particles called microflocs. This marks the beginning of coagulation. Second, because many of these microfloc particles are now positively charged, they begin to attract and hold more negatively charged color-causing and turbidity-causing material. These first two coagulation steps occur very quickly (in microseconds).

Third, the microfloc particles grow into easily visible, "mature" floc particles. This growth occurs partly by the continual attraction of the color and turbidity materials; partly by the adsorption of viruses, bacteria, and algae onto the microfloc; and partly by the random collision of the microflocs that cause particles to stick together. Later, during sedimentation, the large floc particles settle rapidly, leaving the water clear.

The alum dosage to use for best results on a particular water will depend on the various factors listed previously. In general, alum dosages range from 15 to 100 mg/L.

The effective pH range for alum dosing is between 5.5 and 8.5. It is often desirable to adjust pH within this range (by increasing alkalinity) so that alum will perform at its best. The pH for best performance can be found by using the jar test technique.

The rate of floc formation varies with temperature. Lower temperature means slower floc formation. Although there is usually no practical way of adjusting water temperature, one can compensate for lower temperatures by increasing the alum dosage.

From time to time, raw-water quality will change, perhaps as a result of changing seasons, or possibly because of a high runoff of turbid water from a rainstorm. Routine dosage checks that make use of the jar test will help to ensure adequate coagulation–flocculation results.

Ferric Sulfate

Ferric sulfate is a reddish-gray [commercially called Ferric-floc, $Fe_2(SO_4)_3 \cdot 3H_2O$] or grayish-white [commercially called Ferriclear, $Fe(SO_4)_3 \cdot 2H_2O$] granular material. The densities of the different forms vary. Ferric-floc has a density of 60–74 lb/ft^3 and Ferriclear has a density of 78–90 lb/ft^3.

When ferric sulfate is added to water it reacts as follows:

$$\underset{\text{ferric sulfate}}{Fe_2(SO_4)_3} + \underset{\text{natural bicarbonate alkalinity}}{3Ca(HCO_3)_2} \rightarrow \underset{\text{ferric hydroxide floc}}{2Fe(OH)_3\downarrow} + \underset{\text{calcium sulfate}}{3CaSO_4} + \underset{\text{carbon dioxide}}{6CO_2} \quad \text{(C6-8)}$$

The reaction is very similar to that for alum. Notice that as with alum, ferric sulfate requires alkalinity in the water in order to form the floc particle ferric hydroxide [$Fe(OH)_3$]. If natural alkalinity is not sufficient, then alkaline chemicals (soluble salts containing HCO_3^-, CO_3^{-2}, and OH^- ions) must be added.

Ferric sulfate has several advantages compared with alum. The floc particle [$Fe(OH)_3$] is denser and therefore more easily or quickly removed by sedimentation. Ferric sulfate will react favorably over a much wider range of pH, usually 3.5–9.0, sometimes even greater. However, ferric sulfate can stain equipment; it is difficult to dissolve; its solution is corrosive; and it may react with organics to form soluble iron (Fe^{+2}), as discussed later in this chapter. (Soluble iron in water causes red water, as well as staining and taste problems.)

The best ferric sulfate dosage to use in any coagulation application must be decided on a case-by-case basis based on the jar test. However, experience has shown that dosages in water treatment usually run between 5 and 50 mg/L.

Iron and Manganese Removal

Iron and manganese, two troublesome constituents affecting water quality, are found predominantly in groundwater supplies and occasionally in the anaerobic bottom waters of deep lakes. In nature, iron and manganese exist in stable forms — the ferric (Fe^{+3}) form and the manganic (Mn^{+4}) form; these forms are insoluble. Under anaerobic conditions, which can develop in groundwater aquifers and at the bottom of deep lakes, these forms are reduced to the soluble forms — ferrous (Fe^{+2}) and manganous (Mn^{+2}). The objective of most iron and manganese removal processes is to oxidize the reduced forms back into their insoluble forms so they will settle out.

Chlorine

When iron is present in the form of ferrous bicarbonate [$Fe(HCO_3)_2$], it is easily treated for removal by chlorine, as shown in the following reaction:

$$2Fe(HCO_3)_2 + Cl_2 + Ca(HCO_3)_2 \rightarrow 2Fe(OH)_3\downarrow + CaCl_2 + 6CO_2 \quad \text{(C6-9)}$$

This reaction, which is almost instantaneous, works in a pH range of 4–10, working best at pH 7. Removing 1 mg/L of iron (Fe^{+2}) requires 0.64 mg/L of chlorine. The resulting precipitate, ferric hydroxide, is easily recognized as a fluffy, rust-colored sediment. The calcium bicarbonate

[$Ca(HCO_3)_2$] in the reaction represents the bicarbonate alkalinity of the water. The reaction works using either free or combined chlorine residual (free and combined chlorine are discussed later in this chapter).

The reaction with manganese works similarly. The manganese may begin as a salt, such as manganous sulfate ($MnSO_4$).

$$MnSO_4 + Cl_2 + 4NaOH \rightarrow MnO_2\downarrow + 2NaCl + Na_2SO_4 + 2H_2O \quad \text{(C6-10)}$$

Then, with the addition of chlorine and sodium hydroxide, oxidized manganese produces the precipitate manganese dioxide (MnO_2). In order for the reaction to work, 1.3 mg/L of free available chlorine must be added per 1 mg/L of manganese to be removed. Chloramines (combined chlorine residuals) have little effect on manganese.

Sodium hydroxide, NaOH, causes the hydroxyl alkalinity in the sample. The reaction works best in the pH range of 6–10. The speed with which manganese oxidizes varies from a few minutes at pH 10, to 2–3 hours at pH 8, to as much as 12 hours at pH 6.

Aeration

Both ferrous iron (Fe^{+2}) and manganous manganese (Mn^{+2}) can be removed by aeration.

$$\underset{\text{ferrous bicarbonate}}{4Fe(HCO_3)_2} + 10H_2O + O_2 \xrightarrow{\text{aeration}} \underset{\text{ferric hydroxide}}{4Fe(OH)_3\downarrow} + 8H_2CO_3 \quad \text{(C6-11)}$$

$$\underset{\text{manganous sulfate}}{2MnSO_4} + O_2 + \underset{\text{sodium hydroxide}}{4NaOH} \xrightarrow{\text{aeration}}$$

$$\underset{\text{manganese dioxide}}{2MnO_2} + \underset{\text{sodium sulfate}}{2Na_2SO_4\downarrow} + 2H_2O \quad \text{(C6-12)}$$

The removal of iron is actually a two-step reaction. First, the soluble ferrous bicarbonate [$Fe(HCO_3)_2$] is converted to the less soluble form, ferrous hydroxide [$Fe(OH)_2$]. Then, under further aeration, the ferrous hydroxide is converted to the insoluble form, ferric hydroxide [$Fe(OH)_3$], which will filter out or settle out of solution as a fluffy, rust-colored sludge. The reaction works best in the pH range 7.5–8.0. It will take about 15 minutes to complete. To remove 1 mg/L of iron (Fe^{+2}) requires about 0.14 mg/L of oxygen.

The soluble salt of manganese, in this case manganous sulfate ($MnSO_4$), is oxidized to the insoluble form manganese dioxide (MnO_2), which will filter out or settle out of solution. The reaction works best at a high pH, normally greater than 10. Sodium hydroxide (NaOH), shown as part of the reaction, raises the pH to the desired level and provides the hydroxide alkalinity (hydroxyl ions, OH^-) necessary to raise the pH for the reaction. To remove 1 mg/L of manganese (Mn^{+2}) requires about 0.27 mg/L of oxygen. The manganese removal reaction takes place in about 15 minutes.

Potassium Permanganate

Potassium permanganate ($KMnO_4$) is a powerful oxidizing agent and can be used successfully to remove both iron and manganese, as shown by the following equations. This form of chemical treatment is preferable for manganese removal.

$$Fe(HCO_3)_2 + KMnO_4 + H_2O + 2H^+ \rightarrow MnO_2\downarrow + Fe(OH)_3\downarrow + KHCO_3 + H_2CO_3 \quad (C6\text{-}13)$$

$$3Mn(HCO_3) + 2KMnO_4 + 2H_2O \rightarrow 5MnO_2\downarrow + 2KHCO_3 + 4H_2CO_3 \quad (C6\text{-}14)$$

$$3MnSO_4 + 2KMnO_4 + 2H_2O \rightarrow 5MnO_2\downarrow + K_2SO_4 + 2H_2SO_4 \quad (C6\text{-}15)$$

Experience has shown that about 0.6 mg/L of $KMnO_4$ is adequate to remove 1 mg/L of iron (Fe^{+2}). Similarly, it takes about 2.5 mg/L of $KMnO_4$ to remove 1 mg/L of manganese (Mn^{+2}).

Softening

The lime-softening process operates normally at a pH in the range 10–11. Contingent removal of iron and manganese occurs within this range. Iron is eliminated in the form of ferrous hydroxide $Fe(OH)_2$, instead of the familiar ferric hydroxide $Fe(OH)_3$ shown in previous reactions.

Lime–Soda Ash Softening

The two ions most commonly associated with *hardness* in water are calcium (Ca^{+2}) and magnesium (Mg^{+2}). Although aluminum, strontium, iron, manganese, and zinc ions can also cause hardness, they are not usually present in large enough concentrations to produce a hardness problem. Chemical precipitation is one of the more common methods used to soften water. The chemicals normally used are lime [calcium hydroxide, $Ca(OH)_2$] and soda ash [sodium carbonate, Na_2CO_3].

There are two types of hardness: (1) *carbonate hardness*, caused primarily by calcium bicarbonate, and (2) *noncarbonate hardness*, caused by the salts of calcium and magnesium, such as calcium sulfate [$CaSO_4$], calcium chloride [$CaCl_2$], magnesium chloride [$MgCl_2$], and magnesium sulfate [$MgSO_4$]. Lime is used to remove the chemicals that cause carbonate hardness. Soda ash is used to remove the chemicals that cause noncarbonate hardness.

Carbonate Hardness

During treatment for carbonate hardness, lime is the only softening chemical needed, as illustrated in the following reactions:

To remove calcium bicarbonate:

$$\underset{\text{calcium bicarbonate}}{Ca(HCO_3)_2} + \underset{\text{lime}}{Ca(OH)_2} \rightarrow \underset{\text{calcium carbonate}}{2CaCO_3\downarrow} + \underset{\text{water}}{2H_2O} \quad \text{(C6-16)}$$

To remove magnesium bicarbonate:

$$\underset{\text{magnesium bicarbonate}}{Mg(HCO_3)_2} + \underset{\text{lime}}{Ca(OH)_2} \rightarrow \underset{\text{calcium carbonate}}{CaCO_3\downarrow} + \underset{\text{magnesium carbonate}}{MgCO_3} + \underset{\text{water}}{2H_2O} \quad \text{(C6-17)}$$

then

$$\underset{\text{magnesium carbonate}}{MgCO_3} + \underset{\text{lime}}{Ca(OH)_2} \rightarrow \underset{\text{calcium carbonate}}{CaCO_3\downarrow} + \underset{\text{magnesium hydroxide}}{Mg(OH)_2\downarrow} \quad \text{(C6-18)}$$

To remove the magnesium bicarbonate, it takes two separate reactions and twice the lime needed to remove calcium bicarbonate.[1] In Eq C6-16, lime reacts with calcium bicarbonate to form calcium carbonate ($CaCO_3$). Calcium carbonate is relatively insoluble and precipitates.

In Eq C6-17, lime reacts with magnesium bicarbonate to form calcium carbonate, which precipitates, and magnesium carbonate, which does not. In addition to being soluble, magnesium carbonate is a form of carbonate hardness. Therefore, the same amount of lime added in Eq C6-17 is called for in Eq C6-18. Thus, twice the lime required to remove calcium bicarbonate is necessary to remove magnesium bicarbonate and carbonate. In Eq C6-18, the additional lime reacts with magnesium carbonate to form calcium carbonate [$CaCO_3$] and magnesium hydroxide [$Mg(OH)_2$], both relatively insoluble materials that will settle out.

If the water originally had no noncarbonate hardness, then further softening is not needed. However, because $CaCO_3$ and $Mg(OH)_2$ are very slightly soluble, a small amount of hardness remains, usually at least 35 mg/L.

Noncarbonate Hardness

For noncarbonate hardness to be removed, soda ash must be added to remove the noncarbonate calcium compounds, and soda ash together with lime must be added to remove the noncarbonate magnesium compounds.

To remove calcium noncarbonate hardness:

$CaSO_4$	+	Na_2CO_3	→	$CaCO_3\downarrow$	+	Na_2SO_4	(C6-19)
calcium sulfate		soda ash		calcium carbonate		sodium sulfate	

$CaCl_2$	+	Na_2CO_3	→	$CaCO_3\downarrow$	+	$2NaCl$	(C6-20)
calcium chloride		soda ash		calcium carbonate		salt	

In both cases, the calcium noncarbonate hardness is removed by soda ash. The calcium sulfate and the calcium chloride acquire CO_3 from soda ash and precipitate as $CaCO_3$. The compounds that remain after softening, Na_2SO_4 in Eq C6-19 and NaCl in Eq C6-20, are salts and do not cause hardness.

---

[1]Chemistry 7, Chemical Dosage Problems.

To remove magnesium noncarbonate hardness:

$MgCl_2$	+	$Ca(OH)_2$	→	$Mg(OH)_2\downarrow$	+	$CaCl_2$	(C6-21)
magnesium chloride		lime		magnesium hydroxide		calcium chloride, which must also be removed	

$CaCl_2$	+	Na_2CO_3	→	$CaCO_3\downarrow$	+	$2NaCl$	(C6-22)
calcium chloride from Eq C6-21		soda ash		calcium carbonate		salt	

$MgSO_4$	+	$Ca(OH)_2$	→	$Mg(OH)_2\downarrow$	+	$CaSO_4$	(C6-23)
magnesium sulfate		lime		magnesium hydroxide		calcium sulfate, which must also be removed	

$CaSO_4$	+	Na_2CO_3	→	$CaCO_3\downarrow$	+	Na_2SO_4	(C6-24)
calcium sulfate from Eq C6-23		soda ash		calcium carbonate		sodium sulfate	

The removal of magnesium noncarbonate hardness with lime forms calcium noncarbonate hardness, which must then be removed with soda ash.

The final reaction related to softening involves carbon dioxide (CO_2), a gas that is found in dissolved form in most natural waters. Unless CO_2 is removed prior to softening (for example, by aeration), it will consume some of the lime added.

CO_2	+	$Ca(OH)_2$	→	$CaCO_3\downarrow$	+	H_2O	(C6-25)
carbon dioxide		lime		calcium carbonate		water	

If CO_2 is present, then enough lime must be added at the beginning of softening to allow this reaction to take place and still leave enough lime to complete the softening reactions. Whenever CO_2 is present, the CO_2 reaction occurs before the softening reactions take place.

Chemical softening takes place at a high pH. To precipitate calcium carbonate ($CaCO_3$), a pH of about 9.4 is necessary; the precipitation of magnesium hydroxide requires a pH of 10.6. In both cases, the necessary pH is achieved by adding the proper amount of lime.

Recarbonation

Recarbonation is the reintroduction of carbon dioxide into the water either during or after lime–soda ash softening. When hard water is treated by conventional lime softening, the water becomes supersaturated with calcium carbonate and may have a pH of 10.4 or higher. This very fine, suspended calcium carbonate can deposit on filter media, cementing together the individual media grains (encrustation), and depositing a scale in the transmission and distribution system piping (postprecipitation). To prevent these problems, carbon dioxide is bubbled into the water, lowering the pH and removing calcium carbonate as follows:

$$\underset{\text{calcium carbonate (in suspension)}}{CaCO_3} + \underset{\text{carbon dioxide (gas)}}{CO_2(g)} + \underset{\text{water}}{H_2O} \rightarrow \underset{\text{calcium bicarbonate}}{Ca(HCO_3)_2} \qquad \text{(C6-26)}$$

This type of recarbonation is usually performed after the coagulated and flocculated waters are settled but before they are filtered, thereby preventing the suspended $CaCO_3$ from being carried out of the sedimentation basin and cementing the filter media.

When the excess-lime technique is used to remove magnesium, a considerable amount of lime remains in the water. This creates a water that is undesirably caustic and high in pH. Carbon dioxide introduced into the water reacts as follows:

$$\underset{\text{excess lime}}{Ca(OH)_2} + \underset{\text{carbon dioxide (gas)}}{CO_2(g)} \rightarrow \underset{\text{calcium carbonate (precipitate)}}{CaCO_3\downarrow} + \underset{\text{water}}{H_2O} \qquad \text{(C6-27)}$$

This form of recarbonation is performed after coagulation and flocculation but before final settling. Carbon dioxide reacts with the excess lime, removing the cause of the caustic, high-pH condition and, incidentally, removing the calcium that added to the hardness. The product, calcium carbonate, is removed by the filtration process.

It is important to select the correct carbon dioxide dosage.[2] If too much CO_2 is added, the following can happen:

$Ca(OH)_2$	+	$2CO_2(g)$	→	$Ca(HCO_3)_2$	(C6-28)
excess lime		carbon dioxide (gas)		calcium bicarbonate	

Notice that the excess lime combines with the excessive CO_2 to form carbonate hardness. In the excess-lime method, there is too much calcium hydroxide present, and the reaction in Eq C6-28 would significantly increase water hardness. In the methods using lower lime doses, the conversion to calcium bicarbonate may not significantly increase water hardness.

Ion Exchange Softening

The ion exchange process of water softening uses the properties of certain materials (termed *cation exchange materials*) to exchange the hardness-causing cations of calcium and magnesium for non–hardness-causing cations of sodium. The most common cation exchange materials are synthetic polystyrene resins. Each resin particle is a BB-sized, transparent, amber-colored sphere. Each of these insoluble resin spheres contains sodium ions, which are released into the water in exchange for hardness ions of calcium and magnesium. When properly operated, ion exchange is completely effective in removing all hardness, carbonate or noncarbonate.

The two reactions involved in the cation exchange softening process are

Ca^{+2}	+	Na_2X	→	CaX	+	$2Na^{+}$	(C6-29)
hardness cation		cation exchange resin		spent resin		sodium released to treated water in exchange for calcium	

[2]Chemistry 7, Chemical Dosage Problems.

Mg^{+2}	+	Na_2X	→	MgX	+	$2Na^{+}$	(C6-30)
hardness cation		cation exchange resin		spent resin		sodium released to treated water in exchange for magnesium	

The letter X is used to represent the exchange resin. Before softening, the resin with the nonhardness cations of sodium appears in the equations as Na_2X. Although this is not a chemical compound, it does behave somewhat like one. The sodium cations ("Na_2" in the resin, "$2Na^{+}$" after release) are released into the water just as the sodium in Na_2SO_4 would be released when that compound is dissolved in water. Although the resin "X" acts similarly to an anion such as SO_4^{-2}, the resin is actually an insoluble organic material that does not react chemically as SO_4^{-2} would. Instead, it functions more like a "parking lot" for exchangeable cations. The terms CaX and MgX represent the same resin after the exchange has been made.

As shown in the equations, calcium and magnesium hardness ions are removed from the water onto the surface of the resin. In exchange, the resin releases sodium ions. Note that one hardness ion (Mg^{+2} or Ca^{+2}) with a charge of +2 is exchanged for two sodium ions, each having a charge of +1, a total charge of +2. Hence, a +2 charge is exchanged for a +2 charge, an electrically equivalent exchange.

Equations C6-29 and C6-30 can be expanded to show the reactions of the specific hardness-causing compounds (similar to the equations shown for lime–soda ash softening):

$$Ca(HCO_3)_2 + Na_2X \rightarrow CaX + 2NaHCO_3 \quad \text{(C6-31)}$$

$$CaSO_4 + Na_2X \rightarrow CaX + Na_2SO_4 \quad \text{(C6-32)}$$

$$CaCl_2 + Na_2X \rightarrow CaX + 2NaCl \quad \text{(C6-33)}$$

$$Mg(HCO_3)_2 + Na_2X \rightarrow MgX + 2NaHCO_3 \quad \text{(C6-34)}$$

$$MgSO_4 + Na_2X \rightarrow MgX + Na_2SO_4 \quad \text{(C6-35)}$$

$$MgCl_2 + Na_2X \rightarrow MgX + 2NaCl \quad \text{(C6-36)}$$

In each reaction, the anion originally associated with the hardness cation stays in the softened water; after softening, these anions are associated with the sodium cations released by the resin. Hence, the softened water contains sodium bicarbonate ($NaCHO_3$), sodium sulfate (Na_2SO_4), and sodium chloride (NaCl). These compounds do not cause hardness and are present in such small concentrations that they do not cause tastes. Unlike lime–soda ash softening, ion exchange softening operates the same for carbonate and noncarbonate hardness. Both are removed by the same exchange reactions.

After most of the sodium ions are removed from the exchange resin in the softening process, the resin must be *regenerated* in order to restore its softening capacity. That is, the exchange process must be reversed, with the hardness cations of calcium and magnesium being forced out of the resin and replaced by cations of sodium. This reverse exchange is achieved by passing a strong brine solution (a concentrated solution of common table salt) through the resin bed. The two ion exchange regeneration reactions are shown below:

$$CaX + 2NaCl \rightarrow CaCl_2 + Na_2X \quad \text{(C6-37)}$$

$$MgX + 2NaCl \rightarrow MgCl_2 + Na_2X \quad \text{(C6-38)}$$

When sodium is taken back into the exchange resin, the resin is again ready to be used for softening. The calcium and magnesium, released during regeneration, are carried to disposal by the spent brine solution.

Properly maintained and operated, cation exchange removed *all* hardness. Water of zero hardness is corrosive, so the final step in ion exchange softening is to mix a portion of the unsoftened water with the softened effluent to provide water that is still relatively soft, but that contains enough hardness to be noncorrosive (stable).

Scaling and Corrosion Control

Scaling and corrosion are closely related problems in water treatment. They may be thought of as being at opposite ends of a hypothetical stability scale, as shown in Figure C6-1.

The objective of scale and corrosion control is to stabilize the water, thus preventing both scale formation and corrosion. The stable range is relatively

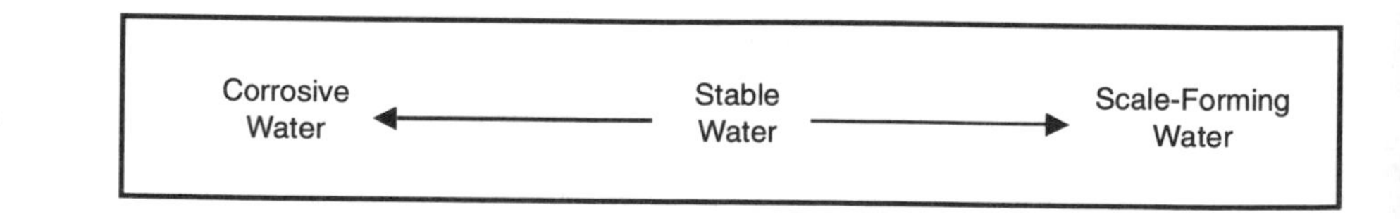

FIGURE C6-1 Hypothetical stability scale

narrow, requiring careful monitoring during treatment in order to avoid under- or overshooting the stable range.

Scaling and Corrosion

Scale is the familiar chalky-white deposit frequently found at the bottom of a tea kettle. It is caused by carbonate and noncarbonate hardness constituents in water.

The exact combination of pH and alkalinity that will result in scale (or corrosion) depends on the overall chemical characteristics of the water. Under conditions of pH 7.0–9.5, with corresponding alkalinities of 300–1,500 mg/L, calcium carbonate ($CaCO_3$) will precipitate and form a scale on interior surfaces of pipes and tanks. One of two chemical reactions will occur:

$$Ca^{+2} + CO_3^{-2} \rightarrow CaCO_3\downarrow \qquad (C6\text{-}39)$$

or

$$Ca^{+2} + 2HCO_3^{-} \rightarrow CaCO_3\downarrow + CO_2 + H_2O \qquad (C6\text{-}40)$$

In controlled amounts, this scale is beneficial, forming a protective coating inside pipelines and tanks. However, excessive scaling can reduce the capacity of pipelines and the efficiency of heat transfer in boilers.

Corrosion is the oxidation of unprotected metal surfaces. In water treatment, a primary concern is the corrosion of iron and its alloys. Corrosion of iron and steel products is easily identified by the familiar red rust that forms. Iron is an important element in many metallic pipe materials and process equipment. It exists naturally as iron ore, in stable forms such as hematite (Fe_2O_3), magnatite (Fe_3O_4), iron pyrite (FeS_2), and siderite ($FeCO_3$). Smelting converts this ore to elemental iron, which is then used in the manufacture of pipeline materials and treatment equipment. Elemental iron is unstable and has a strong tendency to return by the oxidation or corrosion process to the more stable ore forms noted above. Advanced cases of iron corrosion create the problem of "red water."

There are various theories on how iron corrosion occurs and several factors known to affect corrosion. The following simplified discussion of corrosion chemistry highlights the influence of pH, alkalinity, dissolved oxygen, and carbon dioxide in the corrosion process. Remember that corrosion is a complex process that can be influenced by other factors as well. For example, iron bacteria (*Crenothrix* and *Leptothrix*) and sulfate-reducing bacteria can be major causes of corrosion. Increases in water temperature and velocity of flow can accelerate corrosion. The softening process can convert a noncorrosive water to a corrosive one. However, certain constituents in water, such as silica, are believed to protect exposed metal surfaces from corrosion.

When iron corrodes, it is converted from elemental iron (Fe) to ferrous ion (Fe^{+2}).

$$Fe \rightarrow Fe^{+2} + 2 \text{ electrons} \qquad \text{(C6-41)}$$

The electrons that come from the elemental iron build up on metal surfaces and inhibit corrosion. If the water has a very low pH (lower than that of potable water), then hydrogen ions (H^+) in solution will react with electrons to form hydrogen gas ($H_2\uparrow$).

$$2H^+ + 2 \text{ electrons} \rightarrow 2H_2\uparrow \qquad \text{(C6-42)}$$

The hydrogen gas coats the metal surface and could reduce corrosion; however, the coating is removed, partly by the scrubbing action of moving water and partly by combination with oxygen (O_2) normally dissolved in the water:

$$2H_2 + O_2 \rightarrow 2H_2O \qquad \text{(C6-43)}$$

The metal surface is exposed again and corrosion continues. Failure to protect the metal surface or remove corrosion-causing elements will result in destruction of pipes or equipment.

Once the reaction in Eq C6-41 has occurred, subsequent reactions depend on the chemical characteristics of the water. If water is low in pH, is low in alkalinity, and contains dissolved oxygen, then the ferrous ion reacts with water to form ferrous hydroxide [$Fe(OH)_2$].

$$Fe^{+2} + 2H_2O \rightarrow Fe(OH)_2 + H_2\uparrow \qquad \text{(C6-44)}$$

The insoluble ferrous hydroxide immediately reacts with CO_2 present in low-alkalinity water to form soluble ferrous bicarbonate [$Fe(HCO_3)_2$].

$$Fe(OH)_2 + 2H_2CO_3 \rightarrow Fe(HCO_3)_2 + 2H_2O \qquad \text{(C6-45)}$$

Since ferrous bicarbonate is soluble, it detaches from the metal surface and is mixed throughout the water. The dissolved oxygen in the water then reacts with the ferrous bicarbonate to form insoluble ferric hydroxide [$Fe(OH)_3$].

$$4Fe(HCO_3)_2 + 10H_2O + O_2 \rightarrow Fe(OH)_3 + 8H_2CO_3 \qquad \text{(C6-46)}$$

Because this reaction occurs throughout water, ferric hydroxide does not form a protective coating on the metal surface. The ferric hydroxide appears as suspended particles that cause red water.

If water begins with a higher pH and alkalinity (where CO_2 is not present), then the corrosion reaction can be controlled. The ferrous ion shown in Eq C6-41 combines with the hydroxyl alkalinity that is present naturally or induced by lime treatment, forming an insoluble film of ferrous hydroxide on the metal surface.

$$Fe^{+2} + 2(OH^-) \rightarrow Fe(OH)_2 \qquad \text{(C6-47)}$$

If dissolved oxygen is present in the water, it will react with the ferrous hydroxide to form an insoluble ferric hydroxide coating.

$$4Fe(OH)_2 + 2H_2O + O_2 \rightarrow 4Fe(OH)_3 \qquad \text{(C6-48)}$$

Both ferrous and ferric hydroxide are somewhat porous and, although their coatings retard corrosion, they cannot fully protect the pipe. However, the same high-pH and high-alkalinity conditions that cause the rust coating to form also favor the formation of a calcium carbonate coating. Together these coatings protect the pipe from further corrosion.

Chemical Methods for Scale and Corrosion Control

Table C6-1 shows common methods used to control scale and corrosion. The following paragraphs contain brief discussions of each method. Lime, soda ash, and caustic soda arc typically used to raise pH and alkalinity. Carbon dioxide and sulfuric acid are used to lower pH and alkalinity. However, in situations where stabilization is achieved by pH and alkalinity adjustment, the effects of all chemicals used in water treatment must be taken into account. Alum and ferric sulfate (discussed previously) lower pH and alkalinity, as do chlorine (discussed later) and fluosilicic acid (used in fluoridation).

pH and Alkalinity Adjustment With Lime

For each milligram per liter of lime added, approximately 0.56 mg/L of carbon dioxide is removed. Carbon dioxide is in the form of carbonic acid (H_2CO_3) when dissolved in water. The following equation indicates the chemical reaction that takes place:

$$H_2CO_3 + Ca(OH)_2 \rightarrow CaCO_3 + 2H_2O \qquad \text{(C6-49)}$$

TABLE C6-1 Scale and corrosion control methods

Method	For Control of: Scale	Corrosion
pH and alkalinity adjustment with lime	X	X
Chelation	X	
Sequestering	X	
Controlled $CaCO_3$ scaling		X
Other protective chemical coatings		X
Softening	X	

As carbon dioxide (carbonic acid) is removed, pH increases.

For each milligram per liter of lime added, the alkalinity of the treated water will increase by about 1.28 mg/L. The reaction is

$$Ca(HCO_3)_2 + Ca(OH)_2 \rightarrow 2CaCO_3 + 2H_2O \qquad (C6\text{-}50)$$

If the lime dosage is too high, excessive scale will form. If it is too low, the water will be corrosive. The *Langelier saturation index* is one measure used to determine the tendency of water to scale or corrode piping and tanks. The index is based on the assumption that every water has a particular pH value for which the water will neither deposit scale nor cause corrosion. This stable condition is termed *saturation*. The pH value, called saturation pH and abbreviated pH_s, varies, depending on calcium hardness, alkalinity, and temperature. Once the pH_s is calculated, the Langelier saturation index is found as follows:

$$\text{Langelier saturation index} = pH - pH_s$$

If the actual pH of the water is less than the calculated pH_s, then the water has a negative Langelier index and may be corrosive. If the actual pH is greater than the calculated pH_s, then the Langelier index is positive and the water is likely to form scale. In either case the water is unstable. The greater the difference between pH and pH_s, the stronger the tendency for the water to either form scale or cause corrosion. Thus, water with a Langelier index of +0.4 has a stronger scaling tendency than one with an index of +0.1. Similarly, water with a Langelier index of –0.4 has a stronger corrosion tendency than one with an index of –0.2. If the pH and pH_s are equal, then the Langelier

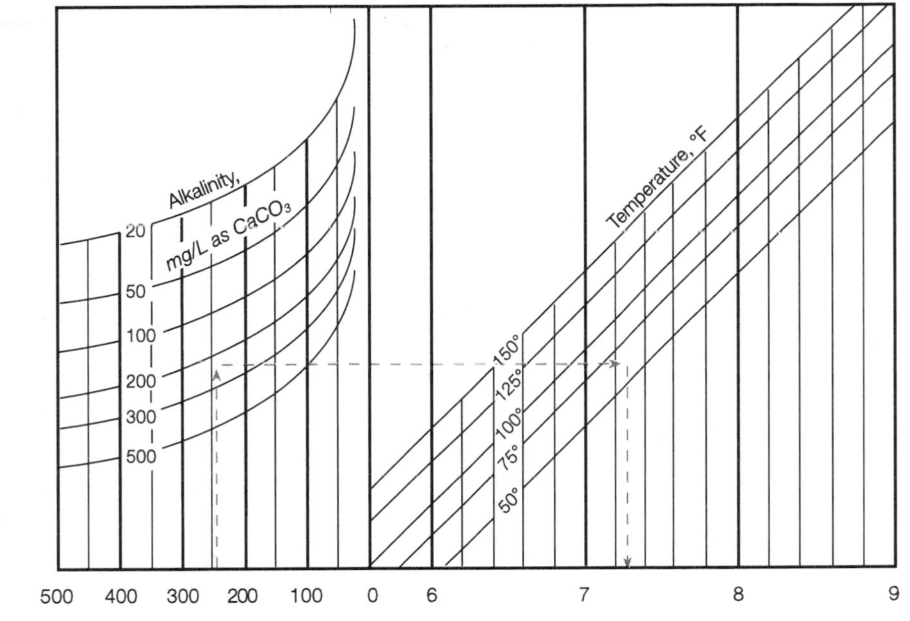

FIGURE C6-2 Determination of pH_s for the Langelier saturation index, given hardness, alkalinity, and temperature (assuming average total dissolved solids of about 500 mg/L)

Reprinted with permission of Nalco Chemical Company, One Nalco Center, Naperville, IL 60563-1198, from Water: The Universal Solvent, *copyright 1977 and 1979.*

saturation index is zero and water is stable. Neither scale formation nor corrosion should occur.

The value of pH_s can be calculated mathematically, but it is simpler to use a graph as illustrated in Figure C6-2. To find pH_s, the calcium hardness in milligrams per liter as $CaCO_3$, alkalinity in milligrams per liter as $CaCO_3$, pH, and water temperature must be determined. For example, the following information may be used to demonstrate reading the graph:

calcium hardness = 240 mg/L as $CaCO_3$

alkalinity = 200 mg/L as $CaCO_3$

pH = 6.8

water temperature = 70°F

Enter the graph at a calcium hardness of 240 mg/L, as shown by the dotted line in Figure C6-2. Proceed upward to the 200-mg/L alkalinity curve. Travel across to locate the temperature of 70°F (estimate the temperature location between the 50°F and 75°F lines). Move downward from the temperature point to the bottom of the graph and identify the pH_s value as 7.3. Finally, calculate the Langelier saturation index as follows:

$$\begin{aligned}\text{Langelier saturation index} &= pH - pH_s \\ &= 6.8 - 7.3 \\ &= -0.5\end{aligned}$$

Since the index is negative, the water is corrosive.

Chelation

Chelation is a chemical treatment process used to control scale formation. The chemical added is known as a chelating agent. It is a water-soluble compound that captures scale-causing ions in solution, preventing precipitation and scale formation. There are several natural organic materials in water that have chelating ability, including humic acid and lignin. When added to water, the chelating agent reacts with calcium ions to keep them in solution and prevent the formation of calcium carbonate scale. In the following equation, the *Y* represents the chelating agent:

$$2Ca^{+2} + Na_4Y \rightarrow Ca_2Y + 4Na^{+} \qquad \text{(C6-51)}$$

Sequestration

Sequestration is a chemical addition treatment process that controls scale. The chemical added to the water *sequesters*, or holds in solution, scale-causing ions such as calcium, iron, and manganese, thus preventing them from precipitating and forming scale. Any one of several polyphosphates may be used in this process. The most commonly used is sodium hexametaphosphate $(NaPO_3)_6$. A common dosage for scale prevention is approximately 0.5 mg/L. Note that sequestration is often related to the deliberate process of adding chemicals to drinking water. Chelation, on the other hand, is often considered a natural process.

Controlled $CaCO_3$ Scaling

One commonly practiced form of corrosion management is controlled $CaCO_3$ scaling. As stated at the beginning of this section, there are three conditions water can have relative to scale formation and corrosion. Water may be corrosive, stable, or scale-forming. By careful control of the pH and alkalinity adjustment, the condition of water can be altered so that it is slightly scale-forming. Eventually the scale will build up beyond the desired thickness. Adjusting pH and alkalinity to a point that is slightly corrosive will then dissolve the excess scale.

Controlled scaling requires careful and continuous laboratory monitoring because the slightest change in the quality of water may require a change in the amount of lime needed. It is also necessary to monitor the thickness of the $CaCO_3$ coating developed in the transmission and distribution system pipeline. The coating should be thick enough to prevent corrosion without obstructing the flow of water.

Other Chemical Protective Coatings

There are two other types of chemicals used to create protective coatings in pipelines: (1) polyphosphates and (2) sodium silicate. Polyphosphates include sodium hexametaphosphate [$(NaPO_3)_6$]; sodium pyrophosphate ($Na_4P_2O_7$); and a group known as bimetallic glassy phosphates. After being fed into the water, polyphosphates form a phosphate film on interior metal surfaces, protecting them from corrosion. Polyphosphates are also effective as sequestering agents for preventing calcium carbonate scale and for stabilizing dissolved iron and manganese. Dosages of 5–10 mg/L are recommended when treatment is initiated. After one to two months worth of protective film is established, the dosages are reduced and maintained at approximately 1 mg/L.

Sodium silicate ($Na_2Si_4O_9$), or water-glass, can also be used to control corrosion in water systems. Sodium silicate combines with calcium to form a hard, dense calcium silicate film ($CaSiO_3$). Dosages vary widely depending on water quality.

Softening

A major problem caused by water hardness is scale formation. Hard waters form calcium carbonate scales in pipelines and boilers. These scales reduce pipeline capacities, lower boiler heat transfer efficiencies, and cause heat exchange tube failures because excessive heating is required to overcome the insulating effects of scale. The results are higher pumping and maintenance costs, longer repair times, shortened equipment life, and higher fuel and power costs.

There are four major water characteristics that control hardness. The tendency to form scale is greater when the hardness, alkalinity, pH, or temperature is increased. Figures C6-3, C6-4, and C6-5 demonstrate how these characteristics are interrelated. The graphs in the figures were determined for one water sample — they should not be applied in general, but they do illustrate the interaction between the indicated variables.

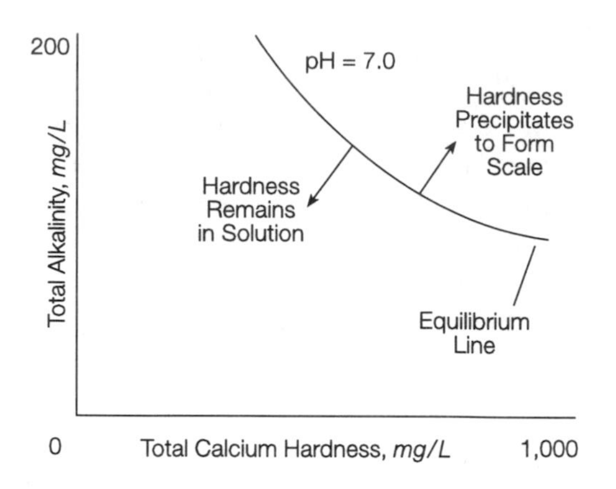

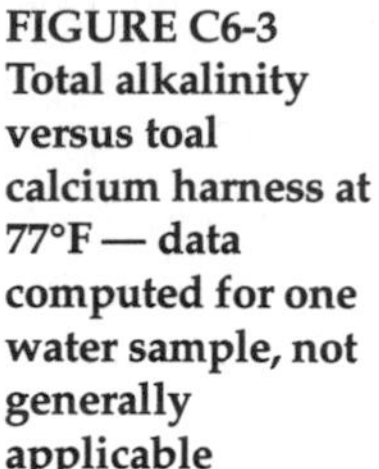
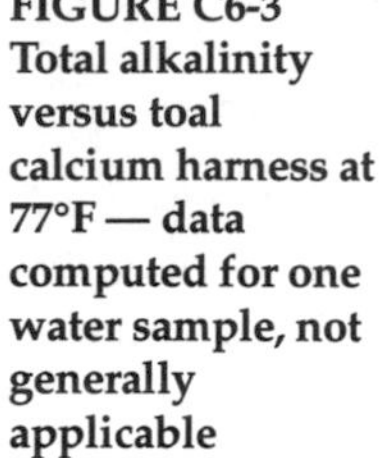

FIGURE C6-3
Total alkalinity versus toal calcium harness at 77°F — data computed for one water sample, not generally applicable

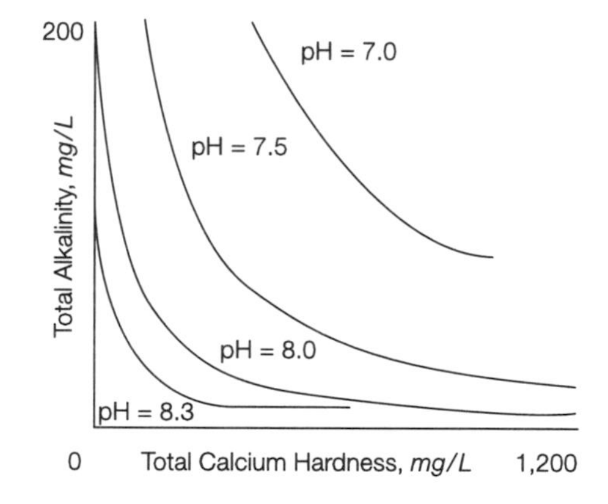

FIGURE C6-4
Effect of total alkalinity and pH on the amount of total calcium hardness that can be kept in solution — data calculated for one water sample, not generally applicable

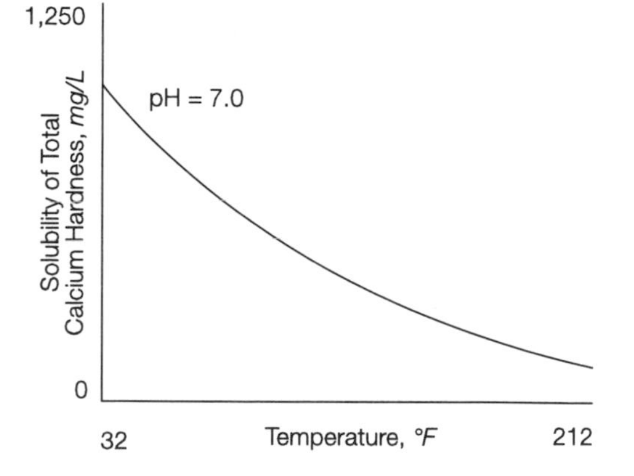

FIGURE C6-5
Effect of temperature on the amount of calcium that will stay in solution — data calculated for one water sample, not generally applicable

The 10 chemical reactions involved in water softening were presented earlier in the discussion of lime–soda ash softening. Refer to that material to review hardness removal using lime and soda ash.

Chlorination

When chlorine is added to water, it reacts to produce various compounds. Some of these compounds are effective disinfectants, whereas others represent the end product of a reaction that has removed an undesirable constituent from the water.

Reaction in Pure Water

When chlorine is added to pure water, it reacts as follows:

$$\underset{\text{chlorine}}{Cl_2} + \underset{\text{water}}{H_2O} \rightarrow \underset{\text{hypochlorous acid}}{HOCl} + \underset{\text{hydrochloric acid}}{HCl} \quad \text{(C6-52)}$$

In this reaction, the chlorine combines with water to produce hypochlorous acid (HOCl). This is one of the two free available chlorine residual forms. Because of the ease with which HOCl penetrates into and kills bacteria, HOCl is the most effective form of chlorine for disinfection. However, some of the HOCl (a weak acid) dissociates as follows:

$$\underset{\text{hypochlorous acid}}{HOCl} \rightarrow \underset{\text{hydrogen}}{H^+} + \underset{\text{hypochlorite ion}}{OCl^-} \quad \text{(C6-53)}$$

As shown by the above equation, this dissociation produces hydrogen (which neutralizes alkalinity or lowers pH) and hypochlorite ion (OCl^-), the second type of free available chlorine residual. The OCl^- is a relatively poor disinfectant compared with HOCl, primarily because of its inability to penetrate into the bacteria.

The following equation shows what happens to the hydrochloric acid (a strong acid) formed in the first reaction:

$$\underset{\text{hydrochloric acid}}{HCl} \rightarrow \underset{\text{hydrogen}}{H^+} + \underset{\text{chloride ion}}{Cl^-} \quad \text{(C6-54)}$$

Notice it also dissociates, forming hydrogen (which neutralizes alkalinity or lowers pH) and chloride ion, one of the same ions formed when common table salt is dissolved in water. Neither the hydrogen nor the chloride ion acts as a disinfectant.

The effectiveness of chlorination is based on five important factors:

- pH
- temperature
- contact time
- concentration
- other substances in water

The pH strongly influences the ratio of HOCl to OCl^-. As shown in Figure C6-6, low pH values favor the formation of HOCl, the more effective free residual, whereas high pH values favor the formation of OCl^-, the less effective free residual form. As pH increases from 7.0 to 10.7, the OCl^- form begins to predominate and the time required for the free residual to effectively disinfect increases. The added time is barely detectable in the pH range 7.0–8.5 but is markedly longer for a pH greater than 8.5.

Very high temperatures (for example, boiling) speed the killing of organisms. Even within the range of temperatures normally found in water,

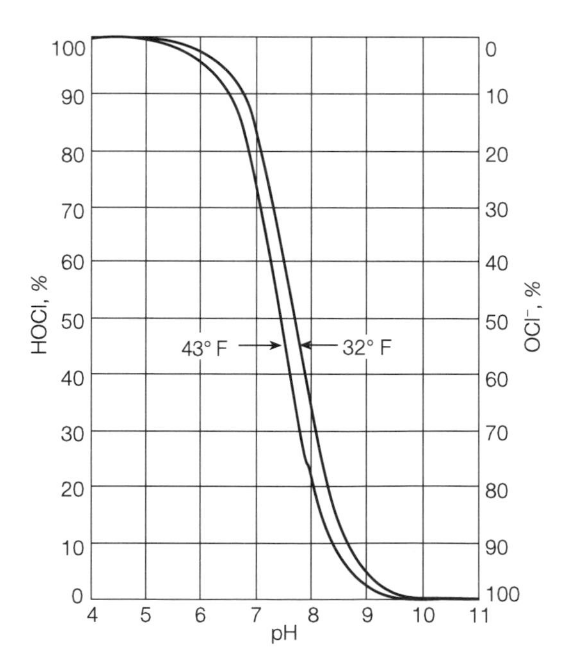

FIGURE C6-6 Relationship among HOCl, OCl^-, and pH

the higher the temperature, the more effective the chlorination. There are several offsetting factors that influence the effectiveness of chlorination. First, as shown in Figure C6-6, lower temperatures slightly favor the formation of HOCl. Lower temperatures also allow chlorine residuals to persist somewhat longer. However, chemical and biochemical reaction rates increase as temperature increases, counteracting these factors, making chlorination more effective at higher temperatures.

The destruction of organisms is directly related to the contact time T and concentration of chlorine C. For example, to accomplish a given kill in a given period of contact time, you might need a certain concentration of chlorine. Provided other conditions remain constant (such as pH and temperature), if the contact time is increased, the chlorine concentration needed to accomplish the same kill is lower; as dosage concentrations are increased, the contact time needed can be decreased.

Equations C6-52 through C6-54 represent what happens when chlorine is added to pure water. However, the water treated at a treatment plant is far from pure. The inorganic and organic materials in raw-water supplies can and do take part in the reaction with chlorine. Some of the most common materials that react with chlorine in water include

- ammonia (NH_3)
- iron (Fe)
- manganese (Mn)
- hydrogen sulfide (H_2S)
- dissolved organic materials

Reaction With Ammonia

One of the most common reactions of chlorine in raw water involves ammonia. Ammonia (NH_3) is an inorganic compound occurring naturally as a result of decaying vegetation or artificially from domestic and industrial wastewater discharges. Chlorine reacts with ammonia to form chloramines, compounds containing both nitrogen and chloride ions. As shown in the following equations, chloramines are formed in three successive steps:

$$\underset{\text{ammonia}}{NH_3} + \underset{\substack{\text{hypochlorous}\\\text{acid}}}{HOCl} \rightarrow \underset{\text{monochloramine}}{NH_2Cl} + H_2O \qquad (C6\text{-}55)$$

$$\underset{\text{monochloramine}}{NH_2Cl} + \underset{\text{hypochlorous acid}}{HOCl} \rightarrow \underset{\text{dichloramine}}{NHCl_2} + H_2O \quad \text{(C6-56)}$$

$$\underset{\text{dichloramine}}{NHCl_2} + \underset{\text{hypochlorous acid}}{HOCl} \rightarrow \underset{\text{trichloramine}}{NCl_3} + H_2O \quad \text{(C6-57)}$$

Whether one chloramine compound or more than one are formed depends on the pH of water and on the presence of enough ammonia. Monochloramine and dichloramine are effective disinfecting agents, but they are often not as effective as free chlorine.

The combined available chlorine is less active as an oxidizing agent than free available chlorine. When water is chlorinated, free available chlorine reacts rapidly with any oxidizable substance. If water contains natural or added ammonia, then the free available chlorine reacts to form combined available chlorine, and the speed of further reactions with oxidizable substances is slowed considerably. The low oxidation potential of combined available chlorine, as compared with free available chlorine, also accounts for slower bactericidal (bacteria-killing) action. Under favorable conditions, equivalent bactericidal action is obtained by using approximately 25 times the combined available chlorine residual as free available chlorine. To obtain equivalent bactericidal action with equal amounts of combined available chlorine residual and free available chlorine residual, a contact period approximately 100 times longer is required.

Chloramines are considerably less effective than HOCl as disinfecting agents. Table C6-2 estimates the effectiveness of the four residual types as compared with HOCl. These numbers are estimates because no method is available to compare effectiveness accurately. Effectiveness can change dramatically, depending on the water's characteristics.

TABLE C6-2 Estimated effectiveness of residual types

Type	Chemical Abbreviation	Estimated Effectiveness Compared with HOCl
Hypochlorous acid	HOCl	1
Hypochlorite ion	OCl^-	1/100
Monochloramine	NH_2Cl	1/150

It is important to place these estimates in proper perspective. Initially, it would appear pointless to use any residual except HOCl. However, from practical experience, it is known that chloramines, particularly monochloramines, do an acceptable job of disinfection if given enough time. Therefore, contact time and concentration are important factors to consider when using monochloramine or combined residual.

Reaction With Iron

Iron is an undesirable element in water, easily removed by chlorination. Iron is often found in groundwater supplies, usually in the form of ferrous bicarbonate [$Fe(HCO_3)_2$]. When chlorine is added, the reaction is

$$\underset{\text{ferrous bicarbonate}}{2Fe(HCO_3)_2} + \underset{\text{chlorine}}{Cl_2} + \underset{\text{calcium bicarbonate}}{Ca(HCO_3)_2} \rightarrow \underset{\text{ferric hydroxide}}{2Fe(OH)_3\downarrow} + \underset{\text{calcium chloride}}{CaCl_2} + \underset{\text{carbon dioxide}}{6CO_2} \quad \text{(C6-58)}$$

Ferric hydroxide [$Fe(OH)_3$] precipitates almost immediately, forming a fluffy, rust-colored sludge. The calcium bicarbonate [$Ca(HCO_3)_2$] in Eq C6-58 represents alkalinity in the water. Iron may be removed using either the free or combined forms of chlorine residual. Each milligram per liter of iron to be removed requires approximately 0.64 mg/L of chlorine.

Reactions With Manganese

Manganese, like iron, is an undesirable constituent often found in groundwaters. It causes similar problems. Just as iron causes red water, manganese may produce brown or black water.

Chlorine reacts with manganese as follows:

$$\underset{\text{manganous sulfate}}{MnSO_4} + \underset{\text{chlorine}}{Cl_2} + \underset{\text{sodium hydroxide}}{4NaOH} \rightarrow \underset{\text{manganese dioxide}}{MnO_2} + \underset{\text{sodium chloride}}{2NaCl} + \underset{\text{sodium sulfate}}{Na_2SO_4} + \underset{\text{water}}{2H_2O} \quad \text{(C6-59)}$$

The manganese in groundwater is normally in the form of a soluble salt, such as manganous sulfate ($MnSO_4$). Chlorination forms the precipitate manganese dioxide (MnO_2), which effectively removes the manganese. The time required to complete the reaction is usually 2 hours.

Each milligram per liter of manganese removed requires 1.3 mg/L of free available chlorine. Unlike iron, manganese is not affected by combined forms of chlorine.

Reaction With Hydrogen Sulfide

Chlorine will also react with hydrogen sulfide (H_2S). Rarely found in surface water, H_2S is frequently found in groundwater supplies. In concentrations as low as 0.05 mg/L, H_2S can give water an unpleasant taste. At slightly higher, yet still small concentrations of 0.5 mg/L, the characteristic odor of rotten eggs is noticeable. Hydrogen sulfide can be *fatal* in a few minutes if inhaled in concentrations equal to or greater than 0.1–0.2 percent by volume in air. At very large concentrations, beginning at 4.3 percent by volume in air, H_2S is flammable. It is one of the more troublesome, obnoxious, and deadly gases that may be encountered in water treatment.

When chlorine is used to remove H_2S, one of two reactions can occur, depending on the chlorine dosage:

$$\underset{\text{chlorine}}{Cl_2} + \underset{\text{hydrogen sulfide}}{H_2S} \rightarrow \underset{\text{hydrochloric acid}}{2HCl} + \underset{\text{sulfur}}{S} \qquad \text{(C6-60)}$$

or

$$\underset{\text{chlorine}}{4Cl_2} + \underset{\text{hydrogen sulfide}}{H_2S} + \underset{\text{water}}{4H_2O} \rightarrow \underset{\text{hydrochloric acid}}{8HCl} + \underset{\text{sulfuric acid}}{H_2SO_4} \qquad \text{(C6-61)}$$

The reaction in Eq C6-60 is instantaneous and occurs at a high pH (the reaction is only about half complete at pH 10). It takes approximately 2.2 mg/L of chlorine to convert 1 mg/L of H_2S to sulfur (S). The sulfur formed by this reaction is a finely divided colloidal-type particle that causes milky-blue turbidity. This turbidity is removed by coagulation and filtration.

In Eq C6-61, the operationally preferred reaction occurs. Chlorine converts the H_2S to sulfuric acid. The H_2SO_4 and the HCl dissociate into hydrogen, chloride, and sulfate ions. Since the initial concentration of H_2S is low in most cases (it can be elevated in groundwater), these ions are not produced in amounts great enough to cause problems. The reaction in Eq C6-61 requires approximately 8.9 mg/L of chlorine for each milligram per liter of H_2S to be removed.

Equation C6-61, like Eq C6-60, is pH dependent. If the pH is less than 6.4, all sulfides are converted to sulfates. At a pH of about 7.0, 70 percent of the hydrogen sulfide changes to hydrogen sulfate, and the remaining 30 percent changes to elemental sulfur. In the pH range of about 9.0–10.0, 50 percent of the hydrogen sulfide is oxidized to hydrogen sulfate, and the remaining 50 percent oxidizes to elemental sulfur.

Sources of Chlorine

There are three types of materials commonly used as a source of chlorine: gaseous chlorine, calcium hypochlorite, and sodium hypochlorite.

Liquid chlorine is a compressed amber-colored gas containing 99.5 percent pure chlorine. At room temperature and pressure, the liquid will expand to approximately 500 times its volume to become a gas. Chlorine gas is greenish-yellow in color and is visible at high concentrations. It is highly toxic, even at concentrations as low as 0.1 percent by volume. Plain chlorine mixed with water produces hypochlorous acid:

$$\begin{array}{ccccccccc} Cl_2 & + & H_2O & \rightarrow & HOCl & + & HCl & & (C6\text{-}62) \\ \text{chlorine} & & \text{water} & & \text{hypochlorous} & & \text{hydrochloric} & & \\ \text{gas or liquid} & & & & \text{acid} & & \text{acid} & & \end{array}$$

Chlorine liquid or gas is not spontaneously combustible, but it will support combustion. The liquid changes easily to a gas at normal temperatures and pressures. If chlorine remains dry, it will not corrode metal. However, when mixed with some moisture, it is extremely corrosive. Chlorine liquid is approximately 1.5 times the weight of water. Gas is approximately 2.5 times the weight of air.

Calcium hypochlorite, $Ca(OCl)_2$, is a dry, white or yellow-white, granular material available in tablets weighing about 0.01 lb. The granular material contains 65 percent available chlorine by weight. This means that when 1 lb of calcium hypochlorite is added to water only 0.65 lb of chlorine is added. Or, in order to add 1 lb of chlorine, 1.54 lb of $Ca(OCl)_2$ must be added.

When added to water, $Ca(OCl)_2$ reacts as follows:

$$\begin{array}{ccccccccc} Ca(OCl)_2 & + & 2H_2O & \rightarrow & 2HOCl & + & Ca(OH)_2 & & (C6\text{-}63) \\ \text{calcium} & & \text{water} & & \text{hypochlorous} & & \text{lime} & & \\ \text{hypochlorite} & & & & \text{acid} & & & & \end{array}$$

TABLE C6-3 Available chlorine in NaOCl solution

Percent Available Chlorine	Available Chlorine, *lb/gal*
10.0	0.833
12.5	1.04
15.0	1.25

Notice that HOCl is produced, just as it was when pure chlorine was added to water. If there are other materials in the water (Fe, Mn, or H_2S), the HOCl will react with them as described previously.

Calcium hypochlorite should be stored carefully to avoid contact with easily oxidized organic material. This type of chlorine can cause fires when brought into contact with many types of organic compounds.

Sodium hypochlorite (NaOCl) is a clear, greenish-yellow liquid chlorine solution normally used in bleaching. Normal household bleach is an example of sodium hypochlorite. It contains 5 percent available chlorine, which is equivalent to 0.42 lb/gal. Commercial bleaches are stronger, containing 9–15 percent available chlorine. Table C6-3 lists the weight of available chlorine in various strengths of NaOCl solution.

Sodium hypochlorite reacts with water to produce the desired HOCl as follows:

$$\underset{\substack{\text{sodium}\\\text{hypochlorite}}}{NaOCl} + \underset{\text{water}}{H_2O} \rightarrow \underset{\substack{\text{hypochlorous}\\\text{acid}}}{HOCl} + \underset{\substack{\text{sodium}\\\text{hydroxide}}}{NaOH} \quad \text{(C6-64)}$$

Sodium hypochlorite solution can be purchased in 5-gal rubber-lined steel drums and in railroad tank cars. Because it is popular for use in small water systems, it is often purchased in refillable plastic jugs packaged four to a box. Unlike calcium hypochlorite, there is no fire hazard connected with NaOCl storage. It is quite corrosive, however, and should be separated from equipment susceptible to corrosion damage.

Of the types of chlorine discussed above, gaseous chlorine is more commonly used. Sodium hypochlorite is preferred for small systems, because it is easy to handle and meter without risk of residues clogging pipes and equipment. Since NaOCl is in liquid form, it may be quite expensive to ship to remote areas. Therefore, where transportation costs are significant, dry granular calcium hypochlorite is usually selected.

CHEMISTRY 7

Chemical Dosage Problems

One of the more common uses of mathematics in water treatment practices is for performing chemical dosage calculations. As a basic- or intermediate-level operator, there are generally seven types of dosage calculations that you may be required to perform:

- milligrams-per-liter to pounds-per-day conversions
- milligrams-per-liter to percent conversions
- feed rate conversions
- calculations for chlorine dosage, demand, and residual
- percent strength calculations
- solution dilution calculations
- reading nomographs

Percent strength calculations,[1] solution dilution calculations,[2] and nomographs[3] are discussed in other sections of this book. The remaining four types of calculations are explained in this chapter.

In addition to the general types of calculations, the operator may be required to perform calculations dealing with specific treatment processes. The last part of this chapter deals with the mathematics specifically needed for lime–soda ash softening, ion exchange softening, recarbonation, and fluoridation.

---

[1]Chemistry 4, Solutions.

[2]Chemistry 4, Solutions.

[3]Mathematics 12, Graphs and Tables.

Milligrams-per-Liter to Pounds-per-Day Conversions

The formula for converting milligrams per liter to pounds per day is derived from the formula for converting parts per million (ppm) to pounds per day, which is as follows:

$$\underset{(\text{lb/d})}{\text{feed rate}} = \underset{(\text{ppm})}{\text{dosage}} \times \underset{(\text{mgd})}{\text{flow rate}} \times \underset{(8.34\ \text{lb/gal})}{\text{conversion factor}} \tag{C7-1}$$

In the range of 0–2,000 mg/L, milligrams per liter are approximately equal to parts per million. For example, 150 mg/L of calcium is approximately equal to 150 ppm of calcium. Therefore, "mg/L" can be substituted for "ppm" in the equation just given. The substitution yields the following equation, which is used to convert between milligrams per liter and pounds per day:

$$\underset{(\text{lb/d})}{\text{feed rate}} = \underset{(\text{mg/L})}{\text{dosage}} \times \underset{(\text{mgd})}{\text{flow rate}} \times \underset{(8.34\ \text{lb/gal})}{\text{conversion factor}}$$

Converting milligrams per liter to pounds per day is a common water treatment calculation. Therefore, you should *memorize the conversion formula.* The following examples illustrate how the formula is used.

Example 1

The dry alum dosage rate is 12 mg/L at a water treatment plant. The flow rate at the plant is 3 mgd. How many pounds per day of alum are required?

This is a milligrams-per-liter to pounds-per-day conversion problem. From Eq C7-1, we have:

$$\text{feed rate} = (\text{dosage})\ (\text{flow rate})\ (\text{conversion factor})$$

Now fill in the information given in the problem and solve for the unknown value.[4]

---

[4]Mathematics 4, Solving for the Unknown Value.

$$x \text{ lb/d} = (12 \text{ mg/L}) (3 \text{ mgd}) (8.34 \text{ lb/gal})$$

$$300.24 \text{ lb/d} = x \text{ lb/d}$$

$$= 300.24 \text{ lb/d}$$

Example 2

Fluoride is added at a concentration of 1.5 mg/L. The flow rate at the treatment plant is 2 mgd. How many pounds per day of fluoride are added?

First write the conversion equation:

feed rate = (dosage) (flow rate) (conversion factor)

Fill in the information given in the problem, and solve for the unknown value:

$$x = (1.5)(2)(8.34)$$

$$= 25.02 \text{ lb/d}$$

Example 3

The chlorine dosage rate at a water treatment plant is 2 mg/L. The flow rate at the plant is 700,000 gpd. How many pounds per day of chlorine are required?

First write the conversion formula:

feed rate = (dosage) (flow rate) (conversion factor)

Before the information given in the problem can be filled in, 700,000 gpd must be converted to million gallons per day. To do this, locate the position of the "millions comma" and move the decimal to there. For example, in the number 2,400,000, the millions comma is between the 2 and 4. To express 2,400,000 gpd as million gallons per day, replace the millions comma with a decimal point:

2,400,000 gpd = 2.4 mgd

In this example, to convert 700,000 gpd to million gallons per day,

,700,000 gpd

↑

millions comma

700,000 gpd = 0.7 mgd

Now write the information into the equation and solve for the unknown value:

$$x = (2)\,(0.7)\,(8.34)$$

$$= 11.68 \text{ lb/d}$$

Example 4

A pump discharges 0.025 m^3/s. What chlorine feed rate (in kilograms per day) is required to provide a dosage of 2.5 mg/L?

This problem involves a flow rate in cubic meters per second, a dosage in milligrams per liter, and a feed rate in kilograms per day. First, write Eq C7-1:

$$\text{feed rate} = (\text{dosage})\,(\text{flow rate})\,(\text{conversion factor})$$

Fill in the equation and solve for the unknown value:

$$x = (2.5)\,(0.025)\,(86{,}400 \text{ s/d})\,(1000\text{L})$$

$$= 5.4 \text{ kg/d}$$

In the four preceding examples, the unknown value was always pounds or kilograms per day. In Eq C7-1, however, any one of the three variables may be the unknown value:

$$\text{feed rate} = (\text{dosage})\,(\text{flow rate})\,(\text{conversion factor})$$

variables

Occasionally, for example, you may know how many pounds per day of chemicals are added and what the plant flow rate is, and need to know what chemical concentration in milligrams per liter this represents.

The next two examples illustrate the procedure used.

Example 5

In a treatment plant, 250 lb/d of dry alum is added to a flow of 1,550,000 gpd. What is this dosage in milligrams per liter?

From the statement of the problem, you can see that this is a milligrams-per-liter and pounds-per-day problem. First write the conversion equation that relates the two terms:

$$\text{feed rate} = (\text{dosage})\,(\text{flow rate})\,(\text{conversion factor})$$

Next, the flow rate in gallons per day must be converted to million gallons per day:

1,550,000 gpd
↑
millions comma

1,550,000 gpd = 1.55 mgd

Now complete the problem by filling in the information given and solving for the unknown value:

$$250 = (x)(1.55)(8.34)$$

$$\frac{250}{(1.55)(8.34)} = x$$

$$19.34 \text{ mg/L} = x$$

Example 6

On one day at a treatment plant, the coagulation process was operated for 15 hours. The average flow during the period was 0.013 m^3/s; 15 kg of alum were fed into the water. What was the alum dosage in milligrams per liter during that period?

The problem involves flow rate, concentration, and feed rate. Write the equation that relates the terms:

feed rate = (dosage) (flow rate) (conversion factor)

Now write the information into the equation and solve for the unknown:

$$15 = (x)(0.013)(86.4)$$

$$\frac{15}{(0.013)(86.4)} = x$$

$$13.35 \text{ mg/L} = x$$

In some problems the flow variable might be the unknown value. The following example illustrate the procedure used to solve such a problem:

Example 7

If 100 lb/d of dry alum were fed into the flow at a treatment plant to achieve a chemical dosage of 20 mg/L, what was the flow rate at the plant in million gallons per day?

First write Eq C7-1:

$$\text{feed rate} = (\text{dosage})\ (\text{flow rate})\ (\text{conversion factor})$$

Then fill in the information given and solve for the unknown value.

$$100 = (20)\ (x)\ (8.34)$$

$$\frac{100}{(20)\ (8.34)} = x$$

$$0.6 \text{ mgd} = x$$

One variation of the mg/L to lb/d problem involves the calculation of hypochlorite dosage. Calcium hypochlorite (usually 65 or 70 percent available chlorine) or sodium hypochlorite (usually 5 to 15 percent available chlorine) is sometimes used for chlorination instead of chlorine gas (100 percent available chlorine).

To provide the same disinfecting power as chlorine gas, a greater weight of calcium hypochlorite is required, and an even greater weight of sodium hypochlorite is needed. For example, assume that water flowing through a treatment plant requires a chlorine dosage of 300 lb/d. If chlorine gas were used for disinfection, the proper dosage rate would be 300 lb/d. If hypochlorites were used, however, about 430 lb/d of calcium hypochlorite or 3,000 lb/d of sodium hypochlorite would be required to provide the same disinfecting power.

In calculating the amount of hypochlorite required, first calculate the amount per day of chlorine (100 percent available) required, and then determine the amount of hypochlorite required. The following examples illustrate the calculations.

Example 8

Disinfection at a treatment plant requires 280 lb/d of chlorine. If calcium hypochlorite (65 percent available chlorine) is used, how many pounds per day will be required?

Since there is only 65 percent available chlorine in the calcium hypochlorite compound, *more* than 280 lb/d of calcium hypochlorite will have to be added to the water to obtain the same disinfecting power as 280 lb/d of 100 percent available chlorine. In fact, *65 percent of some number greater than 280* should equal 280:

$$(65\%)\text{ (greater number of lb/d)} = 280\text{ lb/d}$$

This equation can be restated as

$$(0.65)\ (x\text{ lb/d}) = 280\text{ lb/d}$$

Then solve for the unknown value:

$$x = \frac{280}{0.65}$$

$$x = 430.77\text{ lb/d calcium hypochlorite required}$$

Example 9

A water supply requires 30 lb/d of chlorine for disinfection. If sodium hypochlorite with 10 percent available chlorine is used for the disinfection, how many pounds per day of sodium hypochlorite are required?

Since there is only 10 percent available chlorine in the sodium hypochlorite, considerably more than 30 lb/d will be required to accomplish the disinfection. In fact, 10 percent of some greater number should equal 30 lb/d:

$$(10\%)\text{ (greater number of lb/d)} = 30\text{ lb/d}$$

This equation can be restated as

$$(0.10)\ (x\text{ lb/d}) = 30\text{ lb/d}$$

Then solve for the unknown value:

$$x = \frac{30}{0.10}$$

$$= 300 \text{ lb/d sodium hypochlorite required}$$

Normally, hypochlorite problems as shown in the two examples above are part of a milligrams-per-liter to pounds-per-day calculation. The following example illustrates the combined calculations.

Example 10

How many pounds per day of hypochlorite (70 percent available chlorine) are required for disinfection in a plant where the flow rate is 1.4 mgd and the chlorine dosage is 2.5 mg/L?

First calculate how many pounds per day of 100 percent chlorine are required; then calculate the hypochlorite requirement.

Write the equation to convert milligrams-per-liter concentration to pounds per day:

$$\text{feed rate} = \text{(dosage) (flow rate) (conversion factor)}$$

Fill in the equation with the information given in the problem:

$$x = (2.5)(1.4)(8.34)$$

$$= 29.19 \text{ lb/d chlorine required}$$

Since 70 percent hypochlorite is to be used, *more* than 29.19 lb/d of hypochlorite will be required. In fact, 70 percent of some greater number should equal 29.19 lb/d:

$$(70\%) \text{ (greater number of lb/d)} = 29.19 \text{ lb/d}$$

This equation can be restated as

$$(0.7)\ (x \text{ lb/d}) = 29.19 \text{ lb/d}$$

Then solve for the unknown value:

$$x = \frac{29.19}{0.7}$$

$$= 41.7 \text{ lb/d hypochlorite required}$$

Sometimes chlorine is used to disinfect tanks or pipelines. In such cases, a certain concentration must be achieved by the one-time addition of chlorine to a volume of water. The equation used to convert from milligrams per liter to pounds of chlorine required is similar to the equation used to convert to pounds-per-day feed rate; however, the volume of water in the container to be disinfected replaces the flow rate through the plant. Thus, the equation to calculate pounds of chlorine needed, given the disinfection dosage to be achieved, is as follows:

chlorine weight (lb)	=	dosage (mg/L)	×	volume of container (mil gal)	×	conversion factor (8.34 lb/gal)

The following two examples illustrate this idea.

Example 11

How many pounds of hypochlorite (65 percent available chlorine) are required to disinfect 4,000 ft of 24-in. water line if an initial dose of 40 mg/L is required?

In this problem, the volume of the water line must be calculated:[5]

$$(0.785)\ (2\text{ ft})\ (2\text{ ft})\ (4{,}000\text{ ft})\ (7.48\text{ gal/ft}^3) = 93{,}949\text{ gal}$$

The volume must be expressed in terms of million gallons:

$$93{,}949\text{ gal} = 0.094\text{ mil gal}$$

Next, use the equation that relates milligrams per liter to weight of chlorine required:

$$\text{number of pounds} = (\text{dosage})\ (\text{tank volume})\ (8.34)$$

$$x = (40)\ (0.094)\ (8.34)$$

$$= 31.36\text{ lb/d chlorine required}$$

Since 65 percent hypochlorite is to be used, more than 31.36 lb/d of hypochlorite will be required. In this case, 65 percent of some greater number should equal 31.36 lb/d:

$$(65\%)\ (\text{greater number of lb}) = 31.36\text{ lb}$$

---

[5]Mathematics 10, Volume Measurements.

This equation can be restated as

$$(0.65)(x \text{ lb}) = 31.36 \text{ lb}$$

Then solve for the unknown value:

$$x = \frac{31.36}{0.65}$$

$$x = 48.25 \text{ lb hypochlorite required}$$

Example 12

How many kilograms of chlorine (100 percent available) are required to disinfect a 500,000-L tank if the tank is to be disinfected with 50 mg/L of chlorine?

First write the conversion equation:

$$\text{number of kilograms} = \frac{(\text{dosage})(\text{tank volume})}{1{,}000{,}000}$$

Then fill in the equation and solve:

$$x = \frac{(50)(500{,}000)}{1{,}000{,}000}$$

$$= 25 \text{ kg chlorine required}$$

Milligrams-per-Liter to Percent Conversions

A concentration or dosage expressed as milligrams per liter can also be expressed in terms of percent. Milligrams per liter are approximately equal to parts per million, and percent[6] means "parts per hundred."

$$\text{milligrams per liter} \approx \text{parts per million} = \frac{\text{parts}}{1{,}000{,}000}$$

$$\text{percent} = \text{parts per hundred} = \frac{\text{parts}}{100}$$

---

[6] Mathematics 7, Percent.

Because 1,000,000 ÷ 100 = 10,000, converting from parts per million (or milligrams per liter) to percent is accomplished by dividing by 10,000. The following box diagram[7] can be used:

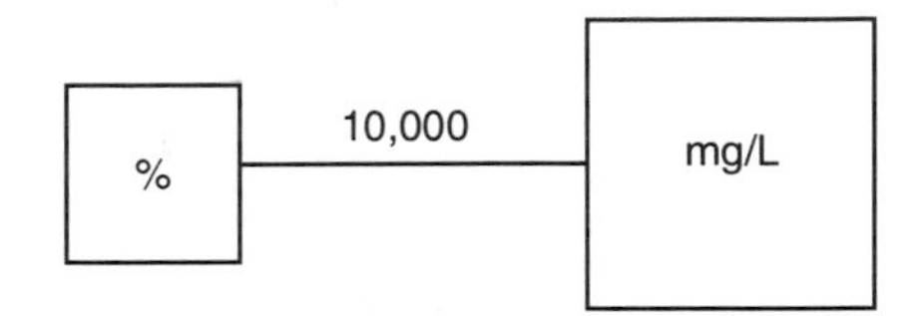

It will help to remember that you can *divide* by 10,000 by moving the decimal point four places to the left (four places because 10,000 has four zeros), yielding a much smaller number. For example,

$$31{,}400 \div 10{,}000 = 3.1400$$
$$= 3.14$$

You can *multiply* by 10,000 by moving the decimal point four places to the right, yielding a much larger number. For example,

$$2.31 \times 10{,}000 = 23100.$$
$$= 23{,}100$$

The following examples illustrate the use of the box diagram:

Example 13

A chemical is to be dosed at 25 mg/L. Express the dosage as percent.

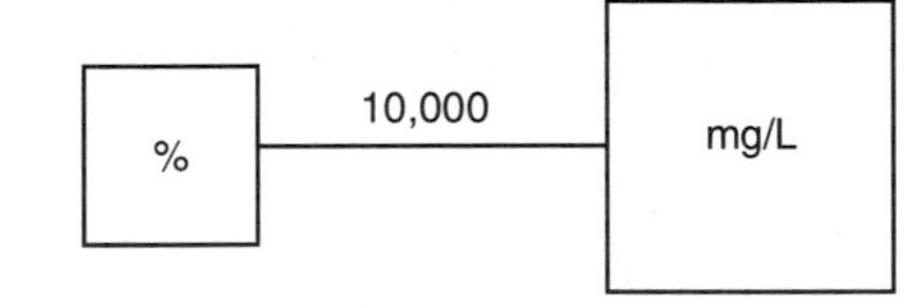

You are moving from a larger to a smaller box, so division by 10,000 is indicated:

$$\frac{25}{10{,}000} = 0.0025\%$$

[7] Mathematics 11, Conversions.

Example 14

Express 120 ppm as percent.

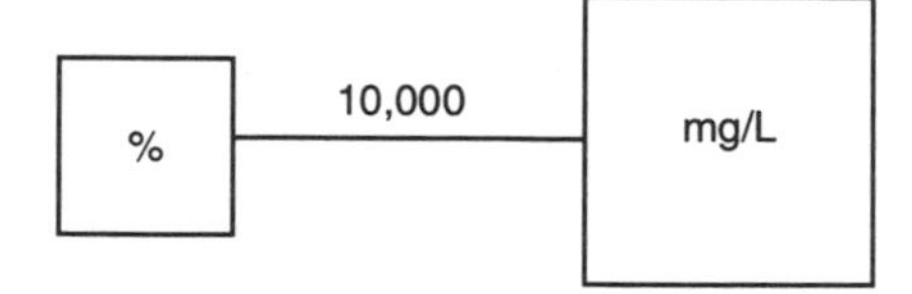

Referring to the box diagram, you are moving from a large box to a small box, so division is indicated:

$$\frac{120}{10{,}000} = 0.0120\%$$

Example 15

The sludge in a clarifier has a total solids concentration of 2 percent. Express the concentration as milligrams per liter of total solids.

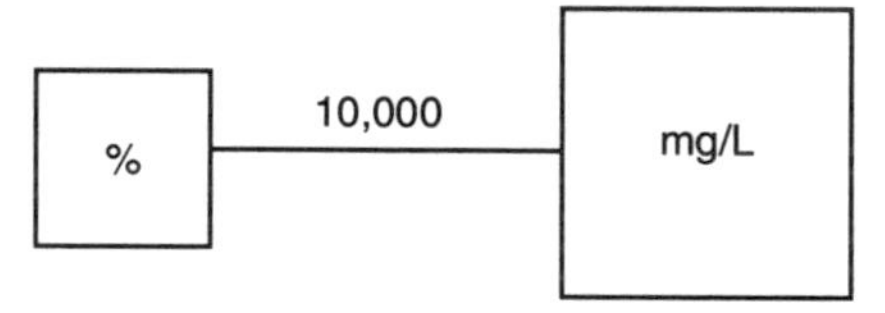

Referring to the box diagram, you are moving from a smaller box to a larger box, so multiply by 10,000:

$$(2)\,(10{,}000) = 20{,}000 \text{ mg/L}$$

Table C7-1 shows the relationship between percent concentration and milligrams-per-liter concentration.

TABLE C7-1 Percent versus milligrams-per-liter concentration

Percent	mg/L or ppm	Percent	mg/L or ppm
100.000	1,000,000	0.100	1,000
50.000	500,000	0.050	500
10.000	100,000	0.010	100
5.000	50,000	0.005	50
1.000	10,000	0.001	10
0.500	5,000	0.0001	1

Feed Rate Conversions

Some chemical dosage problems involve a conversion of feed rates from one term to another, such as from pounds per day to pounds per hour, or from gallons per day to gallons per hour. Many of these conversions are discussed in the Mathematics section.[8] Some types of feed rate conversions not covered in the math section are discussed in the following material:

- gallons-per-hour to gallons-per-day
- gallons-per-hour to pounds-per-hour
- pounds-per-hour to pounds-per-day
- gallons-per-day to pounds-per-day

As with many of the conversions discussed in the math section, the box method can be used in making the conversions. The following box diagram will be used:

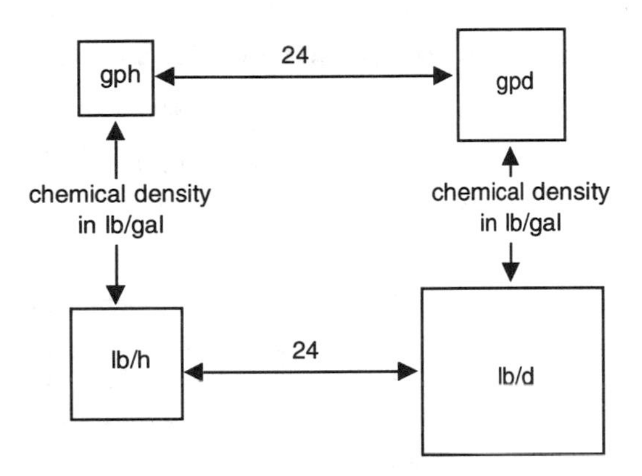

Example 16

The feed rate for a chemical is 230 lb/d. What is the feed rate expressed in pounds per hour?

In moving from pounds per day to pounds per hour, you are moving from a larger box to a smaller box. Therefore, division by 24 is indicated:

$$\frac{230 \text{ lb/d}}{24} = 9.58 \text{ lb/h feed rate}$$

[8]Mathematics 11, Conversions.

Example 17

A chemical has a density of 11.58 lb/gal. The desired feed rate for the chemical is 0.6 gal/h. How many pounds per day is this?

Converting from gallons per hour to pounds per day, you are moving from a smaller box to a larger box (gph → gpd) and then to a still larger box (gpd → lb/d). Therefore, multiplication by the density of the chemical and by 24 is indicated:

$$x = (0.6)\begin{pmatrix}\text{density of}\\ \text{chemical}\end{pmatrix}(24)$$

$$= (0.6)(11.58)(24)$$

$$x = 166.75 \text{ lb/d feed rate}$$

Calculations for Chlorine Dosage, Demand, and Residual

The chlorine requirement, or chlorine dosage, is the sum of the chlorine demand and the desired chlorine residual. This can be expressed mathematically as

$$\begin{matrix}\text{chlorine}\\ \text{dosage}\\ \text{(mg/L)}\end{matrix} = \begin{matrix}\text{chlorine}\\ \text{demand}\\ \text{(mg/L)}\end{matrix} + \begin{matrix}\text{chlorine}\\ \text{residual}\\ \text{(mg/L)}\end{matrix} \qquad \text{(C7-2)}$$

Example 18

A water is tested and found to have a chlorine demand of 6 mg/L. The desired chlorine residual is 0.2 mg/L. How many pounds of chlorine will be required daily to chlorinate a flow of 8 mgd?

$$\begin{matrix}\text{chlorine}\\ \text{dosage}\\ \text{(mg/L)}\end{matrix} = \begin{matrix}\text{chlorine}\\ \text{demand}\\ \text{(mg/L)}\end{matrix} + \begin{matrix}\text{chlorine}\\ \text{residual}\\ \text{(mg/L)}\end{matrix}$$

Fill in the information given in the problem:

$$\text{dosage} = 6 \text{ mg/L} + 0.2 \text{ mg/L}$$

$$= 6.2 \text{ mg/L}$$

Now use Eq C7-1 and Table C7-1 to convert the dosage in milligrams per liter to a feed rate in pounds per day:

$$\text{feed rate} = (\text{dosage})(\text{flow rate})(\text{conversion factor})$$
$$= (6.2)(8)(8.34)$$
$$= 413.66 \text{ lb/d}$$

Example 19

The chlorine demand of a water is 5.5 mg/L. A chlorine residual of 0.3 mg/L is desired. How many pounds of chlorine will be required daily for a flow of 28 mgd?

$$\begin{matrix}\text{chlorine} \\ \text{dosage} \\ \text{(mg/L)}\end{matrix} = \begin{matrix}\text{chlorine} \\ \text{demand} \\ \text{(mg/L)}\end{matrix} + \begin{matrix}\text{chlorine} \\ \text{residual} \\ \text{(mg/L)}\end{matrix}$$

$$\text{dosage} = 5.5 \text{ mg/L} + 0.3 \text{ mg/L}$$
$$= 5.8 \text{ mg/L}$$

Convert dosage in milligrams per liter to feed rate in pounds per day:

$$(5.8)(28)(8.34) = 1{,}354 \text{ lb/d chlorine feed rate}$$

Note that if prechlorination is performed, it is not usually necessary to achieve a residual. Setting the residual equal to zero gives

$$\begin{matrix}\text{chlorine} \\ \text{dosage} \\ \text{(mg/L)}\end{matrix} = \begin{matrix}\text{chlorine} \\ \text{demand} \\ \text{(mg/L)}\end{matrix} + \begin{matrix}\text{chlorine} \\ \text{residual} \\ \text{(mg/L)}\end{matrix}$$

$$\begin{matrix}\text{chlorine} \\ \text{dosage} \\ \text{(mg/L)}\end{matrix} = \begin{matrix}\text{chlorine} \\ \text{demand} \\ \text{(mg/L)}\end{matrix} + 0$$

$$\begin{matrix}\text{chlorine} \\ \text{dosage} \\ \text{(mg/L)}\end{matrix} = \begin{matrix}\text{chlorine} \\ \text{demand} \\ \text{(mg/L)}\end{matrix}$$

Therefore, when no residual is required, as in most prechlorination, the dosage is equal to the chlorine demand of the water.

The equation used to calculate chlorine dosage can also be used to calculate chlorine demand when dosage and residual are given, or to calculate chlorine residual when dosage and demand are given. The following examples illustrate the calculations.

Example 20

The chlorine dosage for a water is 5 mg/L. The chlorine residual after 30 min contact time is 0.6 mg/L. What is the chlorine demand in milligrams per liter?

$$\begin{array}{c} \text{chlorine} \\ \text{dosage} \\ \text{(mg/L)} \end{array} = \begin{array}{c} \text{chlorine} \\ \text{demand} \\ \text{(mg/L)} \end{array} + \begin{array}{c} \text{chlorine} \\ \text{residual} \\ \text{(mg/L)} \end{array}$$

$$5 \text{ mg/L} = x \text{ mg/L} + 0.6 \text{ mg/L}$$

Now solve for the unknown value:

$$5 \text{ mg/L} - 0.6 \text{ mg/L} = x \text{ mg/L}$$

$$4.4 \text{ mg/L} = x$$

Example 21

The chlorine dosage of a water is 7.5 mg/L, and the chlorine demand is 7.1 mg/L. What is the chlorine residual?

$$\begin{array}{c} \text{chlorine} \\ \text{dosage} \\ \text{(mg/L)} \end{array} = \begin{array}{c} \text{chlorine} \\ \text{demand} \\ \text{(mg/L)} \end{array} + \begin{array}{c} \text{chlorine} \\ \text{residual} \\ \text{(mg/L)} \end{array}$$

Fill in the given information and solve for the unknown value:

$$7.5 \text{ mg/L} = 7.1 \text{ mg/L} + x \text{ mg/L}$$

$$7.5 \text{ mg/L} - 7.1 \text{ mg/L} = x \text{ mg/L}$$

$$0.4 \text{ mg/L} = x$$

Sometimes the dosage may not be given in milligrams per liter, but the feed rate setting of the chlorinator and the daily flow rate through the plant will be known. In such cases, feed rate and flow rate should be used to calculate the dosage in milligrams per liter (following the procedure shown in examples 5 and 6). The dosage in milligrams per liter can then be used in Eq C7-2, as in examples 20 and 21.

Lime–Soda Ash Softening Calculations

Two methods for calculating lime and soda ash dosages are

- the conventional method
- the conversion-factor method

The conventional method, although much longer, is helpful in understanding the chemical and mathematical relationships involved in softening. The conversion-factor method is simpler and quicker, and is the more practical method to use in day-to-day operations.

In both methods of calculation, the lime and soda ash dosages required are dependent on the carbonate and noncarbonate constituents in the water. Although hard water will always require lime, it may not always need soda ash. Lime is used to remove carbonate hardness, and both lime and soda ash are used to remove noncarbonate hardness.

To determine whether lime only or both lime and soda ash are needed, total hardness of the water must be compared with total alkalinity. In some areas of the United States, the only alkalinity occurring naturally in water is bicarbonate alkalinity; in these areas, many labs report only the bicarbonate alkalinity concentration or only the total alkalinity concentration, assuming that the operator will be aware that, for the water tested, total alkalinity = bicarbonate alkalinity. However, to be certain of the characteristics of the water tested, an operator should request a detailed lab report showing the total alkalinity as well as the bicarbonate, carbonate, and hydroxyl alkalinity.

Table C7-2 shows how to determine the levels of carbonate and noncarbonate hardness based on the laboratory results for total hardness (TH) and total alkalinity (TA).

A total hardness less than or equal to total alkalinity indicates two important facts about lime and soda ash dosages:

- The noncarbonate concentration is zero, so only lime treatment will be required.
- The amount of lime normally added to treat magnesium noncarbonate hardness (for example, magnesium chloride, $MgCl_2$) is not required.

TABLE C7-2 Noncarbonate and carbonate hardness based on total hardness and total alkalinity

Laboratory Results	Noncarbonate Hardness	Carbonate Hardness
TH less than TA	0	TH
TH = TA	0	TH
TH greater than TA	TH – TA	TA

The following chemical equation is used to determine lime dosage requirements when TH is less than or equal to TA:

$$\text{lime dosage} = [CO_2] + [\text{total hardness}] + [Mg] + [\text{excess}] \qquad (C7\text{-}3)$$

The brackets, [ ], in the equation mean "the concentration of." Therefore, when total hardness is less than or equal to total alkalinity, the lime dosage equals the sum of the concentrations of carbon dioxide, total hardness, and magnesium, plus any excess lime desired.

In most operating situations, more lime will be needed than is indicated by the chemical equations. This additional amount, called "excess lime," is added either to speed the reactions or to force them to react more completely. It is common practice to add from 25 to 50 mg/L as calcium carbonate ($CaCO_3$) of excess lime. An equivalent amount of soda ash (Na_2CO_3) must also be added in order to remove the excess calcium ions added with the lime. To remove 1 mg/L of excess lime requires 1.06 mg/L as Na_2CO_3 of excess soda ash.

When total hardness is greater than total alkalinity, carbonate and noncarbonate hardness constituents are present in the water. This means that both lime and soda ash will be needed in treating for hardness — lime to remove carbonate hardness constituents, and both lime and soda ash to remove noncarbonate hardness constituents in the water. The following chemical equations are used to determine lime and soda ash dosage requirements:

$$\text{lime dosage} = [CO_2] + [HCO_3^-] + [Mg] + [\text{excess}] \qquad (C7\text{-}4)$$

$$\text{soda ash dosage} = [\text{total hardness}] - [HCO_3^-] + [\text{excess}] \qquad (C7\text{-}5)$$

Thus, the lime dosage equals the sum of the concentrations of carbon dioxide, bicarbonate alkalinity, and magnesium, plus the excess lime desired.

The soda ash dosage is equal to the total hardness, minus the bicarbonate alkalinity, plus any excess soda ash desired. Note that when the bicarbonate alkalinity (HCO_3^-) is equal to the total alkalinity (TA), Eq C7-5 can be written

$$\text{soda ash dosage} = [\text{total hardness}] - [\text{TA}] + [\text{excess}] \qquad \text{(C7-6)}$$

And, as shown in Table C7-2, when the total hardness (TH) is greater than the total alkalinity, TH – TA = noncarbonate hardness. Under these conditions, Eq C7-5 can be written

$$\text{soda ash dosage} = [\text{noncarbonate hardness}] + [\text{excess}] \qquad \text{(C7-7)}$$

The two ions that most commonly cause hardness in water are calcium and magnesium. Although aluminum, strontium, iron, manganese, barium, and zinc ions can also cause hardness, they are not usually present in large enough concentrations to cause a significant problem. Consequently it is usually assumed, without much sacrifice in accuracy, that

$$[\text{total hardness}] = [\text{Ca}] + [\text{Mg}] \qquad \text{(C7-8)}$$

When only calcium and magnesium are measured in the laboratory, the total hardness can be found from Eq C7-8. When the laboratory measures both total hardness and calcium, Eq C7-8 can be used to determine the magnesium concentration.

In some cases, total hardness, calcium, and magnesium are all measured in the laboratory. When this occurs, total hardness should be used for Eq C7-3 and Eq C7-4, instead of the sum of calcium and magnesium, because total hardness includes the effects of all hardness-causing ions.

Equations C7-3 through C7-8 are the basis for determining lime and soda ash dosages through either the conventional or the conversion-factor method. Before these equations can be used, the concentrations of CO_2, total hardness, magnesium, HCO_3^-, and excess lime must all be expressed in "mg/L as $CaCO_3$." Usually, when a laboratory test is made for CO_2, the results are reported in "mg/L as CO_2." Similarly, a magnesium concentration would be reported in "mg/L as Mg," and a bicarbonate alkalinity would be reported in "mg/L as HCO_3^-" or in "mg/L as $CaCO_3$." Since these concentrations are not all expressed in the same units of measure, they cannot be added together. However, each concentration can be converted[9] to "mg/L as $CaCO_3$" (the

---

[9]Chemistry 4, Solutions.

units commonly used to express hardness). After all laboratory results are converted to these units, they can be used in Eq C7-3 through Eq C7-8.

The conventional and conversion-factor methods are based on the theory that all reactions occur in definite proportions to the compounds involved. More specifically, the reactions occur in definite proportion to the equivalent weights of the reacting ions or compounds. The difference between the two methods is that, in the conversion-factor method, some of the conversion calculations (such as conversion from "mg/L as HCO_3^-" to "mg/L as $CaCO_3$") are shortened by using conversion factors for which part of the calculation has been completed.

Conventional Method

Given the total hardness, calcium, magnesium, bicarbonate alkalinity, total alkalinity, and free carbon dioxide concentrations from laboratory test results, the conventional method uses equivalent weights to convert all hardness data to milligrams per liter as $CaCO_3$, and then uses the $CaCO_3$ concentrations to calculate the quantity of lime and soda ash needed. This consists of six steps, which are summarized below and explained in detail in the following examples:

Step 1: Convert all data to concentrations as $CaCO_3$, using the following equivalent weights:

$$Ca = 20$$

$$Mg = 12$$

$$HCO_3^- = 61$$

$$CO_2 = 22$$

$$CaCO_3 = 50$$

For example,

$$5 \text{ mg/L as } CO_2 = (5)\left(\frac{50}{22}\right)$$

$$= 11.36 \text{ mg/L as } CaCO_3$$

Step 2: Determine whether lime only or both lime and soda ash will be required in treating the hardness constituents. (Use Table C7-2.)

Step 3: Calculate the lime dosage or lime and soda ash dosage (in milligrams per liter as $CaCO_3$). When total hardness is *less than or equal to* total alkalinity, use Eq C7-3:

$$\text{lime dosage} = [CO_2] + [\text{total hardness}] + [Mg] + [\text{excess}]$$

When total hardness is *greater than* total alkalinity, use Eq C7-4 and Eq C7-5:

$$\text{lime dosage} = [CO_2] + [HCO_3^-] + [Mg] + [\text{excess}]$$

$$\text{soda ash dosage} = [\text{total hardness}] - [HCO_3^-] + [\text{excess}]$$

If noncarbonate hardness is given, Eq C7-7 can be used for the soda ash dosage:

$$\text{soda ash dosage} = [\text{noncarbonate hardness}] + [\text{excess}]$$

Step 4: Convert lime and soda ash dosages expressed as $CaCO_3$ to dosages expressed as CaO and as Na_2CO_3, respectively, using the following equivalent weights:

$$CaCO_3 = 50$$

$$CaO = 28$$

$$Na_2CO_3 = 53$$

For example,

$$227 \text{ mg/L lime as } CaCO_3 = (227)\left(\frac{28}{50}\right) = 127.12 \text{ mg/L as CaO}$$

Step 5: Adjust dosages for the purity of the lime or soda ash used. For example,

$$(88\%)\ (\text{actual dosage of 88\% pure CaO}) = 127 \text{ mg/L of 100\% pure CaO}$$

$$\text{actual dosage of 88\% pure CaO} = 144.5 \text{ mg/L}$$

Step 6: Convert these adjusted dosages to pounds per million gallon or pounds per day. For example:

$$(\text{mg/L})\ (8.34 \text{ lb/gal}) = \text{lb/mil gal}$$

$$(mg/L)\ (mgd\ flow)\ (8.34\ lb/gal) = lb/d$$

The following examples illustrate softening calculations using the conventional method.

Example 22

The laboratory returned the following test results:

calcium	=	100 mg/L as Ca
total hardness	=	300 mg/L as $CaCO_3$
bicarbonate alkalinity	=	250 mg/L as HCO_3^-
total alkalinity	=	205 mg/L as $CaCO_3$
carbon dioxide	=	25 mg/L as CO_2

Express each concentration as $CaCO_3$.

Using the equivalent weight procedure,[10] you can make the conversions as follows:

calcium: $(100\ mg/L)\left(\frac{50}{20}\right) = 250\ mg/L\ as\ CaCO_3$

total hardness: (No conversion needed.) = 300 mg/L as $CaCO_3$

bicarbonate alkalinity: $(250\ mg/L)\left(\frac{50}{61}\right) = 204.92\ mg/L\ as\ CaCO_3$

total alkalinity: (No conversion needed.) = 205 mg/L as $CaCO_3$

carbon dioxide: $(25\ mg/L)\left(\frac{50}{22}\right) = 56.82\ mg/L\ as\ CaCO_3$

Example 23

The laboratory returned the following test results:

magnesium	=	45 mg/L as Mg
total hardness	=	280 mg/L as $CaCO_3$
bicarbonate alkalinity	=	245 mg/L as HCO_3^-
total alkalinity	=	202 mg/L as $CaCO_3$
carbon dioxide	=	20 mg/L as CO_2

[10] Chemistry 4, Solutions.

(a) Express each concentration as $CaCO_3$.

(b) According to step 2 in the softening calculations (and Table C7-2),will lime only or lime and soda ash be required in treating the hardness constituents?

(a) Using the equivalent weight procedure, as in example 22, make the conversion as follows:

magnesium: $(45 \text{ mg/L})\left(\frac{50}{12}\right) = 187.5 \text{ mg/L as } CaCO_3$

total hardness: (No conversion needed.) $= 280 \text{ mg/L as } CaCO_3$

bicarbonate alkalinity: $(245 \text{ mg/L})\left(\frac{50}{61}\right) = 200.82 \text{ mg/L as } CaCO_3$

total alkalinity: (No conversion needed.) $= 202 \text{ mg/L as } CaCO_3$

carbon dioxide: $(20 \text{ mg/L})\left(\frac{50}{22}\right) = 45.45 \text{ mg/L as } CaCO_3$

(b) In this example, total hardness (280 mg/L as $CaCO_3$) is greater than total alkalinity (202 mg/L as $CaCO_3$). As shown in Table C7-2, the water will have both noncarbonate and carbonate hardness constituents and will therefore require both lime and soda ash treatment.

Example 24

The laboratory test results for a water sample were as follows:

magnesium = 32 mg/L as Mg
total hardness = 345 mg/L as $CaCO_3$
bicarbonate alkalinity = 156 mg/L as HCO_3^-
total alkalinity = 128 mg/L as $CaCO_3$
carbon dioxide = 5 mg/L as CO_2

(a) Determine whether lime alone or lime and soda ash will be required in treating the hardness constituents.

(b) Calculate the lime dosage or lime and soda ash dosages in (milligrams per liter as $CaCO_3$). (Assume no excess lime or soda ash is to be added.)

(a) The total hardness (345 mg/L as $CaCO_3$) is greater than the total alkalinity (128 mg/L as $CaCO_3$). Therefore, according to Table C7-2, water will have both carbonate and noncarbonate hardness constituents and will require both lime and soda ash treatment.

(b) Before calculating the lime and soda ash dosages, express all concentrations in terms of $CaCO_3$:

magnesium: $(32 \text{ mg/L})\left(\frac{50}{12}\right) = 133.33 \text{ mg/L as } CaCO_3$

total hardness: (No conversion needed.) $= 345 \text{ mg/L as } CaCO_3$

bicarbonate alkalinity: $(156 \text{ mg/L})\left(\frac{50}{61}\right) = 127.87 \text{ mg/L as } CaCO_3$

total alkalinity: (No conversion needed.) $= 128 \text{ mg/L as } CaCO_3$

carbon dioxide: $(5 \text{ mg/L})\left(\frac{50}{22}\right) = 11.36 \text{ mg/L as } CaCO_3$

Now use the dosage equations shown in step 3:

$$\text{lime dosage} = [CO_2] + [HCO_3^-] + [Mg] + [\text{excess}]$$

$$= 11.36 \text{ mg/L} + 128 \text{ mg/L} + 133.33 \text{ mg/L} + 0$$

$$= 272.69 \text{ mg/L as } CaCO_3$$

$$\text{soda ash dosage} = [\text{total hardness}] - [HCO_3^-] + [\text{excess}]$$

$$= 345 \text{ mg/L} - 128 \text{ mg/L} + 0$$

$$= 217 \text{ mg/L as } CaCO_3$$

If, instead of a magnesium ion concentration, a calcium ion concentration is given in the test results, then the magnesium ion concentration can still be calculated (by the use of Eq C7-7 and the total hardness). Then the lime dosage can be determined in the usual manner.

Example 25

The laboratory returned the following results of testing:

calcium	=	50 mg/L as Ca
total hardness	=	150 mg/L as $CaCO_3$
bicarbonate alkalinity	=	149 mg/L as HCO_3^-
total alkalinity	=	130 mg/L as $CaCO_3$
carbon dioxide	=	10 mg/L as CO_2

Calculate the lime dosage in milligrams per liter as $CaCO_3$ and, if needed, the soda ash dosage. (Assume no excess lime or soda ash are to be added.)

Since the total alkalinity (130 mg/L) is less than the total hardness, the water has both carbonate and noncarbonate hardness constituents. Therefore, both lime and soda ash will be required.

Before the lime and soda ash dosages can be determined, all the concentrations must be expressed in similar terms (as $CaCO_3$).

Use the equivalent weight procedure to make the conversions to "as $CaCO_3$":

calcium: $$(50 \text{ mg/L})\left(\frac{50}{20}\right) = 125 \text{ mg/L as } CaCO_3$$

total hardness: (No conversion needed.) = 150 mg/L as $CaCO_3$

bicarbonate alkalinity: $$(149 \text{ mg/L})\left(\frac{50}{61}\right) = 122.13 \text{ mg/L as } CaCO_3$$

total alkalinity: (No conversion needed.) = 130 mg/L as $CaCO_3$

carbon dioxide: $$(10 \text{ mg/L})\left(\frac{50}{22}\right) = 22.73 \text{ mg/L as } CaCO_3$$

Next, determine the concentration of magnesium. Total hardness is basically calcium ion concentration plus magnesium ion concentration. Mathematically this is

$$[\text{total hardness}] = [\text{Ca}] + [\text{Mg}] \qquad \text{(C7-8)}$$

In this example, two of these factors are known, total hardness and calcium. So, write the known information into the equation:

$$[\text{total hardness}] = [\text{Ca}] + [\text{Mg}]$$

$$150 \text{ mg/L} = 125 \text{ mg/L} + x$$

Now solve for the unknown value, making sure that all concentrations in the above equation are expressed in terms of the same constituent (as $CaCO_3$). Since calcium represents 125 mg/L of the total 150 mg/L hardness, magnesium represents the balance of the hardness, or

$$\text{Mg} = 150 \text{ mg/L} - 125 \text{ mg/L}$$

$$= 25 \text{ mg/L as } CaCO_3$$

Keep in mind that this calculation, though accurate enough, is not exact. There may be small amounts of hardness-producing cations other than calcium and magnesium, so it is likely that if the magnesium ion concentration were measured, it would be slightly less than the 25 mg/L calculated above.

Now the lime and soda ash dosages can be calculated:

$$\text{lime dosage} = [CO_2] + [HCO_3^-] + [\text{Mg}] + [\text{excess}]$$

$$= 22.73 \text{ mg/L} + 122.13 \text{ mg/L} + 25 \text{ mg/L} + 0$$

$$= 169.86 \text{ mg/L as } CaCO_3$$

$$\text{soda ash dosage} = [\text{total hardness}] - [HCO_3^-] + [\text{excess}]$$

$$= 150 \text{ mg/L} - 122.13 \text{ mg/L} + 0$$

$$= 27.87 \text{ mg/L as } CaCO_3$$

Dosages expressed as $CaCO_3$ are useful in order to simplify and standardize the calculation process. However, these dosages are not useful operationally until they are expressed as lime or as soda ash. Once they have been expressed in terms of their own equivalent weight, the operator can use the dosages to dispense the correct amount of chemical. Examples 26 and 27 illustrate this calculation.

Example 26

Using the lime and soda ash dosages given in example 25, determine these dosages expressed in terms of quicklime (lime as CaO) and soda ash (Na_2CO_3), respectively.

To solve this problem, make the equivalent weight conversions. Do this by multiplying the dosages by the equivalent weight of the chemical being changed *to*, and dividing that result by the equivalent weight of the chemical being changed *from*.

$$\text{lime dosage} = (169.86 \text{ mg/L})\left(\frac{28}{50}\right)$$

$$= 95.12 \text{ mg/L as CaO}$$

$$\text{soda ash dosage} = (27.87 \text{ mg/L})\left(\frac{53}{50}\right)$$

$$= 29.54 \text{ mg/L as } Na_2CO_3$$

Example 27

The most current laboratory data sheet shows the following water hardness characteristics:

calcium	=	140 mg/L as $CaCO_3$
total hardness	=	180 mg/L as $CaCO_3$
bicarbonate alkalinity	=	220 mg/L as $CaCO_3$
total alkalinity	=	220 mg/L as $CaCO_3$
carbon dioxide	=	6 mg/L as $CaCO_3$

Find the concentration of quicklime (lime as CaO) and soda ash (Na_2CO_3), if needed. (Assume that no excess lime or soda ash is to be added.)

Since the laboratory has reported the results as $CaCO_3$, the problem can be solved by beginning with step 2. In this example, total hardness (180 mg/L as $CaCO_3$) is less than total alkalinity (220 mg/L as $CaCO_3$). Based on Table C7-2, lime alone must be added. The equation used to calculate lime dosage as $CaCO_3$ (Eq C7-3) utilizes magnesium concentration information. Therefore, calculate the magnesium concentration by using total hardness and calcium concentration information:

$$[\text{total hardness}] = [\text{Ca}] + [\text{Mg}]$$

Insert the known information in the equation:

$$180 \text{ mg/L} = 140 \text{ mg/L} + x$$

Solving for the unknown gives a magnesium concentration of

$$180 \text{ mg/L} - 140 \text{ mg/L} = 40 \text{ mg/L as } CaCO_3$$

Because total hardness is less than total alkalinity, Eq C7-3 should be used in determining lime dosage requirements:

$$\text{lime dosage} = [CO_2] + [\text{total hardness}] + [Mg] + [\text{excess}]$$

$$= 6 \text{ mg/L} + 180 \text{ mg/L} + 40 \text{ mg/L} + 0$$

$$= 226 \text{ mg/L as } CaCO_3$$

Note that if Eq C7-4 and Eq C7-5 are used with data indicating a total hardness less than the total alkalinity, then the soda ash dosage calculation will result in a zero or negative number. Thus, Eq C7-3 is the equation that must be used.

To complete this problem, use the equivalent weight calculation to convert milligrams per liter as $CaCO_3$ to milligrams per liter as CaO:

$$\text{lime dosage} = (226 \text{ mg/L})\left(\frac{28}{50}\right)$$

$$= 126.56 \text{ mg/L as CaO}$$

Once the milligrams-per-liter concentrations of lime and soda ash have been calculated, the information is expressed in the more practical dosage terms of pounds per million gallons or pounds per day, as noted in step 6 of the chemical softening calculations. The following examples illustrate this calculation.

Example 28

How many pounds of lime per day would be needed to soften the raw water in example 27 at the rate of 640,000 gpd?

This is a typical milligrams-per-liter to pounds-per-day calculation. To convert 126.56 mg/L CaO to pounds per day, insert the known information in Eq C7-1:

$$\text{feed rate} = (\text{dosage})\ (\text{flow rate})\ (\text{conversion factor})$$

$$= (126.56)\ (0.64)\ (8.34)$$

$$= 675.53 \text{ lb/d}$$

The lime and soda ash used in water softening treatment are not always 100 percent pure (lime is most often not 100 percent pure, whereas soda ash is usually, but not always, 100 percent pure). Suppose, for example, that a certain powdered quicklime is only 95 percent pure (contains only 95 percent CaO). This would mean that if a dosage of 60 mg/L of quicklime were required, more than 60 mg/L (in fact, 63 mg/L) would actually have to be added to the water to compensate for the impurity of the powder.

To calculate the actual dosage needed of a chemical that is less than 100 percent pure, remember that the general equation for percent problems is[11]

$$\text{percent} = \frac{\text{part}}{\text{whole x 100}}$$

In calculations for dosages of chemicals with less than 100 percent purity, the "whole" is the amount of impure, commercial chemical actually measured into the water; the "percent" is the purity of the commercial chemical; and the "part" is the calculated dosage of pure chemical that is required for the treatment process. This is summarized in the following equation:

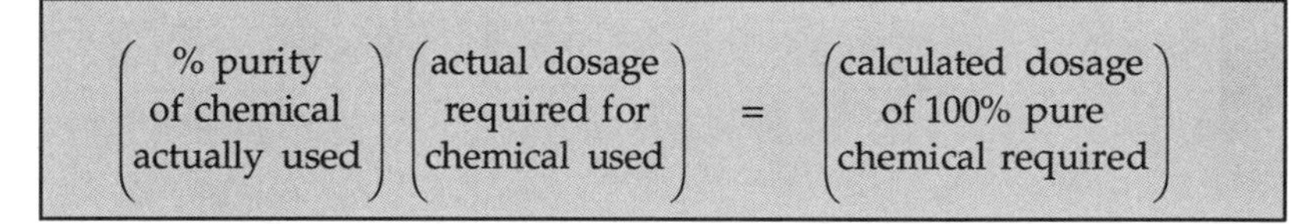

Example 29

Assume that 675.53 lb/d CaO is required for treating a certain water, as calculated in the previous example. How many pounds per day of 90 percent pure quicklime will actually be required?

Keep in mind that the answer to this problem will be slightly more than 675.53 lb/d. To determine the required dosage of the 90 percent pure chemical that is actually used, insert the information in the equation:

$$(\%\ \text{purity})\ (\text{actual dosage}) = (\text{calculated dosage})$$

Express the percent as a decimal:

$$(0.90)\ (\text{actual dosage}) = 675.53\ \text{lb/d}$$

---

[11]Mathematics 7, Percent.

Then solve for the unknown number:

$$\text{actual dosage} = \frac{675.53}{0.90}$$

$$= 750.58 \text{ lb/d CaO required (90\% pure)}$$

Remember that, whenever the purity of the chemical used is less than 100 percent, the actual dosage will be some greater number than the calculated dosage of 100 percent pure chemical.

Example 30

It has been calculated that 95.12 mg/L quicklime as CaO and 29.54 mg/L soda ash as Na_2CO_3 are required in treating a certain water. The quicklime to be used is 94 percent pure, the soda ash is 100 percent pure, and the plant flow at the time is 69.4 L/s. How many kilograms per day of quicklime and soda ash should be used?

First, calculate the kilograms-per-day requirement for the lime and soda ash:

lime:

$$\text{feed rate} = \text{(dosage) (flow rate) (conversion factor)}$$

$$= (95.12)\ (69.4)\ (0.0864)$$

$$= 570 \text{ kg/d CaO (100\% pure)}$$

soda ash:

$$\text{feed rate} = (29.54)\ (69.4)\ (0.0864)$$

$$= 177 \text{ kg/d } Na_2CO_3 \text{ (100\% pure)}$$

Now calculate the adjustment to compensate for the impurity of the chemicals:

lime:

$$\text{(94\%) (actual dosage)} = 570 \text{ kg/d}$$

$$\text{(0.94) (actual dosage)} = 570 \text{ kg/d}$$

$$\text{actual dosage} = \frac{570}{0.94}$$

$$= 606 \text{ kg/d CaO required (94\% pure)}$$

soda ash:

Since the chemical is 100 percent pure, 177 kg/d Na_2CO_3 is the actual dosage required.

It is common practice to add more lime than determined by calculation. Doing so increases the speed and improves the completeness of the reaction. The following two examples illustrate the use of excess-lime information in the calculation of lime and soda ash dosages.

Example 31

A raw-water sample has the following water quality characteristics:

calcium	=	152 mg/L as Ca
magnesium	=	29 mg/L as Mg
bicarbonate alkalinity	=	317 mg/L as HCO_3^-
carbon dioxide	=	9 mg/L as CO_2

It has been decided to include an excess lime dose of 35 mg/L as $CaCO_3$. Including this excess, calculate the lime (89 percent pure) and soda ash (100 percent pure) dosage (a) in milligrams per liter as lime or soda ash; (b) in pounds per million gallons; and (c) in pounds per day. Assume a plant flow of 479,500 gpd.

(a) First, express all data as $CaCO_3$:

calcium: $$(152)\left(\frac{50}{20}\right) = 380 \text{ mg/L as } CaCO_3$$

magnesium: $$(29)\left(\frac{50}{12}\right) = 120.83 \text{ mg/L as } CaCO_3$$

bicarbonate alkalinity: $$(317)\left(\frac{50}{61}\right) = 259.84 \text{ mg/L as } CaCO_3$$

carbon dioxide: $$(9)\left(\frac{50}{22}\right) = 20.45 \text{ mg/L as } CaCO_3$$

Now calculate the lime and soda ash requirements (based on Eq C7-4 and Eq C7-5):

$$\text{lime dosage} = [CO_2] + [HCO_3^-] + [Mg] + [\text{excess}]$$

$$= 20.45 + 259.84 + 120.83 + 35$$

$$= 436.12 \text{ mg/L as } CaCO_3$$

$$\text{soda ash as } CaCO_3 = [\text{total hardness}] - [HCO_3^-] + [\text{excess}]$$

$$[\text{total hardness}] = [Ca] + [Mg]$$

$$= 380 + 120.83$$

$$= 500.83 \text{ mg/L as } CaCO_3$$

$$\text{soda ash dosage} = 500.83 - 259.84 + 35$$

$$= 275.99 \text{ mg/L as } CaCO_3$$

Next, express the dosages as lime or soda ash:

$$\text{lime dosage} = (436.12)\left(\frac{28}{50}\right)$$

$$= 244.23 \text{ mg/L as CaO (pure)}$$

$$\text{soda ash dosage} = (275.99)\left(\frac{53}{50}\right)$$

$$= 292.55 \text{ mg/L as } Na_2CO_3 \text{ (pure)}$$

Finally, adjust for purity:

$$\text{actual lime dosage} = \frac{244.23}{0.89}$$

$$= 274.42 \text{ mg/L as CaO (89\% pure)}$$

No purity adjustment is needed for soda ash, so

soda ash dosage = 292.55 mg/L as Na_2CO_3 (100% pure)

(b) Converting from pounds per million gallons to a dosage is found by multiplying the dosage by 8.34 as follows:

lb/mil gal = (mg/L) (8.34 lb/gal)

lime dosage = (274.42 mg/L) (8.34 lb/gal)

= 2,288.66 lb/mil gal of 89% CaO

soda ash dosage = (292.55 mg/L) (8.34 lb/gal)

= 2,439.87 lb/mil gal of 100% Na_2CO_3

(c) Similarly, the pounds per day can be found as follows:

(mg/L) (mgd flow) (8.34 lb/gal) = lb/d

lime dosage = (274.42 mg/L) (0.4795 mgd) (8.34 lb/gal)

= 1,097.41 lb/d of 89% CaO

soda ash dosage = (292.55) (0.4795 mgd) (8.34 lb/gal)

= 1,169.92 lb/d of 100% Na_2CO_3

Example 32

The test results for a raw-water sample were returned from the laboratory as follows:

total hardness	=	220 mg/L as $CaCO_3$
calcium	=	66 mg/L as Ca
magnesium	=	13 mg/L as Mg
bicarbonate alkalinity	=	156 mg/L as HCO_3^-
carbon dioxide	=	5 mg/L as CO_2

The plant flow on January 4 averaged 52 L/s. How many kilograms of quicklime (88 percent pure) and soda ash (100 percent pure) should have been used that day? It is common practice at this plant to add 26 mg/L excess lime as $CaCO_3$.

Begin by converting all the constituent concentrations to concentrations as $CaCO_3$.

$$\text{total hardness:} \quad = 220 \text{ mg/L as } CaCO_3$$

$$\text{calcium:} \quad (66)\left(\frac{50}{20}\right) = 165 \text{ mg/L as } CaCO_3$$

$$\text{magnesium:} \quad (13)\left(\frac{50}{12}\right) = 54.17 \text{ mg/L as } CaCO_3$$

$$\text{bicarbonate alkalinity:} \quad (156)\left(\frac{50}{61}\right) = 127.87 \text{ mg/L as } CaCO_3$$

$$\text{carbon dioxide:} \quad (5)\left(\frac{50}{22}\right) = 11.36 \text{ mg/L as } CaCO_3$$

Next calculate lime dosage as $CaCO_3$, according to Eq C7-4:

$$\begin{aligned} \text{lime dosage} &= [CO_2] + [HCO_3^-] + [Mg] + [\text{excess}] \\ &= 11.36 + 127.87 + 54.17 + 26 \\ &= 219.4 \text{ mg/L as } CaCO_3 \end{aligned}$$

In a similar manner, calculate the soda ash dosage according to Eq C7-5:

$$\begin{aligned} \text{soda ash dosage} &= [\text{total hardness}] - [HCO_3^-] + [\text{excess}] \\ &= 220 - 127.87 + 26 \\ &= 118.13 \text{ mg/L as } CaCO_3 \end{aligned}$$

Notice that total hardness, calcium, and magnesium were all measured in the laboratory. In such cases, it is best to use total hardness in the soda ash calculation because the sum of calcium and magnesium can be slightly smaller (220 mg/L as $CaCO_3$, versus 219.17 mg/L as $CaCO_3$).

The dosages calculated so far must now be expressed in terms of the equivalent weight of the softening chemicals used (as CaO or as Na_2CO_3).

$$\text{lime dosage} = (219.4)\left(\frac{28}{50}\right)$$

$$= 122.86 \text{ mg/L as CaO (pure)}$$

$$\text{soda ash dosage} = (118.13)\left(\frac{53}{50}\right)$$

$$= 125.22 \text{ mg/L as } Na_2CO_3 \text{ (pure)}$$

These are the concentrations needed if the quicklime and soda ash are 100 percent pure. But the quicklime used is 88 percent pure; therefore, adjust the quicklime dosage by dividing the 100 percent pure dosage by the actual purity to be used.

$$\text{actual lime dosage} = \frac{122.86}{0.88}$$

$$= 139.61 \text{ mg/L as CaO (88\% pure)}$$

The soda ash used was 100 percent pure, so no adjustment need be made.

Finally, based on the average flow rate for the day, given as 52 L/s, determine the total kilograms of lime and soda ash to be used per day, as follows:

$$\text{lime feed rate} = (139.61)(52)(0.0864)$$

$$= 627.24 \text{ kg/d of 88\% pure CaO}$$

$$\text{soda ash feed rate} = (125.22)(52)(0.0864)$$

$$= 562.59 \text{ kg/d of 100\% pure } Na_2CO_3$$

Conversion-Factor Method

All of the equivalent weight conversions required in the conventional method can be combined into conversion factors, as shown in Table C7-3. These factors, multiplied by the concentration of the corresponding constituent, will give the lime or soda ash dosage needed to remove that constituent, in units of either milligrams per liter or pounds per million gallons. To find the total lime dosage, sum the amounts of lime needed to remove carbon dioxide, bicarbonate alkalinity, and magnesium; then add that sum to the excess lime required (if any). The total soda ash dosage is found similarly: Add the amount of soda ash for removing noncarbonate hardness to any excess soda ash required. The final step required is to adjust the dosage based on the actual purity of the lime and soda ash.

Example 33

The following test results were provided by the laboratory:

CO_2	=	25 mg/L as CO_2
HCO_3^-	=	205 mg/L as $CaCO_3$
Mg	=	9 mg/L as Mg
noncarbonate hardness	=	95 mg/L as $CaCO_3$

Assuming no excess lime is added, find the correct dosages in milligrams per liter for lime (90% CaO) and soda ash (99% Na_2CO_3) required to remove all hardness.

First, check to see that the data are expressed in the same terms noted in Table C7-3. In this case they are. If the terms had disagreed, it would have been necessary to convert to the appropriate terms by

TABLE C7-3 Factors for converting constituent concentrations to softening chemical dosages

	Conversion Factor* to Determine Required Amount of Lime (100% Pure CaO)	
Constituent Concentration	*mg/L*	*lb/mil gal*
Carbon dioxide, mg/L as CO_2	1.27	10.63
Bicarbonate alkalinity, mg/L as $CaCO_3$	0.56	4.67
Magnesium, mg/L as Mg	2.31	19.24
Excess lime, mg/L as $CaCO_3$	0.56	4.67
	Conversion Factor* to Determine Required Amount of Soda Ash (100% Pure Na_2CO_3)	
	mg/L	*lb/mil gal*
Noncarbonate hardness, mg/L as $CaCO_3$	1.06	8.83
Excess soda ash, mg/L as $CaCO_3$	1.06	8.83

*Multiply constituent concentration by conversion factor to determine softening chemical dosage in units noted.

using the equivalent weight method. Proceed with the calculations, beginning with the lime calculations. Multiply the constituent concentrations by the respective conversion factors to obtain the lime dosage for each constituent:

$$
\begin{aligned}
CO_2{:}\quad (25)(1.27) &= 31.75 \text{ mg/L as CaO} \\
HCO_3^-{:}\quad (205)(0.56) &= 114.8 \text{ mg/L as CaO} \\
Mg{:}\quad (9)(2.31) &= \underline{20.79 \text{ mg/L as CaO}} \\
\text{total lime dosage} &= 167.34 \text{ mg/L as CaO}
\end{aligned}
$$

Now adjust the lime dosage for purity.

$$\text{actual lime dosage} = \frac{167.34}{0.9}$$

$$= 185.93 \text{ mg/L as CaO (90\% pure)}$$

Next find the soda ash dosage:

$$\text{soda ash dosage} = (95)\ (1.06)$$

$$= 100.70 \text{ mg/L as } Na_2CO_3$$

Then, adjust for purity:

$$\text{actual soda ash dosage} = \frac{100.70}{0.99}$$

$$= 101.72 \text{ mg/L as } Na_2CO_3 \text{ (99\% pure)}$$

Example 34

The following laboratory test results show the hardness of the raw-water supply:

$$
\begin{aligned}
CO_2 &= 20 \text{ mg/L as } CaCO_3 \\
\text{total hardness} &= 200 \text{ mg/L as } CaCO_3 \\
\text{magnesium} &= 50 \text{ mg/L as } CaCO_3 \\
\text{total alkalinity} &= 100 \text{ mg/L as } CaCO_3
\end{aligned}
$$

Note that, for this water supply, the total alkalinity is the same as the bicarbonate alkalinity. The objective is to remove all CO_2 and carbonate hardness while reducing magnesium to 5 mg/L as $CaCO_3$ and noncarbonate hardness to 75 mg/L as $CaCO_3$. Calculate the required dosages in pounds per million gallons. The quicklime used is 88 percent pure, and the soda ash is 98 percent pure. No excess lime is added.

The first step is to determine the carbonate hardness and the noncarbonate hardness (NCH). The total hardness is greater than the total alkalinity, so Table C7-2 gives

$$\text{carbonate hardness} = \text{total alkalinity}$$
$$= \text{bicarbonate } (HCO_3^-) \text{ alkalinity}$$
$$= 100 \text{ mg/L as } CaCO_3$$
$$\text{NCH} = \text{total hardness} - \text{total alkalinity}$$
$$= 200 \text{ mg/L} - 100 \text{ mg/L}$$
$$= 100 \text{ mg/L as } CaCO_3$$

Now list the amount of each constituent to be removed:

CO_2	20 mg/L as $CaCO_3$
carbonate hardness	100 mg/L as $CaCO_3$
Mg	50 – 5 = 45 mg/L as $CaCO_3$
NCH	100 – 75 = 25 mg/L as $CaCO_3$

Now convert as needed to the units shown in Table C7-3:

$$CO_2\text{:} \quad (20)\left(\frac{22}{50}\right) = 8.8 \text{ mg/L as } CO_2$$

$$HCO_3^-\text{:} \quad = 100 \text{ mg/L as } CaCO_3$$

$$\text{Mg:} \quad (45)\left(\frac{12}{50}\right) = 10.8 \text{ mg/L as Mg}$$

$$\text{NCH:} \quad = 25 \text{ mg/L as } CaCO_3$$

Then calculate the required dosages:

$$CO_2\text{:} \quad (8.8)(10.63) = 93.54 \text{ lb/mil gal as CaO}$$
$$HCO_3^-\text{:} \quad (100)(4.67) = 467 \text{ lb/mil gal as CaO}$$
$$\text{Mg:} \quad (10.8)(19.24) = \underline{207.79 \text{ lb/mil gal as CaO}}$$
$$\text{lime dosage} = 768.33 \text{ lb/mil gal as CaO}$$

NCH: (25) (8.83) = 220.75 lb/mil gal as Na_2CO_3

soda ash dosage = 220.75 lb/mil gal as Na_2CO_3

Finally, adjust for purity:

$$\text{actual lime dosage} = \frac{768.33}{0.88}$$

= 873.1 lb/mil gal as CaO (88% pure)

$$\text{actual soda ash dosage} = \frac{220.75}{0.98}$$

= 225.26 lb/mil gal as Na_2CO_3 (98% pure)

Example 35

Based on the following hardness data, calculate the lime and soda ash requirements for complete removal at an average daily flow rate of 49.8 L/s. Excess lime and soda ash are added at the rate of 27 mg/L as $CaCO_3$. The soda ash used is 100 percent pure. The quicklime is 90 percent pure. Give results in milligrams per liter. For this water, total alkalinity and bicarbonate alkalinity are the same.

total hardness	=	215 mg/L as $CaCO_3$
Mg	=	15.8 mg/L as Mg
HCO_3^-	=	185 mg/L as $CaCO_3$
CO_2	=	25.8 mg/L as CO_2

These lab values are in the correct units for use with Table C7-3.

First, determine NCH:

NCH = TH − TA

= 215 mg/L − 185 mg/L

= 30 mg/L as $CaCO_3$

Then use the factors from Table C7-3 to determine the dosages:

CO_2:	(25.8) (1.27)	=	32.77 mg/L as CaO
HCO_3^-:	(185) (0.56)	=	103.6 mg/L as CaO
Mg:	(15.8) (2.31)	=	36.5 mg/L as CaO
excess lime:	(27) (0.56)	=	15.12 mg/L as CaO
	lime dosage	=	187.99 mg/L as CaO

NCH:	(30) (1.06)	=	31.8 mg/L as Na_2CO_3
excess soda ash:	(27) (1.06)	=	28.62 mg/L as Na_2CO_3
	soda ash dosage	=	60.42 mg/L as Na_2CO_3

Next, adjust for purity as follows:

$$\text{actual lime dosage} = \frac{187.99}{0.9}$$

$$= 208.88 \text{ mg/L as CaO (90\% pure)}$$

$$\text{actual soda ash dosage} = 60.42 \text{ mg/L as } Na_2CO_3 \text{ (100\% pure)}$$

Recarbonation Calculations

As discussed in the Chemistry section, chapter 6, there are three possible reactions involved when carbon dioxide is used to stabilize a lime-softened water.[12] Equation C6-26 shows that calcium carbonate, in suspension, will react with carbon dioxide to form calcium bicarbonate. Equations C6-27 and C6-28 show that excess lime will also react with carbon dioxide, forming either calcium carbonate precipitate or soluble calcium bicarbonate.

$CaCO_3$	+	$CO_2(g)$	+	H_2O	→	$Ca(HCO_3)_2$	(C6-26)
calcium carbonate (in suspension)		carbon dioxide (gas)		water		calcium bicarbonate (in suspension)	

[12]Chemistry 6, Chemistry of Treatment Processes.

$Ca(OH)_2$	+	$CO_2(g)$	→	$CaCO_3\downarrow$	+	H_2O	(C6-27)
excess lime		carbon dioxide (gas)		calcium carbonate (precipitate)		water	

$Ca(OH)_2$	+	$2CO_2(g)$	→	$Ca(HCO_3^-)_2$	(C6-28)
excess lime		carbon dioxide (gas)		calcium bicarbonate (in suspension)	

The conditions described by these equations can and do occur together. Consequently, the carbon dioxide dosage needed for recarbonation is the amount needed to reduce calcium carbonate to a desired level and to eliminate excess lime completely. The amount of calcium carbonate present is measured by the alkalinity test and is assumed to be equal to the carbonate alkalinity. The amount of excess lime present, also measured by the alkalinity test, is assumed equal to the hydroxide alkalinity.

The correct carbon dioxide dosage involves adding 0.44 mg/L of CO_2 for every 1 mg/L of carbonate or hydroxide alkalinity expressed as $CaCO_3$. That is, removing 5 mg/L as $CaCO_3$ of carbonate alkalinity and 5 mg/L as $CaCO_3$ of hydroxide alkalinity will require (5 + 5) (0.44) = 4.4 mg/L of CO_2 as CO_2. Consider the following example:

Example 36

The following results of water quality analyses were reported for water in the process of treatment. Calculate the correct CO_2 dosage.

	Concentration, *mg/L as $CaCO_3$*		
Type of Alkalinity	For Raw Water	After Lime–Soda Ash Treatment	Desired After Recarbonation
Bicarbonate	250	0	5
Carbonate	0	35	30
Hydroxide	0	40	0

First, find out how much carbonate and hydroxide alkalinity must be removed. For carbonate alkalinity, subtract the desired concentration from the concentration resulting from the lime–soda ash treatment:

$$\text{carbonate alkalinity to be removed} = 35 \text{ mg/L} - 30 \text{ mg/L}$$

$$= 5 \text{ mg/L as } CaCO_3$$

Do the same for hydroxide alkalinity:

$$\text{hydroxide alkalinity to be removed} = 40 \text{ mg/L} - 0 \text{ mg/L}$$

$$= 40 \text{ mg/L as } CaCO_3$$

Then add the two alkalinities to find the combined amount to be removed:

$$\text{combined alkalinity to be removed} = 5 \text{ mg/L} + 40 \text{ mg/L}$$

$$= 45 \text{ mg/L as } CaCO_3$$

Finally, find the carbon dioxide dosage as follows:

$$CO_2 \text{ dosage} = (\text{alkalinity to be removed})(0.44)$$

$$= (45 \text{ mg/L})(0.44)$$

$$= 19.8 \text{ mg/L}$$

Ion Exchange Softening Calculations

The operation of an ion exchange water softener (more correctly called a cation exchange water softener) involves four types of calculations:

- hardness removal capacity
- volume of water softened per cycle
- length of softening run or cycle
- regeneration salt requirement

All ion exchange materials (the zeolite types and the resin types) have maximum exchange capacities. The usual capacity ranges are summarized below.

Exchange Material	Exchange Capacity, *grains per cubic foot*
Zeolites	2,800–11,000
Resins	11,000–35,000

The use of the unit "grain" is very common in this branch of water treatment. It is used in such terms as grains per cubic foot and grains per gallon (gpg). The following conversion factors will be useful in converting these somewhat old-fashioned units to more familiar units.

$$1 \text{ grain} = 64.80 \text{ mg}$$

$$1 \text{ gpg} = 17.12 \text{ mg/L}$$

$$1 \text{ gpg} = 142.86 \text{ lb/mil gal}$$

Example 37

A treatment unit contains 20 ft^3 of cation exchange material that has a rated removal capacity of 20,000 grains per cubic foot. What is the total hardness removal capacity? Give answer in grains.

Since each cubic foot of material can remove 20,000 grains of hardness and there are 20 ft^3 of material, the answer can be found in one multiplication:

$$\text{Total removal capacity} = (20{,}000 \text{ grain/ft}^3)(20 \text{ ft}^3)$$

$$= 400{,}000 \text{ grains removed per cycle}$$

Example 38

The unit in Example 37 treats water with a hardness of 253 mg/L. How many gallons of water can be softened before the exchange material must be regenerated?

First, convert hardness from milligrams per liter to grains per gallon in order to be consistent with the units of removal capacity (grains per cycle) found in example 37:

$$\text{gpg} = \frac{\text{mg/L}}{17.12}$$

$$\text{hardness} = \frac{253 \text{ mg/L}}{17.12}$$

$$= 14.78 \text{ gpg}$$

Now, find the gallons of water softened by dividing the removal capacity per cycle by the hardness as follows:

$$\text{volume of water softened} = \frac{\text{removal capacity per cycle}}{\text{hardness}}$$

$$= \frac{400{,}000 \text{ grains}}{14.78 \text{ gpg}}$$

$$= 27{,}063 \text{ gal per cycle}$$

Example 39

The area of the exchanger material bed is 5 ft^2 and the softening loading rate is 6 gpm/ft^2. Find the volume of water softened every minute and calculate the length of the softening run cycle, assuming the volume of softened water is 27,063 gal before regeneration is needed.

The volume of water softened each minute is found by multiplying the bed area times the loading rate:

$$\text{volume per minute} = \text{(bed area) (loading rate)}$$

$$= (5 \text{ ft}^2)(6 \text{ gpm/ft}^2)$$

$$= 30 \text{ gpm}$$

Then, divide the total volume treated per cycle by the volume per minute to find the run length.

$$\text{run length} = \frac{\text{volume per cycle}}{\text{volume per minute}}$$

$$= \frac{27{,}063 \text{ gal}}{30 \text{ gpm}}$$

$$= 902.1 \text{ min, or } \frac{902.1}{60} = 15.04 \text{ hours}$$

Example 40

The exchange material in example 37 can be regenerated by adding 0.35 lb of salt for every 1,000 grains of hardness removed. How many pounds of salt are required for this regeneration?

As stated in example 37, the exchange material can remove 400,000 grains of hardness. After the exchange capacity has been exhausted, it takes 0.35 lb of salt to restore each 1,000 grains of removal capacity. Therefore,

$$\text{salt requirement} = (\text{removal capacity})\ (\text{regeneration requirement})$$

$$= (\text{removal capacity})\left(\frac{\text{salt requirement in pounds}}{1{,}000 \text{ grains of hardness}}\right)$$

$$= (400{,}000 \text{ grain})\ \frac{0.35 \text{ lb}}{1{,}000 \text{ grains}}$$

$$= (400)\ (0.35)$$

$$= 140 \text{ lb salt}$$

Example 41

The salt required for exchange material regeneration is often prepared in advance as a solution of salt and water, and then held in storage for use as needed. Suppose a bed of exchange material needs 63.5 kg of salt for regeneration, and the salt solution used contains 0.06 kg of salt per liter. Find the required number of liters of salt solution.

To make the conversion, divide the number of kilograms of salt by kilograms of salt per liter:

$$\text{liters of salt solution} = \frac{\text{kilograms of salt needed}}{\text{kilograms of salt per liter of solution}}$$

$$= \frac{63.5 \text{ kg}}{0.06 \text{ kg/L}}$$

$$= 1{,}058 \text{ L}$$

Example 42

One unit of an ion exchange water softener uses polystyrene resin and has these operating characteristics in a certain plant:

Bed depth	24 in.
Loading rate	5 gpm/ft^2
Brine solution strength	12.5% (1.24 lb/gal)
Surface area	7.5 ft^2
Volume of exchange material	15 ft^3
Rated removal capacity	30,000 grains/ft^3
Average water hardness	180 mg/L
Salt regeneration requirement	0.35 lb/1,000 grains

Calculate the following:

(a) Total removal capacity

(b) Volume treated per cycle

(c) Softening run length

(d) Pounds of salt needed for regeneration

(e) Gallons of salt solution needed for regeneration

(a) total removal capacity = (grains per cubic foot) (cubic feet of resin)

= (30,000 grains/ft^3) (15 ft^3)

= 450,000 grains removed per cycle

(b) $$\text{hardness} = \frac{180 \text{ mg/L}}{17.12}$$

$$= \frac{180}{17.12}$$

$$= 10.51 \text{ gpg}$$

$$\text{volume treated} = \frac{\text{removal capacity per cycle}}{\text{hardness}}$$

$$= \frac{450{,}000 \text{ grains}}{10.51 \text{ gpg}}$$

$$= 42{,}816 \text{ gal per cycle}$$

(c) volume per minute

$$= (\text{bed area})\,(\text{loading rate})$$

$$= (7.5 \text{ ft}^2)\,(5 \text{ gpm/ft}^2)$$

$$= 37.5 \text{ gpm}$$

$$\text{run length} = \frac{\text{volume treated per cycle}}{\text{volume treated per minute}}$$

$$= \frac{42{,}816 \text{ gal}}{37.5 \text{ gpm}}$$

$$= 1{,}142 \text{ min, or } 19 \text{ hours}$$

(d) pounds of salt

$$= (\text{removal capacity})\,(\text{regeneration requirement})$$

$$= (450{,}000 \text{ grain})\left(\frac{0.35 \text{ lb}}{1{,}000 \text{ grain}}\right)$$

$$= 157.5 \text{ lb salt}$$

(e) gallons of salt solution

$$= \frac{(\text{lb salt needed})}{(\text{lb salt/gal of solution})}$$

$$= \frac{157.5 \text{ lb}}{1.24 \text{ lb/gal}}$$

$$= 127.02 \text{ gal of solution}$$

Fluoridation Calculations

There are three chemical compounds commonly used in water fluoridation:

- sodium fluoride, NaF
- sodium fluorosilicate, Na_2SiF_6
- fluosilicic acid, H_2SiF_6

Since each compound is used for its fluoride ion content, it is important to determine the exact amount of fluoride ion available in a given fluoride compound. This amount depends on two factors:

- the weight of fluoride relative to the total weight of the compound
- the purity of the compound

Example 43

Find the percentage of fluoride ion contained in sodium fluoride, NaF.[13]

Find the molecular weight of sodium fluoride by adding the atomic weights of the individual elements:

$$\text{molecular weight of NaF} = \text{molecular weight of Na} + \text{molecular weight of F}$$

$$= 23 + 19 = 42$$

Then divide the weight of the fluoride ion by the total molecular weight of the compound and multiply by 100, as follows:

$$\text{percent F} = \left(\frac{19}{42}\right)(100)$$

$$= 45\% \text{ fluoride ion (rounded from 45.24)}$$

Example 44

The commercial strength of the sodium fluoride in example 43 is between 95 and 98 percent. That is, each 100-lb bag of chemical compound delivered to the plant contains 95–98 lb of sodium fluoride and 2–5 lb of impurities. How much fluoride ion is contained in each 100-lb bag?

From example 43 you know that if it contained pure sodium fluoride, each 100-lb bag would have 45 percent or (0.45) (100) = 45 lb fluoride ion. However, the bags contain slightly impure NaF. Therefore, each bag holds only 95–98 percent of what it would if the NaF were 100 percent pure, or from (45) (0.95) = 42.75 lb of F (43%) to (45) (0.98) = 44.1 lb of F (44%)

---

[13]Chemistry 3, Valence, Chemical Formulas, and Chemical Equations.

Example 45

Find the percentage of fluoride ion contained in sodium fluorosilicate (Na_2SiF_6).

Find the molecular weight of sodium fluorosilicate by adding the atomic weights of the individual elements, as follows:

$$
\begin{aligned}
Na_2: &\quad (2)(23) &= \quad 46 \\
Si: &\quad &\quad 28 \\
F_6: &\quad (6)(19) &= \quad \underline{114} \\
& & 188 = \text{molecular weight of } Na_2SiF_6
\end{aligned}
$$

Then divide the total weight of the fluoride ion by the molecular weight of the compound and multiply by 100, as follows:

$$\text{percent F} = \left(\frac{114}{188}\right)(100)$$

$$= 61\% \text{ fluoride ion (rounded from 60.64)}$$

Example 46

The commercial strength of the sodium fluorosilicate in example 45 is normally 98.5 percent. That is, each 100-lb bag contains 98.5 lb of the compound and 1.5 lb of impurities. How much fluoride ion does each 100-lb bag contain?

From example 45, you know that if it contained pure sodium silicofluoride, each bag would hold 61 percent, or 61 lb, fluoride ion. However, the bags contain slightly impure Na_2SiF_6, with only 98.5 percent of the weight of each bag being composed of Na_2SiF_6. Therefore, the actual weight of fluoride ion in a 100-lb bag is

$$
\begin{aligned}
\text{weight of F in 100-lb bag} &= (98.5\%)(61 \text{ lb}) \\
&= (0.985)(61 \text{ lb}) \\
&= 60.09 \text{ lb}
\end{aligned}
$$

That is, 60 lb (rounded from 60.09 lb) — or 60 percent — of every 100-lb bag is fluoride ion.

Example 47

Find the percentage of fluoride ion contained in fluosilicic acid, H_2SiF_6.

Find the molecular composition of fluosilicic acid, by weight, by adding the atomic weights of the individual elements:

$$\begin{array}{llcr} H_2: & (2)(1) & = & 2 \\ Si: & & & 28 \\ F_6: & (6)(19) & = & \underline{114} \\ & & & 144 = \text{molecular weight of } H_2SiF_6 \end{array}$$

Then divide the total weight of the fluoride ion by the molecular weight of the compound and multiply by 100 as follows:

$$\text{percent F} = \left(\frac{114}{144}\right)(100)$$

$$= 79\% \text{ fluoride ion (rounded from 79.17)}$$

Example 48

The commercial strength of the fluosilicic acid in example 47 is quite variable, ranging from 20 to 35 percent pure. That is, each pound of commercial acid contains 0.2 to 0.35 lb of pure acid. The remainder of the liquid, 0.65 to 0.8 lb, consists of impurities and water. The density of the commercial acid is about 10.5 lb/gal. How much fluoride ion is contained in each gallon of acid?

From example 47, you know that if the fluosilicic acid were 100 percent pure, each pound would contain 79 percent, or 0.79 lb, fluoride ion. However, commercial fluosilicic acid is very dilute, containing only 20 to 35 percent pure acid. To determine the weight of fluoride ion contained in 1 lb of commercial (dilute) acid, multiply the *percentage of fluoride ion per pound of pure acid* times the *percentage of pure acid per pound of commercial acid*:

The weight of fluoride ion in 1 lb of commercial acid ranges from

$$(0.79)(0.20) = 0.16 \text{ lb, or 16\% pure fluoride ion}$$

to

$$(0.79)(0.35) = 0.28 \text{ lb, or 28\% pure fluoride ion}$$

Remember that the acid has a density of 10.5 lb/gal. Each pound of acid contains from 0.16 to 0.28 lb fluoride ion; therefore, 1 gal of acid contains 10.5 times that amount:

The weight of fluoride ion in 1 gal of commercial acid ranges from

$$(10.5)\,(0.16\text{ lb}) = 1.68\text{ lb}$$

to

$$(10.5)\,(0.28\text{ lb}) = 2.94\text{ lb}$$

Example 49

A water treatment plant fluoridates by direct addition of commercial fluosilicic acid. The strength of the acid is 25 percent and the density is 10.5 lb/gal. The treated water flow rate is 1 mgd. What is the dosage rate needed in order to fluoridate at 1 mg/L?

Example 47 shows the fluoride ion content of pure fluosilicic acid to be 79 percent. To adjust for the dilution, multiply the ion content by the strength of the commercial acid, as in example 48.

$$(0.79)\,(0.25) = 0.197\text{ lb fluoride ion per lb of commercial acid}$$

The acid density is 10.5 lb/gal; therefore, find the weight of fluoride ion per gallon of commercial acid as follows:

$$\frac{(10.5\text{ lb/gal})\,(0.197\text{ lb F ion})}{(1\text{ lb commercial acid})} = 2.07\text{ lb/gal commercial acid}$$

For each gallon of commercial acid added to the water, 2.07 lb of fluoride ion is added.

The desired fluoride dosage of 1 mg/L is equivalent to

$$(1\text{ mg/L})\,(8.34\text{ lb/gal})\,(1\text{ mgd}) = 8.34\text{ lb/mil gal}$$

Since the flow rate is 1 mgd, the weight of pure fluoride ion needed is 8.34 lb/day. The number of gallons of commercial acid needed to achieve this dosage is

$$\frac{8.34\text{ lb/day}}{2.07\text{ lb F/gal commercial acid}} = 4.03\text{ gal commercial acid/day}$$

Example 50

How many pounds of NaF are required to fluoridate 1 mgd to a level of 1 mg/L? The NaF used is 95 percent pure.

The fluoride ion content of 95 percent pure NaF is 43 percent (see example 44). Since the flow rate is 1 mgd, the desired dosage of 1 mg/L is equivalent to 8.34 lb/mil gal or 8.34 lb/d. Therefore, to solve the problem, determine the amount of NaF needed to achieve an 8.34-lb/d dosage of fluoride ion. Note that since only 43 percent of each pound of NaF is fluoride ion, to get 8.34 lb of F will require the addition of a little more than 16 lb of NaF. More precisely,

$$(43\%)\ (\text{feed rate of NaF, 95\%}) = 8.34\ \text{lb/d F}$$

$$(0.43)\ (\text{feed rate of NaF, 95\%}) = 8.34\ \text{lb/d F}$$

$$\text{feed rate of NaF, 95\%} = \frac{8.34}{0.43}$$

$$= 19.4\ \text{lb/d NaF, 95\%}$$

Example 51

An important feature of NaF is its relatively constant solubility. The strength of an NaF solution is consistently 4 percent, by weight, within the range of temperatures common in water treatment (0–25°C). How much of such a solution would have to be added each day to meet a dosage requirement of 8.8 kg/d? What would this dosage requirement be in milliliters per second?

You know that the desired dosage is 8.8 kg/d NaF. Because of NaF's relatively constant solubility, 8.8 kg must be 4 percent of the total weight of chemical solution added. Therefore,

$$(4\%)\ (\text{solution added}) = 8.8\ \text{kg}$$

$$\text{solution added} = \frac{8.8}{0.04}$$

$$= 220\ \text{kg/d NaF solution}$$

The density of a 4 percent NaF solution is about 1.03 kg/L. So the dosage rate of 4 percent NaF solution is

$$\frac{220\ \text{kg/d}}{1.03\ \text{kg/L}} = 213.6\ \text{L/d}$$

$$= 213{,}600 \text{ mL/d}$$

$$= 2.47 \text{ mL/s}$$

Example 52

A shipment of 98.5 percent pure sodium fluorosilicate has an average fluoride content of 60 percent (see example 46). How many pounds of Na_2SiF_6 should be added daily to treat a 1-mgd flow to a level of 1 mg/L?

Follow the same procedure described in example 50:

$$(60\%) \text{ (desired dosage)} = 8.34 \text{ lb/d F}$$

$$(0.60) \text{ (desired dosage)} = 8.34 \text{ lb/d F}$$

$$\text{desired dosage} = \frac{8.34 \text{ lb/d F}}{0.60}$$

$$= 13.91 \text{ lb/d } Na_2SiF_6$$

Example 53

The Na_2SiF_6 from example 52 is mixed with water to make a solution of 0.5 percent by weight. How much solution must be added daily to treat 1 mgd to the level of 1 mg/L?

Follow the same type of procedure described in example 51:

$$(0.5\%) \text{ (solution added)} = 13.9 \text{ lb/d}$$

$$\text{solution added} = \frac{13.9 \text{ lb/d}}{0.005}$$

$$= 2{,}780 \text{ lb/d } Na_2SiF_6, \text{ 0.5\% solution}$$

Such a weak solution would be about the same density as water, 8.34 lb/gal. Therefore, the feed rate of the 0.5 percent solution in gallons per day is

$$\frac{2{,}780 \text{ lb/d}}{8.34 \text{ lb/gal}} = 333.33 \text{ gpd of 0.5\% } Na_2SiF_6$$

The feed rate is also equal to 0.25 gpm.

Example 54

The natural fluoride ion concentration of a water supply is 0.2 mg/L. How many kilograms of sodium fluorosilicate must be used each day to treat 1.9 ML/d of water to a level of 1 mg/L? (The sodium fluorosilicate is typically 98.5 percent pure.)

Raising the fluoride level from 0.2 mg/L to 1 mg/L requires increasing the fluoride ion concentration by 0.8 mg/L. Determine feed rate in kilograms per day:

$$\text{required feed rate pure fluoride ion} = \text{(dosage) (flow rate) (conversion factor)}$$

$$= (0.8)\,(1.9)\,(1)$$

$$= 1.52 \text{ kg/d}$$

So 1.52 kg/d of 100 percent pure fluoride ion should be added every day. However, 98.5 percent pure Na_2SiF_6 contains only 60 percent of the total weight of pure fluoride ion that must be added (see example 46). Therefore, to compute the proper dosage,

$$(60\%)\,(\text{feed rate } Na_2SiF_6,\ 98.5\%) = 1.52 \text{ kg/d F}$$

$$(0.60)\,(\text{feed rate } Na_2SiF_6,\ 98.5\%) = 1.52 \text{ kg/d F}$$

$$\text{actual feed rate of } Na_2SiF_6 = \frac{1.52}{0.60}$$

$$= 2.53 \text{ kg/d}$$

Example 55

How many gallons of water are required to dissolve 5.57 lb of Na_2SiF_6, at a temperature of 15.6°C? The solubility of Na_2SiF_6 is 0.62 g/100 mL at 15.6°C.

First convert to English units of measure.[14]

$$\frac{0.62 \text{ g}}{100 \text{ mL}} = \left(\frac{0.62 \text{ g}}{100 \text{ mL}}\right)\left(\frac{3{,}785 \text{ mL}}{1 \text{ gal}}\right)\left(\frac{1 \text{ lb}}{454 \text{ g}}\right)$$

$$= 0.0525 \text{ lb/gal}$$

[14] Mathematics 11, Conversions.

This means that each gallon of water dissolves 0.0525 lb of Na_2SiF_6. Therefore, to determine how many gallons are required to dissolve 5.57 lb of Na_2SiF_6, you need to know how many units of 0.0525 lb there are in 5.57 lb. Stated mathematically, this is

$$\text{solution water required} = \frac{5.57 \text{ lb}}{0.0525 \text{ lb/gal}}$$

$$= 106.1 \text{ gal}$$

Example 56

At what rate should the chemical metering pump be set to pump the fluoride solution of example 55 into the water system? The treatment plant flow rate is constant at a rate of 0.5 mgd. Give answer in gallons per hour.

Since the 106.1-gal supply of fluoride solution is a 1-day supply, and since the plant flow rate is constant, the fluoride solution must be added uniformly throughout the 24-hour period:

$$\text{solution feed setting} = \frac{106.1 \text{ gal}}{24 \text{ h}}$$

$$= 4.42 \text{ gph}$$

Basic Science Concepts
and Applications

Electricity

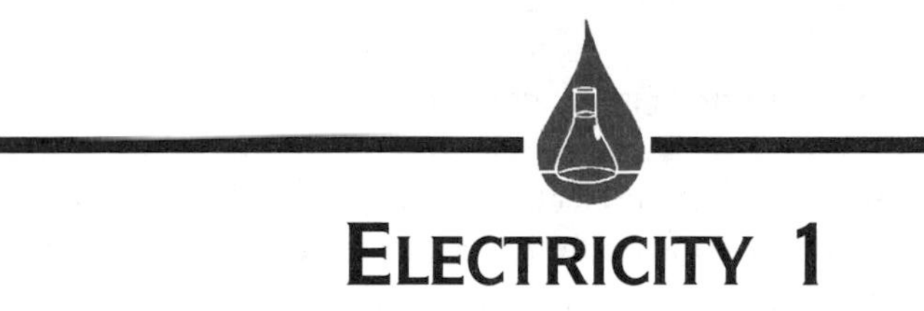

ELECTRICITY 1

Electricity, Magnetism, and Electrical Measurements

All matter is composed of atoms, and the basic particles that make up atoms are protons, neutrons, and electrons.[1] Because electrons are held to atoms by a relatively weak force, the outer electrons of many atoms can easily be removed or transferred to other atoms. This movement may be the result of chemical action, or it may be due to electric or magnetic forces. The study of the movement of large numbers of electrons and their effects is the study of electricity.

Electricity and Magnetism

For purposes of explanation, electricity is often classified as either static or dynamic. Both forms of electricity are composed of large numbers of electrons, and both forms interact with magnetism. The study of the interaction of electricity and magnetism is called electromagnetics.

Static Electricity

Static electricity refers to a state in which electrons have accumulated but are not flowing from one position to another. Static electricity is often referred to as electricity at rest. This does not mean that it is not ready to flow once given the opportunity. An example of this phenomenon is often experienced when one walks across a dry carpet and then touches a doorknob — a spark at the fingertips is likely noticed and a shock is usually felt. Static electricity is often caused by friction between two bodies, as by rotating machinery — conveyor belts and the like — moving through dry air. Another common

[1]Chemistry 1, The Structure of Matter.

example of static electricity is the natural buildup of static electricity between clouds and the earth that results in an electrical discharge — a lightning bolt. In this case, the friction occurs between the air molecules. Static electricity is prevented from building up by properly bonding equipment to ground or earth.

Dynamic Electricity

Dynamic electricity is electricity in motion. Motion can be of two types — one in which the electric current flows continuously in one direction (direct current), and a second in which the electric current reverses its direction of flow in a periodic manner (alternating current).

Direct Current

Current that flows continuously in one direction is referred to as direct current (DC). In this case, the voltage remains at a fixed polarity, or direction, across the path through which the current is flowing. Direct current is developed by batteries and can also be developed by rotating-type DC generators.

Alternating Current

Electric current that reverses its direction in a periodic manner — rising from zero to maximum strength, returning to zero, and then going through similar variations of strength in the opposite direction — is referred to as alternating current (AC). In this case, the voltage across the circuit varies in potential force in a periodic manner similar to the variations in current. Alternating current is generated by rotating-type AC generators.

Induced Current

A change in current in one electrical circuit will induce a voltage in a nearby conductor. If the conductor is configured in such a way that a closed path is formed, the induced voltage causes an induced current to flow. This phenomenon is the basis for explaining the property of a transformer. Since changes of current are a requirement for a transformer to operate, it is obvious that DC circuits cannot use transformers, whereas AC circuits are adaptable to the use of transformers. The function of a transformer is of great value because it allows the voltage of the induced current to be increased or diminished.

Electromagnetics

The behavior of electricity is determined by two types of forces: electric and magnetic. A familiar example of the electric force is the static force that

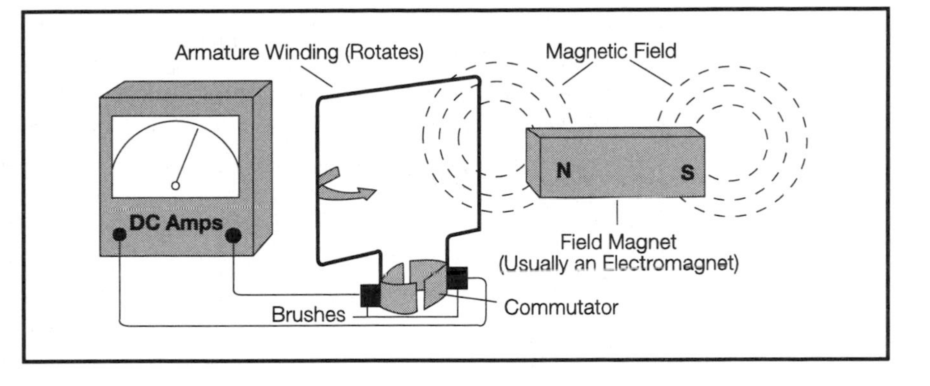

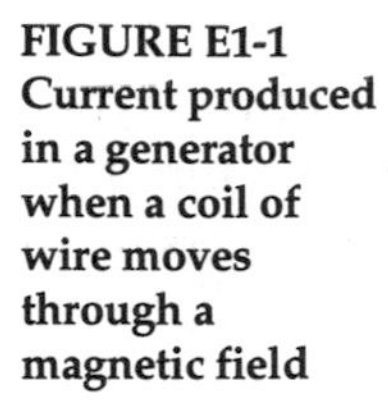
FIGURE E1-1 Current produced in a generator when a coil of wire moves through a magnetic field

attracts dust to phonograph records in dry climates. The magnetic force holds a horseshoe magnet to a piece of iron and aligns a compass with magnetic north. Wherever there are electrons in motion, both forces are created. These electromagnetic forces act on the electrons themselves and on the matter through which they move. Under certain conditions, the motion of electrons sends electromagnetic waves of energy over great distances. Figure E1-1 shows an electric current being produced when a coil of wire is moved through a magnetic field.

Electromagnetic energy is responsible for the transmission of a source of power. Because of electromagnetic energy, power can be transmitted across the air gap between the stator and rotor of a motor. Radio waves are electromagnetic. Radar is electromagnetic. So is light, which unlike most forms of electromagnetic waves, can be seen. The doorbell is an electromagnetic device; so is the solenoid valve on your dishwasher. In fact, much of the electrical equipment with which you are familiar is electromagnetic; your clock motor, the alternator on your car, and the electric typewriter use electromagnetics.

You will find it worthwhile to make your own list of equipment and appliances that are all-electric and that are electromagnetic. You will be surprised at how much you already know about this subject.

In fact, because electricity is such a part of everyday life, even persons who have not studied electricity know more about it today than did early scientific observers. Electricity is often taken for granted, but it can be dangerous and should be respected. Every operator should become familiar with safety rules and practices pertaining to electricity.

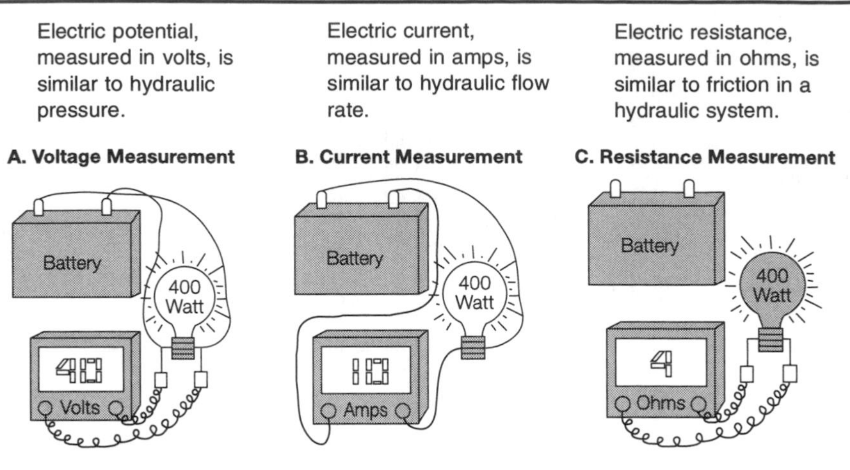

Note: Ohm's law states that voltage drop E across an electrical component equals current I flowing through the component multiplied by resistance R or the component: $E = IR$.

FIGURE E1-2 Representation of Ohm's law

Electrical Measurements and Equipment

The following paragraphs introduce the basic terminology used to describe electricity and electrical equipment. To help the reader understand the new concepts, the behavior of electricity is compared to the behavior of water (Figure E1-2). More technical and detailed descriptions of the terms introduced here are given in the next chapter.

Molecule of liquid ↔ electron of electricity. The smallest unit of liquid is a molecule, whereas the smallest unit of electricity is an electron.

Flow rate (gpm) ↔ current (A). The rate at which water flows through a pipe is expressed as gallons per minute (gpm). It could be expressed as molecules per minute, but the numbers involved would be inconveniently large. Similarly, the rate at which electricity flows through a conductor (called current) could be expressed as electrons per second, but the numbers involved would be too large to be practical. Therefore, the unit of electrical current commonly used is the ampere (A), which represents a flow rate of about 6,240,000,000,000,000,000 electrons per second (also termed 1 coulomb per second). The flow rate of liquid expressed in gallons per minute is, therefore, analogous to the flow rate (current) of electricity expressed in amperes.

Pressure (psi) ↔ *potential (V).* Liquid flow requires a certain head or pressure. The liquid will tend to flow from the high pressure to the low pressure. Electrical potential is similar to head or pressure. Electric current will always tend to flow from high potential to low potential. Electric potential is expressed in terms of volts (V). Whereas liquid pressure is generally expressed in terms of pounds per square inch with reference to atmospheric pressure, electrical potential is generally expressed in terms of voltage with respect to ground or earth, with the general assumption that ground or earth is at zero potential.

Pressure drop ↔ *voltage drop.* The flow of liquid through a pipe is accompanied by a pressure drop as a result of the friction of the pipe. This pressure drop is often referred to as friction head loss. Similarly, the flow of electricity through a wire is accompanied by a voltage drop as a result of the resistance of the wire.

Friction ↔ *resistance.* The flow of water through a pipe is limited by the amount of friction in the pipe. Similarly, the flow of electricity through a wire is limited by resistance.

Pump ↔ *generator.* A pump uses the energy of its prime mover, perhaps a gasoline engine, to move water; the pump creates pressure and flow. Similarly, a generator, powered by a prime mover such as a gasoline engine or a water turbine, causes electricity to flow through the conductor; the generator creates voltage and current.

Turbine ↔ *motor.* A water-driven turbine, like those found in hydroelectric power stations, takes energy from flowing water and uses it to turn the output shaft. An electric motor uses the energy of an electric current to turn the motor shaft.

Turbine-driven pump ↔ *motor-driven generator.* A turbine running off a high-pressure, low–flow-rate stream of water (or hydraulic fluid) can be used to drive a pump to move water at low pressure, but at a high flow rate. (Such an arrangement might be useful for dewatering, for example, although it is not common.) Similarly, a motor operating from a high-voltage, low-current source can be used to run a generator having a low-voltage, high-current output. Note that both the water system and the electrical system could be reversed, taking a source with a low pressure (voltage) and high flow rate (current) and creating an output with a high pressure (voltage) and low flow rate (current). The motor-driven generator (also called a dynamotor) is sometimes used to convert DC battery power to a higher-voltage, lower-current AC power.

Turbine-driven pump ↔ AC transformer. For most electrical applications, the motor-driven generator is replaced with a transformer, which does exactly the same thing — it transforms low-voltage, high-current electricity into high-voltage, low-current electricity; or, it can transform high voltage, low current into low voltage, high current. The transformer will work only with alternating current (AC). It is a very simple device that uses the electromagnetic properties of electricity.

Reservoir ↔ storage battery. Liquids can be stored in tanks and reservoirs, whereas direct-current electricity can be stored in batteries and capacitors.

Flooding ↔ short circuits. The washout of a dam or the break of a water main can cause various degrees of flooding and damage, depending on the pressure and, more important, on the quantity of water that is released. Similarly, an electric fault current, referred to as a short circuit, can cause excessive damage, depending on the voltage and, more important, on the quantity of electric energy that is released. This latter item, the quantity of electric energy that might be released, is referred to as the short-circuit capability of the power system. It is important to recognize that the short-circuit capability determines the physical size and ruggedness of the electrical switching equipment required at each particular plant. In other words, for nearly identical plants, the electrical equipment may be quite different in size because of the different short-circuit capabilities of the power system at the plant site.

Instantaneous transmission of pressure ↔ electricity. When we let water into a pipe that is already filled, the water that we let into the pipe at one end is not the same as that which promptly rushes out of the other end. The water we let in pushes the water already in the pipe out ahead of it. The pressure, however, is transmitted from end to end almost instantaneously. The action of electricity is much the same. The electric current in a wire travels from end to end at high speeds, very close to that of light. However, individual electrons within the wire move relatively slowly and over short distances. When a light switch is turned on, the power starts the electrons on their way.

ELECTRICITY 2

Electrical Quantities and Terms

The following is a glossary of some of the more frequently used units of measurement of electricity, specifically with regard to power systems. Many of these terms are commonly used in describing a system, a piece of electrical equipment, or an amount of electricity required or used. You should become familiar with their usage. The terms defined here are also listed in the glossary at the back of the book.

AC. An abbreviation for *alternating current.*

Alternating current. The term applied to an electric current that is continually reversing its direction of flow at regular intervals. During each interval that it flows in one direction, it starts at zero, rises to maximum flow, then falls back to zero flow. It then starts from this zero flow, rises to a maximum flow in the opposite direction, then again falls back to zero flow. Each repetition from zero to a maximum in one direction, then through zero to a maximum in the other direction, and then back to zero is called a complete cycle. The units referring to cycles per second are called hertz (Hz) in honor of the 19th century German physicist Heinrich Hertz. Thus, an AC power system that goes through 60 cycles per second is a 60-Hz AC power system.

Ampere. A measure of electric current, either AC or DC. One ampere (A) is the amount of current that would flow through a resistance of 1 ohm (Ω) when the voltage on the circuit is 1 volt (V). Most equipment is rated in amperes. If the equipment is operated above its ampere rating, it is overloaded and may be damaged. For example, a motor will have a nameplate that will state how many amperes it will draw when operating fully loaded. If a motor draws more amperes than its nameplate rating, it may be assumed to be overloaded. Amperes are measured with an ammeter.

Ampere-hour (A·h). A measure of a quantity of electricity. One ampere flowing for 1 hour is 1 ampere-hour. This term is used chiefly as a rating of a storage battery, and it simply indicates the amount of energy that a battery can deliver before it needs recharging. The ampere-hour rating of a battery is obtained using standard test times of 4, 8, or 20 hours until total discharge. A battery will have different ampere-hour ratings depending on the rate of discharge current. For example, a battery rated at 100 A·h at a 20-hour discharge rate is discharged at a constant 5 A during the test. If that same battery is discharged at a higher current rate, the battery will not provide a 100 A·h capacity — it will be less. Battery manufacturers provide discharge curves showing the length of time a battery can provide a given rate of current.

Apparent power. The product of AC volts and AC amperes; it is expressed in *volt-amperes* (VA). Although this term is not used frequently, it should be pointed out that this is different from *real power*. Apparent power is normally expressed in units of kVA, meaning thousand volt-amperes. Some of the current that flows in a circuit may not produce any power in the sense of power doing real work. This may seem strange at first, but let's look at it from another point of view. If you tie a rope on your car and try to pull it down the street by hand, would you walk in front of the car or would you go off to the side and walk down the sidewalk? Obviously if you went over to the sidewalk, you would have to pull much harder than if you were directly in front of the car. In this case, you would be pulling more than was really necessary to just move the car forward. The only part of your total pulling that is of any value is the pulling that actually moves the car forward — doing real work, using real power. The pulling you do in trying to pull the car sideways is of no use in moving the car down the street, it only stretches the rope and could be considered useless in moving the car. Relating the pulling example to electrical terms, the pulling that results in moving the car forward down the street is real power. The part of the pulling that only stretches rope is known as reactive power (because the rope stretches and then returns to its original length when the pulling ceases, the power is recovered from the rope). The two types of pulling combined — the total pulling — produce the apparent power, that is, the total pulling that you have to do to move the car. If all your pulling was done directly in front of the car, then the apparent power would be the same as the real power, and the reactive power would be zero.

In electrical systems, the percentage of apparent power that is real power is called the power factor. One hundred kVA at 80 percent power factor represents 80 kW of real power. You will frequently hear the expression "power factor," and it is usually expressed as a decimal, as 0.8, instead of as a percent. Apparent power is measured in volt-amperes; real power is measured in watts; and reactive power is measured in vars. Because these units are quite small with respect to the sizes of electrical loads usually found in water distribution systems, they are generally expressed in thousands. Kilo means thousand. Hence, you will generally find apparent power expressed in kVA (kilovolt-amperes), real power expressed in kW (kilowatts), and reactive power expressed in kvar (kilovars).

Branch circuit. An electrical circuit, which is fed from a protective device, such as a fuse or circuit breaker, that provides electrical power directly to the end load. This load can be a single item of equipment, such as a motor, or can have several loads sharing the same branch circuit, such as lighting fixtures or receptacles. A branch circuit differs from a feeder circuit in that a feeder provides power to a number of branch circuits. A feeder would bring power to a circuit breaker panel, which would, in turn, power multiple branch circuits.

Capacitance. A part of an electrical circuit's total *impedance*, which tends to resist the flow of current. It acts somewhat like inertia in that it tries to resist a sudden change of voltage. For example, your car has inertia, and you will recall that if it is standing still, it takes more force to start it moving than it does to keep it moving. Similarly, after a car is moving, it tends to keep moving and cannot be stopped instantly with ease. The capacitance of a circuit is one of the reasons for the presence of reactive power; capacitance provides storage of electrical energy in the form of electron charge.

Capacitors. Devices that operate on the principle of *capacitance*. They are made of two conductive plates separated by a nonconductor. They are sometimes used to prevent a sudden rise in voltage and sometimes to prevent a sudden drop in voltage. Capacitors are also used in some cases to isolate devices so that no direct-wire physical connection between them is necessary. Frequently, capacitors are used in conjunction with *induction motors* to raise the *power factor* of the electrical system. Capacitors provide a source of leading reactive power, which cancels some of the lagging reactive power of the induction motor. This reduction of total reactive power helps to reduce the value of apparent power in the circuit and make it approach the real power value.

Capacitance and *inductance* may be thought of generally as causing similar but opposite effects on a circuit. Capacitance can be added to cancel the effect of inductance, and vice versa.

Capacitive reactance is measured in ohms.

Conductor. A material or substance that passes electrical current with ease. The most widely used electrical conductor is copper because of its cost-effectiveness and high conductivity.

Continuous duty. A rating on equipment such as motors, engine-generators, or heaters that denotes that the equipment is suitable to operate indefinitely, at its maximum output rating, without interruption. Continuous duty for motors assumes operating without frequent starting and stopping, which can cause excessive heat buildup in the motor windings and shorten the motor's expected life. For electrical code compliance, continuous duty means continuous normal operation for 3 hours or longer.

Cycle. Cycle applies to *alternating current* systems and refers to one complete change in direction of current flow from 0 to maximum to 0 in one direction, then to maximum and back to 0 in the opposite direction.

Demand. A term used frequently in the sale or purchase of electric power. Demand is the amount of electric power expressed in watts (W) or kilowatts (kW) that may be required during a certain time interval (usually a 15-min or a 30-min interval). This is really a measure of how big the power requirement is, i.e., the rate of using the electrical energy, not how much total energy is used. For example, you might have a 1,500-kW (2,000-hp) motor that you start and use only once or twice a year, and then only for a couple of hours each time it is used. Obviously, you wouldn't be buying a lot of electric energy during the year for operation of this motor; but it should be recognized that regardless of how seldom the load is used, the power company has to have that amount of power readily available whenever the load is applied to its system. As a result, the power company may make a separate charge for the size of the load in addition to the charge for energy actually consumed. This is called a demand charge. Demand is measured by a demand meter, which actually measures the maximum kilowatts that have been required during any particular demand interval — 15 min, for instance. The demand meter has a maximum-reading pointer. It is not uncommon that the maximum demand established in any particular month will then apply, as far as billing is concerned, for the following 11 months. It is obvious why it is of chief concern to the operator to prevent establishing a new maximum demand, since this may increase the price of purchased power for the following year. If an operator or maintenance person is not careful, the running of a 500-hp pump motor, for example, for just 20 min could result in a demand charge of several thousand dollars on the electric bill, with only a few cents in energy charges.

This concept of demand is applicable to water distribution systems. You may have a customer who normally uses less than 20 gpm of water but who might very infrequently want several hundred gpm for a short time. In such an instance, pipelines and supply facilities will have to be sized for the maximum usage, even though the usual quantity is not too different from the average residential service.

DC. An abbreviation for *direct current.*

Direct current. An electric current that flows continuously in one direction. Direct current is used quite frequently for controls and instrumentation. It can be generated on-site by an engine-driven DC generator. It is often converted from alternating current by means of an AC, motor-driven DC generator. It may also be directly converted by means of a *rectifier.* The storage battery is a common source of direct current that is kept charged by a battery charger. Of course, dry cells (similar to flashlight batteries) can be used to supply DC current, but they are limited to small loads.

Synchronous motors require a comparatively small amount of direct current in addition to their AC requirements. Magnetic couplings used as variable-speed drives require both AC and DC current. Most electronic instruments use direct current. For example, a signal of 4–20 milliamperes (mA) DC is commonly used with electronic instruments. This is a very small amount of current (milli is a prefix meaning thousandth). Because this DC output is a very small current flow, such an instrument will contain a very small wire size. Direct current from an auxiliary battery supply is frequently used for both power and control of water valves where continued operation is necessary should AC power fail. Many electronic, variable-speed motor controllers employ direct current. For example, there are many types of silicon-controlled rectifier (SCR) drives on the market for comparatively small motors. Many emergency lighting systems are designed for operation on direct current.

Direct current devices will probably play an even greater role in the future as solid-state rectifiers become more fully developed.

Energy. Electric energy is usually expressed in kilowatt-hours (kW•h). For example, 1,000 W used for 1 hour or 500 W used for 2 hours equals 1,000 W•h or 1 kW•h. Note that power is expressed in watts, but that energy is expressed in watt-hours.

One kW•h of electric energy may cost anywhere from 3.5 cents to more than 10 cents, depending on the size of the load and, of course, the location and conditions of the contract for purchasing power. Certainly, one of the primary responsibilities of an operator is to keep the use of power at a minimum because of its cost. For example, by allowing ten 100-W lamps (1,000 W) to burn continuously for one month (720 hours), 720 kW•h of electric energy is used. If it is assumed that energy costs one cent per kW•h, this would amount to $7.20. In a plant where there are hundreds of lamp fixtures and devices, the needless use of electricity becomes an unnecessary operating expense.

It is also interesting to compare the costs of operating the same ten 100-W lamp fixtures with the cost for operating a 1,000-hp motor for 1 hour. One thousand hp equals 746 kW. Therefore, this load is 746 kW•h, which at one cent per kW•h would amount to $7.46. Note that it costs only 26 cents more to operate the motor for 1 hour than it does to leave the 10 lamp fixtures on for a month. This example helps to show why there is usually a *demand* charge in addition to the charge for energy; the electrical system must be many times greater both in size and capacity to start and run a 1,000-hp motor than it needs to be to energize ten 100-W lamps.

Power bills usually have both a demand charge and an energy charge. These charges are based on "how big" and "how much"— concepts that every operator should keep in mind. (See *demand*.)

Frequency. Frequency applies to alternating current and refers to the number of cycles that occur each second. The frequency of most power systems in water distribution facilities will be 60 Hz (i.e., 60 cycles per second). Some power systems in the United States are 25 Hz. In European and African countries, 50 Hz is the norm. Make sure that all electrical equipment is properly rated for the frequency of the power source; otherwise, the equipment could be damaged.

There was once considerable argument about choosing a standard frequency. As frequency increases, the *impedance* to the flow of current in a circuit often also increases, which makes the line losses and voltage drops on electric transmission and distribution systems greater; however, as frequency increases, the physical size of equipment becomes smaller. At 25 Hz, many people could sense a lightbulb dimming and brightening 25 times per second. This, perhaps more than anything else, caused the industry finally to standardize on 60-Hz power. Single-phase and three-phase are most common.

On airplanes it is common to use a power system with a 400-Hz frequency. This allows equipment to be smaller. Tone-transmitting equipment used for telemetering and supervisory control usually covers a range of 300 to 3,000 Hz. Radio equipment operates at much higher frequencies and can use very small components.

Power utilities regulate frequency very closely, making it possible for all of our electric clocks to remain synchronized. If frequency were lowered, our clocks would lose time and all motors would slow down. If frequency were raised, our clocks would gain time and all motors would speed up. There are situations, though, for which varying the frequency can be useful. Equipment is presently available to use frequency-variable motor controllers for variable-speed pump drives.

Frequency can be measured with a frequency meter.

Generator. A device for producing electricity. For water utilities, AC generators are most often used as backup systems in case of a major power outage in the primary electrical grid. AC generators supply power to the critical components of the water utility, including pumps and other control systems. The generators are typically powered by diesel fuel, gasoline, natural gas, or a liquefied petroleum gas (LPG) such as propane.

Ground. This represents an electrical connection to earth or to a large conductor that is known to be at earth's potential. Properly grounded equipment and circuits minimize the hazard of accidental shocks to persons operating electrical equipment. Grounding also serves to facilitate the operation of protective devices used on the equipment. A good ground will generally have a resistance of not more than 10 Ω and preferably less. Motor frames and electrical-equipment enclosures should all be connected together (i.e., bonded) and connected to ground (a process referred to as "grounding"). The word ground is also used to refer to a common reference point in equipment from which voltage measurements are made.

Horsepower. A common term used to denote the rate of doing work. Originally a mechanical term, it is now commonly used to rate electric motors. One horsepower (hp) is equivalent to 746 W of electrical power.

Impedance. The total opposition offered by a circuit to the passage of electrons — that is, to the flow of current. Impedance is made up of three parts: *resistance*, *inductance*, and *capacitance*. It is interesting to note that once a certain flow of direct current is established, the inductive reactance and the capacitive reactance are ineffective — only the resistance of the circuit limits the flow of current because inductance and capacitance act only to oppose *changes* in electrical flow.

Impedance is measured in ohms (Ω) and is made up of a certain number of ohms of inductive reactance, a certain number of ohms of capacitive reactance, and a certain number of ohms of resistance.

A series circuit with 10 Ω inductive reactance, 10 Ω capacitive reactance, and 10 Ω resistance would not, however, have an impedance of 30 Ω. Capacitive reactance and inductive reactance *oppose* one another; in this case they are equal, so the impedance is only 10 Ω, the same as if there were no inductance and capacitance in the circuit. This would be an example of a "tuned" circuit. Neither would a circuit of 10 Ω inductance and 10 Ω resistance have an impedance of 20 Ω. You do not add them directly as one might expect. Instead, you take the square root of the sum of the squares. For example, 10 Ω resistance plus 10 Ω inductance is the square root of $10 \times 10 + 10 \times 10$, or the square root of $100 + 100$; that is, the square root of 200, which is a little over 14. This would be an inductive circuit having an impedance of 14 Ω.

Inductance. Inductance is part of the total *impedance* of an electrical circuit, tending to resist the flow of changing current. Inductance is analogous to inertia in that it tries to prevent a sudden change of current. If a current is flowing in a circuit and a switch is opened, it is the inductance of the circuit that tries to keep the current from instantaneously falling to zero. You have no doubt noticed sparks as you opened a switch or removed a jumper from a car battery. These sparks are simply the electric current jumping across the air gap as the switch is being opened, the current trying to maintain itself. Similarly, when a switch is closed, the inductance of the circuit prevents the current from rising from zero to its maximum amount instantaneously. It takes some time for the current to rise to its full value. You can see this behavior by looking at an ammeter when the switch is opened and closed.

Inductance provides storage of electrical energy in the form of a magnetic field. This is why motors, with their windings and strong internal magnetic fields, have inductive reactance and reactive power (lagging Kvars) in addition to the real power used.

Inductance and *capacitance* may be thought of generally as causing similar but opposite effects on a circuit. Inductance can be added to cancel the effect of capacitance, and capacitance can be added to cancel the effect of inductance.

Inductive reactance is measured in ohms (Ω).

Induction motor. The simplest common AC motor is the three-phase squirrel-cage induction motor. The rotor windings consist of a series of bars placed in slots in the rotor, which are connected together at each end — this gives the rotor the appearance of a squirrel cage. The stator windings, located in the frame, are connected to the three-phase power supply and current flowing through them sets up a rotating magnetic field within the motor. This rotating field induces a current flow in the rotor windings, which generates an opposing magnetic field. The force between the two magnetic fields causes the rotor to turn.

The wound-rotor induction motor has a stator similar to that of the squirrel-cage induction motor, with a number of pairs of *poles* and their associated stator windings connected to the *three-phase power supply*. The rotor has the same number of poles as the stator, and the rotor windings are wired out through slip rings. This allows the *resistance* of the rotor circuit to be controlled while the motor is running, which varies the motor's speed and torque characteristics. The wound-rotor motor offers ease of starting in addition to variable-speed operation. The starting current required by a wound-rotor motor is seldom greater than normal full-load current. In contrast, squirrel-cage and *synchronous motors* may have starting currents from 5 to 10 times their normal full-load currents.

Interrupting current. This is a term applying to the level of current that a fuse or circuit breaker is designed to interrupt or clear to prevent damage to the circuit. A fuse or circuit breaker that interrupts a short-circuit current greater than its designed interrupting current will create a potentially dangerous explosion. A fuse or circuit breaker is rated in amperes-interrupting-current (AIC). Thus, a common 20-A circuit breaker rated at 10,000 AIC could not safely open a circuit with a short-circuit (fault) current of 15,000 A. It could explode and spew out molten metal parts. Short-circuit fault currents can go exceedingly high, depending on the size of the main power source and what type and length of wires are between the power source and the fuse or breaker. Some fuses are rated to interrupt up to 200,000 A of fault current. These fuses may be the only device capable of protecting large electrical services. Other electrical equipment, such as switchboards and motor control centers, have an AIC rating for the busbars and internal construction. High fault currents will produce intense magnetic fields around the equipment conductors. These fields produce large physical forces that push against each other and can rip apart equipment that is not adequately rated for the available fault current. The required interrupting ratings of electrical equipment can be found by doing a fault current study of the power system.

NOTE: The safety issue of interrupting current is becoming a real concern of electrical engineers and inspectors. Many building departments are now routinely requiring fault current studies for all installations.

Inverter. This is a device that changes *direct current* to *alternating current*.

Kilo (k). A prefix meaning 1,000.

Kilowatt (kW). A measure of electric power. Since a unit of power equal to 1 W is so small, in many instances in the water utility it is more convenient to express power in units of 1,000 W, or 1 kW. Remember that 1 hp equals 746 W. (Power is the capacity to do work in a given time.)

Kilowatt-hour (kW·h). A measure of electric energy. Since a unit of energy equal to 1 W·h is so small, it is convenient to express units of electric energy in units of 1,000 W·h, or 1 kW·h.

A kilowatt-hour of electric energy is a measure of the amount of work done. The use of 1 W for 1,000 hours, or 1,000 W for 1 hour, would both be the use of 1 kW·h of electric energy.

One hp-h equals 0.746 kW·h.

Kilovar (kvar). An expression meaning 1,000 reactive volt-amperes. It is a measurement of *reactive power*. It will not be used very often; however, you should have some concept of its meaning. It is used to express the size of power *capacitors*. You may find some *induction motors* that have capacitors connected in *parallel* with the motor. These capacitors will have a nameplate similar to a transformer nameplate, stating size in kilovars.

A capacitor connected to a motor terminal may be used to correct the motor's *power factor*. The capacitor is added to reduce the amount of reactive power (in kilovars) used by the motor circuit. One reason that reactive power is objectionable is that it consumes additional current flow, and current flow causes voltage drop, which results in low voltage at the equipment.

There is another reason to increase power factor by reducing the kilovars. Since power companies have to oversize their equipment to provide reactive power, some companies are now penalizing their customers for low power factor. Power companies may install Kilovar meters and charge for reactive power above a certain level. Thus, it may be very advantageous for the customer to purchase and install power factor improvement capacitors. Engineering studies can determine how many kilovars of capacitors are needed without causing overvoltage problems.

Kilovolt (kV). An expression meaning 1,000 V.

Common voltages used by water utilities in North America are 2.4 kV (2,400 V), 4.16 kV (4,160 V), 15 kV (15,000 V), 33 kV (33,000 V), and 69 kV (69,000 V). Many still higher voltage classes correspond to standard transmission voltages throughout the United States. Usually voltages below 1,000 V are expressed directly, such as 6 V, 24 V, 120 V, 460 V, or 600 V.

kVA. An expression meaning 1,000 volt-amperes, but usually it is referred to in no other way than the three letters *kVA*. The term *kVA* is used in expressing the size of *transformers*, *generators*, and various other pieces of equipment. It is also used to express the amount of *power* that might be released from an electrical system during a fault such as a short circuit. It may also be used to describe the size of a system load. It is useful to recognize that, for a plant that contains mostly squirrel-cage-type *induction motors*, 1 kVA will generally provide adequate capacity to start and run a 1-hp motor. Common sizes of lighting transformers are 50, 75, 100, 225, 500, 750, and 1,000 kVA.

The quantity kVA actually represents a measurement of *apparent power*, which is made up of two parts — real power and reactive power. Real power is expressed in kW, and reactive (wattless) power is expressed in kvar. Apparent power equals real power plus reactive power, or kVA = kW + kvar. It is important to note here, however, that this is not a direct arithmetical addition. The sum is the square root of the sum of the squares of the two parts, similar to the addition described under the discussion of impedance.

Maximum demand. The maximum kilowatt load that occurs and persists for a full demand interval during any billing period, usually a month. It is not uncommon for a maximum demand established during one month to be used as a factor in figuring power bills for the succeeding 11 months, unless of course a still higher demand is established. Demand intervals are usually 15 min or 30 min. This allows large electrical loads to be present for a few minutes during testing and maintenance without a new maximum demand being established for billing purposes. (See *demand*.)

Megohm. This term represents 1,000,000 Ω (ohms). An ohm is a measure of the *impedance*, or opposition, a circuit offers to the flow of current. An ohm is a very small quantity, so impedance is often expressed in megohms. The megohm is particularly useful when speaking of the insulation of high-voltage components in a particular type of equipment (such as *transformers*, *switchgear*, and motors) or in wiring materials (such as cable or insulators or open-wire power lines). For example, the insulation of a high-voltage motor during winding will have an impedance of several thousand megohms.

Ohm. The basic electrical unit used to quantify a material's opposition to the flow of electricity. Obviously, if you want electricity to flow, you must choose a material having little resistance to flow. Likewise, if you do not want electricity to flow, you must choose a material having great resistance to its flow. A material offering little resistance is called a conductor. A material offering great resistance is called an insulator. All materials are really conductors of electricity, but the poorest conductors may make the best insulators.

The practical unit of electric resistance is an ohm. By definition, it is the resistance for which 1 V will maintain a current flow of 1 A. The ohm is used in expressing *resistance*, *reactance*, and *impedance*, all of which are characteristics of a circuit that tend to restrict the flow of current.

Parallel circuits. An arrangement of electrical devices such that the positive poles of each device are connected by a conductor to the positive poles of the other devices, and the negative poles of the devices are similarly connected to another conductor. Each electrical device thereby creates its own "parallel" branch of the circuit; thus, if one device fails, the other devices will continue to operate. The current from the electric power source will divide among the parallel devices and a different current will flow through each device.

Phase. This term has three primary meanings.

(1) *Phase* is a term used to describe the type of electrical service being used. Virtually all systems are either single-phase or three-phase systems. Other types of systems are possible but are not generally found in water utilities.

Single-phase power simply refers to a system that uses a *generator* or *transformer* that has only a single *winding*. This winding requires two conductors (wires), one from each end of the winding, to supply power. (This single winding may be tapped into — that is, center-tapped — so that three conductors are in use. Such a system is called a three-wire, single-phase source.) A single-phase, two-wire system with a 120-V winding may have one end of the winding connected to *ground*; the other end would then be the hot lead. Such a system would be called "a two-wire, grounded, 120-V, single-phase power supply." Similarly, a 240-V winding may have a center tap connected to ground; both ends of the winding would then be hot leads. This system would be a "three-wire, grounded, 240/120-V, single-phase power supply." It would allow you to connect 240-V equipment across both outer leads and 120-V equipment from either outer lead to ground.

Most North American residential small loads use one of the two system types just described. A single-phase motor greater than 2.2–3.7 kW (3–5 hp) is considered a large single-phase motor. Single-phase circuits are most adaptable for lighting and convenience outlets, small fractional-horsepower motors, and appliances.

Three-phase power simply refers to an electric system that has a three-winding generator or transformer. This generator or transformer requires at least three conductors, one from each winding, to supply the power. Three-phase power systems do not have any practical limitation on the voltage that may be delivered to equipment, so all larger load applications (like those found in most industrial plants, treatment plants, and pumping stations) will have three-phase power systems. Some systems will be three-phase–three-wire and some will be three-phase–four-wire, depending on how the three windings are connected inside the generator or transformer. The three windings can be connected in either of two configurations — as a delta or as a wye. The delta-connected transformer has only three corner connections and can be only a three-phase, three-wire system because a wire is connected at each corner. The wye-connected transformer has one end of each of the three windings connected in common so that the other ends are available for a three-phase, three-wire system. It is possible to connect a lead to the common junction (called the neutral) of the three windings, making a fourth wire connection. In such cases, the system is then a three-phase, four-wire system. If the neutral wire (the one connected to the common junction) is also connected to ground, this system is then referred to as a three-phase, four-wire, grounded system.

Single-phase circuits can easily be served from three-phase systems by using the voltage between the two leads of a three-phase, three-wire system, or between the neutral and any of the three hot leads of a three-phase, four-wire system. Three-phase circuits can also be served from a single-phase system, but only in very small loads, by using special solid state or rotary phase converter equipment.

(2) The term *phase* is also used frequently when two or more power supplies are being paralleled (connected together). Such systems may be referred to as "in phase" or "out of phase." They definitely must be in phase when interconnected. The word *phase*, when used in this respect, actually refers to the sequence in which the voltage reaches its maximum value on each of the three power leads. For example, suppose three power leads are connected to a three-phase induction motor and the motor turns in one direction; this is an in-phase system. If you interchange the connection of any two of these three motor leads, the motor will then run in the opposite direction; this is an out-of-phase system. Actually, what you have done is reverse the phase sequence. If you had two power supplies to your plant and these two supplies were out of phase, all of the motors would run in one direction when served by one power feeder and in the opposite direction when served by the second power feeder. When the power supplies are in phase, they are said to be synchronized. The instrument used to show that two systems are in phase is called a syncroscope. A syncroscope is also used to show when two standby generators are in phase. In this case, the term phase not only means that the direction of phase sequence is the same, it also means that the peak voltage of each phase of both generators occurs at the same instant in time. This allows both generators to be connected in parallel to serve the load and act as one larger generator. If the generators are not properly phased before they are paralleled, extensive damage could occur to the generators when they are connected. This would be caused by large currents that would flow between the windings of the two generators without opposition.

Sometimes you will hear the expression that a three-phase motor is "single-phasing." This means that one of the three leads (phases) to the motor has been disconnected by some means and the motor is trying to run with only two leads energized. Obviously, the motor will be overloaded. Loss of phase and reverse phase can be detected by protective devices, called phase protection relays, that will stop equipment when such a condition develops.

(3) The term *phase* is also used in reference to the relationship between AC voltage and the current in a particular circuit. If a circuit has only resistance, the voltage and the current are said to be in phase; that is, the voltage and the current reach their maximum values in both their forward and reverse directions and return to 0 at the same instant. If a circuit has both resistance and a predominate inductive *reactance*, the current will then lag in time behind the voltage. The current is then spoken of as being phase-shifted behind, or lagging, the voltage. If the circuit has both resistance and a predominate capacitive reactance, the current leads the voltage. Most systems are made up of motor loads that are inductive, so it will be usual that the current lags the voltage in any water-utility facility. This is also expressed as a load having a lagging *power factor*. The lagging power factor can be corrected by the addition of capacitors that tend to advance the lagging current of the inductive motors in a leading phase direction.

Pole. This term is used to describe one end of a magnet (or electromagnet) or one electrode of a battery. Magnets are said to have north and south poles; batteries have positive and negative poles.

Power. This term represents the ability of electricity to do a certain amount of work in a certain amount of time. Power is the rate at which work is accomplished. There are actually three types of power in an electrical circuit — apparent power, real power, and reactive power. Real power is the most useful concept, because it represents true work that is accomplished, e.g., pumping a quantity of water in a certain amount of time (such as liters per minute), so it is often referred to simply as "power." The reason for these distinctions in types of power is that some of the current that flows in a circuit may not produce any power. See the example in the definition of apparent power for more information.

Apparent power is given in units of volt-amperes (VA), or more generally *kVA*. Real power is measured in watts (W) or kilowatts (kW), as necessary. Reactive power is expressed in vars or kilovars (kvar), as necessary.

Apparent power is the combination of real and reactive power. To find the number of kVA of apparent power, first add the square of the number of kilowatts of real power to the square of the number of kilovars of reactive power. Then take the square root of the sum to get the number of kVA.

The *power factor* (pf) of the system relates the apparent power to the real (usable) power. It represents the percentage of apparent power that is available as real power. For example, if the apparent power is 100 kVA and the power factor is 80 percent (or 0.8), then the real power is (100)(0.8) = 80 kW.

When we speak of power in a day-to-day sense, we are always referring to real power (in kilowatts). The customary-units equivalent of the kilowatt is the horsepower (hp). One hp equals 0.746 kW, and 1 kW equals 1.341 hp.

Motors are commonly rated in terms of horsepower. A motor rated at 10 hp has the ability to do 7.46 kW of real work. If a 10-hp induction motor operates at 0.8 pf, it will require 7.46/0.8 = 9.3 kVA of apparent power from the electrical system. Thus, approximately 10 kVA of *transformer* capacity would be needed. As a rule of thumb, it may be said that 1 hp of real motor power will require approximately 1 kVA of apparent power (or transformer capacity) to properly start and run the motor.

On DC systems, the number of watts equals the number of volts times the number of amperes (watts = volts × amperes). On single-phase AC systems, the number of watts equals the number of volts times the number of amperes times the power factor (watts = volts × amperes × power factor). On three-phase power systems with relatively balanced phase currents, the number of watts equals the number of volts, measured from phase-to-phase, times the number of amperes in one phase, times the power factor, times 1.73 (the square root of 3).

Power factor (pf). This term refers to the amount by which the current lags or leads the voltage in an *alternating current* circuit (see third definition of *phase*). It is also the ratio of the *real power* to the *apparent power* in a circuit.

The power factor is a useful figure for expressing the characteristics of a particular load, such as a motor or a complete electrical system. For example, during starting, an *induction motor* may have a very low lagging power factor, perhaps 0.3–0.4; then when it is running at full speed, it will have a lagging power factor of approximately 0.8. A *synchronous motor* may have a power factor of 1.0 when running;

in this case, it is usually referred to as having a unity power factor. Some synchronous motors may be designed for operations at 0.8 leading power factor, which provides power factor correction for lagging power factor caused by inductive loads on the same power system.

The power factor of a system is the result of the design and selection of equipment. Usually an operator will adjust a power factor in one of two ways: by operating switches to add or remove *capacitors* or by manually adjusting the field rheostat on synchronous-motor controllers. It may be found that a plant is operating at a very low, lagging power factor. A low power factor can be the cause of low voltage. This low power factor can be corrected by the addition of capacitors or leading–power-factor synchronous motors.

In some cases, a power company may have a power-factor clause in its published rate structure that will require the purchaser of power to pay extra if the power factor of the plant load is below some expressed limit. Usually the power factor of a plant load is approximately 0.8–0.85, which is generally high enough to escape any billing penalty. Generally, power-factor correction by means of modifying the plant design is not undertaken unless there is a low-voltage condition or a significant potential for savings in the cost of power.

Primary. This term is often used in referring to the high-voltage side of a *transformer*. The voltage of a power line that serves a plant site is referred to as the primary voltage. A substation transformer will usually be provided to step the voltage down to a lower voltage, called the secondary voltage, for the customer's use. In a more general sense, the primary side of a transformer is connected to the source of electric power, and the secondary winding(s) is connected to the loads.

It is not uncommon to find that power companies will sell power at either primary voltage or secondary voltage. Usually, the customer will furnish the transformer and substation if power is purchased at primary voltage; likewise, the power company will furnish the transformer and substation if power is furnished at secondary voltage. The rates are generally referred to as primary and secondary rates. Secondary voltage does not necessarily mean low voltage (600 V or lower); it simply means some voltage lower than the power-line distribution voltage. Obviously, power is sold at a somewhat cheaper rate at primary voltage because the users furnish their own substations.

Protective relay. A device used to divert electrical current and/or disconnect electrical equipment for either preplanned or emergency situations. Overcurrent relays are often referred to as overload relays. They sense current surges in the power supply and disconnect the motor if a surge occurs. Frequency relays respond to changes in cycles per second of an AC power supply and are most frequently used on synchronous motor starters to sense when a motor has reached synchronizing speed. Phase-reversal relays are installed to detect a condition whereby the electrical phase sequence is reversed, which would allow the motors to run backward. This is particularly important in deep-well applications because the pump shafting could become unscrewed, which would allow the pump to fall.

Rating. Rating is a term used to designate the limit of output power, current, or voltage at which a material, a machine, a device, or an apparatus is designed to operate under specific normal conditions. Electrical materials and equipment operated in excess of this rating will have a shortened life. Excessive heating or vibration is frequently an indication of operation beyond rated output. Motors are sometimes rated with a "service factor" on the nameplate. This factor means that the motor can provide more horsepower than the nominal continuous rating. However, the motor life will be shortened because it will operate with a higher internal temperature and the winding insulation will deteriorate faster. A typical service factor is 1.15, which means that the motor can provide 15 percent higher horsepower than the nominal rating.

Reactance. The general name given to the other two properties of a circuit, besides *resistance*, that tend to prevent the flow of current. The two reactances are inductive reactance and capacitive reactance. Any piece of electrical equipment will have an impedance of so many ohms, a resistance of so many ohms, an inductive reactance of so many ohms, and a capacitive reactance of so many ohms. The inductive and capacitive reactances have opposite effects, so that they tend to cancel each other. If they are equal, the circuit has a net reactance of zero and is considered a purely resistive circuit. If they are not equal, the circuit's net reactance is either inductive or capacitive. A circuit that has a predominate inductive reactance has a lagging *power factor*, and the current lags the voltage. A circuit that has a predominate capacitive reactance has a leading power factor, and the current leads the voltage.

Reactive power. The term given to describe the stored energy in an electric circuit, device, or equipment. This energy may be stored in the form of a magnetic field in an inductive device (such as a motor or lighting fixture ballast) or it may be stored in the form of an electric field in a capacitive device (such as a power factor correction capacitor). Reactive power does not accomplish any real work, but it may be necessary in order to allow real work to be done. For example, a motor requires a magnetic field to exist between the rotor and stator before the motor can accomplish the real work of turning a pump and moving water. The energy required to produce and maintain this magnetic field is taken from the power line and then returned to the line each time the AC power line reverses polarity. With each cycle of the power line, the magnetic field is created, diminished, then recreated with opposite polarity, and then diminished again. Each time the magnetic field goes through this action, the required energy is taken from the power line to create the field and then returned back to the power line as the field diminishes. Thus, no real energy is permanently expended and no real work is done in the establishment of the magnetic field.

Normally, power companies do not charge for reactive power even though they do have to reserve capacity in their power system to provide the current necessary to supply the reactive power needs of a customer. If the reactive power needs become too high, the power company may charge for reactive power or they may require that the customer reduce the reactive power requirements.

Real power. The form of power most commonly associated with electrical circuits — the kind that can do work. Real power is expressed in watts (W) or kilowatts (kW) and is the product of *kVA* and the *power factor*.

Rectifier. A device that provides unequal resistance to forward and reverse current. A rectifier allows an electric current to flow easily in one direction only and blocks flow in the reverse direction. It is often used for converting alternating current to direct current.

Relay. A device consisting of an electromagnetic coil and one or more electrical contacts that are activated when a current flows through the coil. Common applications are to isolate one electrical circuit from another or to control the flow of a large current in one circuit by using a second, smaller, current in the coil.

Resistance. A part of the total *impedance* of a circuit, tending to resist or restrict the flow of current. The resistance of a circuit causes power losses in the form of heat produced by the current in the resistor. Inductive *reactance* and *capacitive reactance*, the other two components of a circuit, do not consume any *real power*.

Safety factor. The percentage above which a rated device cannot be operated without damage or shortened life. This factor is usually found above the nameplate rating on the device. In some cases, the safety factor implies that the device can be overloaded at some stipulated expense to life of the device. Present-day motors, for example, may have a safety factor of 1.15, which means that they may be operated at 15 percent overload. However, the motor manufacturer may expect the user to recognize that at 15 percent overload, the motor may operate 10°F hotter than the winding insulation is designed for. A general rule of thumb is that for every 10°F above rated temperature, the motor will have its life expectancy cut in half. The safety factor of equipment should be used only for infrequent and abnormal conditions; it should never be used during a continuous or normal operating condition. The term "service factor" is commonly used to mean the safety factor of motors and is found on the motor's nameplate.

Secondary. This term is commonly applied to the low-voltage winding of a *transformer*. It is also used frequently in reference to the voltage of an incoming electric power service. Power can be purchased at either primary or secondary voltage. (See **primary**.)

The term *secondary* is also applied to the rotor windings and controls of wound-rotor *induction motor*. A secondary controller of a wound-rotor motor is the equipment used for varying its speed.

Series circuit. An arrangement of electrical devices such that the devices are connected in a positive pole to negative pole series, so that the same current passes through each device. In a series circuit, if one device fails, the other devices will be inoperable because the current will cease flowing.

Substation. This term refers to any power-switching station that may be accompanied by power *transformer* installations.

Synchronous motor. A motor in which the power-supply leads connect to the stationary windings in such a way that a revolving magnetic field is established, rotating at synchronous speed. The rotor is constructed with poles to match the poles of the stator, and the rotor pole pieces have coil windings that are supplied with direct current, producing a strong constant magnetic field in the rotor poles. This requires that a slip-ring assembly, called a commutator, and brushes be used to connect the direct current to the rotating parts of the motor. Thus, in the synchronous motor, alternating current is supplied to the stator and direct current is supplied to the rotor. When consumers of purchased power pay a penalty for low power-factor conditions, the synchronous motor is often a logical choice, since it has a power factor of 1.0. When speed must be held constant, the synchronous-type motor is necessary.

Transformer. A device that allows energy to be transferred in an *AC* system from one circuit (the primary circuit) to another (the secondary circuit). The two circuits are completely independent and linked together by a common magnetic circuit, comprised of two or more windings on an iron core. The incoming electrical energy is transformed into a magnetic field in the primary winding. The field is concentrated in the iron core and then reconverted into electrical energy when the magnetic field cuts across the secondary winding(s). One of the most common uses of a transformer is to change the "incoming" voltage on the primary side to another voltage on the secondary side. This allows the "incoming" voltage to be changed to a practical level for a particular use. Power transformers are sized for their electrical capacity in kVA. The larger the kVA rating, the larger the physical size of the transformer; the iron core and the windings must be large enough to transform and pass all the energy. (See **primary**.)

Vars. An acronym for volt-amp-reactive. It is computed by multiplying the voltage by the voltage by the reactive current. Kilovars is computed by dividing the vars by 1,000.

Volt. This is the practical unit of voltage, potential, or electromotive force. One volt (V) will send a current of 1 ampere (A) through a resistance of 1 ohm (Ω). The term *voltage* is more often used in place of the terms *electromotive force, potential, potential difference, potential gradient,* or *voltage drop* to designate the electric "pressure" that exists between two points, which is capable of producing a current flow when a closed circuit is connected between the two points. Usually on a three-*phase* system, the voltage is expressed as the voltage from line to line, whereas on a single-phase system, the voltage is expressed as the voltage from line to neutral. A "High Voltage" warning should be respected.

Voltage is measured by a voltmeter.

Volt-Ampere (VA). Computed as the product of the voltage times the current. Also referred to as the apparent power. Kilovolt-ampere, or kVA, is computed by dividing the VA by 1,000.

Watt. A watt (W) is a practical (real) unit of electric *power*. In a DC circuit, watts are equal to volts times amperes. In an AC circuit, watts are equal to volts times amperes times *power factor*. Watts can be measured by a wattmeter. (See **power** and **power factor**.)

Watt-hour (W·h). A watt-hour is a practical unit of electric energy equal to the power of 1 W being used continuously for 1 hour. The rate of doing work determines the number of watts, whereas the amount of work done, regardless of how long it takes, determines the number of watt-hours. One thousand W·h is the equivalent of 1 kilowatt-hour (kW·h). The price of a kilowatt-hour in the United States may vary from approximately 3.5 cents to 10 cents, depending on the size of the load and the conditions of the contract for purchase of power. The price of electric power to the customer will be primarily determined by the source of the energy used to produce the electricity. Hydropower is typically a cheap source, with thermal fuels such as coal, natural gas, diesel fuel, and nuclear power all increasing in cost. The distance from the generating plant to the customer will also affect the customer's cost due to the transmission lines, substations, and distribution equipment that are required to transmit the power.

Winding. The current-carrying coils of a motor, generator, or transformer. A simple motor is formed by placing a rotating magnet near an electromagnet. As current is passed through the electromagnet (the winding of the stator), it creates a magnetic pole that attracts the unlike pole of the other magnet (rotor). In large motors, both the stator and the rotor are electromagnets, each having their own windings. Generators have windings similar to those of motors and operate in a similar manner. Transformers have windings that produce electromagnetic fields when alternating currents flow through them.

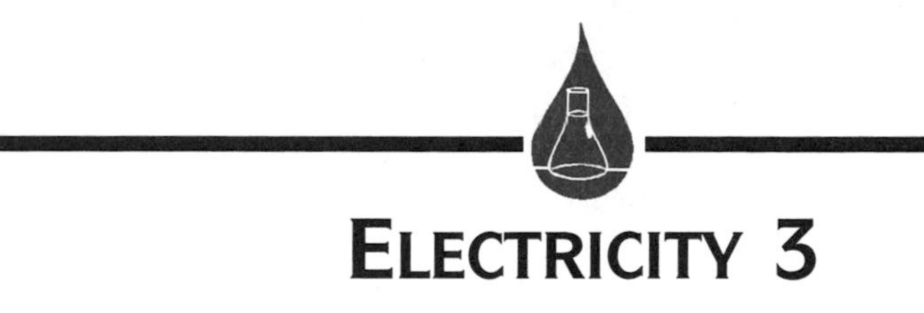

ELECTRICITY 3

Functions and Ratings of Electrical Equipment

Because there are so many kinds of electrical equipment and materials, it is helpful to think of equipment in broad categories or classes based on function or purpose. For example, a reference made to "household appliances" immediately causes one to think of such items as radios, television sets, and toasters, but obviously wouldn't direct one's thoughts or attention to X-ray machines or substation equipment.

Functions of Electrical Equipment

A complete electric power system may also be categorized based on the functions performed by different types of equipment. These different types of equipment are used to

- generate electricity
- store electricity
- change electricity from one form to another
- transport electricity from one location to another
- distribute electricity throughout a plant area
- measure electricity
- convert electricity into other forms of energy
- protect other electrical equipment
- operate and control other electrical equipment
- convert some condition or occurrence into an electric signal (sensors)
- convert some measured variable to a representative electrical signal (transmitters and transducers)

Any assembly of electrical devices that makes up a certain piece of electrical equipment will usually contain elements from several of these categories.

The discussion in this chapter covers the first eight categories listed above. A detailed discussion of certain items in the last three categories is contained in *Water Distribution*, which is also part of this series. The equipment covered in this volume pertains to the power systems commonly used in water supply. There are many more specialized types of equipment used in other fields such as medicine, photography, the military, communications, and exploration.

The operation of a water utility transmission and distribution system will generally be performed by an employee whose title is "operator." The operator of a water transmission and distribution system will generally be responsible for one or perhaps several electrical systems. In effect, the operator operates an electrical system that in turn operates the distribution system. Some of the systems interposed between the operator and the actual service he or she is controlling are power systems, metering systems, control systems, and communications systems, all of which are in some respect specialized electrical systems. Remote control (via a leased telephone circuit) of a pumping station by means of supervisory control equipment and telemetering would be an example. So would the control of a pumping station from a control room using direct wire for control-and-metering circuits.

To be a pump station operator, one need not be an electrician. Neither must one be an automotive mechanic to operate a car or a truck. But some general knowledge about a vehicle is necessary to drive and maintain it properly.

This chapter is not designed to make you an electrician. But it will introduce some general concepts and ideas about electricity and about the electrical equipment and materials that are found in water transmission and distribution systems. Although the installation and repair of electrical equipment should be left to a qualified electrician, the operator should be able to maintain the equipment.

One should make note of the difference in the two words — *repair* and *maintain*. *Repair* means to put back in good condition after damage. *Maintain* means to keep in good condition. Although operators are not required to repair electric failures, they are expected to perform necessary preventive maintenance.

Preventive maintenance may be defined as the periodic inspection of equipment as necessary to uncover conditions that may lead to breakdown or excessive wear, and the upkeep as necessary to prevent or correct such conditions.

Properly applied and correctly installed equipment is obviously one of the first requirements for a well-maintained plant. No one can do a good maintenance job on equipment that is either not appropriate for its task or is installed haphazardly.

The second requirement for good maintenance is an operator who has a general knowledge of what the equipment is for and how it should work.

To operate and maintain an electrical system, one must become acquainted and be familiar with all of its parts. The only way to do this is to pay a great deal of attention to the equipment. Find out what the equipment is supposed to do; then become familiar with how it does it.

How does the equipment normally sound? You use this method to judge the condition of your car engine. The sound of certain electrical equipment under operating conditions is characteristic. There is usually a distinctive hum from coils and transformers. Motor starters and circuit breakers make quite a noise when operated, but this noise is usually the same each time. Motors and rotating equipment make an air noise during operation; this air noise is nearly always the same.

How does the equipment normally feel? Knowledge of the normal operating temperature of a piece of equipment is always helpful in judging its performance.

How does the equipment normally look? Generally, a visual inspection of equipment will show points of unusual wear, corrosion, or overheating. Is it clean and dry? Remember that a piece of equipment's two worst enemies are dirt and moisture.

Most of the maintenance of electrical equipment will be of a mechanical nature (cleaning, lubricating, tightening connections, and recognizing abnormal performance such as excessive vibration, noise, or heating).

Equipment to Generate Electricity

From a practical standpoint, any large block of electricity is generated as alternating current by a rotating-type piece of equipment called a generator. Most generators found in a water utility are AC generators. DC generators, however, are used frequently for battery chargers, synchronous motor field current, and special control systems. Even in DC generators, the initial form of the generated electric current is alternating, but this is changed to DC at the output of the machine by an arrangement of current collectors, or brushes, called the commutator. DC generators are frequently found on AC synchronous motors to supply the motor field with direct current; these DC generators are called exciters. Thus, a synchronous motor with a direct-connected exciter is actually an AC motor with a direct-connected DC generator to supply the motor field current. The motor field current is adjusted to vary the power factor of the synchronous motor.

Power is generally purchased by water utilities, and any on-site generation is provided only for standby power. Generally, it is necessary to reduce the operating load and operate only essential equipment and essential station auxiliaries when the standby power-generating equipment is being used. Generators used for standby power will normally be the gasoline, diesel, or gas-turbine types, because these can be started immediately. It would be unusual to find steam-driven equipment for standby service because the steam equipment would have to remain in continuous operation to be available immediately.

Ratings

Engine-driven generators may be rated in any one of several ways — that is, for continuous duty, primary power duty, continuous standby duty, intermittent duty, and standby duty. Those units rated for intermittent or standby duty should not be operated continuously at full-load nameplate rating. Following are the definitions of various ratings.

The *continuous* rating as applied to an engine generator set indicates that the equipment can deliver its continuous kilowatt rating for the duration of any power outage and is capable of 10 percent overload for a period of 2 hours out of each 24 hours of operation.

The *standby* rating as applied to an engine generator set indicates the 2-hour overload rating at which the unit can operate. For the balance of the time, it would have to meet the continuous rating previously defined.

The *continuous standby* rating as applied to an engine generator set indicates that the equipment can delivery its continuous standby kilowatt rating for the duration of any power outage. No overload is permitted on this rating.

The *primary power* rating is stated as some percentage (always less than 100 percent) of the continuous standby rating.

As the reader can see, the practice of rating engine generator sets is somewhat confusing, if not actually misleading. Different manufacturers do not use the same terms or rating methods for engine–generator sets. The plant operator should remember than an engine–generator set will give its longest life when operated at or below the primary power rating limits. Short-term overload operation is permissible within the overload rating limits. A manufacturer assigns several different duty ratings to the same engine-driven generator set; it is the responsibility of the water treatment plant operator to not exceed these ratings.

Engine-driven generators must be derated for use at altitudes greater than 3,000 ft above sea level or for high ambient temperatures. The derating means that the engine cannot deliver the same mechanical horsepower, or electrical power, that it can at sea level. Two factors affect the derating that each manufacturer assigns its products: (1) the amount of reduced engine output due to a reduction in combustion oxygen in the atmosphere, and (2) the reduction in cooling of the engine and generator due to the reduced density of air. Typically, a unit is derated 4 percent for each 1,000 ft above 3,000 ft. Thus, a 100-kW unit operated at 9,000 ft above sea level will be derated 6 times 4 percent, or 24 percent. The 100-kW unit is then rated at 100 kW minus 24 kW, or 76 kW at 9,000 ft.

Prime Movers

Generating equipment is generally classified by the source of power used to drive the generator's rotating machinery. This machine is usually called the *prime mover*. There are a number of different prime movers used in such generators.

Hydrogenerators. Hydrogenerators are generators driven by water power through the use of water wheels or hydraulic turbines. Some of the earliest uses of water wheels and hydraulic turbines were for the direct drive of pumps. In such cases, the speed of the drive was set by local conditions, so systems of belts and pulleys or gearing were connected to the driven equipment to obtain the required speeds. The conversion of water power to electric power is a good example of the conversion of raw-energy sources to electric energy. Hydrogenerators may be of practically any size. Some water utilities have installed hydrogenerators at water storage reservoirs or dams to produce electrical energy for their own use or for sale to the power company. Some water utilities also install hydrogenerators where a pressure reduction is needed in a water conduit. The hydroturbine extracts the energy from the water during the pressure reduction process. This mechanical energy then turns the electrical generator to convert the mechanical energy into electrical energy.

Steam-Turbine Generators. Both the steam engine and the steam turbine have been used for the conversion to electricity of raw energy found in fossil fuels such as coal, oil, gas, and other combustible substances. In such cases, the heat is used to make steam, which in turn is used to drive the engine or turbine. Steam turbines operate at much higher speeds than engines, so they are used almost exclusively for driving generating equipment. The steam turbine now completely dominates the field of prime movers on large generators in both fossil-fueled and nuclear-fueled power plants. Nuclear power plants simply convert the nuclear energy to heat, which is in turn used to produce steam to drive steam turbine generators. Steam turbine generators are usually any size above 2,500 kVA.

Engine-Driven Generators. Gasoline motors and diesel motors are both used to drive generators. Motor-driven generators are generally referred to as engine-driven units. The smaller engines are usually similar in nearly all respects to car and truck motors, whereas larger units are diesel motors of the large stationary type. Gasoline engines can be adapted to also use natural gas or liquid petroleum gas (LPG) fuels with a special dual-fuel carburetor and controls. Typically, gasoline engines are derated 5 percent when operated on LPG and 10 percent when natural gas is used. Gasoline deteriorates rapidly when stored, and these other fuels overcome this disadvantage. Engine-driven generators are usually any size below 2,500 kVA.

Gas-Turbine Generators. Developed by the aircraft industry, gas turbines are becoming more popular as prime movers because of their light weight, compactness, high speed, and minimum auxiliary requirements. Gas-turbine generators are generally sized from approximately 250 to 2,500 kVA. Gas turbines can burn a variety of light fuel oils. Standard diesel fuel or kerosene are typically used.

Wind-Driven Generators. Wind motors can be used in isolated areas to generate small amounts of electricity. Usually batteries are an adjunct of wind generators so that some power is available during periods when there is a lack of wind. In such cases, the wind generator usually operates a battery charger, which keeps a storage battery charged.

Equipment to Store Electricity

The storage of electricity has been a goal for years, but no practical means has yet been developed to store large amounts of electricity. The nearest approach to this end has been the installation of large pump storage systems wherein water is pumped to a high-level reservoir, then allowed to drain out through a hydroelectric generator. In such cases, a pump and motor combination is designed to be run backward so that the pump acts as a hydraulic turbine and the pump motor acts as a generator. Obviously, this pump storage concept is not a direct means of storing electricity.

All stored electricity is direct current, and only comparatively small amounts of electricity can be stored. Such amounts are so small, in fact, that the measure for stored electricity is the ampere-hour. One thousand A·h of stored electricity may sound like a large amount, but the quantity would power only an ordinary 120 V lighting circuit (approximately 1,500 W) for approximately 75 hours or run a 3,730-W (5-hp) motor approximately 1 day.

The two basic types of electrical equipment that store electricity are the battery and the capacitor (also known as a condenser).

Battery

An electric battery is a device for the direct transformation of chemical energy into electric energy. There are two types of batteries: the primary battery and the storage battery.

Primary Battery. A primary battery is one in which the chemical action is irreversible. In a primary battery, the parts that react chemically are destroyed. A flashlight battery is a primary battery. There are two types of primary batteries, the wet cell and the dry cell. The flashlight battery is a dry-cell battery and has a capacity of approximately 2–3 A·h. Wet-cell batteries have a liquid electrolyte and are therefore not practical for usual industrial use, but they are used frequently as standard cells in laboratory work and in various analytical measuring devices.

Storage Battery. A storage battery is one in which the chemical reactions are almost completely reversible. That is, by returning current to the battery, a chemical action takes place to recondition the parts to their original state. This process is called "charging" the battery. Storage batteries will always be accompanied by battery-charging equipment. Storage batteries are made up of individual cells and can be connected in series and in parallel to develop any practical desired voltage and capacity. Usual voltage ratings vary from 6 to 250 V DC, and capacity ratings generally vary from approximately 100 to 1,000 A·h.

It is interesting to note that the storage battery doesn't really store electricity. It converts electricity to chemical energy during charge and converts chemical energy to electricity during discharge. This is analogous to the pump storage system for storage of water power mentioned earlier.

It is common to find two types of battery-charging equipment used in conjunction with a large-station battery. Automatic chargers are generally applied to maintain the battery at normal charge; manual chargers are provided for a quick charge after an excessive drain, as might occur during an emergency.

Storage batteries should be well ventilated because some will give off hydrogen gas, which if allowed to accumulate can cause an explosion. Battery rooms obviously must be well ventilated.

The most common types of storage battery is the lead-acid battery. This battery uses lead plates and a sulfuric acid–water mixture for the electrolyte. Each of the battery's cells produce an open circuit voltage of approximately 2.1 volts DC. This battery has the lowest cost of the storage batteries; however, it suffers from greatly reduced capacity when the ambient temperatures drop below –20°C. The gel-cell battery is a type of lead-acid battery in which the electrolyte is in paste, or gel, form that helps to retain the gases produced during charging. This type of battery can be sealed and used where a liquid electrolyte would be dangerous or inconvenient.

Another common type of storage battery is the nickel-cadmium battery. This battery is more costly than the lead-acid type, but it provides very high peak currents at very low temperatures. This could be beneficial starting engine–generator sets in cold climates.

Capacitor

A capacitor, also referred to as a condenser, consists of a pair of metallic plates separated by a piece of insulating material called a dielectric. If a capacitor is connected across a voltage source (a potential) and is then disconnected, it will be charged and will hold this voltage for a considerable period. If its terminals are connected together, there will be an immediate discharge causing a current to flow through the connection. The actual amount of electricity stored in a condenser is very small compared with that from even a small storage battery. This amount of electricity, however, can be very dangerous.

The charging of a condenser, or capacitor, is the only true means of actually storing electricity. A condenser will allow alternating current to flow through it, but it will block the flow of direct current. Capacitors vary in size from extremely small, like those used in radio and TV sets, to large power capacitors used in substations as switched capacitors for power-factor correction. They are also used to suppress voltage surges that may be caused by lightning strokes or by large power circuits being switched. In such applications, these condensers are called surge capacitors or surge suppressors.

Capacitors are also used frequently in parallel with motors to correct, or raise, the power factor of an induction motor load.

Equipment to Change Electricity From One Kind to Another

There are many electrical devices that simply change electricity from one type or characteristic to another. This is done for various reasons to make it more suitable for some particular use or specific requirement. Some of the more common types of electrical equipment that perform this function are as follows.

Transformer

A transformer is a device that allows energy to be transferred in an AC system from one circuit, called the *primary*, to a second circuit, called the *secondary*. The primary winding and the secondary winding are essentially two completely independent circuits linked together by a common magnetic circuit. One of the most common uses of the transformer is to change voltage from one level to another. A transformer that raises the voltage in the secondary winding above the voltage in the primary winding is called a step-up transformer. Similarly, a transformer that lowers the voltage in the secondary winding is called a step-down transformer. Transformers can have more than two windings, and in such cases they are referred to as three-winding or four-winding transformers.

Transformers are probably one of the most essential devices for economically distributing electric power. They allow for adjustment of the voltage and current levels to practical values. For example, a circuit of 120 VA of apparent power may have a voltage of 120 V with a current of 1A; or it could have a voltage of 480 V with a current flow of ¼A. When one considers that transmission line power losses are proportional to the square of the current and that voltage drop in the transmission line is directly proportional to the current, one can see that reducing the current flow also reduces the line losses and the voltage drop. Not only does the lower current reduce losses and voltage drop, but it also allows the use of smaller wire size for the circuit, since wire size is selected according to the number of amperes to be carried.

Transformers are made in all sizes, from small control power transformers that can be carried in one's hand, to medium sizes that are pole mounted, to large power transformers like those seen in substations. Transformers can also be constructed as single-phase, three-phase, multiphase, or in any combination. Transformers found at water utilities will generally be single-phase and three-phase.

Substation

The term *substation* is included here because it is used so frequently in place of or as having the same meaning as transformer. A substation more specifically describes any power-switching station that may be (and generally is) accompanied by power transformer installations. Substations may contain high-voltage-line terminal towers, high-voltage power circuit breakers, switches, fuses, power transformers, regulators, lightning arrestors, grounding reactors, couplings for carrier-type communications, batteries, and various other auxiliaries.

The expressions *package substation* and *unit substation* are frequently used to describe substations that are prefabricated to include all of the switchgear, transformer, and auxiliary devices built into one integral assembly.

Voltage Regulator

A regulator can be similar in many respects to a transformer, with the exception that it is designed to maintain automatically a certain voltage on the secondary winding. A voltage regulator may be thought of as analogous to an automatic pressure-regulating valve on a water line. Power voltage regulators are not found frequently in water utilities, but they are used extensively by power utilities on long transmission or distribution circuits.

Small-size voltage regulators are used frequently in a water utility for instrumentation circuits, telemetering, supervisory control, data-handling equipment, and computers.

Converter

Several types of equipment are called converters. Generally, the term *converter* alone refers to an AC-motor-driven DC generator for converting alternating current to direct current.

A frequency converter can be an AC-motor-driven AC generator in which the frequency of the alternating current used to power the motor is different from that which is generated by the generator. Such equipment, for example, might be served by a 60-Hz system to generate power at 400 Hz.

The term *static converter* refers to a nonrotating type of converter that is comprised of modern solid-state electronic power devices such as diodes, silicon-controlled rectifiers, and power transistors. Static converters may convert AC to DC or they may be frequency converters.

Rectifier

A rectifier is a device that changes AC to DC by allowing the current to flow freely in one direction and blocking the flow in the reverse direction. An analogous device in a water system would be a spring-loaded check valve.

Inverter

An inverter is a device that changes DC to AC. Modern inverters incorporate solid-state electronic power devices.

Current Transformer

A current transformer, commonly referred to as a CT, is a special transformer designed to have a 5-A secondary circuit when the primary winding carries full-load current. A 500-A to 5-A current transformer (500-A/5-A CT), for example, would have 5 A in the secondary when the primary current is 500 A, or 4 A in the secondary when the primary current is 400 A. The purpose of the CT is to allow a standard 5-A instrument to be used as, for example, an ammeter suitable for 0–5 A with a scale calibrated at 0–500 A. This type of ammeter and CT is commonly used for electric switchboard meters where large currents must be shown.

The advantage of using a standard CT is that all of the instruments requiring current — such as ammeters, wattmeters, watt-hour meters, demand meters, power-factor meters, var meters, or protective relays — can be standardized for operation over a range of 0 to 5 A. Another advantage is that the CT isolates the high-voltage primary circuit being measured from the secondary metering circuit; thus, all metering and instrumentation are at low voltage and low current.

Potential Transformer

A potential transformer, commonly referred to as a PT, is a special transformer designed to have a 120-V maximum secondary voltage when the primary winding carries full voltage. A 2,400/120-V PT, for example, would have 120 V on the secondary winding when the primary winding voltage is 2,400 V; 100 V on the secondary when the primary voltage is 2,000 V; or 60 V on the secondary when the primary voltage is 1,200 V. The purpose of the PT is to allow a standard 120-V instrument to be used as, for example, a voltmeter suitable for 0–120 V with a scale calibrated at 0–2,400 or 0–5,000 V. This is commonly done on electrical switchboards.

The advantage of using a standard PT is that all instruments requiring voltage — such as voltmeters, wattmeters, watt-hour meters, demand meters, power-factor meters, var meters, or protective relays — can be standardized for operation over a range of 0–120 V. Another advantage is that the PT isolates the high-voltage primary circuit being measured from the secondary metering circuits, so that all metering and instrumentation are at low voltage and low current.

Instrument Transformers

Current transformers and potential transformers are referred to as instrument transformers. CTs and PTs are designed for instrument and metering use and have various accuracy classifications.

Control Power Transformers

Control power transformers are designed to give a secondary voltage, usually of either 120 V or 240 V, when the primary circuit voltage is at full voltage. This allows the use of all control items such as pushbutton stations, selector switches, control relays, timers, and indicating lights to be rated at a standard low voltage. A 4,160-V motor controller, for example, would have a 4,160/240-V control power transformer; similarly, a 2,400-V motor controller would have a 2,400/240-V control power transformer. In both cases, the control items could be identical in style, type, and voltage rating. It is therefore not uncommon to find control power transformers designed for two secondary winding connections — 4,160/240–120, for example — so that both 240-V and 120-V control items may be used in the control wiring as may be required.

Current Regulator

A current regulator is a device that regulates current automatically to stay within certain limits. The current regulator is similar in many respects to the voltage regulator. Current regulators are sometimes referred to as constant-current power sources.

Equipment to Transport Electricity From One Place to Another

Electricity can be transmitted from one place to another by means of controlled or uncontrolled paths. An uncontrolled path would be hazardous because it could cause a flashover (an electrical discharge through the air) or an arc to ground. It is essential to design controlled paths for electric energy to flow from one point to another. The controlled path for electric current is always a conductor, which may be a wire, a bus (which is a bar of conductor with a larger cross section than a wire), or a conducting material of any configuration that is sufficiently insulated to prevent the escape of electricity.

The controlled paths used for the movement of large blocks of power are frequently referred to as transmission systems and distribution systems. These terms are used almost synonymously, but they do carry specific connotations.

Power utilities frequently refer to high-voltage lines, those 69,000 V and greater, as transmission systems and to lines 33,000 V and less as distribution systems. The term *transmission system* can also be interpreted to include components of the high-voltage system such as transformers, substations, regulators, switching stations, steel towers, or poles associated with that particular voltage class. Similarly, reference to the distribution system by a power utility will generally be interpreted to include all of the same components associated with that lower-voltage level.

In the water utility, these terms are used with a somewhat different meaning. A transmission system is generally thought of as the system that supplies power to the water-utility site, regardless of the voltage level. For example, an incoming power line with voltage as low as 2,400 V is referred to as a transmission line. Generally, the expression *transmission system* as used by the water utility seldom relates to any on-site part of the water utility system — but it may.

The term *distribution system* as used by the water utility is generally interpreted to mean the on-site electrical system of the water utility facilities, regardless of whether it is overhead or underground, and regardless of the voltage class.

Most generally, from the water utility standpoint, the term *transmission system* will refer to overhead power lines, whereas the term *distribution system* will refer to on-site underground or in-plant power cables.

The water utility operator will be more concerned and more closely associated with distribution systems, which are discussed in more detail later.

Overhead Transmission

From an electrical standpoint, the design of an overhead power line must take into account voltage selection, conductor size, line regulation, line losses, lightning protection, and grounding.

From a mechanical standpoint, the design must take into account conductor composition, conductor spacing, type of insulators, amount of sag, wind loading, ice loading, and selection of hardware.

From a structural standpoint, the design must take into account the size and type of structures, foundations, guys, and anchoring.

Other features of an overhead line that must be taken into account are line location, acquisition of right of way, road crossings, joint construction with other utilities, and access for line maintenance.

From an operator's standpoint, perhaps the most important aspects of overhead lines are service and maintenance. To maintain an overhead line, it is essential to have all hardware bolts tightened, all grounding connections tightened, cracked insulators replaced, poles inspected and treated, lightning arrestors in order, tree limbs trimmed away from the conductors, and guy wires tightened and protected. Any line work should be done by an electrician.

Weakened lines will generally fail during storms. Loose hardware and cracked insulators can cause radio and television interference. Damaged lightning arrestors can allow lightning to damage equipment served by the power lines. Poor ground connections or opened ground wires can be hazardous to anyone adjacent to poles and guy wires.

Underground Transmission

The trend in some localities is to use underground construction for distribution circuits, particularly in residential areas. The use of high-voltage cable for underground transmission of power has been practiced for years with highly reliable results. An underground cable is less likely to be damaged than an overhead line that is exposed to trafficways or storms. However, it should be noted that should damage occur, an overhead line can be serviced and put back into operation much sooner than an underground power cable can be pulled and replaced. Frequently, it is possible to arrange for two transmission lines to serve a water utility, one overhead and the other underground; then the utility can incorporate the advantages of each.

Underground transmission construction is much more expensive than conventional overhead construction, but in some densely built sections of cities, it is virtually a necessity because there is no space for overhead lines.

Underground construction must take into account many of the same concerns as overhead construction, such as location, routing, acquisition of right of way, location of vaults and access holes, access for services, and flooding. Other features to take into account in the design of underground transmissions are voltage selection, conductor size, regulation, losses, lightning protection, and grounding — the same as for overhead construction.

Unlike overhead construction, there is nothing that an operator can do normally to service and maintain underground lines, with the exception of a periodic high-potential test for suspected deterioration of insulation. Some very high-voltage cables are gas- or oil-filled and require service and attendance from an operator; these types of underground cables, however, are not generally found in water utilities.

Equipment to Distribute Electricity Over a Plant Site

The electrical distribution system of any water utility begins with a source of electric energy that must be distributed to each and every electrical device on the site. This source may be either on-site generation or an interconnection with a power utility.

Sometimes this power source consists of only one incoming power feeder from the power utility; sometimes there are two incoming power feeders, a preferred incoming feeder and a standby. Or it may be that both incoming power feeders are normally in use, with half the load to be served on each feeder, but with switching arrangements so that all of the plant load can be served from either feeder. In other cases, the source of power may be a combination of incoming power services from the power utility and on-site generation for emergency standby power. Obviously, there can be various arrangements for any number of service entrances and any number of on-site generators. The following discussion will review some of the types of distribution systems and their major components.

Power Utility Circuits

The power utility circuit can be called the incoming feeder, power service, service, service entrance, or power supply. In each case, its design will be adapted to meet certain requirements of the water utility. Each utility will have specific standards with which the service entrance must comply. These standards will vary widely, but in most cases will cover such items as voltage, physical arrangement, metering facilities, accessibility, grounding provisions, testing facilities, and type of circuit.

These circuits must, for billing purposes, include the metering equipment (which may consist of watt-hour meters for measuring energy used), demand meters for measuring the maximum amount of power used during any given time interval (usually 15- or 30-min periods), power-factor meters, kvar meters, and any associated current transformers and potential transformers. Such metering equipment is normally the property of the power utility but is generally housed in compartments furnished by the customer.

These compartments, suitable for padlocking or sealing, are designed and located to meet the power utility's requirements. Frequently, these compartments are located within the customer's incoming service switchgear and must be compatible with the switchgear installation.

In addition to the metering facilities, an incoming feeder will have an automatic disconnect that will open in case of an excessive load as a result of a short circuit or fault in the customer's equipment. It is not uncommon for an incoming feeder circuit breaker to have associated relaying equipment that must be coordinated with the power utility's transmission system protection. Power utilities aim to provide reliable service to all customers. To accomplish this, the power utility usually requires that whenever a fault occurs on the user's system, that system will be disconnected automatically from the utility's transmission system. On small services, this automatic disconnect may be a fuse or a plastic molded-case circuit breaker that is usually a part of a distribution panel or switchboard. On large systems, this disconnect device consists of a power circuit breaker activated by protective relays that are usually a part of the switchgear.

Protective relays are used frequently to open incoming service breakers under conditions such as loss of power, low voltage, or reverse phase. In such instances, the user is without power until the service can be automatically or manually restored. Frequently, causes of power outages are of very short duration and the problem is self-restoring. In such cases, it is to the advantage of the user to have additional relay equipment to cause the incoming feeder breaker to reclose automatically. Power utilities also use automatic reclosing equipment to improve the customer's service.

Continuity of Service. Continuity of power service is extremely important to water utilities in many instances, particularly in cases where pumping equipment is supplying a closed water distribution system. In such cases, an "automatic throwover" is employed, wherein two incoming services are used. In this instance, in order to ensure the greatest continuity of power service, it is absolutely necessary that the two incoming feeders be completely independent. Completely independent power feeders could, of course, be supplied from independent transmission systems. But since it usually isn't feasible to acquire services from completely independent transmission systems, generally the two services are obtained from separate power utility substations. The whole purpose of the two services is to ensure continuous power at the site; it is presumed that when power is not available from one source, it will be available from the other. Even the reliability of service provided by two incoming feeders is not considered satisfactory in many instances; it is then necessary to arrange for three or even more services. For example, a high-service pumping station has two 15-kV underground power feeders and two 69-kV overhead power feeders to ensure reliability of power. Naturally, each incoming service must be equipped with the necessary metering, relaying, and disconnect means, so it is not uncommon in such instances to find a good portion of the main switchgear made up of only those items concerning incoming power circuit facilities.

Arrangement of Incoming Service. The arrangement of incoming services can take several forms. Generally, if two services are available, it is common practice to designate one as a preferred and the other as a standby incoming service. Sometimes these are referred to as the normal and alternate, or emergency, sources. It is now common for a power utility to charge the customer a monthly capacity reserve charge for the standby service feeder. Since this feeder may be called on at any time, the power utility must reserve adequate capacity at the substation serving this feeder. This reduces the revenue that they can receive from other customers supplied by the same substation. For any particular site, the power utility may or may not allow the two incoming feeders to be parallel. A parallel connection means that both incoming feeders may be connected to the same load at the same time.

If the two incoming feeders may be connected in parallel, and if either incoming feeder can handle the full load, then the load may be transferred at will from one service to another. This type of switching is called a *closed transition*.

If the two incoming feeders may not be connected in parallel, and if either incoming feeder can handle the full load, then it is necessary to disconnect one service before switching on, or closing, the second service. This type of switching is called an *open transition*.

Generally, a closed transition is not permitted if the two incoming services are independent. Obviously, an open-transition switching arrangement results in a period during which the station load is completely without a power service connection to the power utility. If this period of transfer time is very long, it is obvious that all of the load will be turned off; the motors will stop and lights will go out. It is possible, however, to have this transfer occur automatically and so quickly that it will be hardly noticeable; only a flicker of the lights is seen, and the motors are kept running. This fast-switching operation is the function of the automatic throwover equipment. This equipment and necessary accessories monitor the loss of power, the condition of the standby source, the preference of sources, the transfer time, and control power provisions for operation of the switches. Hopefully, an automatic throwover scheme will function within a transfer time of only a few electric-energy cycles, a few one-sixtieths of a second, so that for all practical purposes, the station continues to operate as if there had been no interruption of power supply.

But what happens when power is restored from the normal source? Sometimes it really doesn't make any difference to the power utility which service is supplying the load; in other cases, the power utility desires that service is taken normally from the preferred source and that the standby source is used only when necessary. There are several different types of automatic throwover schemes — the nonpreferential type, the fixed-preferential type, and the selective-preferential type — to accommodate this possibility.

The *nonpreferential* automatic throwover scheme allows the load to remain in operation on whichever power source it was last connected. In this case, either incoming power source can serve as normal or as standby, so they are usually designated as "alternate source #1" and "alternate source #2." The automatic throwover equipment does not switch the load back to the original incoming source when that source has recovered. This scheme is the most desirable because it reduces the switching operations by eliminating the need to switch back to a preferred source.

The *fixed-preferential* automatic throwover scheme allows the load to remain on the standby supply only until power is restored on the preferred source; then it transfers the load back to the preferred source. After a power failure, when the condition of the preferred source has become normal, the automatic return transfer of the load to the preferred source may be immediate, it may be delayed until after some adjustable time delay, or it may be only partially automatic, being delayed indefinitely until released manually.

The *selective-preferential* scheme is similar in all respects to the fixed-preferential scheme except that the choice of the feeder designated as preferred may be selected by means of a manually operated control transfer switch. This scheme is usually used whenever the incoming services are essentially the same in all respects and the selection is simply a matter of flexibility or convenience that can easily be incorporated in the control scheme. Selective-preferential controls are used when the utility wishes to use the two sources alternately for a reasonably long time on the selected preference to equalize the work load and maintain the electrical equipment in good condition.

In some cases, throwover equipment is not designed to perform the return to normal switching automatically; return switching is done manually. The return-to-normal switching is never urgent; hence, a prolonged time may be reasonable and allows the operator time to make sure that the disturbance that caused the normal source to fail has been cleared up.

Sources of power that can be paralleled are referred to as synchronous sources. Local generation would never be considered a synchronous source with respect to a source from a power utility. Most utilities will require that local generating equipment be both mechanically and electrically interlocked so that the sources can never become parallel. Likewise, two incoming circuits from the power utility that are not synchronous sources must be interlocked so that they cannot be inadvertently connected in parallel.

It is desirable in some instances to use both of the incoming services continuously, and this can be arranged. To do so, the station load is designed to be split essentially in half, with each half of the station normally operating from its respective incoming feeder. A load tie breaker is then employed, so that when one of the incoming feeders is disconnected, its load is connected to the remaining incoming feeder, which will then serve the entire load until power is restored to the original feeder. One great advantage of this circuit arrangement is that only half of the station load is interrupted by the failure of an incoming service. The switchgear in this case not only has the two incoming feeder circuit breakers, but also a third breaker usually called a tie breaker or bus tie breaker. Under normal conditions, the two incoming feeder breakers are closed and the bus tie breaker is open. During outage of either incoming feeder, the bus tie breaker and the incoming feeder breaker for the "live" circuit are closed. Under any condition, normal or emergency, this scheme allows only two of the three breakers to be closed. This is a very effective scheme to keep part of the station in service because the automatic throwover facilities need to transfer only half instead of all of the load, thus causing considerably less shock to the transmission system. The switchgear is usually designed to have two main power buses, sometimes referred to as a sectionalized bus; these buses are connected by the bus tie breaker, with each bus being supplied normally by its respective incoming feeder. Return switching is supplied in one of the regular ways, such as immediately upon restoration of power, after a fixed or adjustable delay, or manually.

In addition to the electrical controls that provide the interlocking functions for the three breakers, some utilities require a mechanical interlock that the operator must use to open a main breaker and close the tie breaker. This mechanical interlock provides electrical safety by assuring that all three breakers cannot be closed at the same time. If this were inadvertently done, the energized feeder would backfeed the dead power utility feed and possibly kill someone working on the line. One type of mechanical interlock is a Kirk-Key interlock system.

It must be clear that the electrical equipment required to meter, protect, and switch two incoming power circuits can require a considerable amount of space and be quite complex and expensive. It is important to note that although it is the power utility circuits that are involved, the equipment discussed herein must usually be provided and maintained by the water utility — the power user. The use of additional incoming power sources does increase the reliability of the station at the expense of more equipment and controls. Although this volume does not discuss the requirement of control power for operation of these breakers, it must be understood that some form of reliable energy must be available for this service. Usually the control power is derived from some mechanical stored-energy mechanism or battery.

Particular care in design, installation, and maintenance must be put into these incoming power-circuit facilities because their emergency function will be called upon at an unexpected moment, usually after a very long period of idleness. Water utility operators should arrange to test all service throwover equipment at least once a year to assure proper function. Protective relays should be current injected and tested for proper timing, sequence, and overall operation at least every three years.

User's Circuits

Beyond the power utility's metering equipment, the distribution system belongs to the user. Any duplication of the power utility's metering equipment must be done independently at the user's expense. Usually the water utility does not duplicate such metering facilities, but it accepts their accuracy and reliability because these characteristics can be verified by calibration.

The plant electrical system can accommodate many arrangements for the distribution of electric power to the various pieces of equipment at the plant site. Immediately after passing through the metering facilities, electricity is connected to a main bus. If there are two main power supplies, there may be two main buses, and if the two main buses can be interconnected, the main bus is referred to as a sectionalized main bus.

The simplest and most common electrical distribution system consists of a main bus with various branch circuits routed from the main to various loads. This type of distribution system is called a simple radial system. The name refers to the point source from which paths radiate outward. The radial system can be used with either the single or sectionalized main bus.

Another type of distribution system consists of a main bus with branch circuits that are connected at the outer ends so that a loop is formed. This is called a loop system and allows a particular load to be served from either of the two branch circuits. Usually a loop system has a number of sectionalizing switches arranged so that the loop may be opened or so that particular loads may be bypassed or taken out of service.

Other types of distribution systems are combinations of radial feeders and loops arranged to form a network. In such systems, there are various paths through which electric energy can flow to a particular load. Networks are generally not found in water utilities.

The user's circuits can take many forms, including any of the following:

- insulated single-conductor cables in conduit
- insulated multiconductor cables in conduit
- insulated single-conductor cables in cable trays or open raceways
- insulated multiconductor cables in cable trays or open raceways
- armored, insulated multiconductor cables either exposed or carried in cable trays
- underground cables of various types in buried duct banks
- direct buried underground cable of various types
- bundled single-conductor cables supported overhead
- preformed cables supported overhead
- open, uninsulated, bare-wire overhead construction

User's Equipment

In addition to the circuits involved, equipment will be provided for protecting and disconnecting each circuit. Since each path that electric energy must flow through to get from the main bus to any particular piece of equipment is a power circuit, there must be many protective and disconnecting devices on the distribution system. A fuse, for example, is a protective device, and a switch is a disconnect device. The fuse will cause the circuit to be disconnected if the current flow becomes greater than normal, whereas the switch is simply a manual means of opening the circuit.

It is natural to assume that in place of a switch and a fuse, a single device might be designed to accomplish both functions. The circuit breaker is such a device.

Circuit breakers have two current ratings: the normal current rating and an interrupting rating. The normal rating of a circuit breaker depends entirely on the size of the normal load that is connected to the circuit. The interrupting rating, however, depends on the size of the electrical system to which the circuit is connected. In one case, for example, a circuit breaker may be rated at 15 A continuous at 240 V and require an interrupting rating of 10,000 A. Yet, for an identical load at another location, the breaker would be rated at 15 A continuous at 240 V, but would require an interrupting rating of 75,000 A. Obviously, these two circuit breakers would not be identical, the latter being larger and more expensive. Similarly, a different shutoff valve would be used on a 1,000-psi (6,900-kPa) constant-pressure system than on a 50-psi (340-kPa) system.

The voltage of a distribution system is generally selected to allow the most economical size of circuit conductors and load equipment such as large horsepower motors. For example, a 37.5-kVA transformer at 208 V could sustain a continuous current flow of 104 A, and a 750-kVA transformer at 4,160 V could also have a current flow of 104 A. In both cases, the size of the cable could be the same — approximately a No. 2 AWG (American Wire Gauge) conductor. A 750-kVA transformer at 208 V could sustain a continuous current flow of 2,080 A, requiring several large cables. Similarly, 1,000 gpm (63 L/s) of water can be pumped through a 6-in. (152-mm) pipe under any configuration, but the power consumed would vary with the head in the system.

Distribution voltages in North American water utilities will generally be any of the following, depending on the size of loads and amount of space covered by the plant site: 15,000 V; 4,160 V; 2,400 V; 460 V; 240 V; 208 V; or 120 V.

The physical size of the equipment will also vary with the voltage — the higher the voltage, the larger the equipment. The combination of high-voltage circuits having high-current-interrupting requirements results in the necessary use of very large circuit breakers.

Generally, since a number of circuits are required, it is customary to purchase assemblies of circuit protective and disconnect equipment arranged for a number of circuits. Such assemblies are generally referred to as load centers, fuse panels, lighting panels, or distribution panels on systems of 600 V or less. Such assemblies for systems of more than 600 V are generally referred to as *switchboards* and *switchgear*. The term *switchboard* is seldom used anymore. The term *switchgear* has become the commonly used term for either type of equipment. However, manufacturers, the Institute of Electrical and Electronics Engineers, and the American National Standards Institute still differentiate between the two classifications of equipment. The differences are based on the type of construction, the overall fault current capacity and time duration of fault, and the type of breakers used.

It is important to recognize that one of the main purposes of load-center, distribution-type equipment is to disconnect automatically any faulted branch circuits as rapidly as possible in order to leave the rest of the branch circuits in operation. From the operator's standpoint, it is generally sufficient to have some means by which to monitor the condition of the branch circuit protective and disconnect devices. Usually on switchgear applications, red and green indicating lamps are used to give an indication of the last operating position assumed by the device.

For monitoring a distribution panel or a lighting panel, it is generally sufficient to glance at the position of the operating handles of the individual breakers. There is usually a clearly noticeable position assumed by the handle of a breaker that has tripped automatically. This position is easily distinguished from the handle position of a breaker that has been turned on or off manually (as discussed later in this section). To monitor a fuse box or panel, it is necessary to actually examine each fuse to determine its condition.

The following general classes of equipment are, for the most part, necessary items for the development of a plant distribution system.

Fuse Box. On small loads, such as single-family residences, for example, an incoming 120- or 240-V, single-phase power service is connected directly from the utility distribution system to a fuse box. This fuse box may have a main switch that, when opened, disconnects the power service and, when closed, connects the power service to the fuse-holder assembly. The fuse-holder assembly has several branch circuit fuses, each going to certain appliances or groups of appliances or to groups of lights. The fuse box, in its simplest form, contains fuses sized to prevent each branch circuit from becoming overloaded and provides a means of disconnecting the branch circuit by removing the fuse. Other types of fuse boxes not only have the fuse for each branch circuit, but also a knife-blade-type switch for use in connecting or disconnecting the branch circuit. When a fuse is blown as a result of an excessive load on its branch circuit, it is necessary to replace the fuse. The fuse box may be referred to as the fuse panel, or lighting panel, although it may serve loads other than lights.

Lighting Panel. Lighting panels are similar in function to fuse boxes and fuse panels; however, the term *lighting panel* most generally implies the use of molded-case circuit breakers instead of fuses. The lighting panel for a single-family residence would connect the single-phase branch circuits to the main 120- or 240-V incoming power service. Generally, there is a main breaker for disconnecting the branch circuit-breaker assembly from the power service, and the branch circuits will each serve an appliance, a group of appliances, or a group of lights.

There are two main advantages to the use of molded-case breakers as compared with fuses. Rather than requiring someone to replace a fuse, the circuit breaker can be reset. The other advantage of the breaker is that it is calibrated to trip at a certain overload that cannot be changed readily, as can a fuse, to a higher rating, thus relieving the danger from overloading branch-circuit wiring and devices.

Distribution Panel. The distribution panel performs a function similar to that of the fuse panel or the lighting panel. The main difference is that a distribution panel is designed to handle larger branch-circuit loads than normally found on lighting panels, and the distribution panel can supply three-phase circuits. The distribution panel may consist of fuses or molded-case circuit breakers and may supply either three-phase or single-phase branch loads. Distribution panels are generally rated to handle 240-V, 480-V, and up to 600-V incoming service and branch circuits.

The molded-case circuit-breaker-type distribution panel has the advantages described for the molded-case circuit-breaker-type lighting panel. The molded-case circuit breakers have a handle that assumes three positions: one when the breaker is opened manually, another when the breaker is closed manually, and a third intermediate position when the breaker trips automatically. The current rating of the breaker is normally printed on the side of the operating handle.

Distribution panels are generally constructed in rectangular wall-mounted boxes with hinged front covers. The branch circuit breakers are generally arranged in two rows, with breakers stacked one above the other. Each breaker is generally provided with a nameplate or a number. Where breakers are numbered, a directory is usually mounted inside the front cover. The directory describes the branch circuit load for each breaker by reference to the breaker number.

Switchboard. The switchboard performs the same general functions as the distribution panel. The main difference is in the type of construction. Generally, the switchboard is a free-standing panel with switches, fuses, or circuit breakers mounted on the panel. Old-style switchboards frequently had the open-type knife switches and fuses mounted on the front of the panel. Newer-type switchboards are generally constructed with a "dead front," using molded-case circuit breakers mounted on the back of the panel, with only their operating handles extending through the panel to the front. Switchboards are generally employed only at voltages less than 600 V. The term *switchboard* has been used very broadly and in some cases, such as in power plants, refers to a compete control and instrument panel. Generally this term does not imply such broad coverage when used with reference to water utilities.

Switchgear. Switchgear is a general term covering an assembly of large switching equipment. Generally, the term *switchgear* applies to switching equipment of higher voltage and current ratings than that used in the assembly of a distribution panel. Switchgear usually applies to equipment rated from 600 V through 15,000 V and for current ratings of several hundred to thousands of amperes. Power-circuit breakers are used for connecting and disconnecting the main branch circuits, and relays, instruments, and control devices are used for sensing conditions and directing the operations of the breakers. Old-style switchgear in many cases consisted of lever-operated mechanisms arranged for manual operation. Modern switchgear is electrically controlled and arranged for manual operation by means of a pushbutton or control switch.

Equipment to Measure Electricity

Devices for measuring electricity include a wide range of equipment and apparatus for the specific purpose of metering and control. The sizes of measurements can vary from the detection of a slight charge and millionths of an ampere to extremely high voltages and currents.

Voltmeter

A voltmeter is an instrument for measuring voltage. Its scale may be graduated in microvolts, millivolts, volts, or kilovolts. A microvolt is one one-millionth of a volt, a millivolt is one one-thousandth of a volt, and a kilovolt is 1,000 V. On three-phase power systems, voltmeters are usually used in combination with a voltmeter switch. The voltmeter switch switches the voltmeter to read the voltage between the conductors of phase 1 and phase 2, phase 2 and phase 3, or phase 3 and phase 1. On three-phase four-wire systems, it is common to have the selector switch provide for additional voltage measurements from each individual phase to neutral. Voltmeters are not generally constructed to measure voltages above 1,000 V. When it is desired to measure voltages above 1,000 V, it is necessary to use a voltage transformer, commonly called a potential transformer, or PT.

Ammeter

An ammeter is an instrument for measuring the amount of current. Its scale is generally graduated in microamperes, milliamperes, or amperes. A microampere is one one-millionth of an ampere. A milliampere is one-thousandth of an ampere. On a three-phase system, an ammeter switch is frequently used with an ammeter. The ammeter switch switches the ammeter to read the current in phase 1, phase 2, or phase 3. Ammeters are generally constructed to measure current only on low-voltage circuits, not more than 600–1,000 V. When it is desired to measure the current on high-voltage circuits, it is necessary to use current transformers, commonly called CTs.

Wattmeter

A wattmeter is an instrument for measuring the amount of real power in watts. Its scale is generally graduated in watts, kilowatts, or megawatts. A kilowatt is 1,000 W. A megawatt is 1,000,000 W. Wattmeters are not designed to measure power on high-voltage circuits. They are designed to operate at a maximum voltage of 120 V and a maximum current of 5 A. On high-voltage circuits, it is therefore necessary to use potential transformers and current transformers with 120-V and 5-A secondary windings, respectively. It should be remembered that determining how much work is done must take time into account. The wattmeter reading on a power service circuit at any given instant is the instantaneous demand.

Demand Meter

A demand meter is an instrument that measures the average power (rate of doing work) of a load during some specific time interval, such as a 5-, 15-, 30-, or 60-min period. Generally, contracts with electric utilities for the purchase of electric power have demand charges based on 15- or 30-min time intervals. The demand is usually spoken of, for example, as a 15-min demand or a 30-min demand. Both the wattmeter and the demand meter are graduated to indicate watts or kilowatts.

An *indicating wattmeter* shows the immediate and instantaneous demand. A *recording wattmeter* records the instantaneous demand. The average of the instantaneous demands over a particular time interval would be equivalent to a demand-meter indication for that same time interval. The maximum demand established during a month is the maximum average load developed over the specified demand time interval, not the peak instantaneous demand. Power bills usually are submitted each month showing, in addition to the charges, the kilowatt-hours of energy consumed and the maximum demand applicable for that month. One word of caution: The demand shown on a power bill for a particular month does not necessarily mean that that demand occurred that month because some contracts are based on the premise that a new maximum demand one month will apply to each of the following 11 months. Power utilities call this a "ratchet" charge.

Demand meters may be of the indicating type, with a pointer and scale arranged so that the pointer remains at its maximum reading until reset to zero when the meter is read. Other types of demand meters may be printing types or recording types. Generally, demand meters are furnished only on the incoming power service feeders to a plant site.

Watt-Hour Meters

A watt-hour meter is used for measuring watt-hours or kilowatt-hours of electric energy. A kilowatt-hour is 1,000 watt-hours. Watt-hour meters are generally provided only on the incoming power feeder; however, it is not uncommon to find them used on large and more important plant motors and loads.

The kilowatt-hour meter generally has a register consisting of four dials, each dial graduated from 0 to 9. The first (rightmost) dial reads clockwise, the second reads counterclockwise, the third reads clockwise, and the fourth (leftmost) dial reads counterclockwise. Adjacent pointers move in opposite directions. Dials are read from left to right, using the figure over which the pointer has passed last. The meter will have a dial register constant (10, for example) that is multiplied by the meter reading to provide the reading in kilowatt-hours.

On most kilowatt-hour meters, a rotating disk can be observed through the glass front or through a small window in the front of the meter case. Revolutions of the disk can be counted by means of a black mark on the disk. The speed of the disk is at any instant proportional to the kilowatt load; it provides a quick and accurate means of measuring kilowatts over short periods of time. The meter will also have a disk constant. This constant expresses the watt-hours (or watt-seconds) per revolution of the meter disk. To measure kilowatt load, determine with a stopwatch the number of seconds required for the disk to make 5 or 10 revolutions, depending on the disk speed. Then, knowing the disk constant, use the following formula to calculate the kilowatt load:

$$\text{kilowatts} = \frac{\text{disk–watt hours constant} \times \text{revolutions} \times 3{,}600}{\text{seconds} \times 1{,}000}$$

When the disk constant is in watt-seconds, you must convert to watt-hours by dividing it by 3,600 for use in the above formula.

You can obtain a great amount of information about a plant load by observing the kilowatt-hour meter. If all the plant motors and lighting circuits are turned off, any rotation of the disk would most likely represent the power required to magnetize the lighting and auxiliary-power transformers. The kilowatt load determined under this condition would be the magnetizing power requirement of the system. Then by turning on all of the plant lights and determining the kilowatt load again, you could determine the amount of lighting load by subtracting the magnetizing power load from this latter kilowatt load.

Similarly, by adding various increments of the plant load in succeeding steps, you could determine the kilowatt load imposed by the various plant auxiliaries.

Varmeter

A varmeter is an instrument for measuring the amount of reactive (wattless) power in vars. Its scale is generally graduated in vars or kilovars. A kilovar is 1,000 vars. Varmeters are not designed to measure power on high-voltage circuits. Varmeters, like wattmeters, are designed to operate on potential and current transformers at maximum voltages of 120 V and a maximum current of 5 A. Generally, varmeters are not found in water utilities unless there are some large synchronous motors at the site.

Power-Factor Meter

Power factor is the ratio of the amount of real power to the total apparent power. Apparent power includes both real and reactive power. A power-factor meter is an instrument that measures this ratio directly. The power-factor meter is generally graduated such that 1.0 power factor is in the center of the scale, with lagging (inductive) power factor values to the left of the 1.0, and leading (capacitive) power factor values to the right of the 1.0. A typical scale would read (from right to left) as: 0.6, 0.7. 0.8, 0.9, 1.0, 0.9, 0.8, 0.7, and finally, 0.6. Thus, a reading of 0.75 to the left of the 1.0 would indicate a lagging power factor of 0.75. Adding power factor capacitors would bring this reading closer to 1.0.

Generally, a water treatment plant with the usual complement of induction motors will have a lagging power factor of approximately 0.85. There is not much that a water-utility plant operator can do to change the power factor or the var flow at a station. These values are quite well determined by the design of the station.

Frequency Meter

A frequency meter is an instrument for measuring the frequency of an AC system. Frequency meters require potential transformers (PTs) for connection to circuits operating above 1,000 V. The frequency of a system depends on the speed of the generators. If the plant operates on an engine generator set, the frequency will be adjusted by adjusting the governor on the engine generator set.

Electronic Metering

Modern water treatment plants may have microprocessor-based meters that combine any or all of the listed electrical meters into a single integrated meter device. The electronic meters use the standard PTs and CTs for the metering signal inputs. The meters electronically calculate, display, and record all the various electrical parameters, such as voltage, current, power, vars, kW·h, etc. Some electronic meters also provide measurement, recording, storage, and latter analysis of several cycles of incoming power when high-voltage spikes (transients) occur. This can help a plant operator to identify and correct problems in the plant power system. Similar electronic metering can also be used for large motors. These meters provide motor temperature measurements, motor overload versus time functions, and other protective relays functions, such as reverse-phase and undervoltage protection.

Equipment to Convert Electricity Into Other Forms of Energy

Electric energy can be converted into many more purposeful types of energy to provide particular services.

Mechanical Energy

The rotating shaft, associated with various types of motors, is perhaps one of the most common forms of electric-to-mechanical energy conversion. Another widely used form of energy is the linear thrust, the pull or push as developed by a solenoid thruster, commonly used for operating a solenoid valve or a doorbell.

Heat Energy

Electric heaters, together with their thermostatic controls, are employed for practically any heating requirement. Electric heaters are in demand because of their cleanliness, convenience, flexibility, accuracy, safety, and in many cases their economy. Other types of special equipment, such as welders and industrial equipment, make use of the direct conversion of electric energy to heat.

Light Energy

Perhaps no other form of conversion is as well known as the conversion to light. The introduction of electric lighting generally preceded all other appliances and equipment.

Chemical Energy

Electrical equipment is used for corrosion-protection equipment. Such equipment, referred to as cathodic-protection equipment, influences the degree and rate of the chemical reactions that take place in the oxidation, rusting, or decomposition that occurs on and within metal structures.

Radio Energy

Electromagnetic waves derived from electrical equipment provide the carrier, commonly referred to as the radio, over which the well-known forms of communication presently take place.

Equipment to Protect Other Electrical Equipment

Certain types of electrical equipment are designed and used only for protecting other types of electrical equipment. Even though these types of equipment are actually nonproductive, the operator should know what they are for, and they should be serviced and maintained for proper operation.

An associated function of protective equipment is to provide personnel safety. Electrical power equipment can be dangerous if protective devices are not properly maintained in service or if their functions are bypassed. If a protective device functions, it means there is a problem in the electrical system that could pose a safety hazard. The size, rating, or adjustment of a protective device should never be changed without consulting a qualified electrical engineer or electrician.

Fuse

A fuse is an overcurrent protective device containing a special metal link that melts when the current through it exceeds some rated value for a definite period of time. Fuse links are made of a special alloy that melts at a relatively low temperature. A fuse is inserted in series with the circuit so that it opens the circuit automatically during an overload and thereby prevents excessive current from damaging other parts of the circuit. Fuses are rated according to voltage, continuous current, interrupting capacity, and speed of response.

The *voltage rating* of a fuse should be equal to or greater than the voltage of the circuit on which it is applied. This voltage is not a measure of its ability to withstand the voltage while carrying current. Rather, it is the ability of the fuse to prevent the open-circuit voltage from restriking and establishing an arc once the fuse link has melted.

The *current rating* of the fuse should be equal to or slightly larger than the current rating of the circuit or device that it protects. The interrupting capacity of a fuse must be such that it can interrupt the inrush of current available during a fault as determined not by the load but by the current capacity of the source of power being supplied through the fuse.

The *speed of response* simply defines how fast the fuse link will melt. Obviously, the faster the fuse link will melt, the lower will be the limit that the fault current can reach before the circuit is opened. Fuses expressly designed for exceedingly fast response are referred to as current-limiting fuses.

Molded-Case Circuit Breaker

A circuit breaker is an electromechanical overcurrent device that opens a circuit automatically when the current rises in excess of a predetermined value. It can be reset by operating a lever to its original position, and it can be used over and over again. There are thermal types, magnetic types, and thermal-magnetic types of circuit breakers.

A *thermal circuit breaker* responds only to temperature change in a bimetallic element. This element is made of two strips of different metals, bonded together. The lengths of the strips increase with temperature — but not equally — so that the composite element bends more and more as its temperature rises. Current flows through the bimetallic element and generates heat. The greater the current, the higher the temperature of the element. The mechanism is adjusted so that the bimetallic element bends just enough to open its contacts at a predetermined value of current. Because this type of breaker must respond to heating, its speed of response may be quite slow.

A *magnetic circuit breaker* responds only to current, and then only if the current is sufficiently great to attract a movable pole-piece by magnetic force. Movement of the pole piece causes a contact to open. In the fully magnetic breaker, there is only one actuating element, the magnetic coil assembly. Its operation is independent of temperature. Because this type of breaker responds only to the magnitude of current, its speed of response may be very fast.

The *thermal-magnetic breaker* operates in exactly the same way as the thermal breaker at low values of current, and in exactly the same way as the magnetic circuit breaker at high values of current. Thus, it essentially combines the good features of both. Slight overloads that persist for a long time will be disconnected by the thermal element, whereas short circuits will be disconnected immediately by the magnetic element.

While molded-case breakers can be reset and reused, this applies only to normal overloads. Any time a molded-case breaker has been subject to a full short-circuit fault current, its proper calibration and operation may be impaired. A breaker that has been subject to severe short-circuit current should be replaced.

Power-Circuit Breaker

A power-circuit breaker differs greatly from the molded-case circuit breaker in that it is much larger and does not have the means for determining when it should operate or what it should do to carry out its protective assignment. Power-circuit breakers generally depend on relays to sense abnormalities and to initiate their operation. The advantage of the power-circuit breaker is its ability to interrupt exceedingly high values of fault current on both low- and high-voltage power circuits. Power-circuit breakers are electromechanical devices used for switching (opening and closing) power circuits under either normal or fault conditions. Power-circuit breakers may be either oil-filled or air-insulated mechanisms. Generally, a row of power-circuit breakers enclosed in a metal housing is referred to as *metal-clad switchgear*.

Protective Relays

A relay is a device activated by either an electrical or physical condition to cause the operation of some other device in an electrical circuit. Although there are all sorts of relays, the protective relay's principal function is to protect power service from interruption and to prevent or limit damage to equipment.

Lightning Arresters

Lightning can produce the most destructive of all types of overvoltages. Overhead power lines exposed to lightning can be elevated to a potential of several million volts by direct stroke and possibly one-half million volts by induction from a near strike in a time interval of only a few millionths of a second. This impressed voltage, called a voltage surge, tends to travel over the complete circuit in the form of steep-front voltage wave. Since power lines and equipment cannot, from a practical standpoint, be insulated to withstand such overvoltages, it is obvious that a breakdown of insulation will occur between the phase conductors and ground. The lightning arrester is designed to break down, thus allowing the other circuit elements to remain in service.

A lightning arrester is a protective device for limiting surge voltages on equipment by discharging a surge current to ground; however, it is designed to prevent a continued flow of follow current after the voltage has returned to normal, and it is capable of restoring itself and repeating the same protective function. The types and ratings of arresters are selected for the particular voltage class and degree of protection desired. Arresters may abe classified as follows:

- station class: for 60,000–276,000-V systems, pad-mounted
- intermediate class: for 3,000–121,000-V systems, pad-mounted
- distribution class: for 3,000–18,000-V systems, pole-mounted
- pellet type: for 1,000–15,000-V systems, pole-mounted
- AC rotating machine: capacitor and arrester combinations for 2,400–13,000-V equipment
- secondary type: for ratings from 0 to 600 V.

With the installation of more and more solid-state control equipment or computers in water treatment plants, there is a need for higher performance protection from power system surge voltages. The traditional lightning protectors provide adequate protection for most electrical power equipment and motors, but other plant circuits now need an advanced form of protection in the form of transient voltage surge suppressor (TVSS) devices. TVSS devices provide protection on low-voltage (less than 600 V) systems. Typical TVSS devices now incorporate a hybrid circuit comprised of metal oxide varistors (MOVs), silicon avalanche diodes, and gas-filled voltage break-over types. This provides a voltage clamping action adequate to protect the sensitive solid-state circuitry. The service entrance equipment at a treatment plant will be fitted with heavy duty IEEE Class C TVSS units. Feeder panels use Class B units, and the branch circuit panels use the smallest Class A suppressors. The differences between the classes is found in the maximum transient energy dissipation capabilities.

Rating Electrical Materials and Equipment

To ensure safe and efficient operation of an electrical system, materials and equipment must be chosen with the proper rating. This is usually the function of the designing engineer, but the operator should be familiar with the factors taken into consideration.

Materials

All materials fall within two general classifications regarding their properties and abilities to allow electric current to flow through them. They are either conductors or insulators. A conductor is a substance that permits the flow of electricity, especially one that conducts electricity with ease. An insulator is a substance that offers very great resistance or hindrance to the flow of electric current.

The path through which current flows has to be a carefully designed circuit from start to finish. This path is made up of a combination of two classes of materials — conductors and insulators.

The windings in a generator are made of copper conductors, but they are wrapped in insulating materials such as dry paper, mica, glass, or rubber so that electricity can flow only in a controlled path. Commercial power is usually generated at 15,000 V and is then transformed to still higher voltage levels for transmission and distribution. The transmission lines usually consist of overhead pole construction using bare copper or aluminum cable for conductance and glass or porcelain and air for insulation. The separation of the conductors is great enough that a flashover between the lines in open air will not occur. At a user's substation, it may be found that the overhead bare wires connect directly to the insulated bushings of a power transformer. The output of the transformer will probably be at lower voltage and connected directly to power cables. Power cables are made up of conductors covered by insulating material such as rubber, varnished cambric, or paper with a lead cover. Modern power cables also use various thermoplastic or thermosetting plastic compounds, such as moisture- and heat-resistant cross-linked synthetic polymer. At each and every point in the electrical system, conductors are provided to keep the flow of current on its intended path, and insulating material is provided everywhere along this path to contain the current.

A good conductor has low resistance and is said to have high conductance. Conductance is just the inverse of resistance and is measured in mhos, which are ohm(s) spelled backward.

A good insulator has high resistance, which is measured in ohms. Insulation is also called *dielectric*. An insulated cable with high dielectric strength is another way of saying the cable is very well insulated.

Following is a list of materials categorized by their ability to conduct electricity.

Good Conductors	Fair Conductors	Poor Conductors	Good Insulators
Silver	Charcoal	Water	Slate
Copper	Carbon	The body	Oils
Aluminum	Acid solutions	Flame	Porcelain
Zinc	Sea water	Linen	Dry paper
Brass	Saline solutions	Cotton	Rubber
Iron	Metallic ores	Wood	Mica
Tin	Vegetation	Fibers	Glass
Lead	Moist earth	Marble	Dry air

Usually, any fault found on an electrical system or a piece of electrical equipment will be either a broken conductor or a breakdown of the insulation. Conductors usually break because of mechanical failure as a result of excessive vibration or movement. Insulation usually fails because of deterioration as a result of excessive heating and aging. Overheating is probably the most common cause of insulation breakdown. Insulation also fails whenever it is subjected to abnormally high voltage, as will occur when it is struck by lightning or subjected to switching surges. Whenever a power switch is opened or closed, surges in system voltage will occur, just as water hammer occurs when a valve is operated quickly.

Equipment

Electrical systems and equipment can be designed for any voltage, any frequency, and any size, but most are standardized. Imagine the deplorable condition that would exist if each of the states within the United States had its own standard voltages and frequencies and its own unique equipment sizes. For example, your 120-V, 60-Hz power tools would be useful only in your state, whereas in a neighboring state the power tools might be rated 300-V, 40-Hz, or 80-V, 100-Hz. Moreover, motors might be manufactured in different watt or horsepower sizes by different companies. There was a period when such a lack of standardization did exist; during that time, equipment was designed at various ratings. There are still systems in operation that use 25 Hz instead of 60 Hz, and other-than-standard voltage classes of equipment are used in some cases.

Standardization of voltage ratings, system frequency, and sizes of equipment has permitted the mass production of equipment at competitive prices, so necessary for the rapid industrial development and expansion that has occurred within the past several decades. This is not to say that the evolved standards were necessarily the best, but that they have become recognized standards for general usage. A knowledge of some of the standard voltage classifications, equipment ratings, and sizes makes it easier to become familiar with a particular system or specific equipment.

Voltage Standards Used in North America

There are two terms that are generally used when expressing voltage: *rated voltage* and *nominal voltage*. Rated voltage is used with reference to the operating characteristics of equipment. Nominal voltage is used in referring to an electrical system that distributes electricity for ultimate use by other equipment. Rated voltage is not applicable here because various pieces of equipment that make up a given system may have different voltage ratings.

For example, a standard system is one having a nominal voltage of 480 V, but the motor used on that system will have a standard rated voltage of 460 V, the motor controller or motor starter may be rated 460 V, and a disconnect switch used on that motor circuit may be rated 600 V.

Table E3-1 shows the standard nominal system voltages and the generally accepted ranges considered as favorable and tolerable within which power may be delivered at each of these nominal voltage levels.

Obviously people may be quite confused when they hear reference to so many voltages in the discussion of a particular electrical system — as, for example, the reference to a 236-V fluorescent lamp ballast, a 220-V motor, a 230-V capacitor, and 240-V switchgear on a specific 240-V system.

Table E3-2 lists only some of the standard nominal voltages and rated voltages for AC equipment, and clearly shows why so many expressions of voltage level may be used.

Reference to any one of the above voltage levels for a particular type of equipment naturally implies the use of the other related voltages for the various associated items of equipment used on that same electrical system.

Current Standards

Because the amperes of electric current vary with the size of electric load and the voltage, there is no standard classification of system nominal

TABLE E3-1 North American standard system voltages, in volts

Minimum Tolerable	Minimum Favorable	Nominal System	Maximum Favorable	Maximum Tolerable	Type (Phase) of System
107	110	120	125	127	1
200	210	240	240	250	3
214/428	220/440	240/480	250/500	254/508	1
244/422	250/434	265/460	227/480	288/500	3
400	420	480	480	500	3
2,100	2,200	2,400	2,450	2,540	3
3,630	3,810	4,160	4,240	4,400	3
6,040	6,320	6,900	7,050	7,300	3
12,100	12,600	13,200	13,800	14,300	3
12,600	13,000	14,400	14,500	15,000	3
30,000		34,500		38,000	3
60,000		69,000		72,500	3
100,000		115,000		121,000	3
120,000		138,000		145,000	3
140,000		161,000		169,000	3

TABLE E3-2 North American standard nominal voltages

Nominal System	Generator Rated	Transformer Secondary	Switch-gear Rated	Capacitor Rated	Motor Rated	Starter Rated	Ballast Rated
Single-Phase Systems							
120	120	120	120	—	115	115	118
120/240	120/240	120/240	240	230	230	230	236
208/120	208/120	208/240	240	230	115	115	118
Three-Phase Systems							
240	240	240	240	230	240	220	236
480/277	480/277	480/277	480	460	460	440	460
480	480/277	480/277	480	480	460	440	460
2,400	2,400/1,388	2,400	2,400	2,400	2,300	2,300	—
4,160	4,160/2,400	4,160/2,400	4,160	4,160	4,000	4,000	—
6,900	6,900/3,980	6,900/3,980	7,200	6,640	6,600	6,600	—
7,200	6,900/3,980	7,200/4,160	13,800	7,200	7,200	7,200	—
12,000	12,500/7,210	12,000/6,920	13,800	12,470	11,000	11,000	—
13,200	13,800/7,970	13,800/7,610	13,800	13,200	13,200	13,200	—
14,400	14,000/8,320	13,800/7,970	14,400	14,400	13,200	13,200	—

TABLE E3-3 Current ratings for low-voltage switches, in amperes

120/240 V	230 V	240 V	600 V
30	30	30	30
60	60	60	60
100	100	100	100
200	200	200	200
	400	400	400
	600	600	600
	800	800	800
	1,200	1,200	1,200

currents as there is for system nominal voltage levels. Only the rated current of the equipment is standardized.

However, for each type of equipment at rated voltage level, there is a set of standard current ratings. Examples of some standard current ratings of equipment are shown in Table E3-3. This does not mean, however, that other ratings are not available — only that they will probably be more expensive than the standards.

The following descriptions are of several types of apparatus that are used to protect systems from electric current overload.

Safety Switches. A safety switch is a disconnect switch. It will have a voltage rating, which means it can be used on systems up to and including its rated voltage. Its current rating means it will carry up to its rated current, regardless of whether or not the voltage of the system is at rated voltage or some lower voltage. From a practical standpoint, the electrical industry has standardized the eight current ratings shown in Table E3-3 for low-voltage switches.

Fuses. Fused disconnect switches are essentially safety switches having a fuse holder and fuse built in series with the switch. Standard, fused, disconnect-switch current ratings are therefore the same as the safety-switch ratings listed in Table E3-3. A fuse holder can accommodate any size fuse below the fuse-holder rating. Standard current ratings of low-voltage fuses are 15, 20, 25, 30, 35, 40, 45, 50, 60, 70, 80, 90, 100, 110, 125, 150, 200, 300, 400, 500, and 600 A. Where fuses larger than 600 A are required, two or more of these fuses can be used in parallel; or a high-capacity fuse, in ratings from 650 to 5,000 A, may be more appropriate. The standard fuses rated 0–600 A have an interrupting rating of only 13,000 A, which is barely equal to the short-circuit or fault current that can be let through a 250-kVA, 200-V transformer. Obviously, whenever a fuse is being selected, not only the load current rating, but also the interrupting current rating of the fuse must be taken into account. The high-capacity-type fuses have current interrupting ratings in the range of 100,000 to 200,000 A. When high-capacity interrupting rated fuses are used, special clips are installed in the fuse holders to permit only the proper fuse type and reject fuses with lower interrupting ratings.

Circuit Breakers. Circuit breakers can do two things: open a circuit automatically when an overload occurs and open a circuit when operated manually. Circuit breakers can be arranged to be operated intentionally, either manually or electrically. The size of a circuit breaker depends on the voltage of the circuit and on the amount of current it must handle under both normal and abnormal conditions. Under normal conditions, the circuit breaker must carry the rated current continuously without overheating. Under abnormal conditions (that is, when the breaker carries the current during a fault or short circuit), the circuit breaker must be able to withstand that amount of intense fault current until it has successfully interrupted that current flow.

Molded-Case Circuit Breakers. Molded-case circuit breakers are used on low-voltage systems, 600 V and less, in locations that would otherwise be held by a safety switch, disconnect switch, fused disconnect switch, or fuse. Single-pole circuit breakers are commonly used in lighting panels. Three-pole circuit breakers are commonly used in distribution panels. Two-pole circuit breakers may be used in either lighting panels or distribution panels. Circuit breakers have a voltage rating, a frame-size current rating, a continuous current rating, and a fault current ampere interrupting rating.

Functionally, circuit breakers serve as disconnecting means or as protective devices. Compared with switches, they have a much higher interrupting capacity. This means that they are capable of interrupting the flow of considerably higher current than could be interrupted by a switch. The size of a circuit breaker is based not only on the maximum continuous current it must carry, but also on its interrupting ability. The interrupting capacity of the breaker must be at least as great as the largest fault current that the power system can cause to flow.

Molded-case circuit breakers are available in certain frame sizes that basically correspond to the maximum continuous current rating of that breaker. A breaker of a certain frame size may be calibrated to carry a continuous current that is only a fraction of the maximum continuous current that the breaker can carry. Such a continuous-current calibration is referred to

TABLE E3-4 Molded-case circuit-breaker design

Frame Size, *A*	Interrupting Capacity, *A*	Continuous Current Ratings, *A*
(240 V AC)		
400	50,000	70, 90, 100, 125, 150, 175, 200, 225, 250, 300, 350, 400
800	50,000	125, 150, 175, 200, 225, 250, 300, 350, 400, 500, 600, 700, 800
1,000	75,000	125, 150, 175, 200, 225, 250, 300, 350, 400, 500, 600, 700, 800, 900, 1,000
(480 V AC)		
100	15,000	5, 8, 10, 12, 15, 20, 25, 30, 35, 40, 50, 70, 90, 100
225	25,000	70, 100, 125, 150, 175, 200, 225
400	35,000	70, 90, 100, 125, 150, 175, 200, 225, 250, 300, 350, 400
800	35,000	125, 150, 175, 200, 225, 250, 300, 350, 400, 500, 600, 700, 800
1,000	40,000	125, 150, 175, 200, 225, 250, 300, 350, 400, 500, 600, 700, 800, 900, 1,000

as the rated-current setting of the breaker. For example, a 100-A frame-size breaker may be adjusted to a rating of 10 A. This means that the breaker will carry up to 10 A continuously. Above 10 A, the breaker would eventually open; how soon would depend on how much the current exceeds the rated current. Currents up to approximately 10 times the rated current will cause the breaker's thermal element to open the breaker. Currents exceeding 10 times the rated current will cause the breaker's magnetic element to open the breaker.

Some of the more common ratings of molded-case circuit breakers are given in Table E3-4.

Low-Voltage Air Circuit Breakers. When breakers with continuous- or interrupting-current ratings greater than those available for molded-case circuit breakers are required, low-voltage air circuit breakers are used. These breakers are mechanical devices, completely metal enclosed and frequently of drawout-type construction for convenience of service and maintenance. An assembly of low-voltage circuit breakers is generally called a lineup of low-voltage switchgear. Low-voltage air circuit breakers have voltage ratings,

TABLE E3-5 Air circuit-breaker design

Frame Size, *A*	Interrupting Capacity, *A*	Continuous Current Ratings, *A*
(240 V AC)		
225	30,000	15, 29, 30, 40, 50, 70, 90, 100, 125, 150, 175, 200, 225
600	50,000	40, 50, 70, 90, 100, 125, 150, 175, 200, 225, 250, 300, 350, 400, 500, 600
1,600	75,000	200, 225, 250, 300, 350, 400, 500, 600, 800, 1,000, 1,200, 1,600
(480 V AC)		
225	25,000	15, 20, 30, 40, 50, 70, 90, 100, 125, 150, 175, 200, 225
600	35,000	40, 50, 70, 90, 100, 125, 150, 175, 200, 225, 250, 300, 350, 400, 500, 600
1,600	60,000	200, 225, 250, 300, 350, 400, 500, 600, 800, 1,000, 1,200, 1,600

frame-size current ratings, continuous-current ratings, and fault-current interrupting ratings similar to those of the molded-case circuit breakers.

Some of the more common ratings of low-voltage air circuit breakers are given in Table E3-5.

High-Voltage Circuit Breakers. High-voltage circuit breakers are used on electrical systems operating above 600 V. These breakers may be oil-filled, contact-type air or modern vacuum bottle breakers. An assembly of high-voltage circuit breakers is generally referred to as a lineup of high-voltage switchgear. Modern high-voltage switchgear for use up to 15,000 V is generally composed of air or vacuum bottle circuit breakers of the drawout type, completely metal enclosed. It is generally an accepted practice to refer to the interrupting capacity of high-voltage switchgear in terms of the apparent power (megavolt-amperes, MVA) that can be interrupted during a fault instead of the fault current in amperes. One MVA represents 1,000,000 volt-amperes. For example, a fault current of 1,000 A on a 2,400-V, single-phase system would represent 2,400,000 VA, or 2.4 MVA. High-voltage switchgear circuit breakers have a voltage rating, a frame-size current rating, a fault-current interrupting rating, and a fault MVA rating. Typical ratings are given in Table E3-6.

TABLE E3-6 High-voltage switchgear design

System Nominal Voltage, *V*	Frame Size, *A*	Momentary Interrupting Capacity, *A*	Interrupting Capacity, *MVA*
2,400	1,200	20,000	30
4,160	2,000	40,000	150
4,160	2,000	60,000	250
4,160	1,200	80,000	350
4,160	2,000	80,000	350
7,200	1,200	70,000	500
7,200	2,000	70,000	500
13,800	1,200	20,000	150
13,800	2,000	35,000	250
13,800	1,200	40,000	500
13,800	2,000	40,000	500
13,800	1,200	60,000	750
13,800	2,000	60,000	750
13,800	1,200	80,000	1,000

Other breakers at higher voltages are available through all voltage classes; however, water utility operators will seldom be required to operate breakers above the 15-kV (15,000-V) class of equipment.

It should be noted that continuous (tripping) current ratings are not listed for the high-voltage breakers. These breakers, in contrast to the molded-case breakers and low-voltage air current breakers, do not have any built-in means of sensing current magnitudes and therefore cannot be individually set at specific current ratings. However, auxiliary current-measuring transformers (current transformers) and protective relays can be set to cause the breakers to operate at any desired magnitude of current.

Basic Science Concepts and Applications

Appendixes

APPENDIX A

Conversion Tables

TABLE A-1 Conversion factors

Conversions		Procedure			Approximations (Actual answer will be within 25% of approximate answer.)
From	To	Multiply number of	by	To get number of	
acres	hectares (ha)	acres	0.4047	ha	1 acre ≈ 0.4 ha
acres	square feet (ft^2)	acres	43,560	ft^2	1 acre ≈ 40,000 ft^2
acres	square kilometers (km^2)	acres	0.004047	km^2	1 acre ≈ 0.004 km^2
acres	square meters (m^2)	acres	4,047	m^2	1 acre ≈ 4,000 m^2
acres	square miles (mi^2)	acres	0.001563	mi^2	1 acre ≈ 0.0015 mi^2
acres	square yards (yd^2)	acres	4,840	yd^2	1 acre ≈ 5,000 yd^2
acre-feet (acre-ft)	cubic feet (ft^3)	acre-ft	43,560	ft^3	1 acre-ft ≈ 40,000 ft^3
acre-feet (acre-ft)	cubic meters (m^3)	acre-ft	1,233	m3	1 acre-ft ≈ 1,000 m^3
acre-feet (acre-ft)	gallons (gal)	acre-ft	325,851	gal	1 acre-ft ≈ 300,000 gal
centimeters (cm)	feet (ft)	cm	0.03281	ft	1 cm ≈ 0.03 ft
centimeters (cm)	inches (in.)	cm	0.3937	in.	1 cm ≈ 0.4 in.
centimeters (cm)	meters (m)	cm	0.01	m	—
centimeters (cm)	millimeters (mm)	cm	10	mm	—
centimeters per second (cm/s)	meters per minute (m/min)	cm/s	0.6	m/min	—
cubic centimeters (cm^3)	cubic feet (ft^3)	cm^3	0.00003531	ft^3	1 cm^3 ≈ 0.00004 ft^3
cubic centimeters (cm^3)	cubic inches ($in.^3$)	cm^3	0.06102	$in.^3$	1 cm^3 ≈ 0.06 $in.^3$
cubic centimeters (cm^3)	cubic meters (m^3)	cm^3	0.000001	m^3	—
cubic centimeters (cm^3)	cubic yards (yd^3)	cm^3	0.000001308	yd^3	1 cm^3 ≈ 0.0000015 yd^3
cubic centimeters (cm^3)	gallons (gal)	cm^3	0.0002642	gal	1 cm^3 ≈ 0.0003 gal
cubic centimeters (cm^3)	liters (L)	cm^3	0.001	L	—

Table continued next page

TABLE A-1 Conversion factors (continued)

Conversions		Procedure			Approximations (Actual answer will be within 25% of approximate answer.)
From	To	Multiply number of	by	To get number of	
cubic feet (ft^3)	acre-feet (acre-ft)	ft^3	0.00002296	acre-ft	1 $ft^3 \approx$ 0.00002 acre-ft
cubic feet (ft^3)	cubic centimeters (cm^3)	ft^3	28,320	cm^3	1 $ft^3 \approx$ 30
cubic feet (ft^3)	cubic inches ($in.^3$)	ft^3	1,728	$in.^3$	1 $ft^3 \approx$ 1,500 $in.^3$
cubic feet (ft^3)	cubic meters (m^3)	ft^3	0.02832	m^3	1 $ft^3 \approx$ 0.03 m^3
cubic feet (ft^3)	cubic yards (yd^3)	ft^3	0.03704	yd^3	1 $ft^3 \approx$ 0.04 yd^3
cubic feet (ft^3)	gallons (gal)	ft^3	7.481	gal	1 $ft^3 \approx$ 7 gal
cubic feet (ft^3)	kiloliters (kL)	ft^3	0.02832	kL	1 $ft^3 \approx$ 0.03 kL
cubic feet (ft^3)	liters (L)	ft^3	28.32	L	1 $ft^3 \approx$ 30 L
cubic feet (ft^3)	pounds (lb) of water	ft^3	62.4	lb of water	1 $ft^3 \approx$ 60 lb of water
cubic feet per second (ft^3/s)	cubic meters per second (m^3/s)	ft^3/s	0.02832	m^3/s	1 $ft^3/s \approx$ 0.03 m^3/s
cubic feet per second (ft^3/s)	million gallons per day (mgd)	ft^3/s	0.6463	mgd	1 $ft^3/s \approx$ 0.6 mgd
cubic feet per second (ft^3/s)	gallons per minute (gpm)	ft^3/s	448.8	gpm	1 ft^3/s = 400 gpm
cubic feet per minute (ft^3/min)	gallons per second (gps)	ft^3/min	0.1247	gps	1 ft^3/min = 0.1 gps
cubic feet per minute (ft^3/min)	liters per second (L/s)	ft^3/min	0.4720	L/s	1 ft^{13}/min = 0.5 L/s
cubic inches ($in.^3$)	cubic centimeters (cm^3)	$in.^3$	16.39	cm^3	1 $in.^3$ = 15 cm^3
cubic inches ($in.^3$)	cubic feet (ft^3)	$in.^3$	0.0005787	ft^3	1 $in.^3$ = 0.0006 ft^3
cubic inches ($in.^3$)	cubic meters (m^3)	$in.^3$	0.00001639	m^3	1 $in.^3$ = 0.00015 m^3
cubic inches ($in.^3$)	cubic millimeters (mm^3)	$in.^3$	16,390	mm^3	1 $in.^3$ = 15,000 mm^3
cubic inches ($in.^3$)	cubic yards (yd^3)	$in.^3$	0.00002143	yd^3	1 $in.^3$ = 0.00002 yd^3
cubic inches ($in.^3$)	gallons (gal)	$in.^3$	0.004329	gal	1 $in.^3$ = 0.004 gal

Table continued next page

TABLE A-1 Conversion factors (continued)

Conversions		Procedure			Approximations (Actual answer will be within 25% of approximate answer.)
From	To	Multiply number of	by	To get number of	
cubic inches (in.^{3})	liters (L)	in.^{3}	0.01639	L	1 in.^{3} = 0.015 L
cubic meters (m^3)	acre-feet (acre-ft)	m^3	0.0008107	acre-ft	1 m^3 = 0.0008 acre-ft
cubic meters (m^3)	cubic centimeters (cm^3)	m^3	1,000,000	cm^3	—
cubic meters (m^3)	cubic feet (ft^3)	m^3	35.31	ft^3	1 m^3 = 40 ft^3
cubic meters (m^3)	cubic inches (in.^{3})	m^3	61,020	in.^{3}	1 m^3 = 60,000 in.^{3}
cubic meters (m^3)	cubic yards (yd^3)	m^3	1.308	yd^3	1 m^3 = 1.5 yd^3
cubic meters (m^3)	gallons (gal)	m^3	264.2	gal	1 m^3 = 300 gal
cubic meters (m^3)	kiloliters (kL)	m^3	1.0	kL	—
cubic meters (m^3)	liters (L)	m^3	1,000	L	—
cubic meters per day (m^3/d)	gallons per day (gpd)	m^3/d	264.2	gpd	1 m^3/d = 300 gpd
cubic meters per second (m^3/s)	cubic feet per second (ft^3/s)	m^3/s	35.31	ft^3/s	1 m^3/s = 40 ft^3/s
cubic millimeters (mm^3)	cubic inches (in.^{3})	mm^3	0.00006102	in.^{3}	1 mm^3 = 0.00006 in.^{3}
cubic yards (yd^3)	cubic centimeters (cm^3)	yd^3	764,600	cm^3	1 yd^3 ≈ 800,000 cm^3
cubic yards (yd^3)	cubic feet (ft^3)	yd^3	27	ft^3	1 yd^3 ≈ 30 ft^3
cubic yards (yd^3)	cubic inches (in.^{3})	yd^3	46,660	in.^{3}	1 yd^3 ≈ 50,000 in.^{3}
cubic yards (yd^3)	cubic meters (m^3)	yd^3	0.7646	m^3	1 yd^3 ≈ 0.8 m^3
cubic yards (yd^3)	gallons (gal)	yd^3	202.0	gal	1 yd^3 ≈ 200 gal
cubic yards (yd^3)	liters (L)	yd^3	764.6	L	1 yd^3 ≈ 800 L
feet (ft)	centimeters (cm)	ft	30.48	cm	1 ft ≈ 30 cm
feet (ft)	inches (in.)	ft	12	in.	—

Table continued next page

TABLE A-1 Conversion factors (continued)

Conversions		Procedure			Approximations (Actual answer will be within 25% of approximate answer.)
From	To	Multiply number of	by	To get number of	
feet (ft)	kilometers (km)	ft	0.0003048	km	1 ft ≈ 0.0003 km
feet (ft)	meters (m)	ft	0.3048	m	1 ft ≈ 0.3 m
feet (ft)	miles (mi)	ft	0.0001894	mi	1 ft ≈ 0.0002 mi
feet (ft)	millimeters (mm)	ft	304.8	mm	1 ft ≈ 300 mm
feet (ft)	yards (yd)	ft	0.3333	yd	1 ft ≈ 0.3 yd
feet (ft) of hydraulic head	kilopascals (kPa)	ft of head	2.989	kPa	1 ft of head ≈ 3 kPa
feet (ft) of hydraulic head	meters (m) of hydraulic head	ft of head	0.3048	m of head	1 ft of head ≈ 0.3 m of head
feet (ft) of hydraulic head	pascals (Pa)	ft of head	2,989	Pa	1 ft of head ≈ 3,000 Pa
feet (ft) of water	inches of mercury (in. Hg)	ft of water	0.8826	in. Hg	1 ft of water ≈ 0.9 in. Hg
feet (ft) of water	pounds per square foot (lb/ft^2)	ft of water	62.4	lb/ft^2	1 ft of water ≈ 60 lb/ft^2
feet (ft) of water	pounds per square inch gauge (psig)	ft of water	0.4332	psig	1 ft of water ≈ 0.4 psig
feet per hour (ft/h)	meters per second (m/s)	ft/h	0.00008467	m/s	1 ft/h ≈ 0.00008 m/s
feet per minute (ft/min)	feet per second (ft/s)	ft/min	0.01667	ft/s	1 ft/min ≈ 0.015 ft/s
feet per minute (ft/min)	kilometers per hour (km/h)	ft/min	0.01829	km/h	1 ft/min ≈ 0.02 km/h
feet per minute (ft/min)	meters per minute (m/min)	ft/min	0.3048	m/min	1 ft/min ≈ 0.3 m/min
feet per minute (ft/min)	meters per second (m/s)	ft/min	0.005080	m/s	1 ft/min ≈ 0.005 m/s
feet per minute (ft/min)	miles per hour (mph)	ft/min	0.01136	mph	1 ft/min ≈ 0.01 mph
feet per second (ft/s)	feet per minute (ft/min)	ft/s	60	ft/min	—
feet per second (ft/s)	kilometers per hour (km/h)	ft/s	1.097	km/h	1 ft/s ≈ 1 km/h
feet per second (ft/s)	meters per minute (m/min)	ft/s	18.29	m/min	1 ft/s ≈ 20 m/min

Table continued next page

TABLE A-1 Conversion factors (continued)

Conversions		Procedure			Approximations (Actual answer will be within 25% of approximate answer.)
From	To	Multiply number of	by	To get number of	
feet per second (ft/s)	meters per second (m/s)	ft/s	0.3048	m/s	1 ft/s ≈ 0.3 m/s
feet per second (ft/s)	miles per hour (mph)	ft/s	0.6818	mph	1 ft/s ≈ 0.7 mph
foot-pounds per minute (ft-lb/min)	horsepower (hp)	ft-lb/min	0.00003030	hp	1 ft-lb/min ≈ 0.00003 hp
foot-pounds per minute (ft-lb/min)	kilowatts (kW)	ft-lb/min	0.00002260	kW	1 ft-lb/min ≈ 0.00002 kW
foot-pounds per minute (ft-lb/min)	watts (W)	ft-lb/min	0.02260	W	1 ft-lb/min ≈ 0.02 W
gallons (gal)	acre-feet (acre-ft)	gal	0.000003069	acre-ft	1 gal ≈ 0.000003 acre-ft
gallons (gal)	cubic centimeters (cm^3)	gal	3,785	cm^3	1 gal ≈ 4000 cm^3
gallons (gal)	cubic feet (ft^3)	gal	0.1337	ft^3	1 gal ≈ 0.15 ft^3
gallons (gal)	cubic inches ($in.^3$)	gal	231.0	$in.^3$	1 gal ≈ 200 $in.^3$
gallons (gal)	cubic meters (m^3)	gal	0.003785	m^3	1 gal ≈ 0.004 m^3
gallons (gal)	cubic yards (yd^3)	gal	0.004951	yd^3	1 gal ≈ 0.005 yd^3
gallons (gal)	kiloliters (kL)	gal	0.003785	kL	1 gal ≈ 0.004 kL
gallons (gal)	liters (L)	gal	3.785	L	1 gal ≈ 4 L
gallons (gal)	pounds (lb) of water	gal	8.34	lb of water	1 gal ≈ 8 lb of water
gallons (gal)	quarts (qt)	gal	4	qt	

Table continued next page

TABLE A-1 Conversion factors (continued)

Conversions		Procedure			Approximations (Actual answer will be within 25% of approximate answer.)
From	To	Multiply number of	by	To get number of	
gallons per capita per day (gpcd)	liters per capita per day (L/d per capita)	gpcd	3.785	L/d per capita	1 gpcd ≈ 4 L/d per capita
gallons per day (gpd)	cubic meters per day (m^3/d)	gpd	0.003785	m^3/d	1 gpd ≈ 0.004 m^3/d
gallons per day (gpd)	liters per day (L/d)	gpd	3.785	L/d	1 gpd ≈ 4 L/d
gallons per day per foot (gpd/ft)	square meters per day (m^2/d)	gpd/ft	0.01242	m^2/d	1 gpd/ft ≈ 0.01 m^2/d
gallons per day per foot (gpd/ft)	square millimeters per second (mm^2/s)	gpd/ft	0.1437	mm^2/s	1 gpd/ft ≈ 0.15 mm^2/s
gallons per day per square foot (gpd/ft^2)	millimeters per second (mm/s)	gpd/ft^2	0.0004716	mm/s	1 gpd/ft^2 ≈ 0.0005 mm/s
gallons per hour (gph)	liters per second (L/s)	gph	0.001052	L/s	1 gph ≈ 0.001 L/s
gallons per minute (gpm)	cubic feet per second (ft^3/s)	gpm	0.002228	ft^3/s	1 gpm ≈ 0.0002 ft^3/s
gallons per minute (gpm)	liters per second (L/s)	gpm	0.06309	L/s	1 gpm ≈ 0.06 L/s
gallons per minute per square foot (gpm/ft^2)	millimeters per second (mm/s)	gpm/ft^2	0.6790	mm/s	1 gpm/ft^2 ≈ 0.7 mm/s
gallons per second (gps)	cubic feet per minute (ft^3/min)	gps	8.021	ft^3/min	1 gpm ≈ 8 ft^3/min
gallons per second (gps)	liters per minute (L/min)	gps	227.1	L/min	1 gps ≈ 200 L/min
grains (gr)	grams (g)	gr	0.06480	g	1 gr ≈ 0.06 g
grains (gr)	pounds (lb)	gr	0.0001428	lb	1 gr ≈ 0.00015 lb
grams (g)	grains (gr)	g	15.43	gr	1 g ≈ 15 gr

Table continued next page

TABLE A-1 Conversion factors (continued)

Conversions		Procedure			Approximations (Actual answer will be within 25% of approximate answer.)
From	To	Multiply number of	by	To get number of	
grams (g)	kilograms (kg)	g	0.001	kg	—
grams (g)	milligrams (mg)	g	1,000	mg	—
grams (g)	ounces (oz), avoirdupois	g	0.03527	oz	1 g ≈ 0.04 oz
grams (g)	pounds (lb)	g	0.002205	lb	1 g ≈ 0.002 lb
hectares (ha)	acres	ha	2.471	acres	1 ha ≈ 2 acres
hectares (ha)	square meters (m^2)	ha	10,000	m^2	—
hectares (ha)	square miles (mi^2)	ha	0.003861	mi^2	1 ha ≈ 0.004 mi^2
horsepower (hp)	foot-pounds per minute (ft-lb/min)	hp	33,000	ft-lb/min	1 hp ≈ 30,000 ft-lb/min
horsepower (hp)	kilowatts (kW)	hp	0.7457	kW	1 hp ≈ 0.7 kW
horsepower (hp)	watts (W)	hp	745.7	W	1 hp ≈ 700 W
inches (in.)	centimeters (cm)	in.	2.540	cm	1 in. ≈ 3 cm
inches (in.)	feet (ft)	in.	0.08333	ft	1 in. ≈ 0.08 ft
inches (in.)	meters (m)	in.	0.02540	m	1 in. ≈ 0.03 m
inches (in.)	millimeters (mm)	in.	25.40	mm	1 in. ≈ 30 mm
inches (in.)	yards (yd)	in.	0.02778	yd	1 in. ≈ 0.03 yd
inches of mercury (in. Hg)	feet (ft) of water	in. Hg	1.133	ft of water	1 in. Hg ≈ 1 ft of water
inches of mercury (in. Hg)	inches (in.) of water	in. Hg	13.60	in. of water	1 in. Hg ≈ 15 in. of water
inches of mercury (in. Hg)	pounds per square foot (lb/ft^2)	in. Hg	70.73	lb/ft^2	1 in. Hg ≈ 70 lb/ft^3
inches of mercury (in. Hg)	pounds per square inch (psi)	in. Hg	0.4912	psi	1 in. Hg ≈ 0.5 psi
inches per minute (in./min)	millimeters per second (mm/s)	in./min	0.4233	mm/s	1 in./min ≈ 0.4 mm/s

Table continued next page

TABLE A-1 Conversion factors (continued)

Conversions		Procedure			Approximations (Actual answer will be within 25% of approximate answer.)
From	To	Multiply number of	by	To get number of	
inches (in.) of water	inches of mercury (in. Hg)	in. of water	0.07355	in. Hg	1 in. of water ≈ 0.07 in. Hg
inches (in.) of water	pounds per square foot (lb/ft^2)	in. of water	5.198	lb/ft^2	1 in. of water ≈ 5 lb/ft^2
inches (in.) of water	pounds per square inch gauge (psig)	in. of water	0.03610	psig	1 in. of water ≈ 0.04 psig
kilograms (kg)	grams (g)	kg	1,000	g	—
kilograms (kg)	pounds (lb)	kg	2.205	lb	1 kg ≈ 2 lb
kiloliters (kL)	cubic feet (ft^3)	kL	35.31	ft^3	1 kL ≈ 40 ft^3
kiloliters (kL)	cubic meters (m^3)	kL	1.0	m^3	—
kiloliters (kL)	gallons (gal)	kL	264.2	gal	1 kL ≈ 300 gal
kiloliters (kL)	liters (L)	kL	1,000	L	—
kilometers (km)	feet (ft)	km	3,281	ft	1 km ≈ 3,000 ft
kilometers (km)	meters (m)	km	1,000	m	—
kilometers (km)	miles (mi)	km	0.6214	mi	1 km ≈ 0.6 mi
kilometers (km)	yards (yd)	km	1,094	yd	1 km ≈ 1,000 yd
kilometers per hour (km/h)	feet per minute (ft/min)	km/h	54.68	ft/min	1 km/h ≈ 50 ft/min
kilometers per hour (km/h)	feet per second (ft/s)	km/h	0.9113	ft/s	1 km/h ≈ 1 ft/s

Table continued next page

TABLE A-1 Conversion factors (continued)

Conversions		Procedure			Approximations (Actual answer will be within 25% of approximate answer.)
From	To	Multiply number of	by	To get number of	
kilometers per hour (km/h)	meters per minute (m/min)	km/h	16.67	m/min	1 km/h ≈ 15 m/min
kilometers per hour (km/h)	meters per second (m/s)	km/h	0.2778	m/s	1 km/h ≈ 0.3 m/s
kilometers per hour (km/h)	miles per hour (mph)	km/h	0.6214	mph	1 km/h ≈ 0.6 mph
kilopascals (kPa)	feet (ft) of hydraulic head	kPa	0.3346	ft of head	1 kPa ≈ 0.3 ft of head
kilowatts (kW)	foot-pounds per minute (ft-lb/min)	kW	44,250	ft-lb/min	1 kW ≈ 40,000 ft-lb/min
kilowatts (kW)	horsepower (hp)	kW	1.341	hp	1 kW ≈ 1.5 hp
kilowatts (kW)	watts (W)	kW	1,000	W	—
liters (L)	cubic centimeters (cm^3)	L	1,000	cm^3	—
liters (L)	cubic feet (ft^3)	L	0.03531	ft^3	1 L ≈ 0.04 ft^3
liters (L)	cubic inches ($in.^3$)	L	61.03	$in.^3$	1 L ≈ 60 $in.^3$
liters (L)	cubic meters (m^3)	L	0.001	m^3	—
liters (L)	cubic yards (yd^3)	L	0.001308	yd^3	1 L ≈ 0.0015 yd^3
liters (L)	gallons (gal)	L	0.2642	gal	1 L ≈ 0.3 gal
liters (L)	kiloliters (kL)	L	0.001	kL	—
liters (L)	milliliters (mL)	L	1,000	mL	—
liters (L)	ounces (oz), fluid	L	33.81	oz (fluid)	1 L ≈ 30 oz (fluid)
liters (L)	quarts (qt), fluid	L	1.057	qt (fluid)	1 L ≈ 1 qt (fluid)
liters per capita per day (L/d per capita)	gallons per capita per day (gpcd)	L/d per capita	0.2642	gpcd	1 L/d per capita ≈ 0.3 gpcd
liters per day (L/d)	gallons per day (gpd)	L/d	0.2642	gpd	1 L/d ≈ 0.3 gpd
liters per minute (L/min)	gallons per second (gps)	L/min	0.004403	gps	1 L/min ≈ 0.004 gps

Table continued next page

TABLE A-1 Conversion factors (continued)

Conversions		Procedure			Approximations (Actual answer will be within 25% of approximate answer.)
From	To	Multiply number of	by	To get number of	
liters per second (L/s)	cubic feet per minute (ft^3/min)	L/s	2.119	ft^3/min	1 L/s ≈ 2 ft^3/min
liters per second (L/s)	gallons per hour (gph)	L/s	951.0	gph	1 L/s ≈ 1000 gph
liters per second (L/s)	gallons per minute (gpm)	L/s	15.85	gpm	1 L/s ≈ 15 gpm
megaliters per day (ML/d)	million gallons per day (mgd)	ML/d	0.2642	mgd	1 ML/d ≈ 0.3 mgd
meters (m)	centimeters (cm)	m	100	cm	—
meters (m)	feet (ft)	m	3.281	ft	1 m ≈ 3 ft
meters (m)	inches (in.)	m	39.37	in.	1 m ≈ 40 in.
meters (m)	kilometers (km)	m	0.001	km	—
meters (m)	miles (mi)	m	0.0006214	mi	1 m ≈ 0.0006 mi
meters (m)	millimeters (mm)	m	1,000	mm	—
meters (m)	yards (yd)	m	1.094	yd	1 m ≈ 1 yd
meters (m) of hydraulic head	feet (ft) of hydraulic head	m of head	3.281	ft of head	1 m of head ≈ 3 ft of head
meters (m) of hydraulic head	pounds per square inch gauge (psig)	m of head	1.422	psig	1 m of head ≈ 1.5 psig
meters per minute (m/min)	centimeters per second (cm/s)	m/min	1.667	cm/s	1 m/min ≈ 1.5 cm/s
meters per minute (m/min)	feet per minute (ft/min)	m/min	3.281	ft/min	1 m/min ≈ 3 ft/min
meters per minute (m/min)	feet per second (ft/s)	m/min	0.05468	ft/s	1 m/min ≈ 0.05 ft/s
meters per minute (m/min)	kilometers per hour (km/h)	m/min	0.06	km/h	—
meters per minute (m/min)	miles per hour (mph)	m/min	0.03728	mph	1 m/min ≈ 0.04 mph
meters per second (m/s)	feet per hour (ft/h)	m/s	11,810	ft/h	1 m/s ≈ 10,000 ft/h
meters per second (m/s)	feet per minute (ft/min)	m/s	196.8	ft/min	1 m/s ≈ 200 ft/min

Table continued next page

TABLE A-1 Conversion factors (continued)

Conversions		Procedure			Approximations (Actual answer will be within 25% of approximate answer.)
From	To	Multiply number of	by	To get number of	
meters per second (m/s)	feet per second (ft/s)	m/s	3.281	ft/s	1 m/s ≈ 3 ft/s
meters per second (m/s)	kilometers per hour (km/h)	m/s	3.6	km/h	1 m/s ≈ 4 km/h
meters per second (m/s)	miles per hour (mph)	m/s	2.237	mph	1 m/s ≈ 2 mph
miles (mi)	feet (ft)	mi	5,280	ft	1 mi ≈ 5,000 ft
miles (mi)	kilometers (km)	mi	1.609	km	1 mi ≈ 1.5 km
miles (mi)	meters (m)	mi	1,609	m	1 mi ≈ 1,500 m
miles (mi)	yards (yd)	mi	1,760	yd	1 mi ≈ 2,000 yd
miles per hour (mph)	feet per minute (ft/min)	mph	88	ft/min	1 mph ≈ 90 ft/min
miles per hour (mph)	feet per second (ft/s)	mph	1.467	ft/s	1 mph ≈ 1.5 ft/s
miles per hour (mph)	kilometers per hour (km/h)	mph	1.609	km/h	1 mph ≈ 1.5 km/h
miles per hour (mph)	meters per minute (m/min)	mph	26.82	m/min	1 mph ≈ 30 m/min
miles per hour (mph)	meters per second (m/s)	mph	0.4470	m/s	1 mph ≈ 0.4 m/s
milligrams (mg)	grams (g)	mg	0.001	g	—
milliliters (mL)	liters (L)	mL	0.001	L	—
millimeters (mm)	centimeters (cm)	mm	0.1	cm	—
millimeters (mm)	feet (ft)	mm	0.003281	ft	1 mm ≈ 0.003 ft
millimeters (mm)	inches (in.)	mm	0.03937	in.	1 mm ≈ 0.04 in.
millimeters (mm)	meters (m)	mm	0.001	m	—
millimeters (mm)	yards (yd)	mm	0.001094	yd	1 mm ≈ 0.001 yd
millimeters per second (mm/s)	gallons per day per square foot (gpd/ft^2)	mm/s	2,121	gpd/ft^2	1 mm/s ≈ 2,000 gpd/ft^2

Table continued next page

TABLE A-1 Conversion factors (continued)

Conversions		Procedure			Approximations (Actual answer will be within 25% of approximate answer.)
From	To	Multiply number of	by	To get number of	
millimeters per second (mm/s)	gallons per minute per square foot (gpm/ft^2)	mm/s	1.473	gpm/ft^2	1 mm/s ≈ 1.5 gpm/ft^2
millimeters per second (mm/s)	inches per minute (in./min)	mm/s	2.362	in./min	1 mm/s ≈ 2 in./min
million gallons per day (mgd)	cubic feet per second (ft^3/s)	mgd	1.547	ft^3/s	1 mgd ≈ 1.5 ft^3/s
million gallons per day (mgd)	megaliters per day (ML/d)	mgd	3.785	ML/d	1 mgd ≈ 4 ML/d
ounces (oz), avoirdupois	grams (g)	oz	28.35	g	1 oz ≈ 30 g
ounces (oz), avoirdupois	pounds (lb)	oz	0.0625	lb	1 oz ≈ 0.06 lb
ounces (oz), fluid	liters (L)	oz	0.02957	L	1 oz ≈ 0.03 L
pascals (Pa)	feet (ft) of hydraulic head	Pa	0.0003346	ft of head	1 Pa ≈ 0.0003 ft of head
pascals (Pa)	pounds per square inch (psi)	Pa	0.0001450	psi	1 Pa ≈ 0.00015 psi
pounds (lb)	grains (gr)	lb	7,000	gr	
pounds (lb)	grams (g)	lb	453.6	g	1 lb ≈ 500 g
pounds (lb)	kilograms (kg)	lb	0.4536	kg	1 lb ≈ 0.5 kg
pounds (lb)	ounces (oz), avoirdupois	lb	16	oz	—
pounds (lb) of water	cubic feet (ft^3)	lb of water	0.01603	ft^3	1 lb of water ≈ 0.015 ft^3
pounds (lb) of water	gallons (gal)	lb of water	0.1199	gal	1 lb of water ≈ 0.1 gal

Table continued next page

TABLE A-1 Conversion factors (continued)

Conversions		Procedure			Approximations (Actual answer will be within 25% of approximate answer.)
From	To	Multiply number of	by	To get number of	
pounds per square foot (lb/ft^2)	feet (ft) of water	lb/ft^2	0.01603	ft of water	1 $lb/ft^2 \approx 0.015$ ft of water
pounds per square foot (lb/ft^2)	inches of mercury (in. Hg)	lb/ft^2	0.01414	in. Hg	1 $lb/ft^2 \approx 0.015$ in. Hg
pounds per square foot (lb/ft^2)	inches (in.) of water	lb/ft^2	0.1924	in. of water	1 $lb/ft^2 \approx 0.2$ in. of water
pounds per square inch gauge (psig)	feet (ft) of water	psig	2.31	ft of water	1 psig ≈ 2 ft of water
pounds per square inch (psi)	inches of mercury (in. Hg)	psi	2.036	in. Hg	1 psi ≈ 2 in. Hg
pounds per square inch gauge (psig)	inches (in.) of water	psig	27.70	in. of water	1 psig ≈ 30 in. of water
pounds per square inch gauge (psig)	meters (m) of hydraulic head	psig	0.7034	m of head	1 psig ≈ 0.7 m of head
pounds per square inch (psi)	pascals (Pa)	psi	6,895	Pa	1 psi ≈ 7,000 Pa
quarts (qt)	gallons (gal)	qt	0.25	gal	—
quarts (qt)	liters (L)	qt	0.9464	L	1 qt ≈ 0.9 L
square centimeters (cm^2)	square inches ($in.^2$)	cm^2	0.1550	$in.^2$	1 $cm^2 \approx 0.15$ $in.^2$
square centimeters (cm^2)	square millimeters (mm^2)	cm^2	100	mm^2	—
square feet (ft^2)	acres	ft^2	0.00002296	acres	1 $ft^2 \approx 0.00002$ acre
square feet (ft^2)	square inches ($in.^2$)	ft^2	144	$in.^2$	1 $ft^2 \approx 150$ $in.^2$
square feet (ft^2)	square meters (m^2)	ft^2	0.09290	m^2	1 $ft^2 \approx 0.09$ m^2

Table continued next page

TABLE A-1 Conversion factors (continued)

Conversions		Procedure			Approximations (Actual answer will be within 25% of approximate answer.)
From	To	Multiply number of	by	To get number of	
square feet (ft^2)	square millimeters (mm^2)	ft^2	92,900	mm^2	1 $ft^2 \approx$ 90,000 mm^2
square feet (ft^2)	square yards (yd^2)	ft^2	0.1111	yd^2	1 $ft^2 \approx$ 0.1 yd^2
square inches ($in.^2$)	square centimeters (cm^2)	$in.^2$	6.452	cm^2	1 $in.^2 \approx$ 6 cm^2
square inches ($in.^2$)	square feet (ft^2)	$in.^2$	0.006944	ft^2	1 $in.^2 \approx$ 0.007 ft^2
square inches ($in.^2$)	square meters (m^2)	$in.^2$	0.0006452	m^2	1 $in.^2 \approx$ 0.0006 m^2
square inches ($in.^2$)	square millimeters (mm^2)	$in.^2$	645.2	mm^2	1 $in.^2 \approx$ 600 mm^2
square inches ($in.^2$)	square yards (yd^2)	$in.^2$	0.0007716	yd^2	1 $in.^2 \approx$ 0.0008 yd^2
square kilometers (km^2)	acres	km^2	247.1	acres	1 $km^2 \approx$ 200 acres
square kilometers (km^2)	square miles (mi^2)	km^2	0.3861	mi^2	1 $km^2 \approx$ 0.4 mi^2
square meters (m^2)	acres	m^2	0.0002471	acres	1 $m^2 \approx$ 0.0002 acre
square meters (m^2)	hectares (ha)	m^2	0.0001	ha	—
square meters (m^2)	square feet (ft^2)	m^2	10.76	ft^2	1 $m^2 \approx$ 10 ft^2
square meters (m^2)	square inches ($in.^2$)	m^2	1,550	$in.^2$	1 $m^2 \approx$ 1,500 $in.^2$
square meters (m^2)	square miles (mi^2)	m^2	0.0000003861	mi^2	1 $m^2 \approx$ 0.0000004 mi^2
square meters (m^2)	square yards (yd^2)	m^2	1.196	yd^2	1 $m^2 \approx$ 1 yd^2
square meters per day (m^2/d)	gallons per day per foot (gpd/ft)	m^2/d	80.53	gpd/ft	1 $m^2/d \approx$ 80 gpd/ft
square miles (mi^2)	acres	mi^2	640	acres	1 $mi^2 \approx$ 600 acres
square miles (mi^2)	hectares (ha)	mi^2	259.0	ha	1 $mi^2 \approx$ 300 ha
square miles (mi^2)	square kilometers (km^2)	mi^2	2.590	km^2	1 $mi^2 \approx$ 3 km^2
square miles (mi^2)	square meters (m^2)	mi^2	2,590,000	m^2	1 $mi^2 \approx$ 3,000,000 m^2

Table continued next page

TABLE A-1 Conversion factors (continued)

Conversions		Procedure			Approximations (Actual answer will be within 25% of approximate answer.)
From	To	Multiply number of	by	To get number of	
square millimeters (mm^2)	square centimeters (cm^2)	mm^2	0.01	cm^2	—
square millimeters (mm^2)	square feet (ft^2)	mm^2	0.00001076	ft^2	1 $mm^2 \approx$ 0.00001 ft^2
square millimeters (mm^2)	square inches ($in.^2$)	mm^2	0.001550	$in.^2$	1 $mm^2 \approx$ 0.0015 $in.^2$
square millimeters per second (mm^2/s)	gallons per day per foot (gpd/ft)	mm^2/s	6.958	gpd/ft	1 $mm^2/s \approx$ 7 gpd/ft
square yards (yd^2)	acres	yd^2	0.0002066	acres	1 $yd^2 \approx$ 0.0002 acre
square yards (yd^2)	square feet (ft^2)	yd^2	9	ft^2	—
square yards (yd^2)	square inches ($in.^2$)	yd^2	1,296	$in.^2$	1 $yd^2 \approx$ 1,500 $in.^2$
square yards (yd^2)	square meters (m^2)	yd^2	0.8361	m^2	1 $yd^2 \approx$ 0.8 m^2
watts (W)	foot-pounds per minute (ft-lb/min)	W	44.25	ft-lb/min	1 W ≈ 40 ft-lb/min
watts (W)	horsepower (hp)	W	0.001341	hp	1 W ≈ 0.0015 hp
watts (W)	kilowatts (kW)	W	0.001	kW	—
yards (yd)	feet (ft)	yd	3	ft	—
yards (yd)	inches (in.)	yd	36	in.	1 yd ≈ 40 in.
yards (yd)	kilometers (km)	yd	0.0009144	km	1 yd ≈ 0.0009 km
yards (yd)	meters (m)	yd	0.9144	m	1 yd ≈ 0.9 m
yards (yd)	miles (mi)	yd	0.0005681	mi	1 yd ≈ 0.0006 mi
yards (yd)	millimeters (mm)	yd	914.4	mm	1 yd ≈ 900 mm

TABLE A-2 Temperature conversions, Celsius to Fahrenheit: °F = 9/5 (°C) + 32

°C	°F	°C	°F	°C	°F	°C	°F	°C	°F	°C	°F
–29	–20.2	–9	15.8	11	51.8	31	87.8	51	123.8	71	159.8
–28	–18.4	–8	17.6	12	53.6	32	89.6	52	125.6	72	161.6
–27	–16.6	–7	19.4	13	55.4	33	91.4	53	127.4	73	163.4
–26	–14.8	–6	21.2	14	57.2	34	93.2	54	129.2	74	165.2
–25	–13.0	–5	23.0	15	59.0	35	95.0	55	131.0	75	167.0
–24	–11.2	–4	24.8	16	60.8	36	96.8	56	132.8	76	168.8
–23	–9.4	–3	26.6	17	62.6	37	98.6	57	134.6	77	170.6
–22	–7.6	–2	28.4	18	64.4	38	100.4	58	136.4	78	172.4
–21	–5.8	–1	30.2	19	66.2	39	102.2	59	138.2	79	174.2
–20	–4.0	0	32.0	20	68.0	40	104.0	60	140.0	80	176.0
–19	–2.2	1	33.8	21	69.8	41	105.8	61	141.8	81	177.8
–18	–0.4	2	35.6	22	71.6	42	107.6	62	143.6	82	179.6
–17	+1.4	3	37.4	23	73.4	43	109.4	63	145.4	83	181.4
–16	3.2	4	39.2	24	75.2	44	111.2	64	147.2	84	183.2
–15	5.0	5	41.0	25	77.0	45	113.0	65	149.0	85	185.0
–14	6.8	6	42.8	26	78.8	46	114.8	66	150.8	86	186.8
–13	8.6	7	44.6	27	80.6	47	116.6	67	152.6	87	188.6
–12	10.4	8	46.4	28	82.4	48	118.4	68	154.4	88	190.4
–11	12.2	9	48.2	29	84.2	49	120.2	69	156.2	89	192.2
–10	14.0	10	50.0	30	86.0	50	122.0	70	158.0	90	194.0

Table continued next page

TABLE A-2 Temperature conversions, Celsius to Fahrenheit: °F = 9⁄5 (°C) + 32 (continued)

°C to °F		°C to °F		°C to °F		°C to °F		°C to °F		°C to °F	
91	195.8	101	213.8	111	231.8	121	249.8	131	267.8	141	285.8
92	197.6	102	215.6	112	233.6	122	251.6	132	296.6	142	287.6
93	199.4	103	217.4	113	235.4	123	253.4	133	271.4	143	289.4
94	201.2	104	219.2	114	237.2	124	255.2	134	273.2	144	291.2
95	203.0	105	221.0	115	239.0	125	257.0	135	275.0	145	293.0
96	204.8	106	222.8	116	240.8	126	258.8	136	276.8	146	294.8
97	206.6	107	224.6	117	242.6	127	260.6	137	278.6	147	296.6
98	208.4	108	226.4	118	244.4	128	262.4	138	280.4	148	298.4
99	210.2	109	228.2	119	246.2	129	264.2	139	282.2	149	300.2
100	212.0	110	230.0	120	248.0	130	266.0	140	284.0	150	302.0

TABLE A-3 Temperature conversions, Fahrenheit to Celsius: °C = 5/9 (°F – 32)

°F to °C		°F to °C		°F to °C		°F to °C		°F to °C		°F to °C	
–19	–28.3	1	–17.2	21	–6.1	41	5.0	61	16.1	81	27.2
–18	–27.8	2	–16.7	22	–5.6	42	5.6	62	16.7	82	27.8
–17	–27.2	3	–16.1	23	–5.0	43	6.1	63	17.2	83	28.3
–16	–26.7	4	–15.6	24	–4.4	44	6.7	64	17.8	84	28.9
–15	–26.1	5	–15.0	25	–3.9	45	7.2	65	18.3	85	29.4
–14	–25.6	6	–14.4	26	–3.3	46	7.8	66	18.9	86	30.0
–13	–25.0	7	–13.9	27	–2.8	47	8.3	67	19.4	87	30.6
–12	–24.4	8	–13.3	28	–2.2	48	8.9	68	20.0	88	31.1
–11	–23.9	9	–12.8	29	–1.7	49	9.4	69	20.6	89	31.7
–10	–23.3	10	–12.2	30	–1.1	50	10.0	70	21.1	90	32.2
–9	–22.8	11	–11.7	31	–0.6	51	10.6	71	21.7	91	32.8
–8	–22.2	12	–11.1	32	0.0	52	11.1	72	22.2	92	33.3
–7	–21.7	13	–10.6	33	+0.6	53	11.7	73	22.8	93	33.9
–6	–21.1	14	–10.0	34	1.1	54	12.2	74	23.3	94	34.4
–5	–20.6	15	–9.4	35	1.7	55	12.8	75	23.9	95	35.0
–4	–20.0	16	–8.9	36	2.2	56	13.3	76	24.4	96	35.6
–3	–19.4	17	–8.3	37	2.8	57	13.9	77	25.0	97	36.1
–2	–18.9	18	–7.8	38	3.3	58	14.4	78	25.6	98	36.7
–1	–18.3	19	–7.2	39	3.9	59	15.0	79	26.1	99	37.2
0	–17.8	20	–6.7	40	4.4	60	15.6	80	26.7	100	37.8

Table continued next page

TABLE A-3 Temperature conversions, Fahrenheit to Celsius: °C = 5/9 (°F – 32) (continued)

°F to °C		°F to °C		°F to °C		°F to °C		°F to °C		°F to °C	
101	38.3	121	49.4	141	60.6	161	71.7	181	82.8	201	93.9
102	38.9	122	50.0	142	61.1	162	72.2	182	83.3	202	94.4
103	39.4	123	50.6	143	61.7	163	72.8	183	83.9	203	95.0
104	40.0	124	51.1	144	62.2	164	73.3	184	84.4	204	95.6
105	40.6	125	51.7	145	62.8	165	73.9	185	85.0	205	96.1
106	41.1	126	52.2	146	63.3	166	74.4	186	85.6	206	96.7
107	41.7	127	52.8	147	63.9	167	75.0	187	86.1	207	97.2
108	42.2	128	53.3	148	64.4	168	75.6	188	86.7	208	97.8
109	42.8	129	53.9	149	65.0	169	76.1	189	87.2	209	98.3
110	43.3	130	54.4	150	65.6	170	76.7	190	87.8	210	98.9
111	43.9	131	55.0	151	66.1	171	77.2	191	88.3	211	99.4
112	44.4	132	55.6	152	66.7	172	77.8	192	88.9	212	100.0
113	45.0	133	56.1	153	67.2	173	78.3	193	89.4	213	100.6
114	45.6	134	56.7	154	67.8	174	78.9	194	90.0	214	101.1
115	46.1	135	57.2	155	68.3	175	79.4	195	90.6	215	101.7
116	46.7	136	57.8	156	68.9	176	80.0	196	91.1	216	102.2
117	47.2	137	58.3	157	69.4	177	80.6	197	91.7	217	102.8
118	47.8	138	58.9	158	70.0	178	81.1	198	92.2	218	103.3
119	48.3	139	59.4	159	70.6	179	81.7	199	92.8	219	103.9
120	48.9	140	60.0	160	71.1	180	82.2	200	93.3	220	104.4

Table continued next page

TABLE A-3 Temperature conversions, Fahrenheit to Celsius: $°C = 5/9\ (°F - 32)$ (continued)

°F to °C		°F to °C		°F to °C		°F to °C		°F to °C		°F to °C	
221	105.0	241	116.1	261	127.2	281	138.3	301	149.4	321	160.6
222	105.6	242	116.7	262	127.8	282	138.9	302	150.0	322	161.1
223	106.1	243	117.2	263	128.3	283	139.4	303	150.5	323	161.7
224	106.7	244	117.8	264	128.9	284	140.0	304	151.1	324	162.2
225	107.2	245	118.3	265	129.4	285	140.6	305	151.7	325	162.8
226	107.8	246	118.9	266	130.0	286	141.1	306	152.2	326	163.3
227	108.3	247	119.4	267	130.6	287	141.7	307	152.8	327	163.9
228	108.9	248	120.0	268	131.1	288	142.2	308	153.3	328	164.4
229	109.4	249	120.6	269	131.7	289	142.8	309	153.9	329	165.0
230	110.0	250	121.1	270	132.2	290	143.3	310	154.4	330	165.6
231	110.6	251	121.7	271	132.8	291	143.9	311	155.0	331	166.1
232	111.1	252	122.2	272	133.3	292	144.4	312	155.6	332	166.7
233	111.7	253	122.8	273	133.9	293	145.0	313	156.1	333	167.2
234	112.2	254	123.3	274	134.4	294	145.6	314	156.7	334	167.8
235	112.8	255	123.9	275	135.0	295	146.1	315	157.2	335	168.3
236	113.3	256	124.4	276	135.6	296	146.7	316	157.8	336	168.9
237	113.9	257	125.0	277	136.1	297	147.2	317	158.3	337	169.4
238	114.4	258	125.6	278	136.7	298	147.8	318	158.9	338	170.0
239	115.0	259	126.1	279	137.2	299	148.3	319	159.4	339	170.6
240	115.6	260	126.7	280	137.8	300	148.9	320	160.0	340	171.1

Appendix B

Periodic Table and List of Elements

Periodic Table of the Elements

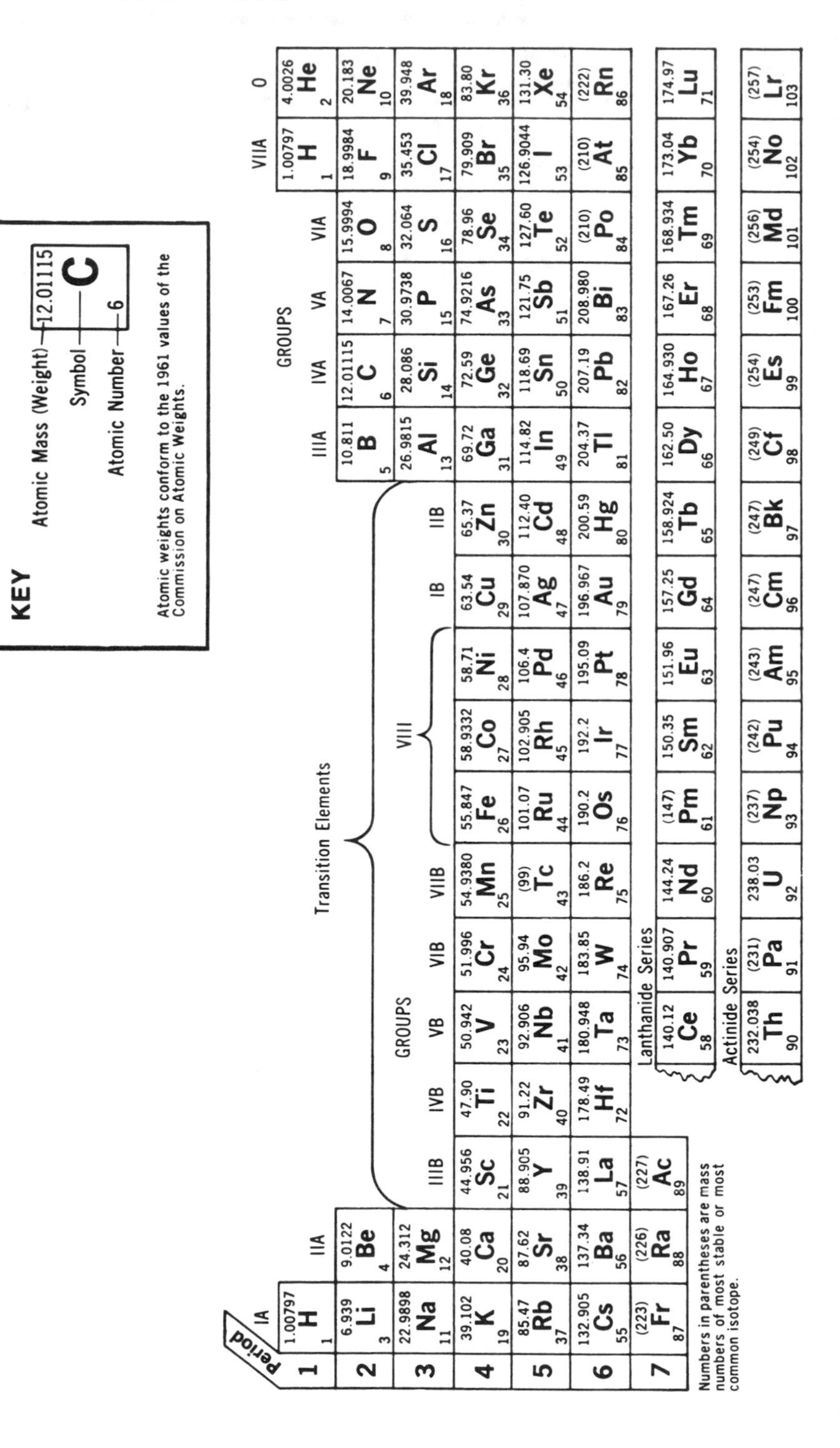

KEY

Atomic Mass (Weight)	12.01115
Symbol	C
Atomic Number	6

Atomic weights conform to the 1961 values of the Commission on Atomic Weights.

Each cell: atomic number, symbol, atomic mass (weight).

Period	IA	IIA	IIIB	IVB	VB	VIB	VIIB	VIII	VIII	VIII	IB	IIB	IIIA	IVA	VA	VIA	VIIA	0
1	1 H 1.00797																1 H 1.00797	2 He 4.0026
2	3 Li 6.939	4 Be 9.0122											5 B 10.811	6 C 12.01115	7 N 14.0067	8 O 15.9994	9 F 18.9984	10 Ne 20.183
3	11 Na 22.9898	12 Mg 24.312											13 Al 26.9815	14 Si 28.086	15 P 30.9738	16 S 32.064	17 Cl 35.453	18 Ar 39.948
4	19 K 39.102	20 Ca 40.08	21 Sc 44.956	22 Ti 47.90	23 V 50.942	24 Cr 51.996	25 Mn 54.9380	26 Fe 55.847	27 Co 58.9332	28 Ni 58.71	29 Cu 63.54	30 Zn 65.37	31 Ga 69.72	32 Ge 72.59	33 As 74.9216	34 Se 78.96	35 Br 79.909	36 Kr 83.80
5	37 Rb 85.47	38 Sr 87.62	39 Y 88.905	40 Zr 91.22	41 Nb 92.906	42 Mo 95.94	43 Tc (99)	44 Ru 101.07	45 Rh 102.905	46 Pd 106.4	47 Ag 107.870	48 Cd 112.40	49 In 114.82	50 Sn 118.69	51 Sb 121.75	52 Te 127.60	53 I 126.9044	54 Xe 131.30
6	55 Cs 132.905	56 Ba 137.34	57 La 138.91	72 Hf 178.49	73 Ta 180.948	74 W 183.85	75 Re 186.2	76 Os 190.2	77 Ir 192.2	78 Pt 195.09	79 Au 196.967	80 Hg 200.59	81 Tl 204.37	82 Pb 207.19	83 Bi 208.980	84 Po (210)	85 At (210)	86 Rn (222)
7	87 Fr (223)	88 Ra (226)	89 Ac (227)															

Transition Elements: groups IIIB through IIB.

Lanthanide Series

58 Ce 140.12	59 Pr 140.907	60 Nd 144.24	61 Pm (147)	62 Sm 150.35	63 Eu 151.96	64 Gd 157.25	65 Tb 158.924	66 Dy 162.50	67 Ho 164.930	68 Er 167.26	69 Tm 168.934	70 Yb 173.04	71 Lu 174.97

Actinide Series

90 Th 232.038	91 Pa (231)	92 U 238.03	93 Np (237)	94 Pu (242)	95 Am (243)	96 Cm (247)	97 Bk (247)	98 Cf (249)	99 Es (254)	100 Fm (253)	101 Md (256)	102 No (254)	103 Lr (257)

Numbers in parentheses are mass numbers of most stable or most common isotope.

TABLE B-1 List of elements

Name	Symbol	Atomic Number	Atomic Weight
Actinium	Ac	89	227*
Aluminum	Al	13	26.98
Americium	Am	95	243*
Antimony	Sb	51	121.75
Argon	Ar	18	39.95
Arsenic	As	33	74.92
Astatine	At	85	210*
Barium	Ba	56	137.34
Berkelium	Bk	97	247*
Beryllium	Be	4	9.01
Bismuth	Bi	83	208.98
Boron	B	5	10.81
Bromine	Br	35	79.90
Cadmium	Cd	48	112.40
Calcium	Ca	20	40.08
Californium	Cf	98	249*
Carbon	C	6	12.01
Cerium	Ce	58	140.12
Cesium	Cs	55	132.91
Chlorine	Cl	17	35.45
Chromium	Cr	24	52.00
Cobalt	Co	27	58.93
Copper	Cu	29	63.55
Curium	Cm	96	247*
Dysprosium	Dy	66	162.50
Einsteinium	Es	99	254*
Erbium	Er	68	167.26
Europium	Eu	63	151.96
Fermium	Fm	100	253*
Fluorine	F	9	19.00
Francium	Fr	87	223*
Gadolinium	Gd	64	157.25
Gallium	Ga	31	69.72
Germanium	Ge	32	72.59
Gold	Au	79	196.97

*Mass number of most stable or best-known isotope.

†Mass of most commonly available, long-lived isotope.

Table continued next page

TABLE B-1 List of elements (continued)

Name	Symbol	Atomic Number	Atomic Weight
Hafnium	Hf	72	178.49
Helium	He	2	4.00
Holmium	Ho	67	164.93
Hydrogen	H	1	1.01
Indium	In	49	114.82
Iodine	I	53	126.90
Iridium	Ir	77	192.22
Iron	Fe	26	55.85
Krypton	Kr	36	83.80
Lanthanum	La	57	138.91
Lawrencium	Lr	103	257*
Lead	Pb	82	207.2
Lithium	Li	3	6.94
Lutetium	Lu	71	174.97
Magnesium	Mg	12	24.31
Manganese	Mn	25	54.94
Mendelevium	Md	101	256*
Mercury	Hg	80	200.59
Molybdenum	Mo	42	95.94
Neodymium	Nd	60	144.24
Neon	Ne	10	20.18
Neptunium	Np	93	237.05†
Nickel	Ni	28	58.71
Niobium	Nb	41	92.91
Nitrogen	N	7	14.01
Nobelium	No	102	254*
Osmium	Os	76	190.2
Oxygen	O	8	16.00
Palladium	Pd	46	106.4
Phosphorus	P	15	30.97
Platinum	Pt	78	195.09
Plutonium	Pu	94	242*
Polonium	Po	84	210*
Potassium	K	19	39.10
Praseodymium	Pr	59	140.91

*Mass number of most stable or best-known isotope.

†Mass of most commonly available, long-lived isotope.

Table continued next page

TABLE B-1 List of elements (continued)

Name	Symbol	Atomic Number	Atomic Weight
Promethium	Pm	61	147*
Protactinium	Pa	91	231.04†
Radium	Ra	88	226.03†
Radon	Rn	86	222*
Rhenium	Re	75	186.2
Rhodium	Rh	45	102.91
Rubidium	Rb	37	85.47
Ruthenium	Ru	44	101.07
Samarium	Sm	62	150.4
Scandium	Sc	21	44.969
Selenium	Se	34	78.96
Silicon	Si	14	28.09
Silver	Ag	47	107.87
Sodium	Na	11	22.99
Strontium	Sr	38	87.62
Sulfur	S	16	32.06
Tantalum	Ta	73	180.95
Technetium	Tc	43	98.91†
Tellurium	Te	52	127.60
Terbium	Tb	65	158.93
Thallium	Tl	81	204.37
Thorium	Th	90	232.04†
Thulium	Tm	69	168.93
Tin	Sn	50	118.69
Titanium	Ti	22	47.90
Tungsten	W	74	183.85
Uranium	U	92	238.03
Vanadium	V	23	50.94
Xenon	Xe	54	131.30
Ytterbium	Yb	70	173.04
Yttrium	Y	39	88.91
Zinc	Zn	30	65.38
Zirconium	Zr	40	91.22

*Mass number of most stable or best-known isotope.

†Mass of most commonly available, long-lived isotope.

APPENDIX C

Chemical Equations and Compounds Common in Water Treatment

Chemical Equations Common in Water Treatment

Taste, Odor, and Color Removal

$4Fe(OH)_2 + O_{2(g)} + 2H_2O \rightarrow 4Fe(OH)_3\downarrow$

$2MnSO_4 + O_{2(g)} + 4NaOH \rightarrow 2MnO_2\downarrow + 2Na_2SO_4 + 2H_2O$

$2H_2S + O_{2(g)} \rightarrow 2H_2O + 2S\downarrow$

Alum Coagulation

$Al_2(SO_4)_3 + 3Ca(HCO_3)_2 \rightarrow 2Al(OH)_3\downarrow + 3CaSO_4 + 6CO_2$

$Al_2(SO_4)_3 + 3Na_2CO_3 + 3H_2O \rightarrow 2Al(OH)_3\downarrow + 3Na_2SO_4 + 3CO_2$

$Al_2(SO_4)_3 + 3Ca(OH)_2 \rightarrow 2Al(OH)_3\downarrow + 3CaSO_4$

Ferric Sulfate Coagulation

$$Fe_2(SO_4)_3 + 3Ca(HCO_3)_2 \rightarrow 2Fe(OH)_3\downarrow + 3CaSO_4 + 6CO_2$$

Iron and Manganese Removal

$$2Fe(HCO_3)_2 + Cl_2 + Ca(HCO_3)_2 \rightarrow 2Fe(OH)_3\downarrow + CaCl_2 + 6CO_2$$

$$MnSO_4 + Cl_2 + 4NaOH \rightarrow MnO_2\downarrow + 2NaCl + Na_2SO_4 + 2H_2O$$

$$4Fe(HCO_3)_2 + O_2 + 2H_2O \rightarrow 4Fe(OH)_3\downarrow + 8CO_2$$

$$2MnSO_4 + O_2 + 4NaOH \rightarrow 2MnO_2\downarrow + 2Na_2SO_4 + 2H_2O$$

$$3Fe(HCO_3)_2 + KMnO_4 + 7H_2O \rightarrow MnO_2\downarrow + 3Fe(OH)_3 + KHCO_3 + 5H_2CO_3$$

$$3Mn(HCO_3)_2 + 2KMnO_4 + 2H_2O \rightarrow 5MnO_2\downarrow + 2KHCO_3 + 4H_2CO_3$$

$$3MnSO_4 + 2KMnO_4 + 2H_2O \rightarrow 5MnO_2\downarrow + K_2SO_4 + 2H_2SO_4$$

Hardness Removal

$$Ca(HCO_3)_2 + Ca(OH)_2 \rightarrow 2CaCO_3\downarrow + 2H_2O$$

$$Mg(HCO_3)_2 + Ca(OH)_2 \rightarrow CaCO_3\downarrow + MgCO_3 + 2H_2O$$

$$MgCO_3 + Ca(OH)_2 \rightarrow CaCO_3\downarrow + Mg(OH)_2$$

$$CaSO_4 + Na_2CO_3 \rightarrow CaCO_3\downarrow + Na_2SO_4$$

$$CaCl_2 + Na_2CO_3 \rightarrow CaCO_3\downarrow + 2NaCl$$

$$MgCl_2 + Ca(OH)_2 \rightarrow Mg(OH)_2\downarrow + CaCl_2$$

$MgSO_4 + Ca(OH)_2 \rightarrow Mg(OH)_2\downarrow + CaSO_4$

$CO_2 + Ca(OH)_2 \rightarrow CaCO_3\downarrow + H_2O$

Corrosion Control

$Fe(OH)_2 + 2H_2CO_3 \rightarrow Fe(HCO_3)_2 + 2H_2O$

$4Fe(HCO_3)_2 + 10H_2O + O_2 \rightarrow 4Fe(OH)_3\downarrow + 8H_2CO_3$

$4Fe(OH)_2 + 2H_2O + O_2 \rightarrow 4Fe(OH)_3\downarrow$

Chlorination

$Cl_2 + H_2O \rightarrow HOCl + HCl$

$NH_3 + HOCl \rightarrow NH_2Cl + H_2O$

$NH_2Cl + HOCl \rightarrow NHCl_2 + H_2O$

$NHCl_2 + HOCl \rightarrow NCl_2 + H_2O$

$Ca(OCl)_2 + 2H_2O \rightarrow 2HOCl + Ca(OH)_2$

$NaOCl + H_2O \rightarrow HOCl + NaOH$

TABLE C-1 Compounds common in water treatment

Chemical Name	Common Name	Chemical Formula
Aluminum hydroxide	Alum floc	$Al(OH)_3$
Aluminum sulfate	Filter alum	$Al_2(SO_4)_3 \cdot 14H_2O$
Ammonia	Ammonia	NH_3 (Ammonia gas)
Calcium bicarbonate	—	$Ca(HCO_3)_2$
Calcium carbonate	Limestone	$CaCO_3$
Calcium chloride	—	$CaCl_2$
Calcium hydroxide	Hydrated lime (slaked lime)	$Ca(OH)_2$
Calcium hypochlorite	HTH	$Ca(OCl)_2$
Calcium oxide	Unslaked lime (quicklime)	CaO
Calcium sulfate	—	$CaSO_4$
Carbon	Activated carbon	C
Carbon dioxide	—	CO_2
Carbonic acid	—	H_2CO_3
Chlorine	—	Cl_2
Chlorine dioxide	—	ClO_2
Copper sulfate	Blue vitriol	$CuSO_4 \cdot 5H_2O$
Dichloramine	—	$NHCl_2$
Ferric chloride	—	$FeCL_3 \cdot 6H_2O$
Ferric hydroxide	Ferric hydroxide floc	$Fe(OH)_3$
Ferric sulfate	—	$Fe_2(SO_4)_3 \cdot 3H_2O$
Ferrous bicarbonate	—	$Fe(HCO_3)_2$
Ferrous hydroxide	—	$Fe(OH)_2$
Fluosilicic acid (hydrofluosilicic acid)	—	H_2SiF_6
Hydrochloric acid	Muriatic acid	HCl
Hydrofluosilicic acid (fluosilicic acid)	—	H_2SiF_6
Hydrogen sulfide	—	H_2S
Hypochlorous acid	—	$HOCl$
Magnesium bicarbonate	—	$Mg(HCO_3)_2$
Magnesium carbonate	—	$MgCO_3$
Magnesium chloride	—	$MgCl_2$
Magnesium hydroxide	—	$Mg(OH)_2$
Manganese dioxide	—	MnO_2
Manganous bicarbonate	—	$Mn(HCO_3)_2$

Table continued next page

TABLE C-1 Compounds common in water treatment (continued)

Chemical Name	Common Name	Chemical Formula
Manganous sulfate	—	$MnSO_4$
Monochloramine	—	NH_2Cl
Potassium bicarbonate	—	$KHCO_3$
Potassium permanganate	—	$KMnO_4$
Sodium bicarbonate	Soda	$NaHCO_3$
Sodium carbonate	Soda ash	Na_2CO_3
Nitrogen trichloride (trichloramine)	—	NCl_3
Sodium chloride	Salt	$NaCl$
Sodium chlorite	—	$NaClO_2$
Sodium fluoride	—	NaF
Sodium fluosilicate (sodium silicofluoride)	—	Na_2SiF_6
Sodium hydroxide	Lye	$NaOH$
Sodium hypochlorite	—	$NaOCl$
Sodium phosphate	—	$Na_3PO_4 \cdot 12H_2O$
Sodium silicofluoride (sodium fluosilicate)	—	Na_2SiF_6
Sodium sulfate	—	Na_2SO_4
Sulfuric acid	Oil of vitriol	H_2SO_4
Trichloramine (nitrogen trichloride)	—	NCl_3

APPENDIX D

Algae Color Plates

The six color plates on the following pages illustrate how some of the more common types of algae encountered in water treatment would appear under a microscope. For positive identification, a biologist familiar with algae should be consulted.

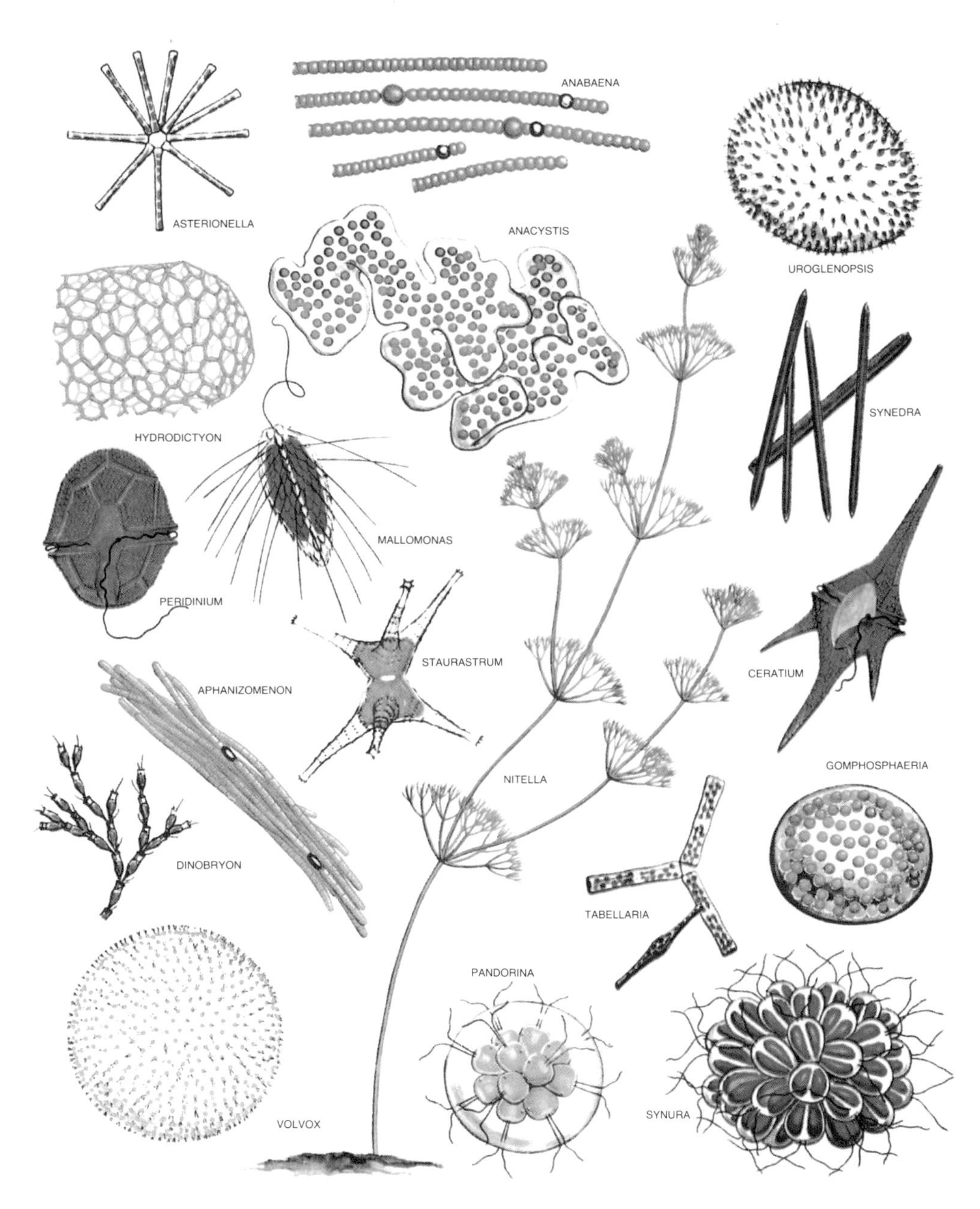

PLATE A
Taste-and-odor algae

Source: Standard Methods for the Examination of Water and Wastewater.

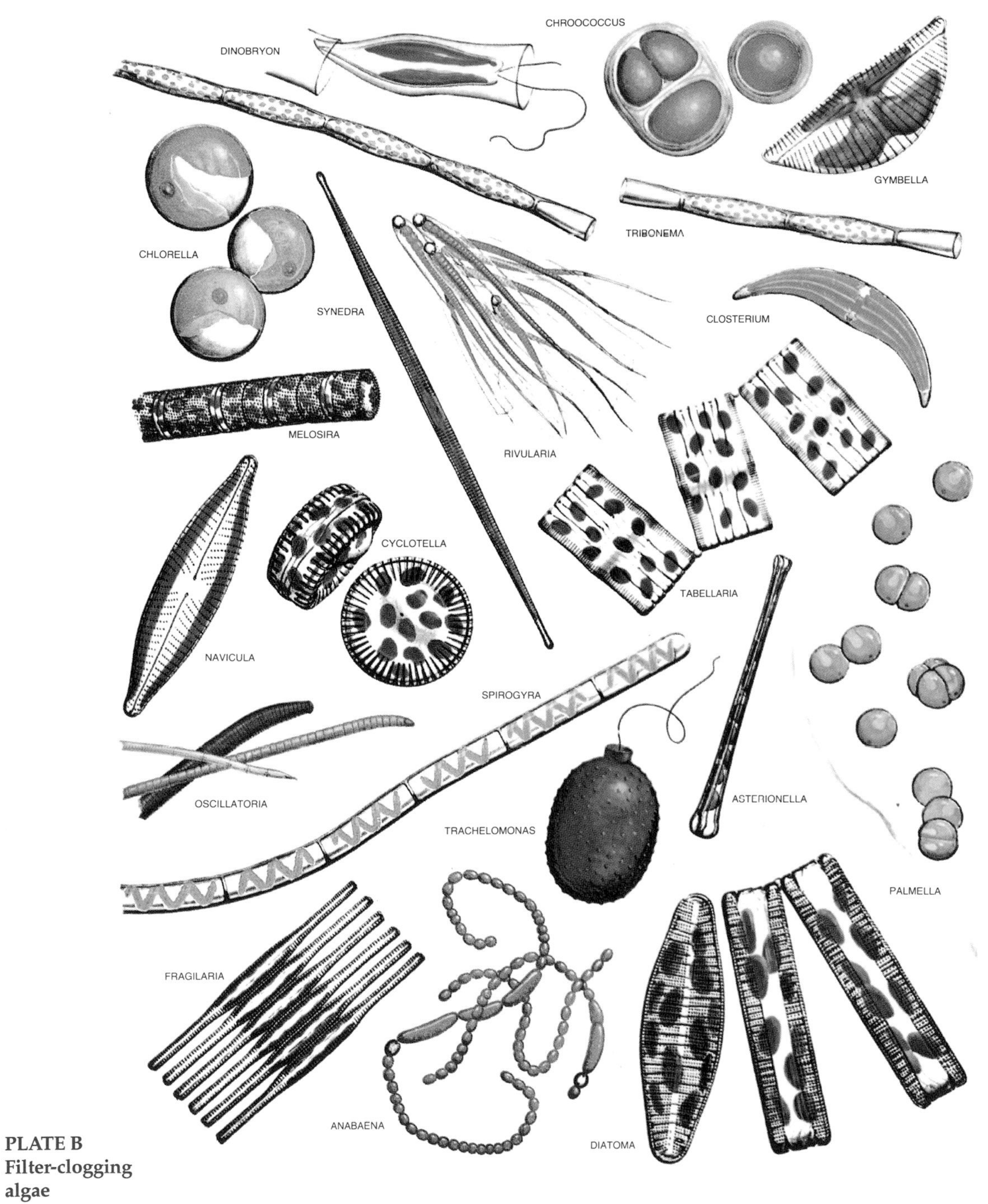

PLATE B
Filter-clogging algae

Source: Standard Methods for the Examination of Water and Wastewater.

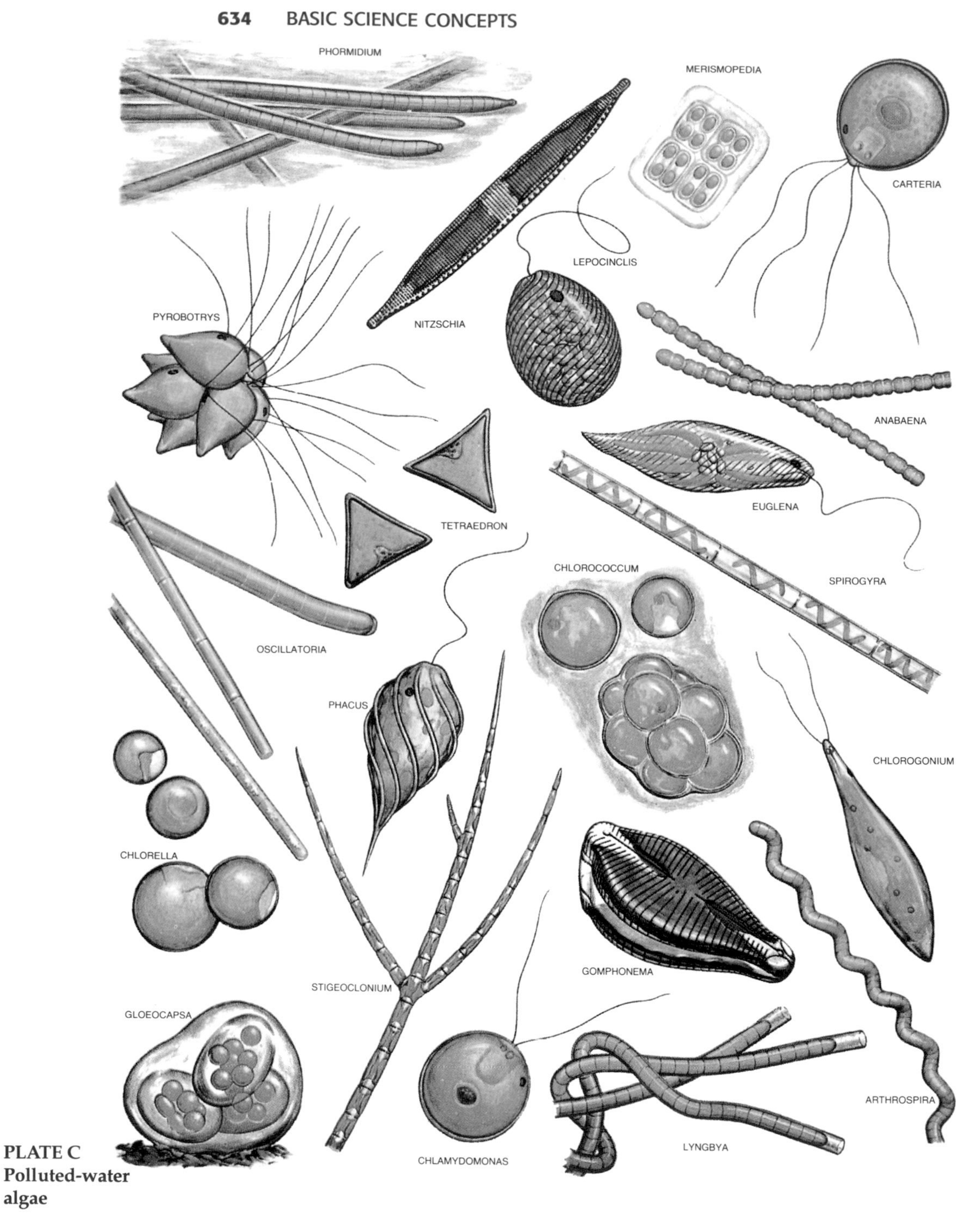

PLATE C
Polluted-water algae

Source: Standard Methods for the Examination of Water and Wastewater.

PLATE D Clean-water algae

Source: Standard Methods for the Examination of Water and Wastewater.

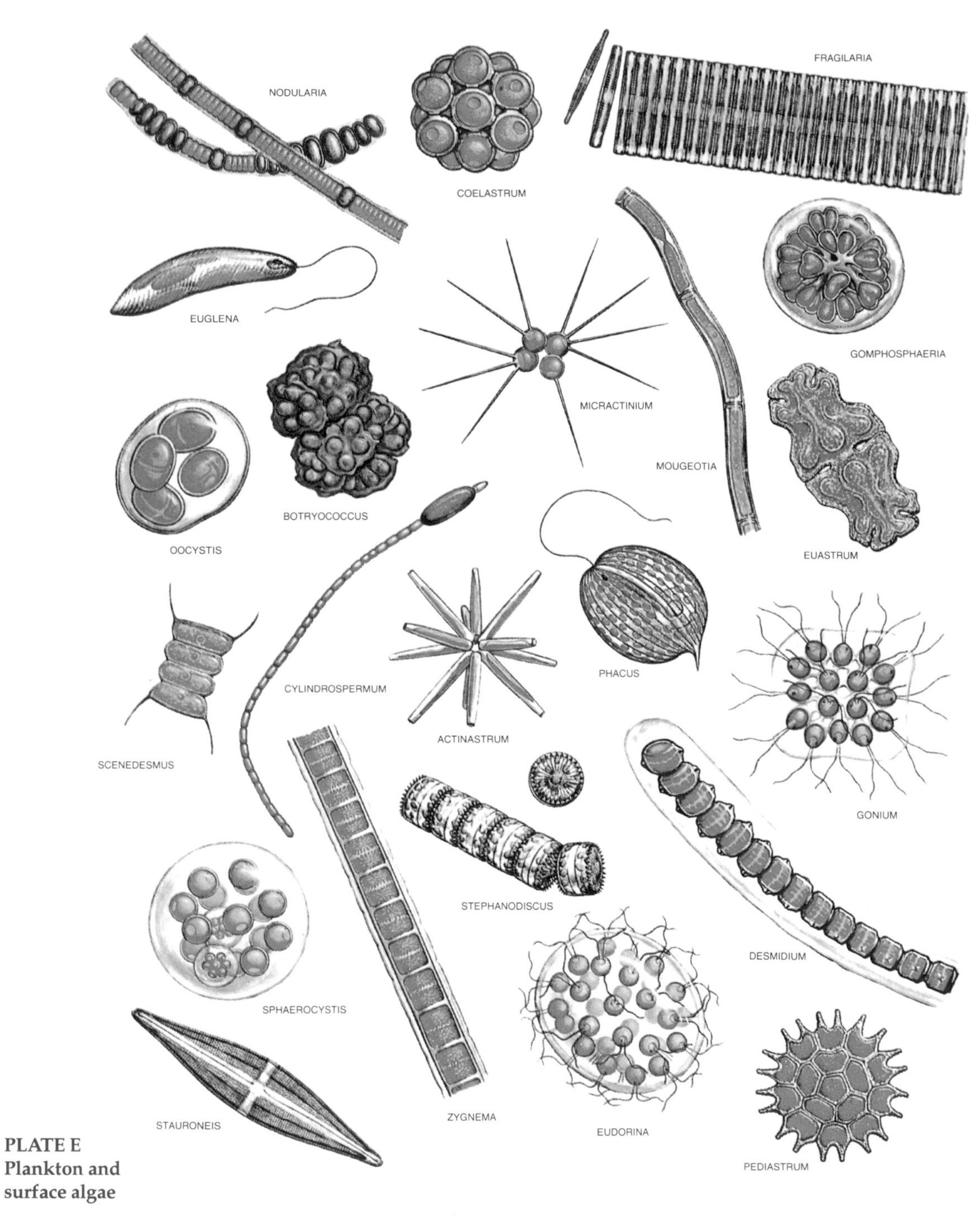

PLATE E
Plankton and surface algae

Source: Standard Methods for the Examination of Water and Wastewater.

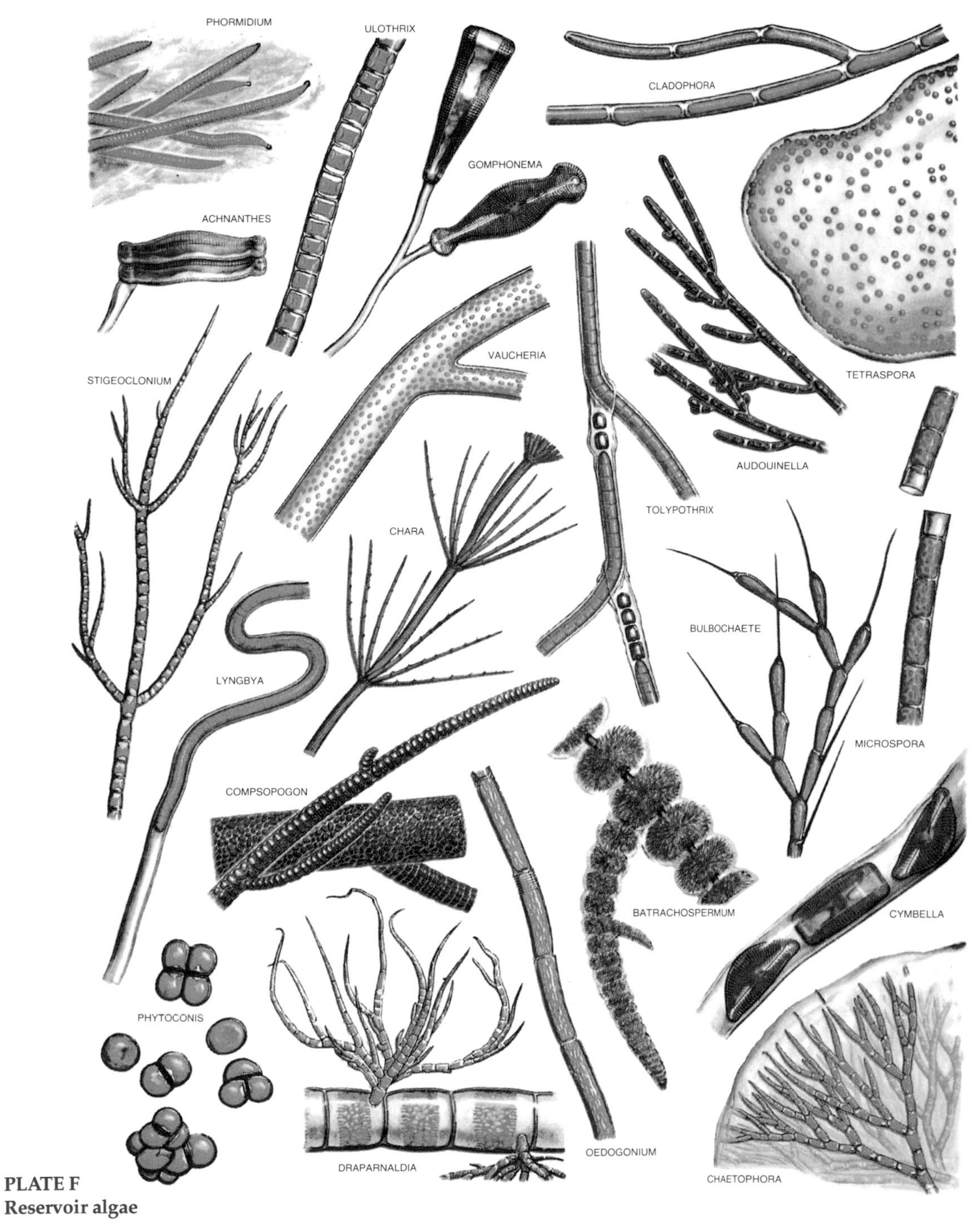

PLATE F
Reservoir algae

Source: Standard Methods for the Examination of Water and Wastewater.

Basic Science Concepts and Applications

Glossary

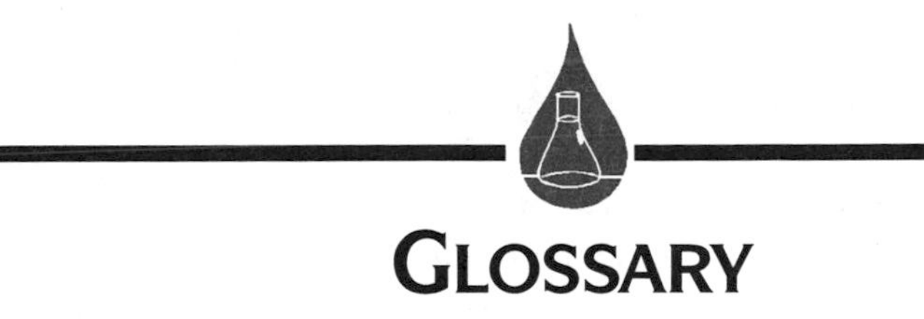

GLOSSARY

absolute pressure The total pressure in a system, including both the pressure of water and the pressure of the atmosphere (about 14.7 psi, at sea level). Compare with *gauge pressure.*

AC See Electricity chapter 2.

acid Any substance that releases hydrogen ions (H^+) when it is mixed into water.

acidic solution A solution that contains significant numbers of H^+ ions.

alkaline solution A solution that contains significant numbers of OH^- ions. A basic solution.

alkalinity A measurement of water's capacity to neutralize an acid. Compare *pH.*

alternating current (AC) See Electricity chapter 2.

ammeter An instrument for measuring amperes.

ampere (amp or A) See Electricity chapter 2.

ampere-hour (A·h) See Electricity chapter 2.

anion A negative ion.

annual ADF (or AADF) The average daily flow calculated using one year of data.

apparent power See Electricity chapter 2.

arithmetic mean A measurement of average value, calculated by summing all terms and dividing by the number of terms.

arithmetic scale A scale is a series of intervals (marks or lines), usually marked along the side and bottom of a graph, that represents the range of values of the data. When the marks or lines are equally spaced, it is called an arithmetic scale. Compare with *logarithmic scale.*

atom The smallest particle of an element that still retains the characteristics of that element.

atomic number The number of protons in the nucleus of an atom.

atomic weight The sum of the number of protons and the number of neutrons in the nucleus of an atom.

average daily flow (ADF) A measurement of the amount of water treated by a plant each day. It is the average of the actual daily flows that occur within a period of time, such as a week, a month, or a year. Mathematically, it is the sum of all daily flows divided by the total number of daily flows used.

average flow rate The average of the instantaneous flow rates over a given period of time, such as a day.

balanced A chemical equation is balanced when, for each element in the equation, as many atoms are shown on the right side of the equation as are shown on the left side.

base Any substance that releases hydroxyl ions (OH^-) when it dissociates in water.

basic solution A solution that contains significant numbers of OH^- ions.

battery A device for producing DC electric current from a *chemical reaction*. In a storage battery, the process may be reversed, with current flowing into the battery, thus reversing the chemical reaction and recharging the battery.

bicarbonate alkalinity Alkalinity caused by bicarbonate ions (HCO_3^-).

bond See *chemical bond*.

brake horsepower The power supplied to a pump by a motor. Compare with *water horsepower* and *motor horsepower*.

buffer A substance capable in solution to resist a reduction in pH as acid is added.

bulk density The weight per standard volume (usually pounds per cubic foot) of material as it would be shipped from the supplier to the treatment plant.

calcium carbonate saturation index See *Langelier saturation index*.

capacitance See Electricity chapter 2.

capacitor See Electricity chapter 2.

capacity The flow rate that a pump is capable of producing.

carbonate alkalinity Alkalinity caused by carbonate ions (CO_3^{-2}).

carbonate hardness Hardness caused primarily by bicarbonate. Compare with *noncarbonate hardness*.

cation A positive ion.

cation exchange materials Materials that release nontroublesome ions into water in exchange for hardness-causing ions.

chelation A chemical process used to control scale formation, in which a chelating agent "captures" scale-causing ions and holds them in solution, thus preventing them from precipitating out and forming scale.

chemical bond The force that holds atoms together within molecules. A chemical bond is formed when a chemical reaction takes place. Two types of chemical bond are ionic bonds and covalent bonds.

chemical equation A shorthand way, using chemical formulas, of writing the reaction that takes place when chemicals are brought together. The left side of the equation indicates the chemicals brought together (the reactants); the arrow indicates in which direction the reaction occurs; and the right side of the equation indicates the results (the products) of the chemical reaction.

chemical formula See *formula*.

chemical reaction A process that occurs when atoms of certain elements are brought together and combine to form molecules, or when molecules are broken down into individual atoms.

circuit breaker A device that functions both as a current overload protective device and as a switch.

circumference The distance measured around the outside edge of a circle.

compounds Two or more elements bonded together by a chemical reaction.

concentration In chemistry, a measurement of how much solute is contained in a given amount of solution. Concentrations are commonly measured in milligrams per liter (mg/L).

condensor See *capacitor*.

conductor A substance that permits the flow of electricity, especially one that conducts electricity with ease.

converter Generally, a DC generator driven by an AC motor.

covalent bond A type of chemical bond in which electrons are shared. Compare with *ionic bond*.

cross multiplication A method used to determine if two ratios are in proportion. In this method, the numerator of the first ratio is multiplied by the denominator of the second ratio. Similarly, the denominator of the first ratio is multiplied by the numerator of the second ratio. If the products of both multiplications are the same, the two ratios are in proportion to each other.

current The "flow rate" of electricity, measured in amperes. Compare with *potential*.

current regulator A device that automatically holds electric current within certain limits.

cycle See Electricity chapter 2.

daily flow The volume of water that passes through a plant in one day (24 hours). More precisely called "daily flow volume."

DC See Electricity chapter 2.

demand See Electricity chapter 2.

demand meter An instrument that measures the average power of a load during some specific interval.

denominator The part of a fraction below the line. A fraction indicates division of the numerator by the denominator.

density The weight of a substance per a unit of its volume; for example pounds per cubic foot or pounds per gallon.

design point The mark on the H–Q (head–capacity) curve of a pump characteristics curve that indicates the head and capacity at which the pump is intended to operate for best efficiency in a particular installation.

detention time The average length of time a drop of water or a suspended particle remains in a tank or chamber. Mathematically, it is the volume of water in the tank divided by the flow rate through the tank. The units of flow rate used in the calculation are dependent on whether the detention time is to be calculated in minutes, hours, or days.

diameter The length of a straight line measured through the center of a circle from one side to the other.

digit Any one of the 10 arabic numerals (0 through 9) by which all numbers may be expressed.

direct current See Electricity chapter 2.

drawdown The amount the water level in a well drops once pumping begins. Drawdown equals static water level minus pumping water level.

dynamic discharge head The difference in height measured from the pump center line at the discharge of the pump to the point on the hydraulic grade line directly above it.

dynamic head See *total dynamic head*.

dynamic suction head The distance from the pump center line at the suction of the pump to the point of the hydraulic grade line directly above it. Dynamic suction head exists only when the pump is below the piezometric surface of the water at the pump suction. When the pump is above the piezometric surface, the equivalent measurement is dynamic suction lift.

dynamic suction lift The distance from the pump center line at the suction of the pump to the point on the hydraulic grade line directly below it. Dynamic suction lift exists only when the pump is above the piezometric surface of the water at the pump suction. When the pump is below the piezometric surface, the equivalent measurement is called dynamic suction head.

dynamic water system The description of a water system when water is moving through the system.

effective height The total feet of head against which a pump must work.

efficiency The ratio of the total energy output to the total energy input, expressed as percent.

electromagnetics The study of the combined effects of electricity and magnetism.

electron One of the three elementary particles of an atom (along with protons and neutrons). An electron is a tiny, negatively charged particle that orbits around the nucleus of an atom. The number of electrons in the outermost shell is one of the most important characteristics of an atom in determining how chemically active an element will be and with what other elements or compounds it will react.

element Any of more than 100 fundamental substances that consist of atoms of only one kind and that constitute all matter.

elevation head The energy possessed per unit weight of a fluid because of its elevation above some reference point (called the "reference datum"). Elevation head is also called position head or potential head.

energy See Electricity chapter 2.

energy grade line (EGL) (Sometimes called "energy-gradient line" or "energy line.") A line joining the elevations of the energy heads; a line drawn above the hydraulic grade line by a distance equivalent to the velocity head of the flowing water at each section along a stream, channel, or conduit.

equivalent weight The weight of an element or compound that, in a given chemical reaction, has the same combining capacity as 8 g of oxygen or as 1 g of hydrogen. The equivalent weight for an element or compound may vary with the reaction being considered.

exponent An exponent indicates the number of times a base number is to be multiplied together. For example, a base number of 3 with an exponent of 5 is written 3^5. This indicates that the base number is to be multiplied together five times: $3^5 = 3 \times 3 \times 3 \times 3 \times 3$.

filter backwash rate A measurement of the volume of water flowing upward (backwards) through a unit of filter surface area. Mathematically, it is the backwash flow rate divided by the total filter area.

filter loading rate A measurement of the volume of water applied to each unit of filter surface area. Mathematically, it is the flow rate into the filter divided by the total filter area.

flow rate A measure of the volume of water moving past a given point in a given period of time. Compare *instantaneous flow rate* and *average flow rate.*

formula Using the chemical symbols for each element, a formula is a shorthand way of writing what elements are present in a molecule and how many atoms of each element are present in each of the molecules. Also called a chemical formula.

formula weight See *molecular weight.*

free water surface The surface of water that is in contact with the atmosphere.

frequency See Electricity chapter 2.

friction head loss The head lost by water flowing in a stream or conduit as the result of (1) the disturbance set up by the contract between the moving water and its containing conduit and (2) intermolecular friction.

fuse A protective device that disconnects equipment from the power source when current exceeds a specified value.

gauge pressure The water pressure as measured by a gauge. Gauge pressure is not the total pressure. Total water pressure (absolute pressure) also includes the atmospheric pressure (about 14.7 psi at sea level) exerted on the water. However, because atmospheric pressure is exerted everywhere (against the outside of the main as well as the inside, for example), it is generally not written into water system calculations. Gauge pressure in pounds per square inch is expressed as "psig."

gallons per capita per day (gpcd) A measurement of the average number of gallons of water used by the average person each day in a water system. The calculation is made by dividing the total gallons of water used each day by the total number of people using the water system.

generator A piece of equipment used to transform rotary motion (for example, the output of a diesel engine) to electric current.

grains per gallon (gpg) A measure of the concentration of a solution. One gpg equals 17.12 mg/L.

gram-mole See *mole.*

ground See Electricity chapter 2.

groups The vertical columns of elements in the periodic table.

hardness A characteristic of water, caused primarily by the salts of calcium and magnesium. Causes deposition of scale in boilers, damage in some industrial processes, and sometimes objectionable taste. May also decrease the effectiveness of soap.

head (1) A measure of the energy possessed by water at a given location in the water system, expressed in feet. (2) A measure of the pressure or force exerted by water, expressed in feet.

head loss The amount of energy used by water in moving from one location to another.

hertz (Hz) A measurement of frequency, equal to cycles per second.

homogenous A term used to describe a substance with a uniform structure or composition throughout.

horsepower A standard unit of power equal to 746 W (watts), approximately equal to 33,000 ft-lb/min.

hydraulic grade line (HGL) A line (hydraulic profile) indicating the piezometric level of water at all points along a conduit, open channel, or stream. In an open channel, the HGL is the free water surface.

hydroxyl alkalinity Alkalinity caused by hydroxyl ions (OH^-).

impedance See Electricity chapter 2.

inductance See Electricity chapter 2.

instantaneous flow rate A flow rate of water measured at one particular instant, such as by a metering device, involving the cross-sectional area of the channel or pipe and the velocity of the water at that instant.

insulator A substance that offers very great resistance, or hindrance, to the flow of electric current.

interpolation A technique used to determine values that fall between the marked intervals on a scale.

interrupting current See Electricity chapter 2.

inverter See Electricity chapter 2.

ion An atom that is electrically unstable because it has more or fewer electrons than protons. A positive ion is called a cation. A negative ion is called an anion.

ionic bond A type of chemical bond in which electrons are transferred. Compare with *covalent bond.*

isotopes Atoms of the same element, but containing varying numbers of neutrons in the nucleus. For each element, the most common naturally occurring isotope is called the principal isotope of that element.

kill The destruction of organisms in a water supply.

kilo (k) A prefix meaning 1,000.

kilovolt See Electricity chapter 2.

kilowatt (kW) See Electricity chapter 2.

kilowatt-hour (kW·h) See Electricity chapter 2.

kVA See Electricity chapter 2.

kvar See Electricity chapter 2.

Langelier saturation index A measure of the tendency of water to scale or corrode pipes, based on the pH and alkalinity of the water.

logarithmic scale (log scale) A scale is a series of intervals (marks or lines), usually marked along the side and bottom of a graph, that represents the range of values of the data. When the marks or lines are varied logarithmically (and are therefore not equally spaced), the scale is called a logarithmic, or log, scale. Compare with *arithmetic scale.*

maximum demand See Electricity chapter 2.

megohm See Electricity chapter 2.

milligrams per liter (mg/L) A measure of the concentration of a solution, equal to one milligram weight of solute in every liter volume of solution. Generally interchangeable with parts per million in water treatment calculations.

minor head loss The energy losses that result from the resistance to flow as water passes through valves, fittings, inlets, and outlets of a piping system.

mixture Two or more elements, compounds, or both, mixed together with no chemical reaction (bonding) occurring.

molality A measure of concentration defined as the number of moles of solute per liter of solvent. Not commonly used in water treatment. Compare with *molarity.*

molarity A measure of concentration defined as the number of moles of solute per liter of solution.

mole Used in this text and generally as an abbreviation for gram-mole. A mole is the quantity of a compound or element that has a weight in grams equal to the substance's molecular or atomic weight.

molecular weight The sum of the atomic weights of all the atoms in the compound. Also called formula weight.

molecule Two or more atoms joined together by a chemical bond.

motor horsepower The horsepower equivalent to the watts of electric power supplied to a motor. Compare with *brake horsepower* and *water horsepower.*

neutralization The process of mixing an acid and a base to form a salt and water.

neutralize See *neutralization.*

neutron An uncharged elementary particle that has a mass approximately equal to that of the proton. Neutrons are present in all known atomic nuclei except the lightest hydrogen nucleus.

nomograph A graph in which three or more scales are used to solve mathematical problems.

noncarbonate hardness Hardness caused by the salts of calcium and magnesium.

normality A method of expressing the concentration of a solution. It is the number of equivalent weights of solute per liter of solution.

nucleus (plural: nuclei) The center of an atom, made up of positively charged particles called protons and uncharged particles called neutrons.

numerator The part of a fraction above the line. A fraction indicates division of the numerator by the denominator.

ohm See Electricity chapter 2.

Ohm's law An equation expressing the relationship between the potential (E) in volts, the resistance (R) in ohms, and the current (I) in amperes for electricity passing through a metallic conductor. Ohm's law is $E = I \times R$.

organic compounds Generally, compounds containing carbon.

organics See *organic compounds*.

parts per million (ppm) A measure of the concentration of a solution, meaning one part of solute in every million parts of solution. Generally interchangeable with milligrams per liter in water treatment calculations.

pascal (Pa) A unit of pressure in the metric system, equal to 0.000145 psi.

per capita Per person.

percent (%) The fraction of the whole expressed as parts per one hundred.

perimeter The distance around the outer edge of a shape.

periodic table A chart showing all elements arranged according to similarities of chemical properties.

periods The horizontal rows of elements in a periodic table.

pH A measurement of how acidic or basic a substance is. The pH scale runs from 0 (most acidic) to 14 (most basic). The center of the range (7) indicates the substance is neutral, neither acidic or basic.

phase See Electricity chapter 2.

pi (π) The ratio of the circumference of a circle to the diameter of that circle, approximately equal to 3.14159, or about $^{22}/_{7}$.

piezometer An instrument for measuring pressure head in a conduit, tank, or soil, by determining the location of the free water surface.

piezometric surface An imaginary surface that coincides with the level of the water in an aquifer, or the level to which water in a system would rise in a piezometer.

pole One end of a magnet (the north or south pole).

potential The "pressure" of electricity, measured in volts. Compare with *current*.

pounds per square inch A measurement of pressure. The comparable metric unit is the pascal (Pa); 1 psi = 6,895 Pa.

pounds per square inch absolute (psia) The sum of gauge pressure and atmospheric pressure. See *absolute pressure*. Compare with *pounds per square inch gauge*.

pounds per square inch gauge (psig) Pressure measured by a gauge and expressed in terms of pounds per square inch. See *gauge pressure*. Compare with *pounds per square inch absolute*.

power (in hydraulics or electricity) The measure of the amount of work done in a given period of time. The rate of doing work. Measured in watts or horsepower.

power (in mathematics) See *exponent*.

power factor (pf) See Electricity chapter 2.

pressure The force pushing on a unit area. Normally pressure can be measured in pascals (Pa), pounds per square inch (psi), or feet of head.

pressure head A measurement of the amount of energy in water due to water pressure.

primary See Electricity chapter 2.

principal isotopes See *isotopes*.

products The results of a chemical reaction. The products of a reaction are shown on the right side of a chemical equation.

proportion (proportionate) When the relationship between two numbers in a ratio is the same as that between two other numbers in another ratio, the two ratios are said to be in proportion, or proportionate.

proton One of the three elementary particles of an atom (along with neutrons and electrons). The proton is a positively charged particle located in the nucleus of an atom. The number of protons in the nucleus of an atom determines the atomic number of that element.

psi See *pounds per square inch.*

psig See *pounds per square inch gauge.*

pump center line An imaginary line through the center of a pump.

pump characteristic curve A curve or curves showing the interrelation of speed, dynamic head, capacity, brake horsepower, and efficiency of a pump.

pumping water level (PWL) The water level measured when the pump is in operation.

radicals Groups of elements chemically bonded together and acting like single atoms or ions in their ability to form other compounds.

radius The distance from the center of a circle to its edge. One half of the diameter.

rating See Electricity chapter 2.

ratio A relationship between two numbers. A ratio may be expressed using colons (for example, 1:2 or 3:7), or it may be expressed as a fraction (for example, $\frac{1}{2}$ or $\frac{3}{7}$).

reactance The combined effect of capacitance and inductance.

reactants The chemicals brought together in a chemical reaction. The chemical reactants are shown on the left side of a chemical equation.

reactive power See Electricity chapter 2.

real power See Electricity chapter 2.

recarbonation The reintroduction of carbon dioxide into the water, either during or after lime–soda ash softening.

rectifier See Electricity chapter 2.

regeneration The process of reversing the ion exchange softening reaction of ion exchange materials, removing the hardness ions from the used materials and replacing them with nontroublesome ions, thus rendering the materials fit for reuse in the softening process.

relay See Electricity chapter 2.

resistance See Electricity chapter 2.

rule of continuity The rule states that the flow (Q) that enters a system must also be the flow that leaves the system. Mathematically, this rule is generally stated as $Q_1 = Q_2$ or (since $Q = AV$), $A_1V_1 = A_2V_2$.

safety factor The percentage above which a rated electrical device cannot be operated without damage or shortened life.

salts Compounds resulting from acid–base mixtures.

saturation A stable condition of water in which the water will neither deposit scale nor cause corrosion.

saturation index (SI) See Langelier saturation index.

scientific notation A method by which any number can be expressed as a number between 1 and 9 multiplied by a power of 10.

secondary See Electricity chapter 2.

sequestration A chemical process used to control scale formation, in which a sequestering agent holds scale-causing ions in solution, preventing them from precipitating out and forming scale.

side water depth (SWD) The depth of water measured along a vertical interior wall.

solute The substance dissolved in a solution. Compare with *solvent*.

solution A liquid containing a dissolved substance. The liquid alone is called the solvent, the dissolved substance is called the solute. Together they are called a solution.

solvent The liquid used to dissolve a substance. See *solution*.

specific capacity A measurement of the well yield per unit (usually per foot) of drawdown. Mathematically, it is the well yield divided by the drawdown.

specific gravity The ratio of the density of a substance to a standard density. For solids and liquids, the density is compared with the density of water (62.4 lb/ft^3). For gases the density is compared with the density of air (0.075 lb/ft^3).

standard solution A solution with an accurately known concentration, used in the lab to determine the properties of unknown solutions.

static discharge head The difference in height between the pump center line and the level of the discharge free water surface.

static suction head The difference in elevation between the pump center line and the free water surface of the reservoir feeding the pump. In the measurement of static suction head, the piezometric surface of the water at the suction side of the pump is higher than the pump; otherwise, static suction lift is measured.

static suction lift The difference in elevation between the pump center line of a pump and the free water surface of the liquid being pumped. In a static suction lift measurement, the piezometric surface of the water at the suction side of the pump is lower than the pump; otherwise, static suction head is measured.

static water level (SWL) The water level in a well measured when no water is being taken from the aquifer, either by pumping or by free flow.

static water system The description of a water system when water is not moving through the system.

substation See Electricity chapter 2.

surface overflow rate A measurement of the amount of water leaving a sedimentation tank per unit of tank surface area. Mathematically, it is the flow rate from the tank divided by the tank surface area.

switch A device to manually disconnect electrical equipment from the power source.

thrust A force resulting from water under pressure and in motion. Thrust pushes against fittings, valves, and hydrants and can cause couplings to leak or to pull apart entirely.

thrust anchor A block of concrete, often a roughly shaped cube, cast in place below a fitting to be anchored against vertical thrust, and tied to the fitting with anchor rods.

thrust block A mass of concrete, cast in place between a fitting to be anchored against thrust and the undisturbed soil at the side or bottom of the pipe trench.

total alkalinity The combined effect of hydroxyl alkalinity (OH^-), carbonate alkalinity ($CO_3^=$) and bicarbonate alkalinity (HCO_3^-).

total dynamic head The difference in height between the hydraulic grade line (HGL) on the discharge side of the pump and the HGL on the suction side of the pump. This head is a measure of the total energy that a pump must impart to the water to move it from one point to another.

total static head The total height that the pump must lift the water when moving it from one point to another. The vertical distance from the suction free water surface to the discharge free water surface.

transformer See Electricity chapter 2.

trihalomethanes (THMs) Certain organic compounds, sometimes formed when water containing natural organics is chlorinated. Some THMs, in large enough concentrations, may be carcinogenic.

valence One or more numbers assigned to each element, indicating the ability of the element to enter into chemical reactions with other elements.

valance electrons The electrons in the outermost electron shells. These electrons are one of the most important factors in determining which atoms will combine with which other atoms.

velocity head A measurement of the amount of energy in water due to its velocity, or motion.

volt (V) See Electricity chapter 2.

voltmeter An instrument for measuring volts.

water hammer The potentially damaging slam, bang, or shudder that occurs in a pipe when a sudden change in water velocity (usually as a result of too rapidly starting a pump or operating a valve) creates a great increase in water pressure.

water horsepower (WHP) The portion of the power delivered to a pump that is actually used to lift water. Compare with *brake horsepower* and *motor horsepower*.

watt See Electricity chapter 2.

watt-hour See Electricity chapter 2.

wattmeter An instrument for measuring real power in watts.

weir overflow rate A measurement of the flow rate of water over each foot of weir in a sedimentation tank or circular clarifier. Mathematically, it is the flow rate over the weir divided by the total length of the weir.

well yield The volume of water that is discharged from a well during a specified time period. Mathematically, it is the total volume discharged, divided by the time during which the discharge was monitored.

whole numbers Any of the natural numbers, such as 1, 2, 3, etc.; the negative of these numbers, such as –1, –2, –3, etc.; and zero. Also called "integers" or "counting numbers."

wire-to-water efficiency The ratio of the total power input (electric current expressed as *motor horsepower*) to a motor and pump assembly, to the total power output (*water horsepower*); expressed as a percent.

work The operation of a force over a specific distance.

Basic Science Concepts and Applications

Index

INDEX

NOTE: *f.* indicates figure; *t.* indicates table.

A

B

D

E

K

L

M

T

U

V

W